Beltsville Symposia in Agricultural Research

A Series of Annual Symposia Sponsored by
THE BELTSVILLE AGRICULTURAL RESEARCH CENTER
Agricultural Research Service
United States Department of Agriculture

[11] Research Instrumentation for the 21st Century

SYMPOSIA IN THIS SERIES

[1] Virology in Agriculture
May 10–12, 1976
Published, 1977

[2] Biosystematics in Agriculture
May 8–11, 1977
Published, 1978

[3] Animal Reproduction
May 14–17, 1978
Published, 1979

[4] Human Nutrition Research
May 8–9, 1979
Published, 1981

[5] Biological Control in Crop Production
May 18–21, 1980
Published, 1981

[6] Strategies of Plant Reproduction
May 17–20, 1981
Published, 1983

[7] Genetic Engineering: Applications to Agriculture
May 16–19, 1982
Published, 1983

[8] Agricultural Chemicals of the Future
May 16–19, 1983
Published, 1984

[9] Frontiers of Membrane Research
May 20–24, 1984
Published, 1985

[10] Biotechnology for Solving Agricultural Problems
May 5–9, 1985
Published, 1986

[11] Research Instrumentation for the 21st Century
May 4–8, 1986

FORTHCOMING SYMPOSIUM

[12] Biomechanisms Regulating Growth and Development: Keys to Progress
May 3–7, 1987

Beltsville Symposia in Agricultural Research

[11] Research Instrumentation for the 21st Century

Gary R. Beecher, editor

Invited papers presented at a symposium held May 4–8, 1986, at the Beltsville Agricultural Research Centre (BARC), Beltsville, Maryland

Organized by the BARC SYMPOSIUM XI COMMITTEE
Gary R. Beecher, Essex E. Finney, Jr. and
Karl H. Norris, Co-chairpersons

Sponsored by
THE BELTSVILLE AGRICULTURAL RESEARCH CENTRE
Agricultural Research Service
United States Department of Agriculture

1988 **MARTINUS NIJHOFF PUBLISHERS**
a member of the KLUWER ACADEMIC PUBLISHERS GROUP
DORDRECHT / BOSTON / LANCASTER

Distributors

for the United States and Canada: Kluwer Academic Publishers, 190 Old Derby Street, Hingham, MA 02043, USA
for the UK and Ireland: Kluwer Academic Publishers, MTP Press Limited, Falcon House, Queen Square, Lancaster LA1 1RN, UK
for all other countries: Kluwer Academic Publishers Group, Distribution Center, P.O. Box 322, 3300 AH Dordrecht, The Netherlands

Library of Congress Cataloging-in-Publication Data

Research instrumentation for the 21st century.

(Beltsville symposia in agricultural research ; 11)
1. Agriculture--Research--Technique--Congresses.
2. Agriculture--Research--Instruments--Congresses.
I. Beecher, Gary R. II. BARC Symposium XI Committee.
III. Beltsville Agricultural Research Center.
IV. Title: Research instrumentation for the twenty-first century. V. Series.
S539.7.R465 1987 630'.72 87-23945

ISBN 90-247-3595-5

PRINTED IN THE NETHERLANDS

CONTENTS

FOREWORD

The Annual Beltsville Symposium serves as a forum for presenting recent developments in basic research that contribute solutions to agricultural problems. The eleventh in this series focuses on instrumentation techniques which provide powerful new research tools. These tools will provide information that can lead to a better understanding of biological and physical processes critical to the solution of today's agricultural problems. The tools discussed include isotope technology, separation techniques, microscopy, electro-magnetic spectroscopy, resonance and mass spectrometry and microcomputers.

This Symposium brought together scientists who are at the forefront of the development of many of these tools and those who are applying them to problems directly related to agricultural research.

WALDEMAR KLASSEN, Director
Beltsville Area

CONTRIBUTORS AND THEIR AFFILIATIONS

Norman G. Anderson
Proteus Technologies, Inc.
Rockville, MD 20852
U.S.A.

Edwin D. Becker
Office of Research Services
NIH
Bethesda, MD 20892
U.S.A.

Klaus Biemann
Department of Chemistry
Massachusetts Institute of Technology
Cambridge, MA 02139
U.S.A.

B.A. Bolton
Development Quality Laboratory
Western Regional Research Center
Agricultural Research Service USDA
Albany, CA 94710
U.S.A.

Bruce R. Buchanan
Center for Process Analytical Chemistry
University of Washington
Seattle, WA 98195
U.S.A.

Karin D. Caldwell
Department of Bioengineering and Center for Biopolymers at Interfaces
University of Utah
Salt Lake City, UT 84112
U.S.A.

J.T. Clerc
Pharmacy Department
University of Bern
Bern
Switzerland

Henry M. Fales
Laboratory of Chemistry
NHLBI, NIH
Bethesda, MD 20892
U.S.A.

Youhanna Fares
Phytokinetics Inc.
College Station, TX 77840
U.S.A.

Cherie L. Fisk
Office of Research Services
NIH
Bethesda, MD 20892
U.S.A.

John D. Goeschl
Phytokinetics, Ltd.
College Station, TX 77840
U.S.A.

R.J. Griesbach
Florist and Nursery Crops Laboratory
Horticultural Science Institute
Agricultural Research Service USDA
Beltsville, MD 20705
U.S.A.

Peter R. Griffiths
Department of Chemisty
University of California
Riverside, CA 92521
U.S.A.

James S. Holler
Toxicology Branch
Centers for Disease Control
Atlanta, GA 30333
U.S.A.

David E. Honigs
Centre for Process Analytical Chemistry
University of Washington
Seattle, WA 98195
U.S.A.

Gregory Hook
Surgical Neurology Branch
NINCDS, NIH
Bethesda, MD 20892
U.S.A.

C.H. Jaeger
Botany Department and Phytotron
Duke University
Durham, NC 27706
U.S.A.

Machael J. Kesler
Packard Instrument Co.
Downers Grove, IL 60515
U.S.A.

Paul Lauterbur
Department of Medical Information Science and Chemistry
University of Illinois at Urbana – Champaign
Urbana, IL 61801
U.S.A.

Kreg A. Leymaster
Breeding and Genetics Research Unit
U.S. Meat Animal Research Center
Agricultural Research Service USDA
Clay Center, NE 68933
U.S.A.

C.E. Magnuson
Phytokinetics Inc.
College Station, TX 77840
U.S.A.

C.E. Nelson
Division of Clinical Radiation Oncology
Radiation Oncology Center
School of Medicine
East Carolina University
Greenvile, NC 27834
U.S.A.

T.C. O'Haver
Department of Chemistry and Biochemistry
University of Maryland
College Park, MD 20742
U.S.A.

Gilbert L. Pariente
Department of Chemistry
University of California
Riverside, CA 92521
U.S.A.

Jonathan Perkins
Center for Process Analytical Chemistry
University of Washington
Seattle, WA 98195
U.S.A.

Philip E. Pfeffer
Plant and Soil Biophysics Laboratory
Eastern Regional Research Center
Agricultural Research Service USDA
Philadelphia, PA 19118
U.S.A.

Daniel Pinkel
Biomedical Sciences Division
Lawrence Livermore National Laboratory
Liverman, CA 94550
U.S.A.

Timothy Poston
Crump Institute for Medical Engineering
University of California
Los Angeles, CA 90024
U.S.A.

John C. Price
Remote Sensing Research Laboratory
Agricultural Systems Research Institute
Agricultural Research Service USDA
Beltsville, MD 20705
U.S.A.

H.W. Scheld
Phytokinetics Inc.
College Station, TX 77840
U.S.A.

J.R. Scherer
Development Quality Laboratory
Western Regional Research Center
Agricultural Research Service USDA
Albany, CA 94710
U.S.A.

Boyd R. Strain
Botany Department and Phytotron
Duke University
Durham, NC 27706
U.S.A.

Claude Veillon
Vitamin and Mineral Nutrition Research Laboratory
Beltsville Human Nutrition Research Center
Agricultural Research Service USDA

Beltsville, MD 20705
U.S.A.

William P. Wergin
Plant Stress Laboratory
Plant Physiology Institute
Agricultural Research Service USDA
Beltsville, MD 20705
U.S.A.

John J. Windle
Development Productivity Laboratory
Western Regional Research Center
Agricultural Research Service USDA
Albany, CA 94710
U.S.A.

F. Eugene Yates
Crump Institute for Medical Engineering
University of California
Los Angeles, CA 90024
U.S.A.

SYMPOSIUM ORGANIZATION

Beltsville Agricultural Research Center (BARC)
Waldemar Klassen, Director, Beltsville Area

BARC SCIENCE SEMINAR COMMITTEE
Patricia C. Augustine, Chairman
Steven Berberich
Gordon Carpenter
Dwight Glotfelty
Donald Keister
Alan Lefcourt
Barbara Leonhardt
Betty Li
Otho Michaelis
John Moseman
David Nickle
Karl Norris
Stephen Rawlins
Richard Sayre
Steven Sinden
John Vandenberg
William Van Der Woude
Lila Vodkin

BARC SYMPOSIUM XI COMMITTEE
Gary R. Beecher
Essex E. Finney, Jr.
Karl H. Norris
Joanna R. Weirman, Secretary

PROGRAM COMMITTEE
Judith A. Abbott
Patricia C. Allen
Charles R. Caldwell
Galen F. Hart
Steven R. Heller
Ken R. Hill
M.D. Huettel
Alan M. Lefcourt
William R. Lusby

REGISTRATION COMMITTEE
Betty Li
Joanna R. Weirman

FINANCE COMMITTEE
John D. Vandenberg

LOCAL ARRANGEMENTS
Ralston R. Rhodes
Gordon T. Carpenter, Sr.

LOGO
S. North

PUBLICITY
Steven M. Berberich
V. Mazzolla

POSTER SESSION
Harry Danforth
John D. Vandenberg

EDITORIAL
Gary R. Beecher
Essex E. Finney, Jr.
Karl H. Norris

Recent advances in detectors for radioisotopes

MICHAEL J. KESSLER
Packard Instrument Company, 2200 Warrenville Road, Downers Grove, Illinois, IL 60515, U.S.A.

Abstract With the increasing use of radioisotopes in agricultural research and the corresponding increase in instrumentation development, several different types of radioactivity detection instruments have been designed. These instruments are being used now and will be used into the 21st century to quantitate radioisotopes used in plants and animals. Five different types of basic radioisotope detectors are available. First, the sample oxidizer which burns the sample in an oxygen atmosphere, then separates and recovers the radiolabeled $^{14}CO_2$ and ^3H_2O for quantitation. Second, the liquid scintillation counter accurately quantitates beta-and low energy gamma-emitting radioisotopes in a homogeneous solution. Third, the imaging scanner quantitates radioisotope components separated on a solid support (TLC, paper, PAGE, etc.). Fourth, radiogas chromatography quantitates radioactive substances separated by a gas chromatograph. Fifth, radioactivity flow detector quantitates on-line, radioactive peaks separated by high performance liquid chromatography (HPLC). The basic theory of operation, special features, and applications each of these radioisotope detectors in agricultural research will be discussed.

Introduction

Five different types of instrumentation for the detection of radioisotopes as applied in agricultural research will be discussed. These instruments are the sample oxidizer, liquid scintillation analyzer, imaging scanner, radiogas chromatography detector, and radioactivity flow detector.

The basic theory of operation and application for each of the detectors will be discussed briefly initially. Then, this chapter will present an in depth description of each of these five radioisotope detectors. The sample oxidizer is used to quantitate radioisotopes in plant and animal samples which are not suitable, because of quenching or solubility problems, for quantitation in a liquid scintillation counter. This is accomplished by burning the sample and quantitating the radioisotope, $^{14}CO_2$ and ^3H_2O, which are formed during the oxidation process. The liquid scintillation counter (LSC), which is used to accurately quantitate radioisotopes in a homogeneous solution, has demonstrated major advances in counting radioisotopes in the areas of quantitation accuracy, automatic efficiency control, chemiluminescence correction, heterogeneity monitoring, pulse indexing, spectral indexing, quench correction parameters, and full spectrum display of single and double labeled isotopes. The linear imaging scanner uses the technique of position sensitive gas flow proportional counting to be able to quantitate radioisotopes on any solid support material. The linear scanner is able to quantitate lanes (one dimensional) on thin layer plates (TLC) or two dimensional separations (TLC or polyacrylamide gels-PAGE) with a resultant topography map, of various radioactive materials on the solid support. The radiogas chromatographic instrument uses two separate reaction tubes and a gas proportional counter to quantitate radioactive compounds separated on a gas chromatograph (GC). Two reaction tubes convert the GC separated compounds to $^{14}CO_2$ and $^{3}H_2$ which can easily be quantitated by a gas proportional counter. This is accomplished by first oxidizing the sample with copper oxide followed by the reduction of sample with a reducing agent. The radioactivity flow detector quantitates, in real time, radioisotopically labeled compounds as they elute from the HPLC system. The radio-HPLC system can use three different methods to quantitate radioisotopes. This quantitation is accomplished by using different types of detector flow cells (liquid, solid, high energy). The isotope specificity, sensitivity, and applications of each of these five radioactivity detectors will be discussed.

Sample oxidation

Basic theory

The sample oxidizer has been used extensively in the area of agricultural research for the quantitation of radioactivity in various samples. This is especially important in agricultural research since plant samples, animal tissues, feces, and other tissues are used in radioisotope tracer studies [1–3]. These experiments require actual quantitation of either single and/or dual labeled samples. The major advantages of using the sample oxidation technique are presented in Table 1. Many agricultural samples are heavily colored (green/plants, red/blood, yellow/urine) and difficult to quantitate directly using the liquid scintillation counting technique. By using the instrument the samples are oxidized in an oxygen atmosphere to produce tritiated water and/or ^{14}C labeled CO_2. These two radioisotopes can then be individually quantitated accurately in a liquid scintillation counter. The oxidation process eliminates chemicals, colored pigments, bleaches used to decolorize samples, organic components (which cause chemiluminescence), and tissue solubilizers (used to make a homogeneous solution for counting by the liquid scintillation technique). The oxidation process allows an accurate comparison of blood, urine, organs, feces, and other body parts because all of the samples are oxidized to two common substances, water and CO_2. The quantitation of each of these radioisotopes can be accurately compared because the quenching factors are the same for each sample because only trapped condensed CO_2 and H_2O are being quantitated. The sample

Table 1. Main advantages of the sample oxidation technique

1.	Elimination or reduction of color and chemical quenching
2.	Elimination of chemiluminescence
3.	Eliminates the need for chemical solubilization of sample
4.	Large sample size can be quantitated (~ 1.5 gm)
5.	Complete separation of tritium and ^{14}C for accurate analysis of both isotopes in same sample (< 0.05 % spillover)
6.	Common counting samples are converted to water and CO_2 (Recovery 99 ± 1 %)

oxidizer also has the distinct advantage of being able to accurately quantitate a double labeled ($^3H/^{14}C$) sample, since the two isotopes are completely separate from each other. An improvement in sensitivity results because no spillover correction and/or quench correction is required to quantitate both radioisotopes.

Instrumentation

The dynamic sample oxidation instrumentation is available in two different combustion types: catalytic [4] and noncatalytic [5]. The catalytic oxidation process involves placing the solid or liquid sample into a tube furnace and burning it in a stream of oxygen. The combustion products and gases are passed over a copper oxide catalyst. This process ensures the complete oxidation of the sample to H_2O and CO_2. Any sulfur or halogens are retained by the catalyst. The water vapor is condensed and the CO_2 is trapped by using a CO_2 absorber (ethanolamine, phenethylamine, etc.). The noncatalytic oxidation type (Fig. 1) can be divided into three sections: oxidation, tritiated H_2O condensation and CO_2 trapping. The sample, either wet or dry, is placed in a platinum-rhodium wire basket. After setting of the proper burn time, an automatic program, microprocessor controlled, process is initiated. The sample in the wire basket is raised into the combustion flask. The oxygen stream is turned on as the wire basket is heated to ignite the sample. The continuous flow of oxygen sweeps the resultant radiolabeled gaseous $^{14}CO_2$ and ^3H_2O through the system. The gaseous products are first passed through a condenser, in the ^3H section of the oxidizer, where the water vapor (tritiated) is condensed and drips into a scintillation vial. In addition, a small amount of water vapor is trapped in the primary and secondary exchange columns. Next, the entire ^3H section is repetitively washed with a scintillation solution to rinse out any residual tritiated water from this section. If the sample is known to contain ^{14}C, the ^{14}C section of the oxidizer is preloaded with a CO_2 absorber, which is soluble in a toluene based scintillation cocktail, prior to the burning of the sample. As the sample is oxidized, the gaseous products (other than water vapor) are passed through a convoluted reaction column where the

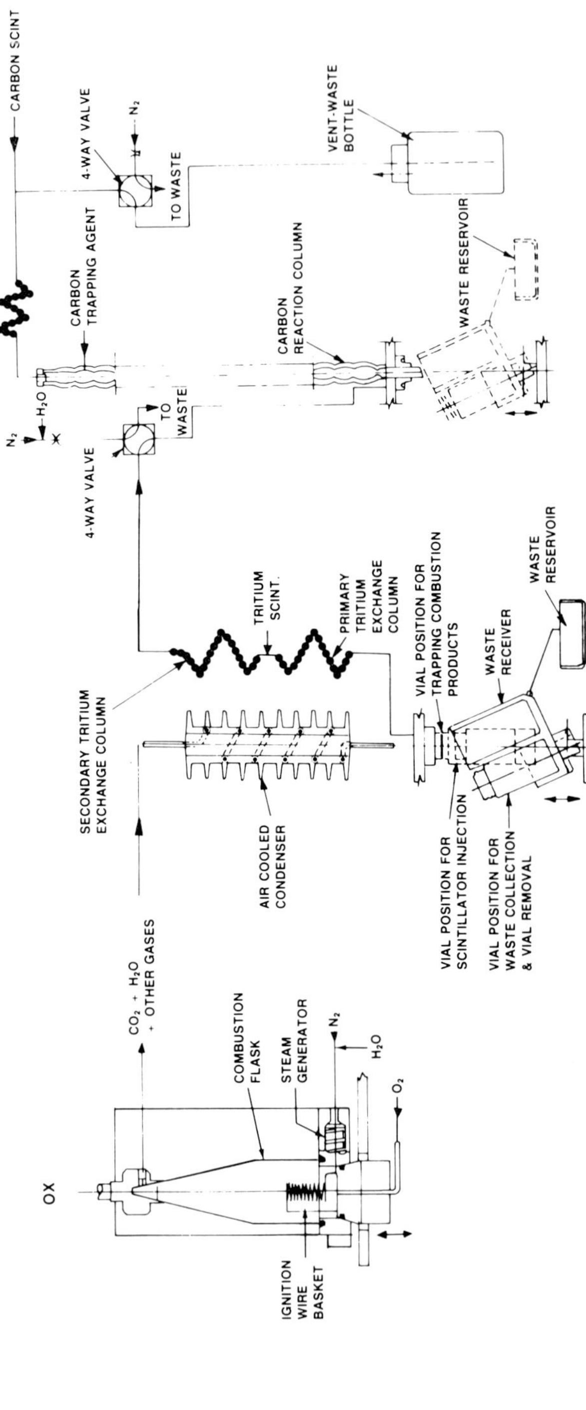

Figure 1. Packard 306 sample oxidizer.

CO_2 is trapped. Any untrapped CO_2 will be recovered in a secondary carbon trapping exchange column (Fig. 1). After the oxidation of the sample, the CO_2 trapping agent is dispensed into a scintillation vial. The entire CO_2 section is rinsed with a toluene based scintillation cocktail. At the end of the oxidation of the sample, the reaction flask is purged with steam and nitrogen. This process purges the water trapped in the condenser, thereby increasing recovery and decreasing memory (radioactivity carryover between samples). The next step is the injection of scintillation cocktail through the primary exchange column and flushing the trapped water into the tritium vial. Finally, scintillation cocktail is injected into the ^{14}C section through the carbon exchange and reaction columns. This, ultimately, washes the trapped $^{14}CO_2$ into a second scintillation vial. The samples are manually capped, and removed, and finally placed into a liquid scintillation counter for quantitation. Thesc steps can also be done by a specially designed robotics system.

Liquid scintillation counters

Basic theory

In order to understand the method of quantitation radioactivity in a liquid scintillation counter, it is essential that the basic theory of liquid scintillation counting [6–9] be investigated. The detection and quantitation of radioactivity, beta and low energy gamma emitting, nuclides is shown in Fig. 2. This quantitative process involves a series of steps. In the first step, the radioisotope emits a beta particle. The energy of the beta particle transfers its energy to an organic solvent molecule. This solvent becomes excited to a higher energy level. The next step is the solvent transferring its energy to a fluor (scintillator). As this molecule releases its energy, by returning to its ground state, a photon of light is released. The intensity (pulse height) of this light is directly dependent on the energy of the beta particle initially released. The released light is then detected and quantitated by a photomultiplier tube (PMT) which measures the number and the energy (intensity) of the photons released by the fluor. Two major factors (quenching agents) can influence the quantitation of the light produced in this multistep process. The first factor (Fig.

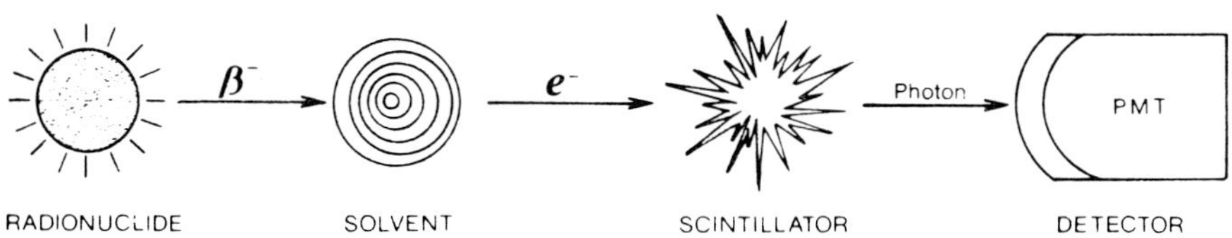

Figure 2. The basic theory of the liquid scintillation counting process.

2) is called chemical quenching and normally interferes in the transfer of the energy of the beta particle to the solvent molecule. The best example of this is the presence of chloroform in the sample, which acts like a sponge to absorb the beta particle (electrons) and prevent the transfer of its energy to the solvent molecule. The second factor (color quenching) is due to the presence of color in the sample.

This interferes with the quantitation of the photon by the photomultiplier released by the scintillator. The color in the sample acts like a colored filter. For example, red and yellow colored samples absorb blue light, the color of the light produced by the scintillator. Therefore, color quenched samples result in an apparent reduction in the amount and intensity of the photons released by the scintillator as seen by the photomultiplier tubes.

Instrumentation

The detection of the photon of light, by the PMT, is the first step in a series of steps involved in actually quantitating the radioactivity and determining the energy of the beta particle (Fig. 3). The photomultiplier converts the photon of light to an electrical pulse. The electrical pulse is first amplified by a preamplifier followed by either a linear or logarithmic amplifier. As can be seen from Fig. 3, two phototubes are used to quantitate the photons emitted by the scintillation process. The two PMTs are necessary in order to eliminate extraneous pulses (single photon events) and thus quantitate only the radioactive pulses produced by the sample. This is accomplished by a coincidence circuit. If both phototubes detect the pulse at the same time then it is a true radioactive

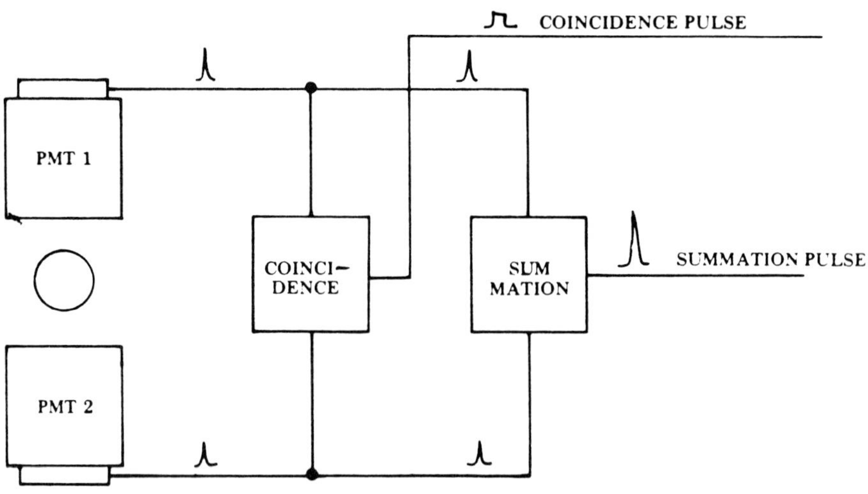

Figure 3. General glock diagram LSC.

pulse, but if only one PMT detects the pulse it could be from cosmic radiation, electronic noise, or extraneous source. The pulses are next summed (pulses from both tubes added together) in order to increase the signal to noise ratio (S/N). The pulses are then sorted and quantitated on the basis of their number and intensity. This is done by one of two methods depending on the liquid scintillation counter being used. The first and older method, is pulse height analysis (PHA). This method uses specific energy window discriminators and collects the pulses which only fall within a predefined energy region. The second method (used in most modern LSC instruments) is the use of a multichannel analyzer (MCA) which collects data from all pulses and creates an energy spectrum which is displayed on a monitor and/or a printer. The complete spectral data can also be stored on a disk for future analysis when required.

Many of the modern features of a liquid scintillation counter are shown in Table 2. Detailed presentations of many of these features can be found in several excellent references [10–12]. Three of the more recent developments will be discussed in more detail in this chapter (3-D spectralyzer, spectral display, and DPM calculations). The 3-D spectralyzer spectrum analyzer allows the instrument to discriminate background counts from the true radioactivity present in the sample. This can be accomplished because background pulses demonstrate a different 3-dimensional pulse height distribution pattern than these beta particles. From the pulse index discrimination plot most background pulses can be excluded from the sample spectrum. This results in a higher performance of the instrument in addition to a lower background level. The second development is that of spectral display (Fig. 4) on a CRT monitor, hard copy on a printer, and spectral data storage via a computer. This spectral display (4096 channels) is ideal for visualizing sample activity, visual optimization of sample counting discriminator setting, interference monitoring, radiation safety checks, environmental and low level counting. The real time display allows the user to perform dynamic studies of such phenomenon as luminescence decay and radioactivity monitoring.

Table 2. Advances in liquid scintillation counting technology

1. Coincidence Counting (1953)
2. Pulse Summation (1963)
3. Automatic External Standard (1963)
4. Automatic Window Tracking (1963)
5. Integrated DPM Calculation (1967)
6. Small Vial Counter (1976)
7. Spectrum Analysis (1979)
8. Luminescence Correction (1980)
9. Three Dimensional Spectrum Analysis (1985)
10. Efficiency Tracing Technique (1985)

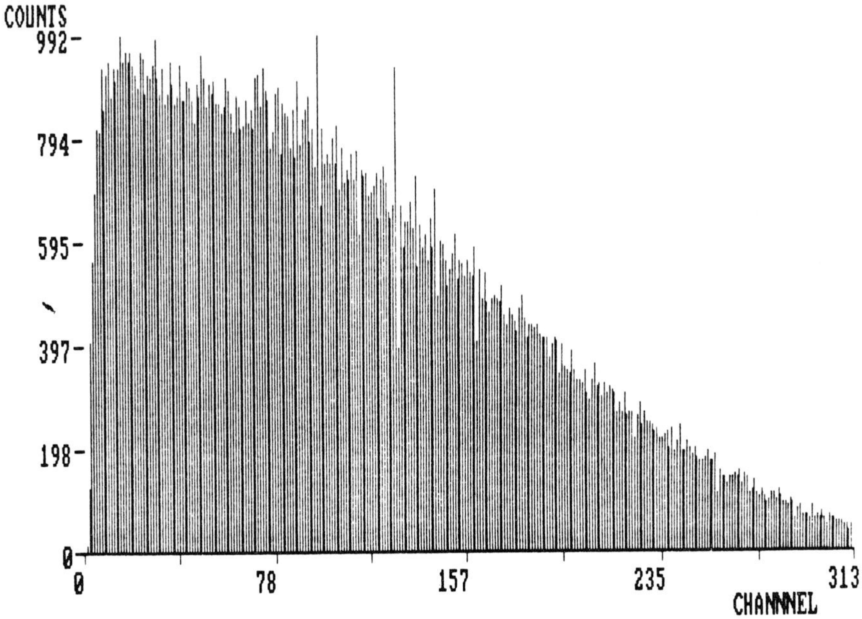

Figure 4. Multichannel analysis plot of ^{14}C (4096 Channels).

Another development is the automatic correction of the sample for quenching (color and chemical), elimination of vial wall effect, elimination of vial size and sample volume dependence using a single correction factor for calculating DPM, specific activity in the sample. The most modern method is that of Transformed Spectral Index using an external standard sample spectrum (tSIE). This feature calculates the DPM present in the sample giving results independent of the factor mentioned above. This allows a rapid and accurate DPM measurement for each sample independent of most factors which can influence the calculation of DPM in the sample.

Using these modern features, liquid scintillation counting has become an accurate method of quantitating radioactivity in soluble or homogeneous sample mixture. Many examples for the use of liquid scintillation counting in agricultural research are available for both plant and animal studies [13–15].

Imaging scanners

Basic theory

Several techniques are available for the detection and/or quantitation of radioactivity on a solid support. An evaluation and comparison of these techniques is important because several applications in agricultural

Table 3. Comparison of instrumentation technique for quantitating and detecting radioactivity on a solid support

Characteristic	Techniques				
	Radioauto–graphy	Spark Camera	LSC	Chrom Scanner	Linear Scanner
Sensitivity	L	M	H	L	M
Resolution	H	M	L	M	M
Quantitation	L	L	H	M	H
Cost/Sample	L	L	H	L	L
Health Hazard	L	L	H	L	L
Sample Recovered	Y	Y	N	Y	Y
Speed of Analysis	S	F	M	M	F

L = Low, H = High, M = Medium, Y = Yes, N = No, S = Slow, F = Fast.

research use radioactivity and solid supports. These applications include thin layer chromatography (TLC), a method of separating compounds on a thin layer coated on a solid support. Polyacrylamide gel electrophoresis (PAGE), a method for separating large molecules on the basis of either their size and/or charge. A more recent method is the application of membrane transfer technique in which a molecule is transferred from a gel matrix to a membrane for quantitation – Southern (DNA), Northern (RNA), and Western (protein) blots. As can be ascertained, solid supports and radioactivity are used routinely in the separation and quantitation of various agricultural products [16–18].

Five techniques can be used to either detect and/or quantitate the radioactivity on these solid supports. These include radioautography [19], spark camera [20], liquid scintillation counting [21], radiochromatogram scanners [22] and radiochromatogram imaging systems [23]. The capabilities of each technique is summarized in Table 3. Radioautography involves the use of a photographic or x-ray film to record the ionizations produced by beta particles or x-rays. This technique provides excellent spatial resolution but low sensitivity and quantitation. The spark camera uses an ionizable gas (P10) filled chamber which produces light along an ionizing track which can then be photographed. This technique has high sensitivity but low resolution and quantitation. The liquid scintillation technique, the most sensitive technique, involves scraping or cutting of the solid support at regular intervals (a very time consuming process) and quantitating in a liquid scintillation counter. This provides excellent sensitivity and quantitation but has very poor resolution and is very time consuming. The radiochromatogram scanner plots counts against position using a Geiger or proportional counter. This technique provides moderate sensitivity and resolution but quantitation is difficult. The radiochromatogram imaging system is the most

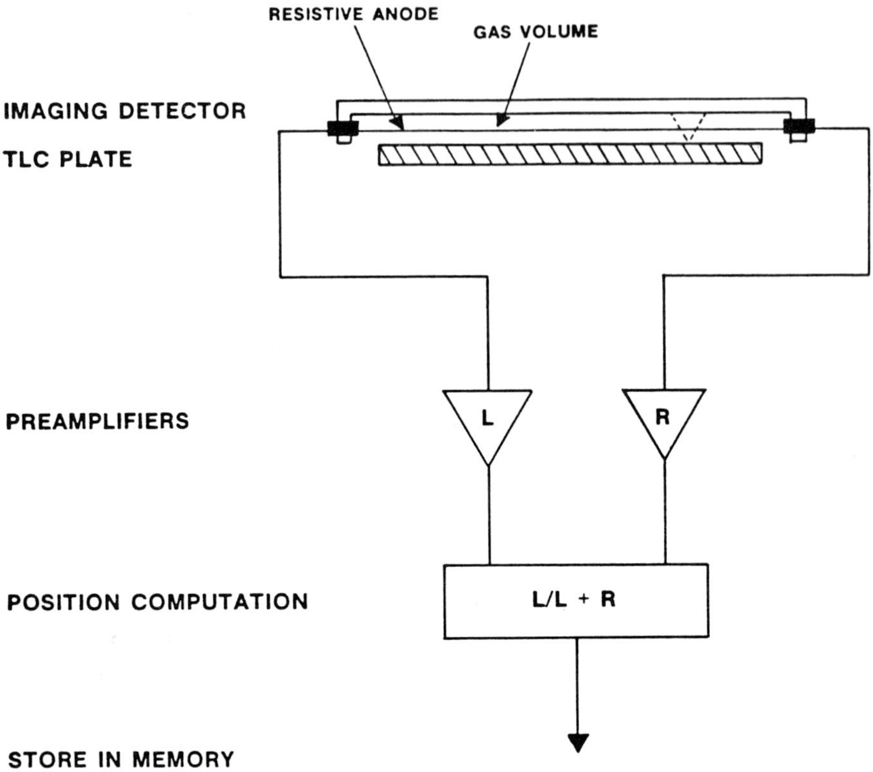

Figure 5. Block diagram of resistive anode imaging scanner.

modern and fastest instrumentation technique. It uses an imaging proportional counter (resistive anode or delay line) to quantitate and locate the radioactive substance on a solid support. This technique, which will be discussed in more detail next, has a high sensitivity, excellent quantitation, moderate spatial resolution, and excellent speed of analysis. Because an entire section ($\sim 1.0 \times 25$ cm) can be analyzed at a time a complete quantitation of a 2-D separation sample (PAGE or TLC) can be done in a matter of minutes or hours.

Instrumentation design – Imaging system

Imaging system. The imaging scanner system usually consists of two parts, a position sensitive proportional counter and a data reduction system. The position sensitive counter can be one of two basic designs. The first design (Fig. 5) called the resistive anode technique, uses a large metal anode with a uniform resistance over its entire length. The radioactivity interacts in the ionizer detector gas (P10, argon/methane, 90 %/10 %) producing free electrons. These electrons are accelerated toward the high voltage anode. A multiplication of the electrons occurs and they are detected as a series of pulse at the anode. If the anode is resistive, the pulse divide and flow toward both ends of the anode. To obtain the position of a radioactive event, the pulse amplitudes at each end of the anode are converted to a digital number by an analog to digital converter. The resultant numbers are sent to a computer for the quantitations and location of the radioactivity. The second design uses the same basic counting technique of an ionizable gas (proportional counting) but instead of a resistive anode employs a delay line. As the electrons are collected by an anode, a positive ion cloud is induced on the delay line. This signal propagates toward the two ends of the delay line with a given velocity. The difference in the arrival time of the signal at both ends of the delay line are sent to a TAC (time-to-amplitude converter). The TAC determines location and the quantitation of the radioactivity on the solid support.

Performance and data presentation. Because of the insensitivity of the gas proportional counter, only low efficiencies can be obtained for low energy beta emitters (~ 1.0 % for ^{3}H) but high efficiency can be obtained for ^{14}C, ^{32}P, ^{125}I, ^{35}S, and ^{99m}Tc (~ 20 %). But because of the low backgrounds (~ 0.20 cpm/mm of solid support) and the fact that the

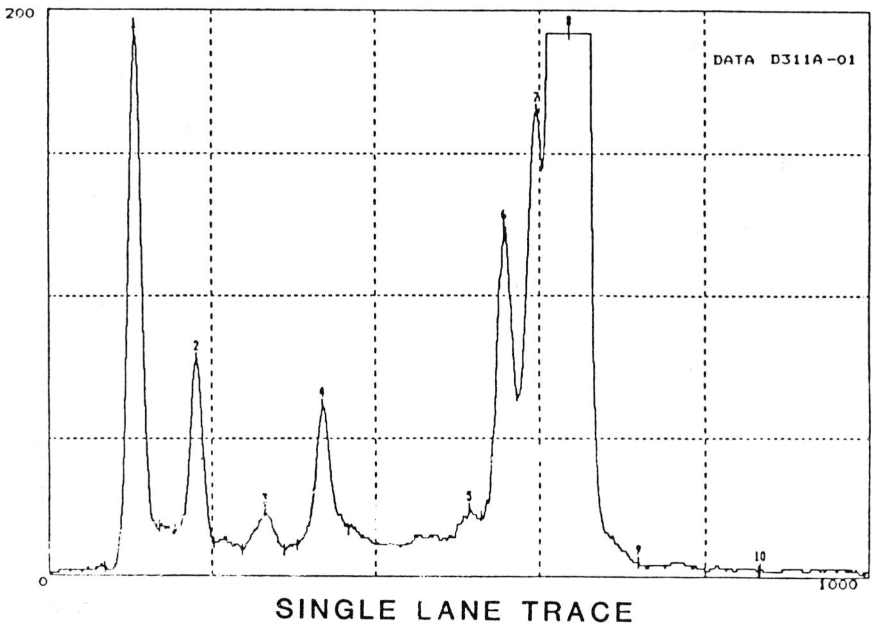

Figure 6. Graphic plot and data presentation of a single dimensional TLC plate with ^{14}C metabolites.

sample can be counted for long periods of time (1–999 minutes), radioactive peaks as low as 100 dpm for ^{14}C and 1000 dpm for ^{3}H can be quantitated on a solid support in less than 10 minutes. The overall sensitivity is dependent on the actual proportional counter used in the instrument. The uniformity of response for the determination of location of radioactivity is dependent on the design of the delay line and resistive anode used in the instrument. If the instrument is functioning properly the response (quantitation and location) of a series of radioactive spots should be uniform over the entire 1.0 × 25 cm counting chamber. The resolution of two radioactive spots using this technique is dependent on the energy of the beta-particle being emitted. The higher the energy, in general, the worse the resolution. Typical resolution values for various isotopes are as follows: ^{3}H, 1–2 mm, ^{14}C, 2–3 mm, ^{32}P, 3—4 mm. The exact resolution, linearity, and uniformity of response depends on the exact design and implementation of the basic principles described earlier.

The data output for an imaging scanner can be presented visually via a CRT and/or a printer. Three basic types of data presentation are available (single lane, multiple lanes, and aerial map). The first, single lane, presents a 1:1 proportional plot of the radioactivity present in a single lane 1.0 × 25 cm) on the solid support. Peak area, peak number, retention times (min), % of total activity can be presented on the 1:1 dimensional plot (Fig. 6). The second, multi-lane (Fig. 7), enables the

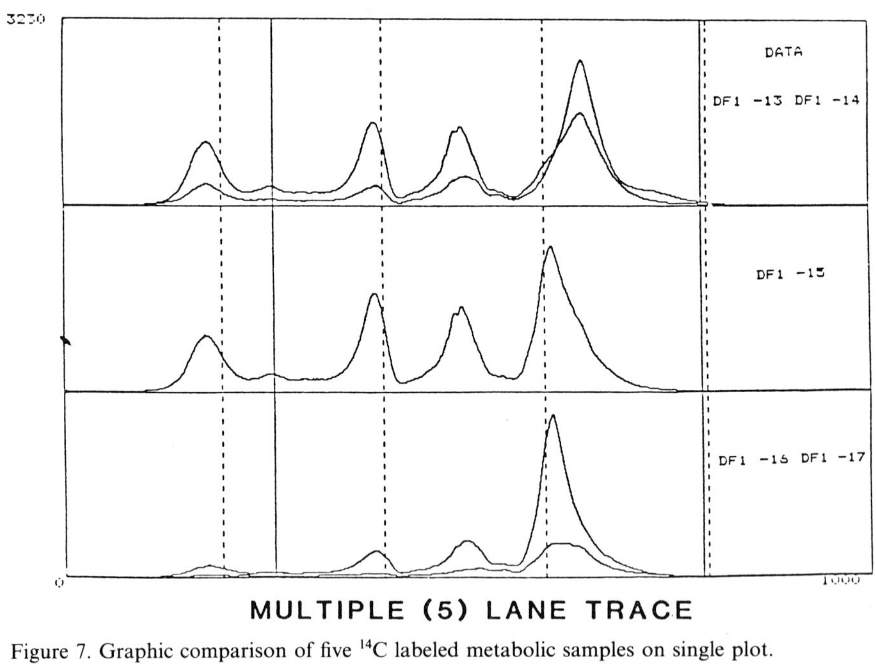

Figure 7. Graphic comparison of five ^{14}C labeled metabolic samples on single plot.

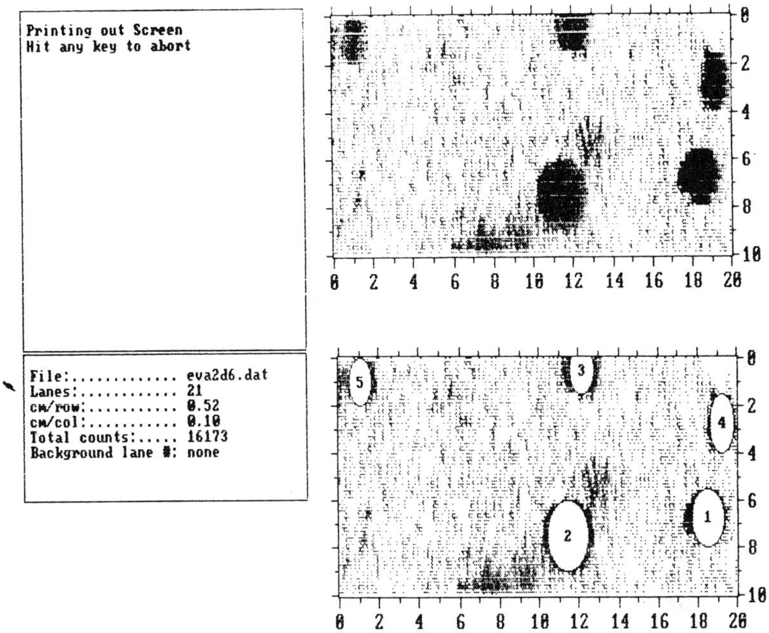

spot	X	Y	% of total	counts	shape
1	18.50	6.89	25.2	4070	Ellipse
2	11.49	7.67	28.3	4582	Ellipse
3	12.18	0.45	13.4	2164	Ellipse
4	19.27	2.74	9.7	1563	Ellipse
5	1.12	0.97	5.1	825	Ellipse

Figure 8. Graphic presentation and data reduction of ^{14}C metabolites chromatographed on 2-D TLC plate.

user to be able to compare various lanes to a standard sample lane. In addition, data from several lanes can be added or subtracted, subtract background, and normalized to a single lane or a single peak in a lane of radioactive peaks. The third, topographic presentation, is becoming a very useful and popular feature of the imaging scanner. This permits the user to be able to scan an entire solid support (2-dimensional – TLC or PAGE) and present a two dimensional map of the radioactive spots using 1:1 dimensional plot (Fig. 8). The radioactive spots can then be quantitated using cursor keys on the computer, a computer mouse, or a light pen. An example of the quantitative results is shown in Fig. 8. Each spot can be quantitated by total count/peak, % of total, and X and Y coordinates of the solid support. This permits the user to easily quantitate and locate the radioactivity on a two-dimensional solid support. The radioactivity can then be removed from the solid support for further

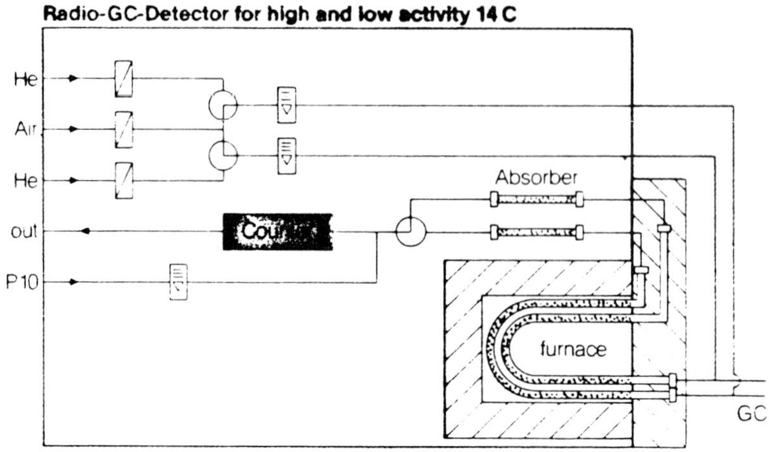

Figure 9. Block diagram of radio G.C. detector system.

analysis and/or quantitated more accurately using a liquid scintillation counter.

Radioactivity gas chromatography detector

Basic theory and instrumentation

The use of gas chromatographs to separate and quantitate various agricultural substances has been a very popular technique [24]. Since tritium and carbon-14, the most frequently used radioisotopes, are both low energy beta emitting isotopes, they require a special type of detector to accurately quantitate them. The original method of gas proportional counters suffered from the problems of insensitivity, quenching of gas amplification, and low electrical resistance which resulted in the destruction of the operation of this type of counter and gave irreproducible results. The latest radioactivity gas chromatography detector (Fig. 9) involves a series of steps and reactions before quantitation of the radioactivity was a gas proportional counter. The first step uses copper oxide to oxidize the sample to CO_2 and H_2O. The second step is the reduction of the H_2O to H_2 gas. The CO_2 and H_2 gases are combined with P10 gas and quantitated using a gas proportional counter. The counter uses a small internal diameter quartz glass tube for the oxidation and reduction processes. The detector has a low background because of the optimization of the detector design and the 5.0 cm lead shielding used. The detector has high sensitivity because of the oxidation/reduction process and the 4π counting geometry of the system. In addition, a low noise, high sensitivity preamplifier is used to ensure that all radioactivity signals are quantitated.

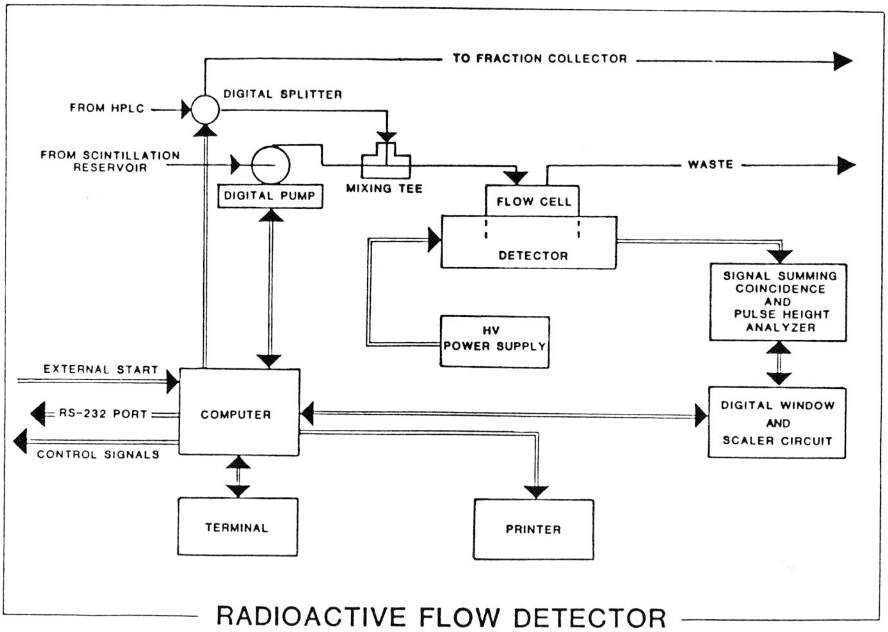

Figure 10. Block diagram of typical radioactivity flow detector.

Radioactivity flow detector for HPLC

Basic theory

The method of detection of radioactivity in a flowing stream, usually HPLC, is that of scintillation counting by one of three different types of methods (solid, high energy, liquid cells). The beta or low energy gamma ray emitted by the nuclide activates a solvent molecule (liquid cell) in the liquid scintillation cocktail, then a resultant fluor is excited. If a solid cell is used the radionuclide directly transfers its energy (no solvent intermediate) to a scintillation molecule (fluor). The fluor converts the energy of each beta particle directly to a flash of light, whose intensity is directly proportional to the energy of the beta particle. Since the energy of the beta particle is extremely small (low intensity), a pair of photomultiplier tubes are used to measure the light flashes. The resultant pulses are sorted by a pulse height analyzer board, with a resultant energy spectrum being analyzed (Fig. 10). This information is then sent to a computer for calculations, analysis and graphic presentation of the resultant radiochromatogram.

Now that the basic theory of radioactivity flow-through detector has been examined, it is essential to differentiate how flow counting differs from standard static scintillation counting. An excellent model uses a train to illustrate the basic concept. This basic model assumes that flashing lights (radioactivity) exist as discrete segment (train cars) on a train travelling down a track at a specific speed (flow rate of HPLC and scintillation fluid). An individual counter (photomultiplier tube) is standing in the train station to quantitate the light flashes seen on the train cars as they pass through the train station. Two basic parameters need to be determined for quantitating in a radioactive flow detector: background counts and sample counts. The flow system determines these two parameters using different methods. The background counts are composed of radioactivity eluting from the HPLC, electronic noise, and cosmic radiation and are independent of the flow rate of the system. This background count can be depicted as a flashing light on a stationary pole in the train station (independent of speed of train). The sample counts in the flowing system (moving train cars) are dependent on two separate parameters: flow rate (train's speed) and cell volume (size of station). These two factors determine the residence time of the sample (train car) in the cell (eqn 1):

$$\text{Residence Time (min)} = \frac{\text{Cell volume (ml)}}{\text{Total flow rate (ml/min)}} \qquad (1)$$

These parameters are used to calculate the residence time of the sample in the flow cell (how long the PMT counts the radioactivity) and subsequently used to calculate the number of CPM of each peak in the sample (eqn 2):

$$\text{CPM} = \frac{\text{counts} - (\text{Background} \times \text{Residence Time})}{\text{Residence Time}} \qquad (2)$$

Finally, the actually DPM in the sample is calculated by one of two methods, depending on whether the counting efficiency is changing or remains constant during the HPLC elution profile (eqn 3):

$$\text{DPM} = \frac{\text{CPM}}{\%\ \text{Efficiency}} \qquad (3)$$

If the efficiency is constant, DPM (isocratic), the DPM in each peak can be simply calculated. If the efficiency changes, DPM (gradient), then a different efficiency is calculated relative to time. Thus, a different efficiency is calculated for each peak with the DPM for each peak properly corrected for the changing efficiency.

Instrumentation

Sensitivity of radioactivity flow detector. Two basic factors contribute to the overall sensitivity of the radioactivity flow detector: cell type and

Table 4. Radioactivity detector flow cells

Cell type	Scintillator type	% Efficiency
Standard (liquid)	Mix HPLC eluent with nongelling scintillation cocktail	$^{3}H > 40\%$ $^{14}C > 80\%$
Solid	Plastic (200 micron) Glass (80 micron) CaF_2 (Eu) (100 micron)	$^{14}C > 20\%$ $^{14}C > 20\%$ $^{14}C > 75\%$ $^{3}H \approx 0.2\%$
High Energy Isotope	No solid support No scintillation cocktail, scintillator plastic window	$^{32}P > 40\%$ $^{99m}Tc > 30\%$ $^{125}I > 20\%$

residence time. First, how does the cell type effect sensitivity? Three different cell types (solid, liquid and high energy isotope) are commonly used in radioactivity flow detectors (Table 4). The liquid is the most commonly utilized cell because it gives the highest sensitivity (% efficiency) but requires mixing the sample with scintillation solution – no sample recovery. The solid cell is most commonly used in quality control analysis or when the exact nature of the radioactivity components is known. This cell type also requires a higher number of DPM than the liquid cell type. This is due to the lower counting efficiency of the solid cell. The major problem with this cell type is that it can become easily contaminated. The high energy isotope cell is normally used to quantitate very high energy beta emitters (^{32}P) and low energy gamma emitters (^{125}I, ^{99m}Tc). It has the advantage of using no solid scintillation (no contamination) and no liquid scintillation cocktail mixing required (sample not lost).

The second factor influencing sensitivity is a factor called residence time. This factor relates directly to the amount of time that each radioactive peak is in the flow cell. The residence time is directly dependent upon the flow cell size and inversely dependent on the total flow rate of either the HPLC element (solid cell) and scintillation cocktail (liquid cell). Thus, the maximum sensitivity is produced by a large flow cell and/or a lower total flow rate.

Radioactivity flow detectors – Data presentation and reduction. The radioactivity flow detector provides a real time on-line graphics presentation of up to two radioactivity channels and two input HPLC detector signals (UV, fluorescence, electrochemical, visible) on a single graphics presentation (Fig. 11). The graphics presentation and data summary calculates net counts, cpm, dpm (isocratic or gradient) corrected for sample quenching. Each radioactive peak is numbered, retention time

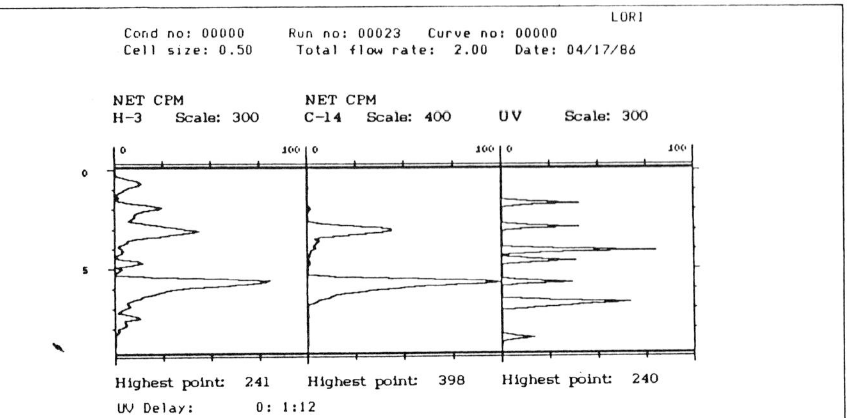

NET CPM
H-3

PEAK	INTEG	FROM	TO	RET	%RUN
1	215	0.20	1.30	0.70	6.58
2	431	1.30	2.60	1.90	13.17
3	786	2.60	3.90	3.10	24.03
4	118	4.40	4.90	4.70	3.60
5	1516	5.30	6.70	5.60	46.36
6	78	6.70	7.20	6.90	2.38
7	127	7.20	7.70	7.50	3.88
--	3269				TOTAL

NET CPM
C-14

PEAK	INTEG	FROM	TO	RET	%RUN
1	849	2.60	3.50	3.00	26.36
2	118	3.50	4.50	3.70	3.65
3	2254	5.30	6.90	5.70	69.99
--	3221				TOTAL

UV

PEAK	INTEG	FROM	TO	RET	%RUN
1	190	1.50	1.90	1.70	10.35
2	190	2.70	3.10	2.90	10.35
3	380	3.90	4.30	4.10	20.71
4	210	4.40	4.80	4.60	11.44
5	200	5.50	5.90	5.70	10.90
6	575	6.50	7.10	6.70	31.34
7	90	8.30	8.70	8.50	4.90
--	1835				TOTAL

RADIOCHROMATOGRAM OF DUAL-LABELED METABOLITES

Figure 11. Graphic presentation of a dual labeled sample eluting from the HPLC with the UV signal plotted on the same graph.

(elution time) reported, total area (defined with start and stop markers) and % that each peak is of the total radioactivity eluting from the HPLC is presented. All of these parameters as well as the conditions used for the experiment are also recorded on the printout of the graphic results. The raw data or processed data from each HPLC run can be stored permanently on a floppy disk (20 hours) or on a 10 megabyte hard drive for data storage (30 days). The stored data can be reprocessed, replotted, various backgrounds subtracted, and the resultant CPM corrected for quench. Some flow detectors provide a special data reduction package which allows for selective background subtraction, skimming of minor peaks, smoothing data, expanded region of interest, with manual and automatic integration markers.

Advantages of radioactivity flow detection

The flow-through radioactivity detector has three advantages over the previous method of fraction collection/liquid scintillation counting. First, the flow-through detector enables the investigator to obtain quantitation of each radioactive peak in real time as it elutes from the HPLC. This includes on-line graphics, data reduction, and integration of each peak automatically as it elutes from the system. Second, the time required for analysis is substantially reduced from approximately 12 hours to only the time of the HPLC run (5–60 minutes). Third, the cost of analysis is substantially decreased using the radioactive flow detector to approximately one-tenth of the fraction collector/scintillation technique, from $3.00/sample for flow detector to $30.00/sample for the alternative method. Therefore, the radioactivity flow detector can substantially assist the user to obtain immediate quantitative results of radioactive molecules eluting from the HPLC in about 5 % of the time and at about 10 % the cost of the alternative method.

The overall sensitivity, resolution, and reproducibility of the radioactive flow detector have been examined in detail [24–27]. In general, the minimum detectable limit was found to be 40 dpm ^{14}C and 80 dpm ^{3}H using a 2.5 ml liquid cell. The reproducibility of peak areas from repetitive injections was found to be 2–5 % for counts above 2500 and 5–10 % for counts from 500–2500 dpm. The resolution for a 2.5 ml flow cell at a flow rate of 5.0 ml/min was found to be 1.5 min for baseline resolution and 15 seconds for a 0.5 ml cell using 5.0 ml/min flow rate. These results are excellent, and in most cases better than the fraction collector/scintillation counter technique.

References

1. Davidson, J.D. and Oliverio, V.T. 1968. In *Advances in Tracer Methodology* (Rothschild, S. ed.), Vol. 4, p. 67. Plenum, New York.
2. Peterson, J.I., Wagner, F., Siegel, S. and Nixon, W. 1969. Anal. Biochem. 31: 189.
3. Peterson, J.I. 1969. Anal. Biochem. 31: 204.
4. Davidson, J.D., Oliverio, V.T. and Peterson, J.I. 1970. *The Current Status of Liquid Scintillation Counting* (Bramsome, E.D., Jr. ed.) p. 222, Grune & Stratton, New York.
5. Sher, D.W., Kaartinen, N., Everett, L.J. and Justes, V. Jr. 1971. *Organic Scintillators and Liquid Scintillation Counting* (Horrocks, D.L. and Peng, C.T. eds.), p. 849, Academic Press, New York.
6. Kobayashi, Y. 1961. *Liquid Scintillation Counting and Some Practical Considerations*, Waltham, Mass., Tracerlab.
7. Kobayashi, Y. and Maudsley, D.V. 1970. *Practical Aspects of Double Isotope Counting.* (In Bransome, E.D. Jr. ed.), The Current Status of Liquid Scintillation Counting, New York and London, Gruna and Stratton, pp. 273–282.
8. Horrocks, D.L. 1969. *Liquid Scintillation Counting*, Surv. Progr. Chem. 5, pp. 185–235. See also Horrocks, D.L., Applications of Liquid Scintillation Counting, London, Academic Press, 1974.
9. Newton, Hayes, F. 1962. *Solutes and Solvents for Liquid Scintillation Counting*, Tech. Bull. No. 1, Rev. ed., Downers Grove, IL., Packard Instrument Company.
10. Ring, J.R., Nguyen, D.C. and Everett, L.J. 1980. *SIS-Spectral Index of a Sample and SIE-Spectral Index of an External Standard* in *Liquid Scintillation Counting: Recent Application and Development*. Ping, C.T., Horrocks, D.C. and Alpin, E.C. (eds.), Vol. 1, Academic Press, New York, p. 89.
11. De Filippis, S.J. and Everett, L.J. 1983. *Determination of Sample Homogeneity by Liquid Scintillation Counting*, in *Advances in Scintillation Counting*, McQuarrie, S.A., Ediss, C. and Wiebe, L.I. (eds.), Univ. Alberta, Edmonton, Alberta, Canada.
12. van Cauter, S. 1985. *Three Dimensional Spectrum Analysis, A New Approach to Reduce Background of Liquid Scintillation Counters*, Packard Instrument Company.
13. Synthesis and Applications of Isotopically Labeled Compounds, 1985. Proceedings of the

Second International Symposium, Kansas City, MO., U.S.A., 3–6 September 1985, Muccino, R.R., (ed.), pp. 219–224.
14. McQuarrie, S.A., Ediss, C. and Wiebe, L.I. 1983. *Advances in Scintillation Counting*, University of Alberta, Edmonton, Alberta, Canada.
15. Peng, Chin-Tzu, Horrocks, D.L. and Alpen, E.L. 1980. *Liquid Scintillation Counting*, Recent Applications and Development, Academic Press, A Subsidiary of Harcourt Braco Jovanovich, Publishers, New York, London, Sydney, Toronto, San Francisco.
16. Baird, W.M., Chern, C.J. and Diamond, L. 1977. Cancer Res. 37: 3190–3197.
17. Daird, W.M., Chemerys, R.A., Chern, C.J. and Diamond, L. 1978. Cancer Res. 38: 3432–3437.
18. Baird, W.M. and Diamond L. 1976. Chem. Biol. Interact. 13: 67–75.
19. Yang, S.K. and Dower, W.V. 1975. Proc. Nat. Acad. Sci. USA, 72: 2601–2605.
20. Gabriel, A. and Bram, S. 1974. FEBS Lett. 39: 307–309.
21. Zanivsky, Yu. V., Chernenko, S.P., Ivanov, A.B., Kaminir, L.B., Peshekhonov, V.D., Senchankov, E.P., Tyapkin, I.A. and Kalinin, V.N. 1978. Nuclear Instruments and Methods, 153: 445–447.
22. Baird, W.M., Diamond, L., Borun, T.W. and Shulman, S. 1979. Analytical Biochem. 99: 165–169.
23. Goulianos, K., Smith, K.K. and White, S.N. 1980. Analytical Biochem., 103: 64–69.
24. Policarpo, A.J.P.L., Alves, M.A.F., Dos Santos, M.C.M. and Carvalho, M.J.T. 1972. Nucl. Instr. and Method, 102: 337.
25. Cazer, F.D. and Kessler, M.J. 1983. In *Synthesis and Applications of Isotopically Labeled Compounds*, Duncan, W.P. and Susan, A.B., (eds.), Elsevier Scientific Publishing Comp., Amsterdam, pp. 223–227.
26. Kessler, M.J. 1983. J. of Chromatography 255: 209–217.
27. Kessler, M.J., Analytical and Chromatographic Techniques in Radiopharmaceutical Chemistry, Wieland, D.M., Tobes, M.C., and Manager, T.J., (eds.), Springer-Verlag, New York, pp. 149–170.

Short-lived isotope kinetics: a window to the inside

JOHN D. GOESCHL, YOUHANNA FARES, C.E. MAGNUSON and H.W. SCHELD
Phytokinetics Inc., Suite 101-E, 707 Texas Avenue, College Station, TX 77840, U.S.A.

BOYD R. STRAIN and C.H. JAEGER
Botany Department and Phytotron, Duke University, Durham, NC 27706, U.S.A.

C.E. NELSON
Division of Clinical Radiation Oncology, Radiation Oncology Center, School of Medicine, East Carolina University, Greenville, NC 27834-4354, U.S.A.

Abstract An integrated approach to studying the effects of environmental factors on plants is described. The central theme of the system is the use of CO_2 labelled with ^{11}C supplied continuously to the plant and following the emitted radiation *in vivo* in the leaf, the phloem, and the various sinks. The system consists of six components and with minor changes can be used with other tracers such as ^{13}N for nitrogen fixation studies. Because of the short half-life of the isotope, several experiments can be carried out on the same plant under the same environmental conditions without disturbing the plant. When this system is coupled with mathematical models of phloem transport and carbon allocation a host of ecologically, agriculturally, and genetically important questions can be answered (redundent). Carbon assimilation and allocation patterns in plant leaves were measured using this system. The concepts and analytical techniques of tracer kinetics of extended square wave (ESW) input of $^{11}CO_2$ were used in calculating the parameters of carbon allocation dynamics in plant systems under different physiological conditions. Examples of the latter include (1) recovery from mechanical agitation, (2) diurnal patterns, and (3) effects of drought stress.

Introduction

Possible interactions between photosynthesis, storage, and the export of photosynthetic products in leaves have long been major topics of scientific and agronomic interest. The question of whether photosynthesis or translocation controls productivity is of great agronomic importance, because many efforts are being directed at selecting plant varieties with high rates of photosynthesis. For example, Neales and Incoll (1968) revived an hypothesis proposed 100 years earlier by Baussingault (1868) i.e., . . . that the accumulation of assimilates in an illuminated leaf may be responsible for a reduction in net photosynthesis rate of the leaf. They concluded (as of 1968) that the hypothesis was supported in general by a considerable body of experimental evidence, but not strictly proven.

Since then, a number of papers e.g. Moorby, Troughton, and Currie, (1974); Fensom, Williams, Aikman, Dale, Scobie, Ledingham, Drinkwater, and Moorby, (1977); Geiger, (1979b); Austin, Bingham, Blackwell, Evans, Ford, Morgan, and Taylor, (1980); Zelitch, (1982); books e.g. Wilson, (1972); Wareing and Patrick, (1975); Mitsui, Miyachi, San Pietro, and Tamura S., (1977), Siegleman & Hind, (1978); Carlson,

(1980); Hesketh & Jones, (1980); Rachie & Lyman, (1981); and conferences e.g. Kende & Zeevaart, (1979), have broadened this concept of feedback control to include the relationships between stomatal behavior, photosynthesis (source strength), photorespiration, transport, sink assimilation (sink strength), and crop productivity. Despite the magnitude of research and publication on this subject, there is still a need for improved conceptual and experimental approaches.

As indicated by Geiger (1979a), a well-adapted plant should be viewed as an integrated system or continuum of sources, translocation paths, and sinks. The study of carbon flow through this continuum should be developed from a rigorous systems analysis approach, both conceptually and technically. Conceptually, carbon is considered as moving through a series of compartments or pools for which there are characteristic kinetics. The mathematical description of the systems kinetics gives a quantitative predictive model. When methods of compartmental analysis are applied, parameters such as carbon exchange rates, relative pool sizes, turnover rates, export rates, and translocation speeds can be calculated and compared to those predicted by detailed mechanistic models (Atkins, (1969); Jacquez, (1972); Moorby and Jarman, (1975); Geiger, Ploeger, Fox and Fondy, (1983)). It is important that both time-dependent and steady-state kinetic parameters are obtained.

Since most other measurements require destructive sampling at various intervals, much of the short term interactive information is lost, and the impact of the experimental conditions and procedures on the plants' behavior may not be known. Real-time non-destructive techniques for measuring many of the dynamic parameters of carbon flow in plants are needed. The use of radioactively labelled $^{11}CO_2$ makes these measurements possible *in vivo*, where it is unnecessary to know either the biochemical pathway or the constituents of each compartment in order to analyze the data (Moorby and Jarman, (1975)). This approach has proven valuable in pharmacokinetics studies (cf. Rescingo and Segre, (1966); Jacquez, (1972)); and in medical studies such as energy metabolism in the brain, e.g. cerebral glucose utilization (Sokoloff, et al., (1977)). Indeed the whole discipline of nuclear medicine is based on quantitative and qualitative analysis of dynamic tracer flow and distribution through animal and human organs.

As mentioned above, the principles of tracer kinetics can be applied to any live biological system, including plants. In this work we use these concepts in the calculations of some of the dynamic parameters of carbon partitioning and allocation in plant leaves. The results of three experiments using $^{11}CO_2$ to study the influence of biological and environmental factors on carbon partitioning will be used as examples of these calculations.

Extended square wave

C-11 was first used in plant photosynthesis studies in 1939 by Ruben, Hassid, and Kamen and as pulsed input of $^{11}CO_2$ in translocation research in 1963 by Moorby, Ebert, and Evans, (1963). In our experimental program we used $^{11}CO_2$ as a tracer for the reasons discussed in the literature (cf. Fares, DeMichele, Goeschl, and Baltuskonis, (1978); and Magnuson, Fares, Goeschl, Nelson, Strain, Jaeger, and Bilpuch, (1982)) and the references therein). The continuous $^{11}CO_2$ production system allows for pulse as well as steady-state labelling, the latter of which has a number of advantages as discussed by Geiger, (1979a); Fisher, (1975a, b); Fares, et al., (1978); and Magnuson, et al., (1982).

In order to obtain both time dependent and steady-state information we use an extended square wave pattern (ESW), which results in a leading edge (buildup), a trailing edge (washout), and a period when the plant approaches isotope steady-state. We will show below how the steady-state and washout data can be used to analyze carbon allocation in leaves.

The central theme of this approach is the utilization of CO_2 labelled with the radioactive isotope ^{11}C. However, before proceeding to describe the system's components, there are two questions that should be answered first.

(a) Why the use of C-11?

The possible use of C-11 was first suggested by workers in the 1930's (see for example Fares et al., 1978 and references therein). As a tracer, C-11 has several advantages. First it decays by positron (β^+) emission followed by positron-electron annihilation with the emission of two oppositely directed gamma rays. The γ-rays have sufficient energy (0.511 MeV) to be detected through several cm of tissue *in vivo* and in time coincidence, which makes it possible to localize the source and to reduce undesired background activity. Second, the short half-life of C-11 (20.3 min) makes it possible to perform several experiments on the same plant under the same set of environmental conditions. Finally, the half-life is comparable to the turnover times of the photosynthetic pool and the velocity of transport, making possible dynamic measurements which cannot be obtained with long-lived tracers.

(b)Is there a unique requirement?

Since the half-life of C-11 is about 20 min, it is preferable if not necessary that the radioactive isotope production facility be as close as

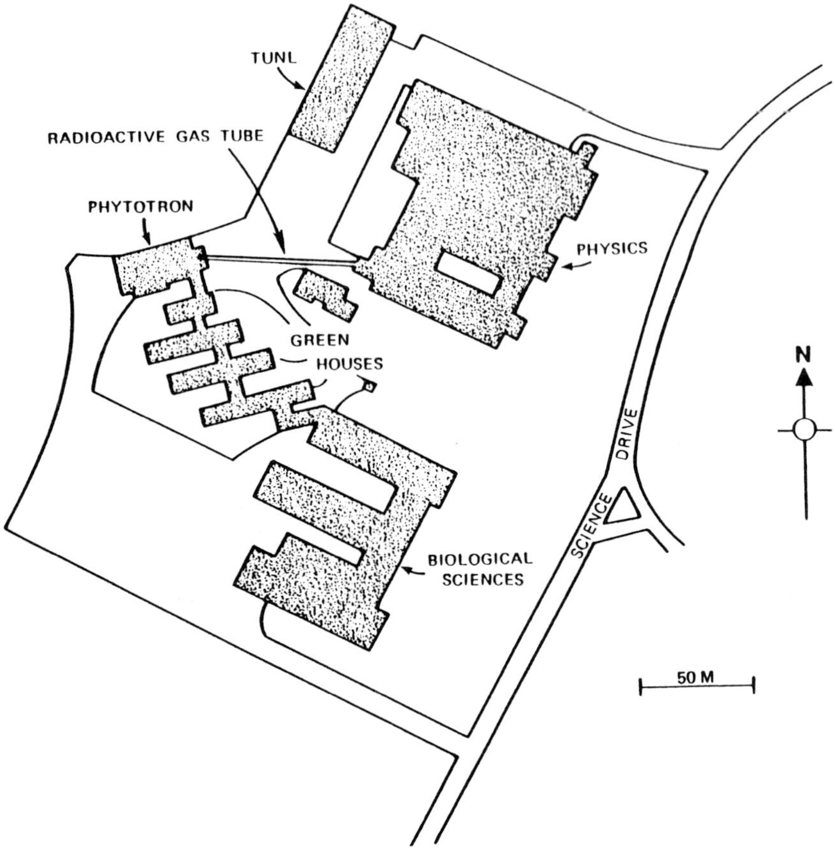

Figure 1. The relative location of the Duke Phytotron with respect to the 4 MeV Van de Graaff of the Duke University Physics Department.

possible to the controlled plant environment facility, where the measurement will be made. This unique situation is found at Duke University where the 4 MeV Van de Graaff accelerator of the physics department is less than 100 m away from the Duke Phytotron, the controlled environment facility, where the tracer studies would be conducted. With reasonable gas flow rates, the transport time of the radioactive gas to the, phytotron can be adjusted to a few minutes so that only a small fraction of the isotope decays. Figure 1 shows the relative locations of the two facilities.

Examination of mathematical models of carbon allocation in plants indicates that much needed information could be obtained by continuous $^{11}CO_2$ labelling methods (see for example Thornley 1976, and De Michele, et al., 1978). Emphasis is placed on the significance of bringing the system to steady-state conditions. Often the analytical treatment of data, e.g., mathematical techniques of tracer kinetics, is used with the implicit assumption that a steady-state exists. However, the technique that will be described here could also be used in pulse labelling as was described by several authors before, though the information obtained in this case is limited (Minchin and Troughton, 1980).

Systems components

The integrated system is made up of six components, including (1) Radioactive Isotope Production, (2) On-line Chemistry, Gas Flow System, and Gas Exchange Measurement, (3) Isotope Uptake Measurement, (4) Plant Detector Assembly and Environmental Chamber, (5) Signal Processing and Counting, and (6) The Processor. Figure 2 shows the relative arrangement of the system's components.

Radioactive isotope production

The maximum charged particle energy of the Duke Van de Graaff accelerator, 4 MeV, and the need for a continuously flowing CO_2 labelled with C-11 dictated the choice of target and projectile of the nuclear

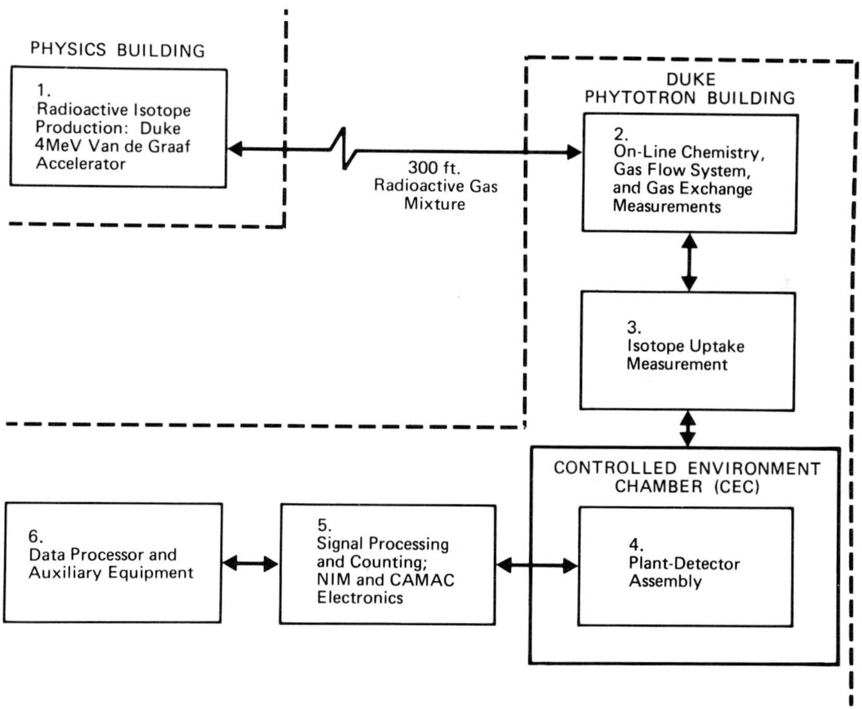

Figure 2. Outline of the integrated system for studying the carbon allocation in plants using radioactive tracers under controlled environment conditions.

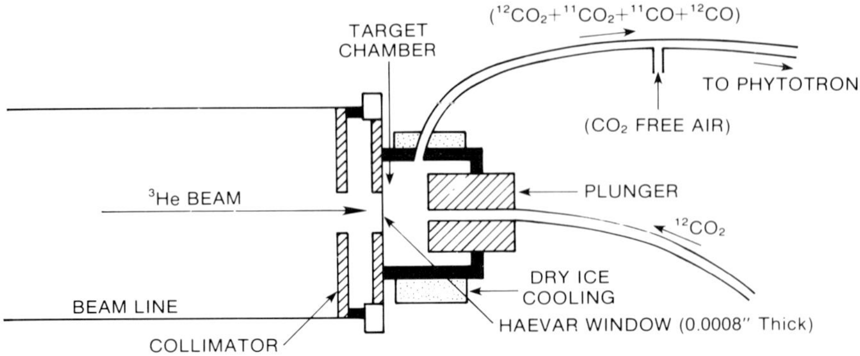

Figure 3. The target chamber mounted on the Duke 4 MeV Van de Graaff beam line. The chamber is made of an aluminum cylinder 1.5 cm in radius. The plunger adjusts the length of the chamber to produce maximum specific activity (see text). The target chamber is cooled with dry ice.

reaction utilized. Briefly, the nuclear reaction ^{12}C (^3He, ^4He) ^{11}C (Cirilov et al., 1966), is used to produce $^{11}CO_2$ continuously and directly by bombarding $^{12}CO_2$ gas flowing at a given rate in the target chamber with a ^3He beam at 4 MeV. Since about 0.7 MeV of beam energy is lost in the window of the target chamber, and the nuclear reaction threshold energy is close to 2 MeV, the $^{12}CO_2$ gas layer in the target chamber is adjusted using a plunger, to 6–8 mm (as shown in Fig. 3) in order to maximize the production of C-11 without reducing the specific activity (see Williamson and Boujot, 1962).

There are two other competing nuclear reactions that take place in the target, $^{12}C(^3He, n)\,^{14}O$ and ^{16}O (^3He, ^4He) ^{15}O, O-14 and O-15 have half lives of 71 and 122 s respectively and both decay by positron emission similar to C-11. To eliminate these two isotopes, the transport time of activity from the target chamber to the phytotron is adjusted to be longer than six half-lives of the longest-lived of the two isotopes, O-15. The transport time is adjusted by either slowing down the CO_2 flow or by the addition of CO_2 free air to the effluent gases at the exit line of the target chamber (see Fig. 3). Another problem encountered with this reaction is that the hot atom chemistry is different than an earlier study with 35 MeV protons as projectiles. In that study, of a total of 100 % C-11 activity, 98 % was in the form of ^{11}CO and 2 % $^{11}CO_2$, which made it possible to separate the activity from the target with only 2 % loss (Fares et al., 1978). The separation of ^{11}CO from the target $^{12}CO_2$ and any $^{11}CO_2$ is necessary in order to control the specific activity and the total carbon dioxide presented to the plant leaf. In the present study, with 4 MeV ^3He, the composition is 10 % ^{11}CO and 90 % $^{11}CO_2$. However, the cross section of the ^3He reaction is three to four times that of the proton reaction, which compensates for some of the loss of activity to the target material and produces enough C-11 labelled carbon dioxide after separation from the target to carry out the experiments without difficulty. An attempt to increase the recovery of ^{11}CO was done by mixing the target, $^{12}CO_2$, with some H_2 gas. The attempt succeeded in reducing some $^{11}CO_2$ to ^{11}CO but at the same time it reduced $^{12}CO_2$ to ^{12}CO, so in the end the overall recovery of C-11 was increased but the specific activity did not change or decreased slightly (the results of this study will be reported elsewhere). What happens in the target chamber can be summarized as follows.

Without Hydrogen

$$^{12}CO_2 \rightarrow \boxed{\text{Target Chamber}} \rightarrow \underset{\text{target}}{^{12}CO_2} + \underbrace{\underset{90\%}{^{11}CO_2} + \underset{10\%}{^{11}CO}}_{100\%\ ^{11}C} + {}^{15}O + {}^{14}O$$

$$\xrightarrow[12-13\ \text{min}]{\text{Transport}} {}^{12}CO_2 + {}^{11}CO_2 + {}^{11}CO$$

{^{15}O, ^{14}O may also be in the form of traces of ($C^{15}O$, $C^{14}O$, $C^{15}O^{16}O$, $C^{14}O$, ^{16}O)}

With Hydrogen

$$^{12}CO_2 + H_2 \rightarrow \boxed{\text{Target Chamber}} \rightarrow {}^{12}CO_2 + \underbrace{\underset{10\%}{^{11}CO_2} + \underset{90\%}{^{11}CO}}_{100\%\ ^{11}C} + {}^{12}CO + H_2O +$$

$$H_2 + {}^{15}O + {}^{14}O \xrightarrow[12-13\ \text{min}]{\text{Transport}} {}^{12}CO_2 + {}^{11}CO_2 + {}^{11}CO + H_2\,.$$

(Note: some ^{15}O and/or ^{14}O would be scrubbed as $H_2{}^{15}O$.) The ratio of $^{11}CO_2$ to ^{11}CO will depend on the amount of H_2 added. The production rate as monitored at the entry station in the phytotron i.e., by the line monitor (see below) is, after correcting for the decay in transport, $0.12\,\mu ci\ s^{-1}\,\mu A^{-1}$.

On-line chemistry, gas flow system, and gas exchange measurements

The objective of the on-line chemistry station positioned in the phytotron is to separate ^{11}CO from the target material and any other contaminates. It then oxidizes ^{11}CO to $^{11}CO_2$ so that when mixed (or diluted) with CO_2 free air and when the right amount of $^{12}CO_2$ is added, air with the desired concentration of CO_2 and known specific activity will be presen-

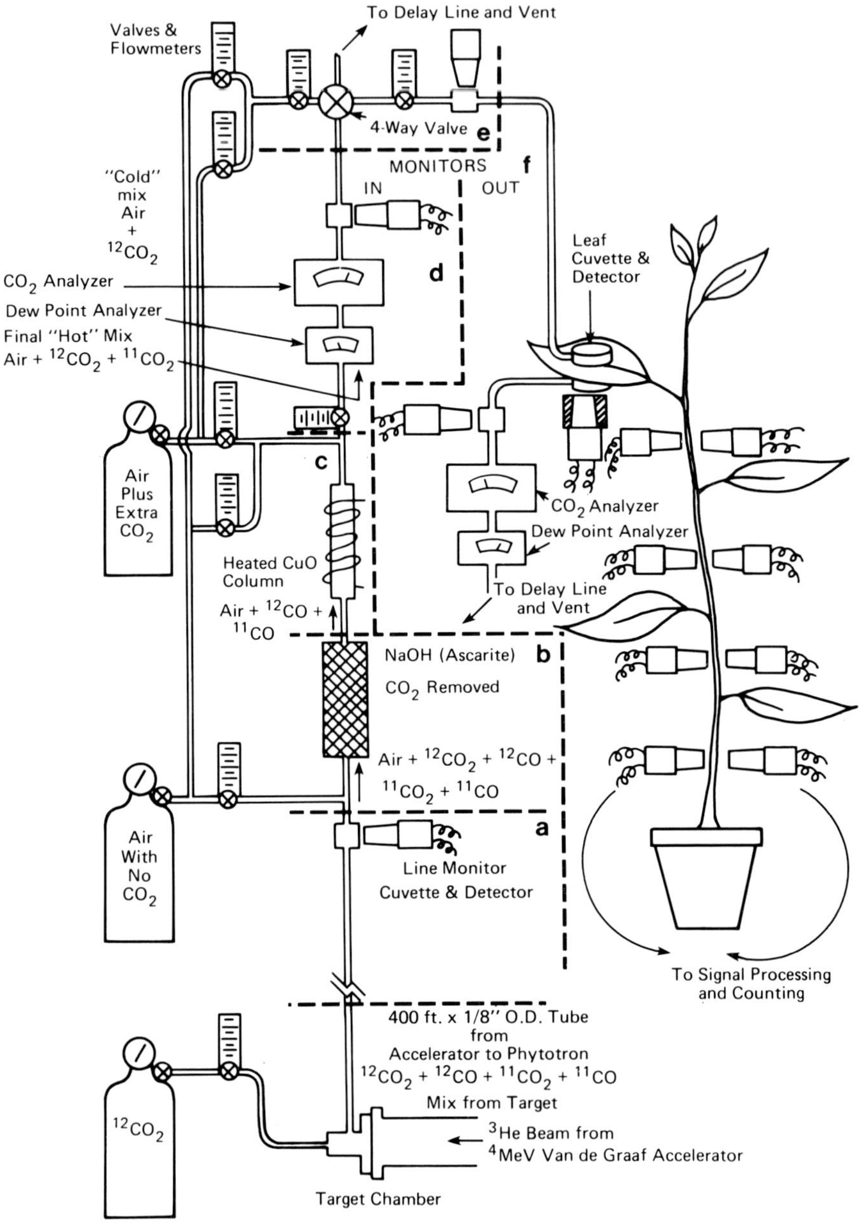

Figure 4. The experimental lay out. (a) Line Monitor (b) Removal of target material (c) Oxidation of ^{11}CO (d) In Line Monitor, CO_2 analyzer and dew point analyzer (e) Final in Monitor (f) Out Monitor, leaf cuvette and plant and detector assembly in the controlled environment chamber (CEC).

ted to the plant leaf. As shown above, part of the chemistry takes place in the target chamber and this dictates what is to be done next. The effluent gases from the target chamber arc transported a distance of 100 m (in 12 min) to the phytotron via a 2.5 mm i.d. teflon tube. The following sequence of steps takes place in the chemistry and gas flow station as shown in Figs. 4a, b, c, and d. This system has been redesigned recently.

(a) On arrival at the phytotron the effluent gases from the target chamber flow at a known rate through a spiral plastic cuvette of known volume positioned in front of a γ-ray detector at a given, fixed detection geometry. The detection efficiency of the detector in the given geometry is calibrated for every experiment with a standard γ-ray source. The details of the structure of all the detectors used will be described in Sect. 4. It is imperative that the detection efficiency of each detector in the system be known for every experiment. This station that monitors the activity of the incoming gases is called the "Line Monitor" as shown in Fig. 4a and it is this measurement that indicates to the production rate of the radioactive isotope.

(b) As the gases exit the line monitor, they flow through an ascarite column. The ascarite column is sodium hydroxide impregnated asbestos, which removes all the $^{11}CO_2$ and $^{12}CO_2$ gases. The small volume of ^{11}CO left (or ^{11}CO, ^{12}CO, H_2) is purged from the column by CO_2 free air added to the gases as they enter the ascarite column (when no air was added to the effluent gases they exited the target chamber as shown in Fig. 3).

(c) The next step is to oxidize ^{11}CO to $^{11}CO_2$ by flowing the mixture of gases exiting the ascarite column (air + ^{11}CO + ^{12}CO + H_2) over a CuO bed heated to 600 °C in a quartz tube in a horizontal muffle as shown in Fig. 4c. Oxygen in the air recovers the used CuO.

(d) To the effluent gases from the CuO bed (air + $^{11}CO_2$ or air + $^{11}CO_2$ + $^{12}CO_2$) either $^{12}CO_2$ free air or air + a controlled amount of $^{12}CO_2$ is added to bring the concentration of total CO_2 in the so-called "Hot Mix" to the desired value. The gases then flow through a dew point analyzer to determine their moisture content, then through an infra-red CO_2 analyzer to determine the total concentration of CO_2 in the flowing gases in ppm as shown in Fig. 4d.

As the gases exit the CO_2 analyzer, they flow through a spiral cuvette positioned in front of a γ-ray detector, as described earlier, to measure the specific activity of the gas mixture that will be presented to the leaf.

(e) Finally, adjustment of the rate of gas flow to be comparable to field flow rates is accomplished by adding either CO_2 free air or air + extra CO_2 via a 4-way valve. This valve allows for venting unwanted gases to the absorbers. If any adjustment is needed, the activity and the final flow rate are measured again as shown in Fig. 4e.

Isotope uptake

An essential measurement is performed at this station. The net isotope uptake by the portion of the leaf exposed to the final "Hot Mix" in the leaf cuvette is determined. It is simply a station where the specific activity of the input and output gas mixtures into and out of the leaf cuvette is measured. The flow rate and specific activity of the input gas mixture are measured in station (e) above and called the "In-Monitor". Similarly, the specific activity of the output gas mixture is measured in a cuvette-detector arrangement called the "Out-Monitor". The effluent gases from the "Out-Monitor" flow through a CO_2 analyzer then through a dew point analyzer before they are vented to the absorbers as shown in Fig. 4f. Thus, net photosynthetic rates can be calculated as well. The final flow rate of the gases is also monitored to adjust for the "Out-Monitor" measurement in case there is any perceptible reduction of flow. Signals from the CO_2 and dew point analyzers are either recorded by a multipen strip chart recorder or digitized and entered into the computer as will be described below.

Plant detector assembly

The plant and the detector used are positioned in the controlled environment chamber. The radioactive gas mixture is presented to a portion of the leaf through a cuvette which consists of a Plexiglas® water jacket on the top and bottom of the leaf. The cuvette is created by a 2-mm thick closed-cell form neoprene gasket on each water jacket. The cuvette is clamped to the leaf in such a way that the gas mixture coming out of the "In-Monitor" flows across the upper surface of the leaf, around the edge, and back across the lower surface. Several such cuvettes are designed to cover different leaf areas to accomidate various experimental requirements. Immediately below the leaf cuvette a γ-ray detector is positioned to monitor the tracer buildup in the leaf. This detector is shielded with a lead collar, thick enough to stop all radiation from other sources (e.g., activity flowing through the phloem).

In order to monitor the movement of the labelled photosynthates in the phloem, 4–6 pairs of detectors are positioned on either side of the plant stem above and below the labelled leaf. The detector pairs are operated in time-coincidence to detect the 2 γ-rays coming off at 180° with respect to each other as explained in the introduction (see Fig. 4).

The distances between the detector pairs and their geometry with respect to the stem and to one another are carefully assigned.

Each detector is made of 5 × 2.8 cm NE102 plastic scintillator crystal light coupled to an RCA 4517, twelve stage photomultiplier tube, which to a voltage divider that is capable of producing fast and slow signal output. The front end of the detector assembly facing the source of activity, could be collimated with a lead collar to give a window of desired width (see Fig. 5). Any other scintillator could be used e.g., NaI (TI), CsF etc. depending on availability of the detector and the necessary related electronics. A highly stabilized high-voltage power supply, positioned outside the environmental chamber was used to bias the detectors.

The controlled environment room CER (or chamber) in which the plant detector station is located is a typical chamber of type B of the Duke Phytotron (see Kramer et al., 1970; and Down, et al., 1972). In such a chamber, temperature, humidity, light intensity, light cycle, CO_2 concentration, and flow speed of air are controlled. The light cycle and CO_2 concentration are adjusted to the conditions of growth of the plant under study.

Signal processing and counting

We have, in this station two sets of electronic units. Standard NIM electronics for processing the signals from the detectors to prepare them for counting and CAMAC electronics for data acquisition and interfacing with the data processor. The fast signal from each detector is processed by the NIM electronics and then presented to the scalars in the CAMAC crate to be counted under the command of the CAMAC unit controller. The CAMAC unit is under control of the processor, as shown in Fig. 6, through the general purpose interface bus (GPIB). Standard LeCroy fast electronics modules are used in the NIM system, and Kinetic Systems CAMAC units are used in the CAMAC system. The CAMAC crate provides the physical mounting, power, cooling, and dataway connections for the modules. The dataway provides all of the power connections to the modules and interconnections (addresses, data, and interrupt signals) between the modules and the crate controller. A normal crate has 25 stations the rightmost of which, station 25, is assigned for the crate controller; the remaining slots are normal stations for modules such as scalars, digitizers, etc. The system simply works by commands and responses from the controller to the modules and back from the controller to the computer as shown in Fig. 6. The controller is the traffic director of the systems signals. All system parameters are carefully defined within the IEEE CAMAC specifications 583. The

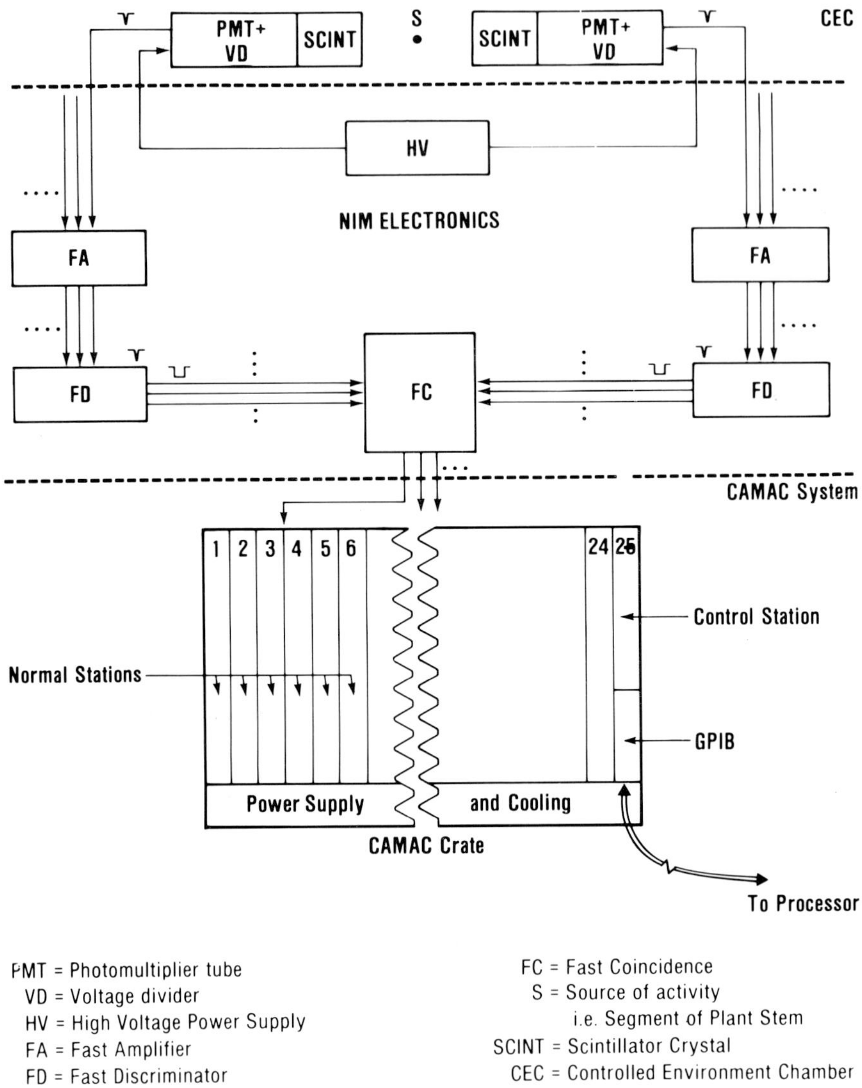

Figure 5. Signal processing and counting using NIM and CAMAC systems. NIM = Nuclear Instrumentation Electronic Modular. CAMAC = Automated Measurement and Control.

modules in the CAMAC crate also receive signals from the CO_2 and dew point analyzers as well as from the mass sensitive flow meters.

The processor

For the above system to be an integrated on-line facility, a processor that can perform the following functions is needed:

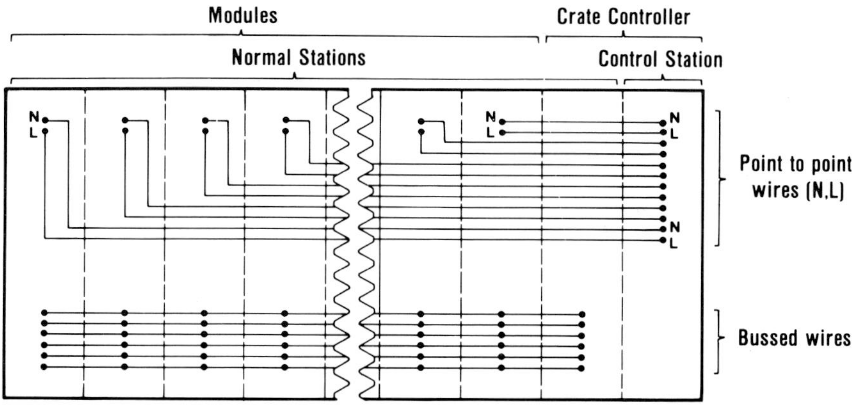

a. Signal flow between modules and crate controller.

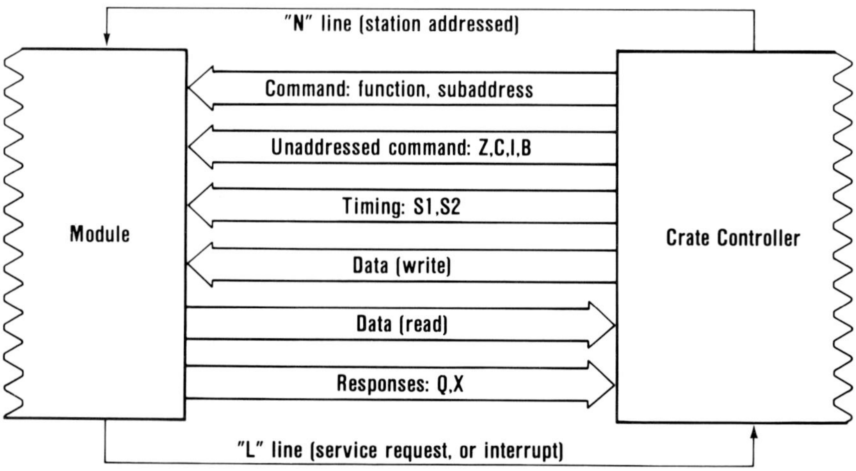

b. Simplified diagram of Dataway layout.

Figure 6. Communication and control paths on the Dataway of the CAMAC Crate.

(i) Communicate with the CAMAC system i.e., send commands to and receive responses from it;
(ii) Implement routines such as calibration and background corrections, (the latter is especially important where successive pulse experiments will have a changing residual background left over from the previous pulse);
(iii) On-line data acquisition, reduction and storage;
(iv) On-line data display and graphic representation;

(v) On-line and off-line testing of acquired data i.e., comparison of acquired data to predicted behavior from stored models of the system;

(vi) Utility programs.

Several computers available in the market, are CAMAC compatible, follow the IEEE 488 specifications, and can do most if not all the above mentioned functions, e.g. PDP-11 and LSI-11. In this system an HP 9845 B (option 250) minicomputer is used. This computer, though termed mini, has 186K byte memory expandable to 500K and uses the extended basic language which is easy to program. It has the additional advantage of graphic capabilities (color graphics if desired) and a large CRT for visual display. It can drive the following accessories for input/output or display options: two magnetic tape drives, floppy disc drives, thermal printer, multipen multicolour recorders, and extra CRT's for display or monitoring. Other important features of this minicomputer can be found in HP publications. Figure 7 is a copy of the graphic display

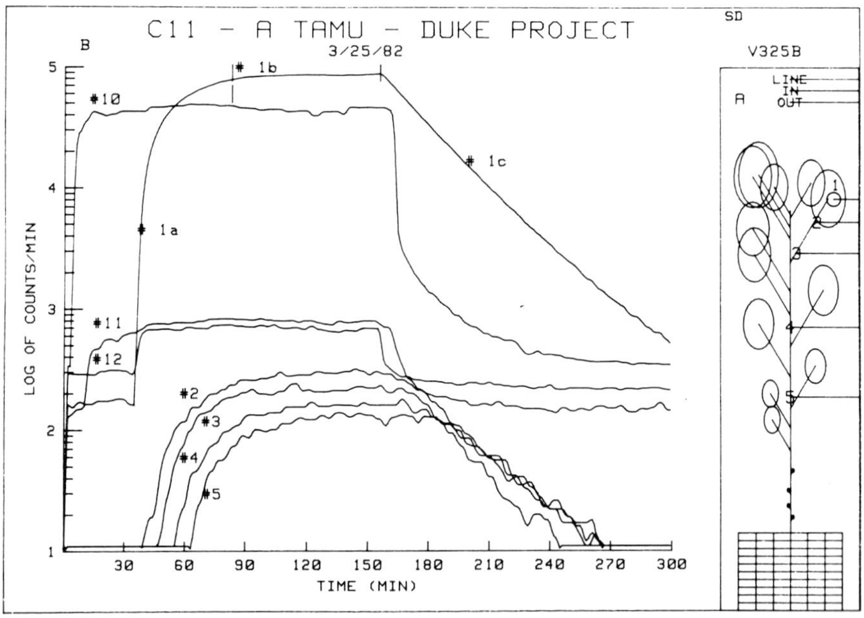

Figure 7. Live-time display of tracer profiles of Velvet Leaf *Abutilon theophrasti* Medic. (a) Plant geometry with detectors # 1 monitoring the leaf and detector pairs # 2, # 3, # 4 and # 5 monitoring the phloem activity. (b) Profile # 1 shows the line activity as recorded by the "Line Monitor", # 2 and # 3 the input and output activities as recorded by the "Input" and "Output" Monitors respectively. Profile # 4-a, # 4-b and # 4-c show the time-dependent, the steady-state and the washout behaviour of the leaf respectively. Profile 5, 6, 7 and 8 show the tracer profiles in the phloem as monitored by detector pairs # 2, # 3, # 4 and # 5 shown in Fig. 7-a.

on the CRT of this computer of an on-line real-time experiment described in the protocol below and reproduced by the multipen recorder. This figure can also be copied directly from the screen by the thermal printer. (Note that the system described above could also be used for the production of N-13 for nitrogen fixation studies with a few changes in the chemistry.)

Experimental protocol

Described here is the simplest experiment possible, which we refer to as "extended square wave" or, in the jargon of the data analysis techniques used, as "step input" to illustrate our daily experimental protocol and the information obtained from a simple experiment using this system. Figure 7 shows the on-line real-time computer output of the data obtained from this experiment for rates of net photosynthesis, carbon storage (e.g., starch), and export (e.g., as sucrose) from a given area of a leaf and the velocity of translocation of Velvet Leaf, *Abutilon theophrasti* Medic.

The plant of choice, grown under the environmental conditions chosen for a particular study, is moved into the controlled environment chamber, CEC, 24–48 h before the start of the experiment. A portion of one of the leaves of the plant is enclosed with a clear plastic Lexan® water-cooled cuvette and is supplied with a continuous flow of air at a given, steady-state concentration of $^{12}CO_2$ (e.g., 400 ppm at a flow which will maintain 350 ppm in the cuvette). Net photosynthesis, transpiration, and other biologically related functions are monitored continuously as explained above. The detectors are arranged on the plant at the desired positions and the geometry of the leaves and the petiole of the leaf on which the experiment is conducted is measured and entered into the computer (see Fig. 7a). On the morning of the experiment, as soon as the lights in the CEC come on, the background activity at each detector is determined and the detectors are calibrated with a standard Na-22 source. This information is automatically entered and stored in the computer which would have already started the data acquisition phase of the program. An updated listing of data by the thermal printer displays the information at predetermined intervals. Radioactive gases flow from the target chamber in the accelerator laboratory to the phytotron, the activity is monitored by the "line monitor" as explained above, and is displayed continuously on the computer CRT (profile # 1 in Fig. 7b) till a steady-state level is reached. The gases then flow through the chemical processing station and are monitored for steady-state level of specific activity (profile 2 Fig. 7b). At the appropriate time the 4-way valve is turned to divert the flow of unlabelled air and substitute an identical flow of the "Hot Mix" at the same total CO_2 concentration as the original. The build up of activity in the leaf and its movement in the transport stream are continuously measured and updated by the computer. A real-time continuous display of that behavior is shown on the CRT as in Fig. 7, profile 4, for the leaf and profiles 5, 6, 7, and 8 in the phloem. The specific activity of the gases exiting and the leaf cuvette is shown in profile #3. All data input into the computer is stored on cassette tapes as well as listed in the updated printout.

Tracer studies of this type must be conducted when the leaf has reached and is expected to maintain steady-state rates of photosynthesis and carbon allocation because of the implicit assumption of steady-state in tracer data analysis. These conditions tend to be reached mid-morning in most plants grown in growth chambers. However, extended square wave, ESW, experiments early in the morning or late afternoon when the rates of carbon metabolism are changing with time are also possible but require non-steady state data analysis. Figure 7 shows typical ESW results for the plant.

Two hours after stopping the tracer input in the leaf, only about 3 % of the activity would be left in the plant and the same experiment (or other experiments) could be repeated on the same plant under the same environmental conditions. However, it is not necessary to wait that long to replicate the experiment or start another one on the same plant, since corrections by the computer for residual activity could be made at the same time data is acquired.

It is obvious that a host of ecologically related environmental and biological factors could be tested with this technique. The technique should lead to improved qualitative and quantitative understanding of plant responses to environmental stimuli.

In any experiment that we conduct, we measure, in addition to all the biological and environmental factors such as light intensity, temperature, humidity, respiration . . . etc., the following parameters: average activity count rate in the line over the ESW period (ACL), overall leaf input count rate (ACI), the average leaf output activity (ACO) over ESW period, total flow rate through the input cuvette to the leaf (TF), the input-output CO_2 mixture to the leaf, the net photosynthetic rate (CER), and specific activity of the input mixture to the leaf (SA). We also calculate the rate of accumulation of C-11 in the leaf at steady-state (RE), the rate of storage in the leaf at steady-state (RS), the turnover time (TE) of the export pool in minutes, the ratio of the rate of export of C-11 to that of assimilation (RE/RT) which we call % E, the relative export pool size, E. Pool, in nci cm^{-2}, and the fractional exchange constant of the export pool λ_E which we call Lam E.

Theory and concepts of tracer analysis

In this section we concisely describe how the plant labelling proceeds throughout any given experiment and what each step means in terms of the theory of compartmental analysis. We will illustrate the various steps with the computer graphics output of a real-time experiment (cf., Magnuson et al., (1982)).

In Fig. 8 the difference between tracer profiles # 2 and # 3 represents the net uptake of isotope in the leaf (and net uptake of all carbon since the specific activity of the input is known) and it takes the shape of a step function input *I*(*t*) where

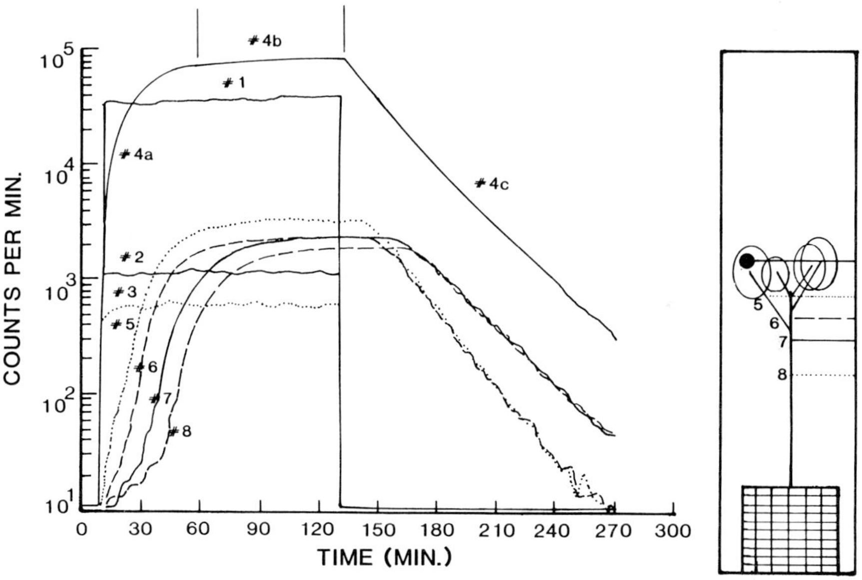

Figure 8. Real time display of activity in various parts of the C-11 system and in a velvet leaf plant during and following a 120 min, period of steadystate labelling. Profiles # 1, # 2 and # 3 show activities flowing through 7 ml cuvettes, respectively, on the line from the accelerator (at 0.5 ml sec^{-1}), on the entrance to the leaf cuvette (after processing and dilution with air to 8 ml sec^{-1}), and on the exit from the leaf cuvette. Activity in the labelled portion of the leaf (dark area on plant diagram, inset right) is shown by trace marked # 4a, # 4b and # 4c which indicate the buildup, steady-state, and washout respectively, and represent the ESW. Profiles # 5, # 6, # 7 and # 8 show activity at points along the petiole and stem as indicated on the plant diagram, at distances 12 cm, 22 cm, 28 cm and 42 cm respectively from the leaf.

$$I(t - t_0) = \begin{cases} 0 & t < t_0 \\ h & t \geqslant t_0 \end{cases} \tag{1}$$

Time-dependent tracer behavior in the leaf

If we consider the leaf as a linear system of n compartments and the state variable of compartment i is x_i the size of the compartment (in gm, mole, . . . etc.) and the f's are positive constant fractional transfer coefficients (in time^{-1}), then Fig. (9) and equation (2) represent the ith compartment in the leaf (Jacquez, (1972))

$$\frac{dx_i}{dt} = -f_{ii}x_i + \sum_{j \neq i} f_{ij}x_j + I_i(t) \tag{2}$$

where $f_{ji}x_i$ is the rate of exchange from the ith to the jth compartment and for simplicity we let $f_{ii} = f_{oi} + \Sigma_{j \neq i} f_{ji}$.

At time t_0 on our time scale (e.g. 10 min in Fig. 10) we start introducing a trace amount of $^{11}CO_2$ continuously at a constant level into one or more compartments of steady-state system, the leaf. Let N_i be the total amount of labelled material in the ith compartment at time t, measured in the same units as x_i, and let k_i be the conversion factor from mass units to radioactivity units for the labelled material in ci mole^{-1} of C-11; then the specific activity is $s_i = k_i N_i/x_i$. By steady-state assumptions x_i, x_j, f_{ij}, f_{ji} and $I(t)s_i/k_i = I'(t)$ are all constant. $N_i = s_i x_i/k_i$, therefore the change in the amount of radioactivity in the ith pool is given by

$$\frac{dN_i}{dt} = -f_{ii}N_i + \sum_{j \neq i} f_{ij}N_j + I_i'(t) \tag{3}$$

and

$$\sum_{i=1}^{n} \frac{dN_i}{dt} = \frac{dN_1}{dt} + \frac{dN_2}{dt} + \ldots + \frac{dN_n}{dt} \tag{4}$$

from which one can estimate the transfer coefficients f_{ji} and f_{ij}, where n is the number of compartments or pools in the leaf. Profile 4-a of Fig. 8 (see also Fig. 10) illustrates the time dependent behavior of the leaf described by equation 4.

We emphasize the significance of the steady-state assumptions because often equations 3 and/or 4 are used with the implicit assumption that a steady-state exists.

There are several methods for solving the set of n simultaneous linear differential equations (4) (cf. Jacquez, (1972); Ackerman and Gatewood, (1979); Finkelstein and Carson, (1979)).

The steady-state tracer distribution in the leaf

Profile 4-b of Fig. 8, and Fig. 10 show the steady-state level of the tracer distribution in the leaf. At tracer steady-state, the tracer influx and efflux in the leaf are equal, and the mean residence time of C-11 in the leaf pool becomes constant. This also means that $I'(t) = f_{oi}N_i$ and $f_{ii} = \Sigma_{j \neq i} f_{ij}$ for all i in equation (3).

Washout curve of the leaf

In work with the radioactive tracers the washout function is obtained by adding labelled material to the input *to a steady-state system* at a *constant level* until a *steady-state in the tracer* is reached, then stopping the administration of tracer, and following the total activity in the system. Profile 4-c of Fig. 8, and Fig. 11 illustrate such a washout function for C-11 tracer in the leaf. Usually the washout function is expressed as a sum of exponentials, $W(t)$, as in equation 5, where λ_i and A_i correspond to the turnover time and the size of the ith compartment;

$$W(t) = \sum_{i=1}^{n} N_i(t) = A_i e^{-\lambda_i t} + \ldots + A_n e^{-\lambda_n t} \tag{5}$$

(cf. Bergner, (1961); Jacquez, (1972); Ackerman and Gatewood, (1979)).

The rate of appearance of material in the excretion or export from the system, $-\dot{W}(t)$, is shown in profile 5 of Fig. 8. Figure 10 depicts the time-dependent uptake, the steady state level, and the washout curve of C-11 tracer from the leaf, i.e. one ESW.

System specification

Frequently the experimental problem includes estimating the number of compartments, n, of the system. Can we tell from a washout curve or the time course of changes in one or more compartments how many exponential terms are present in equations 4 or 5? This is the problem of *system specification.* Washout curves can be resolved by fitting sums of exponentials using computers (cf. Berman, Shahn and Weiss, (1962); and Hearon, (1963)); or by using transform techniques such as Fourier Transform as explained by Gardner, (1963); Gardner, Gardner, Laush, and Meinke, (1959). Transform methods have two advantages over data fitting methods: (a) it is not necessary to have initial estimates of the parameters, and (b) it has the ability to resolve smaller and faster and/or slower turning compartments that, otherwise, are lumped. However, the influence of insufficient data points and the effect of random error should be carefully considered (cf. Gardner, et al., (1959); Gardner, (1963); Callahan and Pitzer, (1966); Cochran, Cooley, Favin, Helms, Kaenel, Lang, Maling, Nelson, Rader, and Welch, (1967).

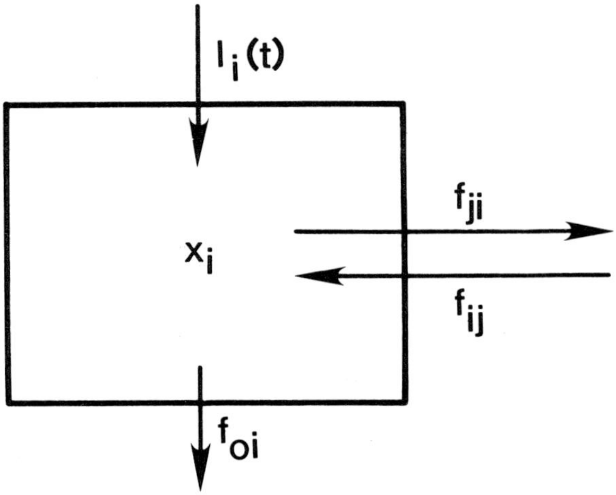

Figure 9. The *i*th member of an *N*-compartment system.

Translocation of photosynthates

Profile #5 in Fig. 8 represents the shape of the photosynthate efflux or export profile in the phloem as measured at the petiole. Curves #6, #7, and #8 show what happens to this profile as the photosynthates

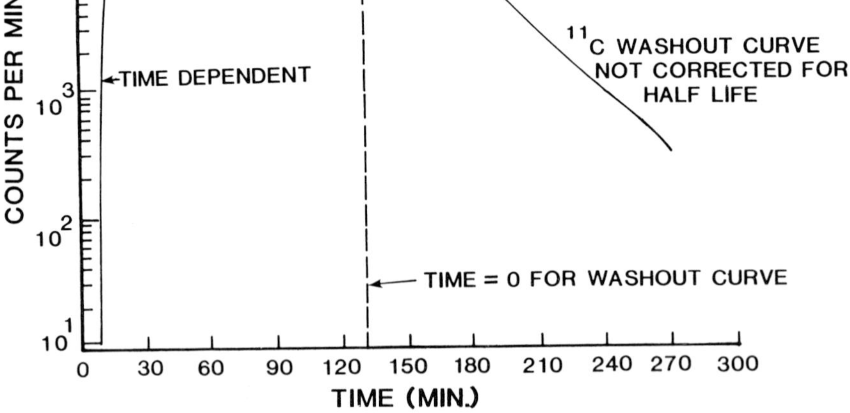

Figure 10. Time-dependent, steady-state, and washout of C-11 in the leaf (c.f. Fig. 8, #4a, 4b, and 4c respectively). Time = 0 for the washout curve is the time of cessation of labeling.

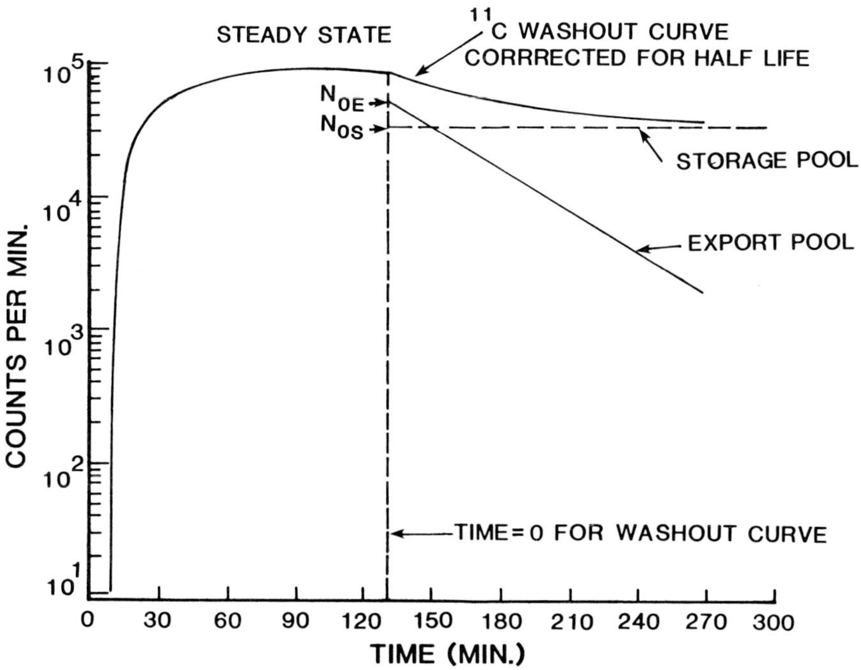

Figure 11. The washout curve resolved by curve fitting the data to eqn. 5 (see text), in two compartments of pools; the storage and the export pools. N_{OS} and N_{OE} and the relative sizes of the storage and the export pools at steady-state respectively, i.e. at the time of cessation of labelling.

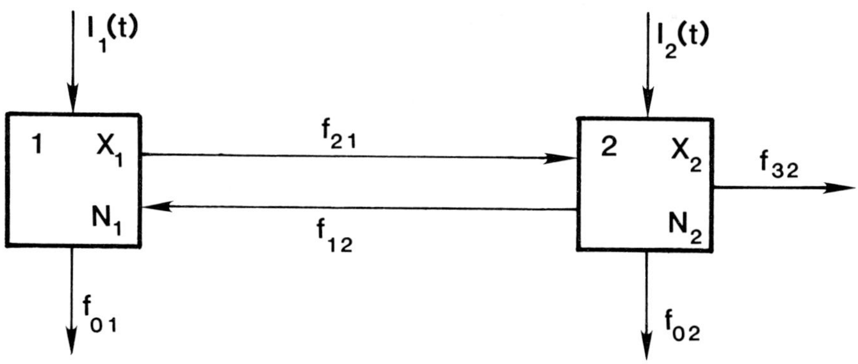

Figure 12. A two compartment model for carbon allocation in the leaf. Compartment 1 represents the stored products, and 2 represents the pool of exportable products of $^{11}CO_2$ incorporated by net photosynthesis. respectively, I_1 and I_2 represent the input rates into compartments 1 and 2, while the f_{ij}'s are the fractional exchange constants as explained in the text. Normally, f_{21}, and f_{12} are believed to be very small, therefore f_{01} is equal to the decay constant of the stored C-11 in compartment 1, i.e. it is equal to the decay constant, of C-11. In compartment 2, f_{32} represents the export constant, while f_{02} is equal to the decay constant of C-11 in the "export pool".

move down the stem towards the roots. Obviously, the shape of the profile has already been set by the loading mechanism of the leaf. If we consider Profile #5 as input into that segment of the stem and #6 as output some distance away from it, then we have both spatial and temporal relationships of the translocating photosynthates. The relationship between input and output is sometimes called the transfer function $h(t)$. This quantity is of great importance, not only because it tells us what happened to the translocates as they travel the distance between input and output, but it is a measure of the transit time or speed of the translocates since it is the probability density function of transit times (Jacquez (1972); Minchin, (1978); (1979); Minchin and Troughton, (1980)). There are two main techniques by which $h(t)$ and the mean speeds of translocation can be calculated: (a) deconvolving the convolution integral of the input and output data points using the discrete Fourier Transform Methods, (cf. Cooley, Lewis, and Welch, (1967); Hung (1970); Ackerman and Gatewood, (1979); and Jacquez, (1972); (b) using discrete time-series methods (cf., Cadzow, (1973); Minchin, (1978); (1979); Minchin and Troughton, (1980)).

Analysis of the washout curve

Using an algorithm for least-square estimation of nonlinear parameters for fitting sums of exponentials as in equation 5, (Marquardt, (1963)), we found that most leaves behaved like a two-compartment system with input into one or both compartments and export from one of the compartments. Figure 11 illustrates the resolution of the leaf pools into storage and export compartment, ($n = 2$). However, the data do not exclude the possibility of other small compartments which turn over much more rapidly or much more slowly. The sampling times and the fractional decay constant of the isotope are such that compartments with biological decay constants greater than $0.7\,\text{min}^{-1}$ ($t_{1/2} = 1\,\text{min}$) or less than $0.002\,\text{min}^{-1}$($t_{1/2} = 6\,\text{hr}$) would not be detected. Based on this information we suggest that storage and export parameters in most leaves can be closely estimated by a simple two compartment model as illustrated in Fig. 12.

Examples and discussion

Three examples are shown here to illustrate the type of phenomena which can be measured and analyzed by the ESW methods explained above. The data shown in Table 1 and Figs. 13 through 15 represent typical cases of the allocation of carbon under different physiological conditions using the washout curve analysis.

Table 1. A comparison of the CER, storage rate, export rate, % export, turnover time, and export pool for the examples cited in the text

Figure	Time or day	CER ($n\,mol\,cm^{-2}\,sec^{-1}$)	Storage rate ($n\,mol\,cm^{-2}\,sec^{-1}$)	Export rate ($n\,mol\,cm^{-2}\,sec^{-1}$)	% Export	Turnover time (min)	Export pool ($n\,mol\,cm^{-2}$)
13-A	First Day	1.80	0.66	1.14	64	30.7	3040
	Second Day	1.84	0.67	1.17	63	32.8	3239
13-B	First Day	1.69	0.93	0.76	45	160	10628
	Second Day	1.81	0.63	1.18	65	37.8	3072
	Third Day	1.86	0.66	1.20	64	30.4	3146
14-A	Morning	1.67	0.11	0.22	93	10.9	1429
	Afternoon	1.54	1.51	1.32	86	13.2	1496
14-B	Morning	1.41	0.20	1.21	86	19.1	2001
	Afternoon	1.37	0.57	0.80	59	33.4	2327
15	A-first day	1.82	0.78	1.04	57	36	3254
	B-third day	1.63	1.12	0.51	31	56	2490

*Where turnover time = $T_{1/2}/0.693$.

Example 1. Plant recovery from agitation

As indicated in the methods section of this and previous papers (Goeschl, Magnuson, Fares, Jaeger, Nelson, and Strain, (1984)), plants to be studied with the C-11 system are moved on carts from a large growth chamber to a smaller one which houses the necessary apparatus. At each step great care is exercised to minimize physical trauma, but some agitation and flexing of the plant is unavoidable.

Attempts to measure photosynthate export and/or translocation on the same day plants were transferred, have revealed a wide range of export rates from essentially none to moderate rates. Within 24 hr after being moved, the majority of plants recover to within 95% of their eventual level of transport activity, while others require longer recovery times. These responses vary not only between species, but also between individual specimens, even where no obvious differences occurred in the manner of handling the plants.

This is exemplified by the patterns seen in Fig. 13-A representing a 5, week old cotton seedling being prepared for a study of water stress. The first two days of measurement were intended to determine its normal baseline behavior before the cessation of watering. In this case moving the plant caused slight decreases in photosynthesis, storage, export rate, and export pool size (cf. Table 1). The cotton seedling shown in Fig. 13-B, however had not recovered by the first day, and similar patterns occurred in two other cotton plants out of approximately 40 subjects over a two year period. Less frequent examples of delayed recovery have been observed in specimens of velvet leaf, sweet potato, and pangola grass.

In the present example, analyses of the washout curves on the first and second days of recovery (see Table 1) confirm and quantify the visual observations.

On the first day the export rate (proportional to the slope of the washout curve) is clearly restricted, and is accompanied by a high rate of storage (proportional to the asymptotic level of activity approached by the washout curve, please refer to Fig. 11). That the low rate of export is not simply a result of decrease, rate of photosynthesis is suggested by the large pool size (i.e. high concentration) of exportable products and the concomitant long turnover time. Thus we might conclude that photosynthesis is being inhibited by its accumulated products.

On the second day the various rates and pool sizes approached those of the third day, indicating an essentially complete recovery to normal levels, and illustrating the repeatability of this method when applied to plants which have fully recovered. It also illustrates the ability of this

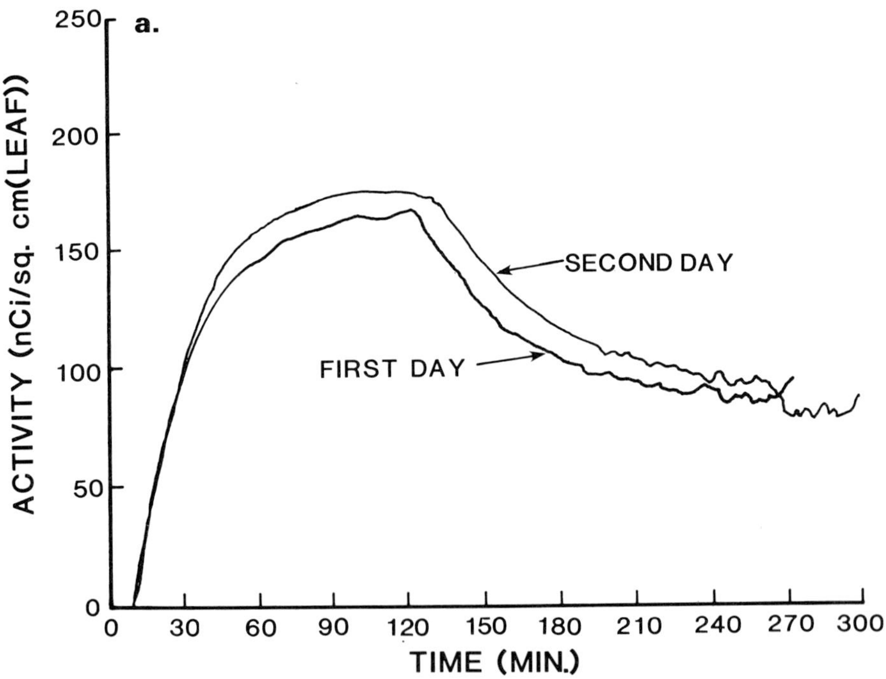

Figure 13-A. C-11 activity profiles in the leaf of a 5 week old cotton plant 24 hr and 48 hr after being placed in the experimental apparatus. The amplitude of the activity, and especially the shape of the washout curve indicate that the plant was almost fully recovered by the first day. (see Table 1 for calculated differences in photosynthesis, storage, export, and export pool size).

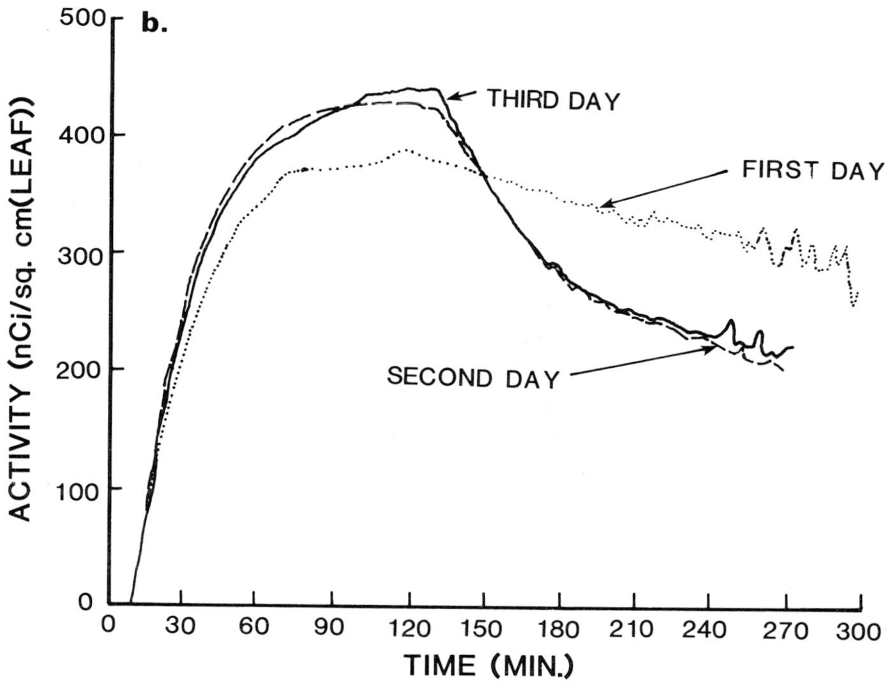

Figure 13-B. Differences in the amplitude and shape of the C-11 tracer profiles on the first and second days indicate that this cotton plant had not recovered on the first day after being moved. Full recovery by the second day is illustrated by its nearly identical profile to that of the third day. (The plant in Fig. 13-A was subjected to other treatments on the third day).

non-destructive method to determine the physiological status of plants before proceeding with experiments which might otherwise yield misleading results.

Example 2. Diurnal effects on carbon flow

During the standard 14 hr photoperiod of our experiments, CER generally rose to a steady level within 1 to 2 hr, remained constant for 6 to 10 hr (these times being dependent upon species), then began to decline gradually until the lights were turned off. In many plants (e.g. corn, velvet leaf, and barnyard grass) this decline was generally not more than 5 %, in others (e.g. sweet potato, or pangola) the decline reached 5 to 15 %, and in cotton it would decline as much as 20%. In most cases these declines in CER were followed closely by lesser declines in transpiration. As revealed by the C-11 tracer data, these overt patterns were accompanied by, or perhaps caused by, various changes in the allocation of photosynthetic products within the leaves (see Table 1).

Plants with a modest decline in CER are typified by the behavior of the C_4 bunch grass, little bluestem. As seen in Table 1, after 9 hr of the photoperiod (afternoon) CER had decreased by about 5 % from that after 4 hr (morning) (these being the respective times at the beginning of the washout curves). Despite the lower net rate of uptake, the level of C-11 activity in the leaf during the 120 min labelling period (Fig. 14-A) was higher at 9 hr (afternoon) than at 4 hr (morning) as was the asymptotic level of the washout curve because the rate of accumulation as storage products had doubled. While the rate of export decreased, the concentration of exportable carbon products increased slightly.

A quantitative extreme in this pattern was found in brownseed paspalum (Fig. 14-B & Table 1) with a substantial decrease in export, a near tripling in storage rate, a noticeable increase in export pool size and turnover rate, but only a small decrease in CER. An important point

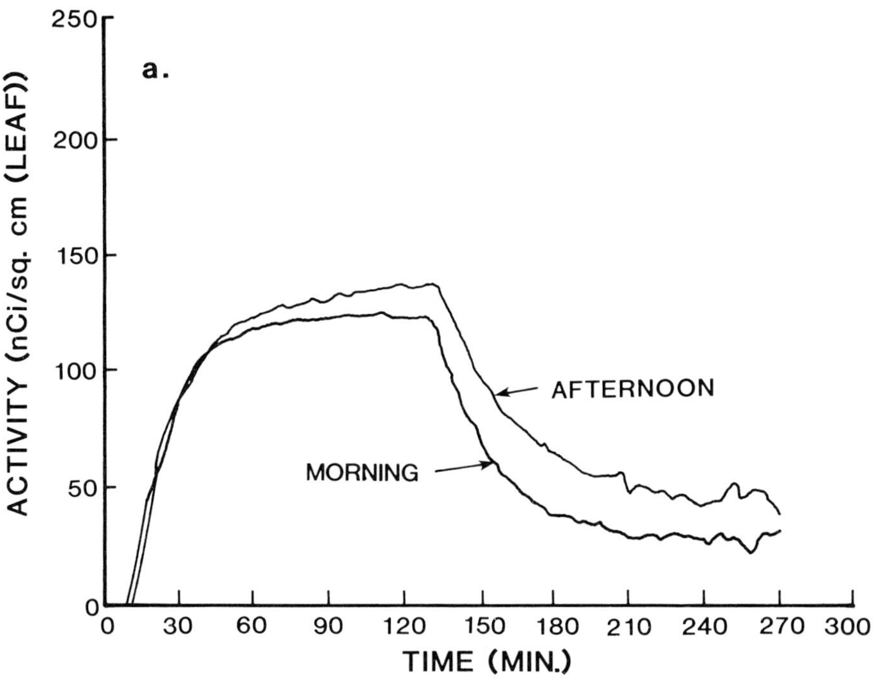

Figure 14-A. Profiles of C-11 activity in the portion of leaf of the grass little bluestem labeled for 120 min. starting 2 hr (morning) and 7 hr (afternoon) after the beginning of the 14 hr photo-period. The afternoon profile indicates a higher rate of storage (proportional to the asymptotic level at the end of the washout curve) and a lower rate of export (based on the slope of the washout curve, see Table 1). There was also a 8 % decrease in net photosynthesis from morning to afternoon (see Table 1). The higher level of C-11 activity in the afternoon resulted from increased storage and export pool size.

here is that allocation patterns which changed only slightly with time would be difficult or impossible to detect with destructive sampling techniques where variability of population samples might have masked the physiological trends.

Example 3. Effects of withholding water

Assimilation and allocation patterns of C-11 on the first and third days after the cessation of watering of a cotton seedling are shown in Fig. 15, lines A and B. Wilting of the lowermost leaves would have occurred on the 5th day. As the stress condition began to develop there was a reduction in the export rate and an increase in the concentration of exportable products and rate of storage (see Table 1). By the third day there were noticeable decreases in CER, transpiration, export rate, and export pool size. Combining the C-11-labelling with conventional gas exchange measurements one can follow a number of carbon flow parameters as affected by an environmental factor over a period of time, and identify which of several responses was earliest response. In this case

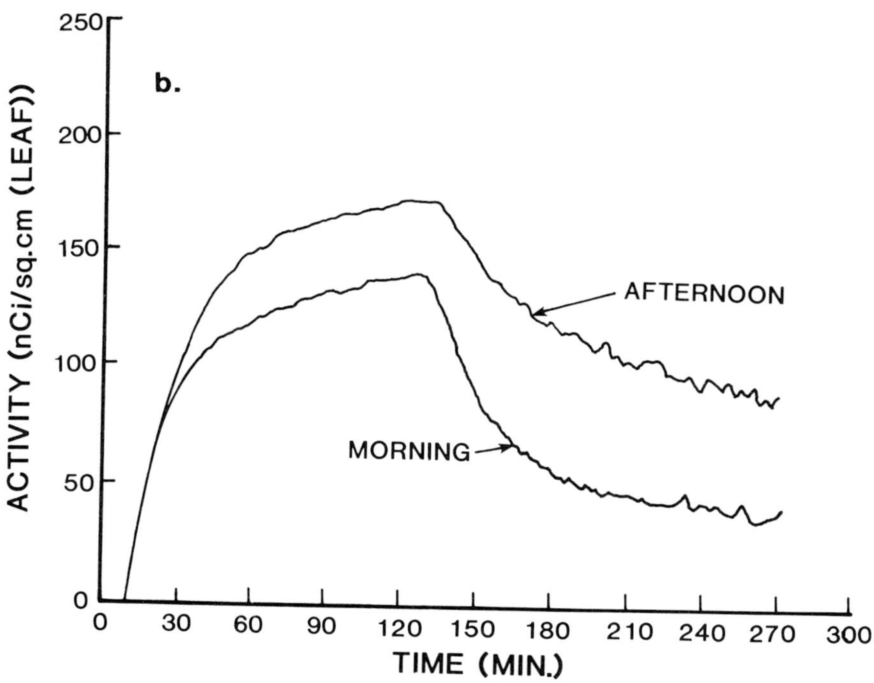

Figure 14-B. Profiles of C-11 activity in the labeled portion of a leaf of the grass brownseed paspalum indicate a much greater increase in storage and decrease in export during the photoperiod. However the decrease in net photosynthesis (11 % see Table 1) was nearly the same as little bluestem.

our purpose is to test the predictions of a mathematical model of phloem transport regarding the maintenance of phloem turgor (see DeMichele, Sharpe, and Goeschl, (1978)). That a decrease in carbon export would precede a decrease in gas exchange is a logical consequence of the hypothesis.

In each of the above studies, other detectors on the plants made it possible to measure the flow of carbon out of the leaves, through the transport system, and into various sink tissues. Similar analyses of the time-dependent and steady-state aspects of the C-11 kinetics in the remaining parts of the plants make it possible to follow a large number of simultaneous parameters and thus avoid overlooking the earliest response to a given stimulus.

Conclusion

An integrated approach to study the effects of environmental factors on plants is described. Though the tracer described is C-11, N-13 could be used for nitrogen fixation studies with minor changes in the system. However, C-11 and N-13 seem to provide very convenient tracers that are natural constituents of plants and their environment. A host of

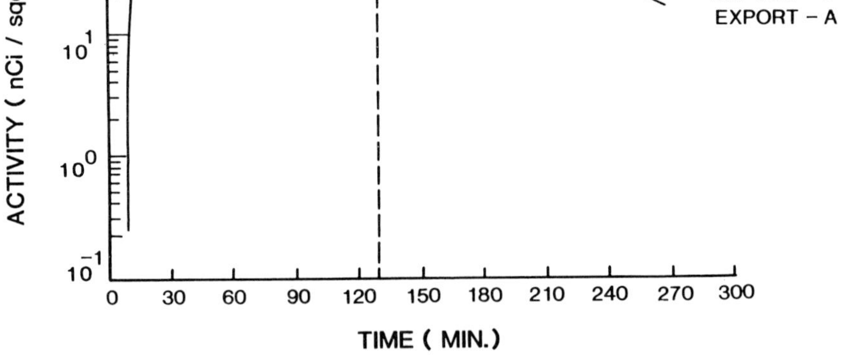

Figure 15. Profiles of C-11 activity in the labeled portion of a cotton leaf on the first day (Washout A) and third day (Washout B) of withholding water. The higher asymptotic level at the end of Washout B indicates a higher rate of storage, while the slope of the calculated export B indicates a much lower rate of export (see Table 1). On the second day of this experiment the export pool size had increased, but decreased on the third day as net photosynthesis began to decrease.

ecologically, agriculturally, and genetically important questions can be answered using this technique. The consequences of elevated CO_2 concentration in the atmosphere, the effects of water stress on growth rates of plants, and the information required to genetically engineer a plant to suit our needs, are but a few of the major scientific questions that can be studied by this method.

The three examples above represent data gathered as part of various experiments, involving usually no more than 6 to 8 individual specimens. More extensive studies of these physiological phenomena are in progress. However, some conclusions about the methodology and its usefulness can be reached.

First, we recognize that photosynthesis, photorespiration, intermediary metabolism, respiration, and transport processes involve considerable compartmentalization and recycling of carbon within the cells and tissue of leaves. This being the case one might have expected that the C-11 washout curves might have revealed more than two compartments, i.e. n in equation 5 would have been larger than 2. However mathematical analyses to date have not revealed any consistent, significant evidence of such compartments (i.e. during the 2 hr period before C-11 decays completely). Some evidence for a smaller, slower turnover or loss of ^{14}C-labelled photosynthates has been reported by Fondy & Geiger, (1980). However, our primary purpose at present is to estimate the net rates of storage and export; thus our treatment of export as a single compartment is both consistent with empirical results and adequate for present purposes. In the future we plan to improve the specific activity of the $^{11}CO_2$ and thus raise the signal-to-noise ratio, extend the period of detectability, and improve the resolution of the washout curves. In addition, we will analyze the buildup curves (equation 4), which again may reveal more compartments.

Second, very subtle differences in carbon allocation patterns can be observed in individual plants as a function of time and/or changes in environmental inputs. For example, Figs. 13A & B and Fig. 15 illustrate labelling patterns, some of which differ by only a few percent. As a further example four specimens of corn, two of velvet leaf, and six of cotton have been subjected to gradual drought stress by withholding water from large pots. In all cases decreases in phloem speed ranging from 4 % to 17 % were observed previous to measurable decreases in CER. The standard deviation in those parameters in small populations (6 to 10 specimens) of these plants ranges from 15 to 30 % of their absolute value. Thus these subtle physiological changes would have been lost in the genetic variability if destructive sampling at intervals from treated and control populations had been required.

Third, since the behavior of each specimen can be measured prior to the application of any treatment, it is possible to eliminate specimens whose base-line behavior is not representative of its population. For example we often choose the "median" seedlings from a small population on the basis of size, color, and configuration. Once detailed physiological measurements of several specimens have been completed it is possible to eliminate any results from specimens which are not representative of the population.

Acknowledgements

This research was supported jointly by the National Science Foundation, Ecosystems Studies Program (DEB 80-22165); US Department of Energy, Office of Carbon Dioxide and Climate Research (DOE 101-81ER60012); Texas Agricultural Experiment Station; Texas Engineering Experiment Station and the Standard Oil Company of Ohio. We also acknowledge the NSF Biological Research Resources program (DEB 80-21312) for support of the Duke University Phytotron.

References

1. Ackerman, E. and Gatewood, L.C. 1979. *Mathematical models in the health sciences, a computer-aided approach.* Univ. of Minnesota Press, Minneapolis, Minn.
2. Atkins, G.L. 1969. *Multicompartment models for biological systems.* Methuen & Co. Ltd., London.
3. Austin, R.B., Bingham, J., Blackwell, R.D., Evans, L.T., Ford, M.A., Morgan, C.L. and Taylor, M. 1980. *Genetic improvements in winter wheat yields since 1900 and associated physiologic changes.* Journal of Agricultural Science. 94: 675–689.
4. Boussingault, J.B. 1868. *Agronomic Chimie agricole et physiologie* 2^e Ed Mallet Bachelier, Paris, 1860–1874, 5 Vols, (pp. 236–312).
5. Bergner, P.E. 1961. *Tracer dynamics. I. A tentative approach and definition of fundamental concepts.* Journal of theoretical Biology 2: 120.
6. Berman, M., Shahn, E. and Weiss, M.F. 1962. *The routine fitting of kinetic data to models, a mathematical formalism for digital computers.* Biophysics Journal 2: 275.
7. Cadzow, J.A. 1973. *Discrete-time systems: an introduction with interdisciplinary applications.* Prentice Hall, Inc., Englewood Cliffs, New Jersey.
8. Callahan, A.B. and Pitzer, S.M. 1966. *The applicability of Fourier Transforms Analysis to biological compartmental models. Natural automata and useful simulations. Proceedings of a symposium on Fundamental Biological Models.* (H.H. Patee, E.A. Edelsack, L. Fein and A.B. Callahan, eds.) Spartan Books, 1250 Connecticut Ave., N.W., Washington, D.C., pp. 149–177.
9. Carlson, P.S., ed., 1980. *The biology of crop productivity.* Academic Press, New York.
10. Cirilov, S.D., Newton, J.D. and Schapira, J.P. 1966. *Total cross sections for the reaction* $^{12}C(^{3}He,{}^{4}He)^{11}C$ *and* $^{12}C(^{3}He,n)^{14}O$. Nucl. Phys. 77: 472–476.
11. Cochran, W.T., Cooley, J.W., Favin, D.L., Helms, H.D., Kaenel, R.A., Lang, W.W., Maling, G.E., Nelson, E.D., Rader, C.M. and Welch, P.D. 1967. *What is the Fast Fourier Transform*? IEEE Transactions on Audio and Electroacoustics, Vol. Au-15, No. 2., pp. 45–79.
12. Cooley, J.W., Lewis, P.A.W. and Welch, P.D. 1967. *Application of the Fast Fourier Transform to computation of Fourier integrals Fourier series and convolution integrals.* IEEE Transactions on Audio and Electroacoustics, Vol. Au-15, No. 2, pp. 79–84.

13. DeMichele, D.W., Sharpe, P.J.H. and Goeschl, J.D. 1978. *Toward the engineering of photosynthetic productivity.* CRC Critical Reviews in Bioengineering 3: 23–91.
14. Down, R.J., Hellmers, H. and Kramer, P.J. 1972. *Engineering problems in the design and operation of phytotrons.* ASHRAE J. June 1972: 47–55.
15. Evans, N.T.S., Ebert, M. and Moorby, J. 1963. *A model for the translocation of photosynthates in the soybean.* Journal of Experimental Botany 14: 221–231.
16. Fares, Y., DeMichele, D.W., Goeschl, J.D. and Baltuskonis, D.A. 1978. *Continuously produced, high specific activity* $^{11}CO_2$ *for studies of photosynthesis, transport, and metabolism.* International Journal of Applied Radiation and Isotopes 29: 431–441.
17. Fensom, D.S., Williams, E.J., Aikman, D.P., Dale, J.E., Scobie, J., Ledingham, K.W.D., Drinkwater, A. and Moorby, J. 1977. *Translocation of* $^{11}CO_2$ *from leaves of* Helianthus: *preliminary results.* Canadian Journal of Botany 55: 1787–1793.
18. Finkelstein, L. and Carson, E.R. 1979. *Mathematical modeling of dynamic biological systems.* D.W. Hill, ed. Research Studies Press, Forest Grove, Oregon.
19. Fisher, D.B. 1970a. *Kinetics of C-14 translocation in soybean. I. Kinetics in the stem.* Plant Physiology 45: 107–113.
20. Fisher, D.B. 1970b. *Kinetics of C-14 translocation in soybean. II. Kinetics in the leaf.* Plant Physiology 45: 114–118.
21. Fisher, D.B. 1970c. *Kinetics of C-14 translocation in soybean III. Theoretical considerations.* Plant Physiology 45: 119–125.
22. Fisher, D.B. and Outlaw, Jr. W.H. 1979. *Sucrose compartmentation in the Palisade Parenchyma of* Vicia faba *L.* Plant Physiology 64: 481–483.
23. Fondy, B.R. and Geiger, D.R. 1980. *Effect of rapid changes in sink-source ratio on export and distribution of products of photosynthesis in leaves of* Beta vulgaris *L. and* Phaseolus vulgaris *L.* Plant Physiology 66: 945–949.
24. Fondy, B.R. and Geiger, D.R. 1982. *Diurnal pattern of translocation and carbohydrate metabolism in source leaves of* Beta vulgaris *L.* Plant Physiology 70: 671–676.
25. Gardner, D.G. 1963. *Resolution of multi-compartment exponential decay curves using Fourier transforms.* Annals of the New York Academy of Science, Vol. 108, pp. 195–203.
26. Gardner, D.G., Gardner, J.C., Laush, G. and Meinke, W.W. 1959. *Method for the analysis of multi-component exponential decay curves.* Journal of Chemical Physiology 31: 978–986.
27. Geiger, D.R. and Fondy, B.R. 1979. *A method for continuous measurement of export from a leaf.* Plant Physiology 64: 361–365.
28. Geiger, D.R. 1979a. *Measurement of Translocation.* Methods of Enzymology. 69: 561–571.
29. Geiger, D.R. 1979b. *Control of partitioning and export of carbon in leaves of higher plants.* Botanical Gazzet. 140 (3): 241–248.
30. Geiger, D.R. 1975. *Phloem Loading in Source Leaves.* In I.F. Wardlaw, J.B. Passioura, eds. *Transport and transfer processes in Plants.* Academic Press, New York, pp. 167–183.
31. Geiger, D.R., Ploeger, B.J., Fox, T.C. and Fondy, B.R. 1983. *Sources of sucrose translocated from illuminated sugar beet source leaves.* Plant Physiology 72: 964–970.
32. Geiger, D.R., Saunders, M.A. and Cataldo, D.A. 1969. *Translocation and accumulation of translocates in sugar beet petiole.* Plant Physiology 44: 1657–1665.
33. Geiger, D.R. and Swanson, C.A. 1965a. *Sucrose translocation in the sugar beet.* Plant Physiology 40: 685–690.
34. Geiger, D.R. and Swanson, C.A. 1965b. *Evaluation of selected parameters in a sugar beet translocation system.* Plant Physiology 40: 942–947.
35. Gifford, R.M. and Evans, L.T. 1981. *Photosynthesis, carbon partitioning and yield.* Annual Rev. Plant Physiol. 32: 485–509.
36. Goeschl, J.D., Magnuson, C.E., DeMichele, D.W. and Sharpe, P.J.H. 1976. *Concentration dependent unloading as a necessary assumption for a closed form mathematical model of osmotically driven pressure flow in phloem.* Plant Physiology 58: 556–562.
37. Goeschl, J.D., Magnuson, C.E., Fares, Y., Jaeger, C.H., Nelson, C.E. and Strain, B.R. 1984. *Spontaneous and induced blocking and unblocking of phloem transport.* Plant, Cell and Environment 7: 89–100.

38. Hearon, J.Z. 1963. *Theorems in Linear Systems*. Annals of the New York Academy of Science, 108.
39. Hesketh, J., Jones, J.W., eds, 1980. *Predicting photosynthesis for ecosystem models, Vol. II.*, CRC Press, Boca Raton, FL.
40. Hofstra, G. and Nelson, C.D. 1969a. *A comparative study of translocation of assimilated ^{14}C from leaves of different species*. Planta (Berl.) 88: 103–112.
41. Hofstra, G. and Nelson, C.E. 1969b. *The translocation of photosynthetically assimilated ^{14}C in corn*. Canadian Journal of Botany 47: 1435–1442.
42. Hunt, B.R. 1970. *In Inverse Problem in Radiography*. Mathematical Biosciences 8: 161–179.
43. Jacquez, J.A. 1972. *Compartmental Analysis in Biology and Medicine*. Amsterdam: Elsevier.
44. Kende, H.J. and Zeevaart, J.A.D. eds. 1979. *Partitioning of assimilates, summary of workshop*. Pub. by Amer. Soc. Plant Physiology, Rockville, Maryland, 20850.
45. Kramer, P.J., Hellmers, H. and Down, R.J. 1970. *SEPEL: New phytotrons for environmental research*. 20 (22): 1201–1208.
46. Liu, Pl, Wallace, D.H. and Ozbun, J.L. 1973. *Influence of translocation of photosynthetic efficiency of* Phaseolus vulgaris *L*. Plant Physiology 52: 412–415.
47. Loomis, R.S., Rabbigne, R. and Ng, E. 1979. *Explanatory models in crop physiology*. Annu. Rev. Plant Physiology 30: 339–67.
48. Lush, W.M. and Evans, L.T. 1974. *Translocation of photosynthetic assimilate from grass leaves, as influenced by environment and species*. Australian Journal of Plant Physiology 1: 417–431.
49. Magnuson, C.E., Fares, Y., Goeschl, J.D., Nelson, C.E., Strain, B.R., Jaeger, C.H. and Bilpuch, E.G. 1982. *An integrated tracer kinetics system for studying carbon uptake and allocation in plants using continuously produced $^{11}CO_2$*. Radiation and Environmental Biophysics 21: 51–65.
50. Magnuson, C.E., Goeschl, J.D., Sharpe, P.J.H. and DeMichele, D.W. 1979. *Consequences of insufficient equations in models of the Much hypothesis of phloem transport*. Plant, Cell and Environment 2: 181–188.
51. Marquardt, D.W. 1963. *An algorithm for least squares estimation of nonlinear parameters*. Journal of the Society of Industrial Applied Mathematics. Vol. 11, pp. 431–441.
52. Minchin, P.E.H. 1978. *Analysis of tracer profiles with applications of phloem transport*. Journal of Experimental Botany. Vol. 29, 113, pp. 1441–1450.
53. Minchin, P.E.H. 1979. *The relationship between spatial and temporal tracer profiles in transport studies*. Journal of Experimental Botany. Vol. 30, No. 119, pp. 1171–1178.
54. Minchin, P.E.H. and Troughton, J.H. 1980. *Quantitative interpretation of phloem translocation data*. Annual Review of Plant Physiology 31: 191–215.
55. Mitsui, A., Miyachi, S., San Pietro, A. and Tamura, S., eds. 1977. *Biological solar energy conversion*. Academic Press, New York.
56. Moorby, J. and Jarman, P.D. 1975. *The use of compartmental analysis in the study of the movement of carbon through leaves*. Planta (Berl.) 122: 155–168.
57. Moorby, J., Ebert, M. and Evans, N.T.S. 1963. *The translocation of ^{11}C-labelled photosynthate in the soybean*. Journal of Experimental Botany 14: 211–220.
58. Moorby, J., Troughton, J.H. and Currie, B.G. 1974. *Investigations of carbon transport in plants. II The effects of light and darkness and sink activity on translocation*. Journal of Experimental Botany 937–944.
59. Neales, T.F. and Incoll, L.D. 1968. *The control of leaf photosynthesis rate by the level of assimilate concentration in the leaf: a review of the hypothesis*. Botanical Review 34: 107–125.
60. Outlaw, Jr. W.H. and Fisher, D.B. 1975. *Compartmentation in* Vicia faba *leaves I. Kinetics of ^{14}C in the tissue following pulse labelling*. Plant Physiology 55: 699–703.
61. Outlaw, Jr. W.H., Fisher, D.B. and Christy, A.L. 1975. *Compartmentation in* Vicia faba *leaves II. Kinetics of ^{14}C-sucrose redistribution among individual tissues following pulse labelling*. Plant Physiology 55: 704–711.
62. Pearson, C.J. 1974. *Daily changes in carbon-dioxide exchange and photosynthate translocation of leaves in* Vicia faba. Planta (Berl.) 119: 59–70.
63. Pickard, W.F., Minchin, P.E.H. and Troughton, J.H. 1978. *Real time studies of carbon-11 translocation in Moonflower I. The effects of cold blocks*. J. exp. Bot. 29(111). 993–1001. *II. The*

effects of metabolic and photosynthetic activity and of water stress. III. Transient inhibition of translocation in Ipomoea alba L. by small temperature reductions. Australian Journal of Plant Physiology. 5: 127–130.

64. Pizer, S.M., Ashare, A.B., Callahan, A.B. and Brownell, G.L. 1969. *Fourier transform analysis of tracer data. Concepts and models of biomathematics* (F. Heinments, ed.), Marcel Dekker, Inc., New York, pp. 105–129.
65. Rescingo, A. and Segre, G. 1966. *Drug and tracer kinetics.* Blaisdell Publishing Co., Wallham, Mass.
66. Rachie, K.O., Lyman, J.M., eds., 1981. *Genetic engineering for crop improvement: A Rockefeller Foundation conference.* The Rockefeller Foundation. New York.
67. Servaites, J.C. and Geiger, D.R. 1974. *Effects of light intensity and oxygen on photosynthesis and translocation in sugar beet.* Plant Physiology 54: 575–578.
68. Siegleman, H.W. and Hind, G. eds. 1978. *Photosynthetic carbon assimilation.* Plenum Press, New York.
69. Smith, K.C., Magnuson, C.E., Goeschl, J.D. and DeMichele, D.W. 1980. *A time-dependent mathematical expression of the Münch hypothesis of phloem transport.* Journal of Theoretical Biology 86: 493–505.
70. Sokoloff, L., Reivich, M., Kennedy, C., Des Rosiers, M.H., Patlak, C.S., Pettigrew, K.D., Sakurada, O. and Shinohara, M. 1977. *The ^{14}C deoxyglucose method for the measurement of local cerebral glucose utilization: theory, procedure and normal values in the conscious and anesthetized albino rat.* Journal of Neurochemistry 28: 897–916.
71. Spanner, D.O. and Prebble, J.N. 1962. *The movement of tracers along the petiole of* Nymphoides peltatum *I. A preliminary study with ^{137}Cs.* Journal of Experimental Botany 13: 194–306.
72. Thornley, J.H.M. 1976. *Mathematical models in plant physiology.* Academic Press, New York.
73. Troughton, J.H., Currie, B.G. and Chang, F.H. 1977. *Relations between light level, sucrose concentration, and translocation of carbon 11 in* Zea mays *leaves.* Plant Physiology 59: 808–820.
74. Troughton, J.H., Moorby, J. and Currie, B.G. 1974. *Investigations of carbon transport in plants. I. The use of carbon-11 to estimate various parameters of the translocation process.* Journal of Experimental Botany 25: 684–694.
75. Vernon, L.P. and Aronoff, S. 1952. *Metabolism of soybean leaves. IV. Translocation from soybean leaves.* Archives of Biochemistry and Biophysics 36: 383–398.
76. Wareing, P.F. and Patrick, J. 1975. *Source-sink relations and the partition of assimilates in the plant.* pp. 481–499 in J.P. Cooper, ed. *Photosynthesis and productivity in different environments. Vol. 3. International Biological Programme.* Cambridge University Press, London.
77. Williamson, C. and Boujot, J.P. 1962. *Rapport CEA 2189. Centre D'Etudes Nucleaires de Saclay. Service de Documentation.* Boite Postale No. 2-G. f. sur. Yvette (Seto) France.
78. Wilson, J.W. 1972. *Control of crop processes.* pp. 7–30 in A.R. Rees, K.E. Coskshull, D.W. Hand and R.G. Hurd, eds., *Crop process in controlled environment.* Academic Press, New York.
79. Zelitch, I. 1982. *The close relationship between net photosynthesis and crop yield.* Bioscience 32: 796–802.

Application of stable isotopes in analysis and metabolism

CLAUDE VEILLON
Vitamin and Mineral Nutrition Laboratory, Beltsville Human Nutrition Research Center, U.S. Department of Agriculture, ARS, Beltsville, MD 20705, U.S.A.

Introduction

Many of the elements in the periodic table consist of more than one isotope, i.e., chemically identical species with differing atomic masses. Most of these isotopes are not radioactive in their natural state and are referred to as *stable* isotopes. The element copper, for example, has an atomic number of 29 and an atomic weight of 63.5, and consists of two stable isotopes: ^{63}Cu and ^{65}Cu. The atomic nucleus of each contains 29 protons, but ^{65}Cu contains 2 more neutrons than ^{63}Cu in its nucleus. The 2 stable isotopes of copper have "atomic weights" of 63 and 65 and thus differ in mass while being chemically identical. Naturally occurring copper is about 69 % ^{63}Cu and about 31 % ^{65}Cu, resulting in the overall elemental atomic weight of 63.5.

The various stable isotopes of most elements can be separated (enriched) in a variety of ways, the most common of which is electromagnetically. Here the element is ionized, accelerated and separated in a magnetic field, in much the same way that a mass spectrometer operates. These enriched stable isotopes are separated by and are commercially available from the Oak Ridge National Laboratory in Tennessee. Actually, for most of the elements, this is the only source in the western world, the only other producer worldwide being the Soviet Union. The equipment used for separations at Oak Ridge is now over 40 years old, which causes some apprehension among those needing enriched stable isotopes in their work or product. Somewhat different demands are placed on the facility by the two main user groups. Researchers like myself need milligram quantities of all of the isotopes, while the medical/nuclear/physics users want gram quantities of a few isotopes.

This discussion will not include the use of stable isotopes as targets in particle accelerators nor the production of radioisotopes by the absorption of neutrons or other particles. Thus, the two primary uses of stable isotopes to be discussed are: as ideal internal standards in analyses; and, as metabolic tags or tracers, particularly for human studies. The discussion will also concentrate on application of stable isotopes of the metallic elements, the so-called "trace elements". This merely reflects the author's experiences and is not meant to reduce in any way the importance of other applications. Non-metals, like ^{2}H, ^{13}C, ^{15}N, ^{18}O, etc. have been and are extensively used in a wide variety of tracer studies. Incorporating these into organic species enables one to study, for example, metabolic pathways and metabolic fate of compounds.

Labeled compounds can be used as ideal internal standards in determinations of the compound by isotope dilution, isotope ratio mass spectrometry. The types of things that can be done, the kinds of information that can be obtained, and the mathematical treatments are very similar for stable isotopes of metals and non-metals.

Application in analysis

If one considers the periodic table through element number 86 (radon), and neglecting the lanthanide and actinide elements, all of the elements essential to life are included. True, this also includes non-essential and even downright dangerous ones as well. Of these 72 elements, 4 are unstable (radioactive), 16 are mononuclidic (only one isotope), 20 consist of only 2 stable isotopes, and the remaining 32 consist of more than 2 stable isotopes. Some of the mononuclidic trace elements are of considerable biological interest, such as manganese, cobalt, aluminum, arsenic, fluorine and iodine, and this precludes their determination by stable isotope dilution. Actually, iodine had been measured by isotope ratio mass spectrometry employing a *radioactive* isotope as the internal standard, a perfectly valid procedure but not one to be recommended for everyone.

Measurement

To employ enriched stable isotopes for analysis or as tracers, a means of measuring them quantitatively is needed. Since the isotopes of an element are chemically identical, differing only in atomic mass, a measurement system must utilize properties affected by mass differences or properties of the atomic nucleus. Spectroscopic techniques can be used in some cases. For example, infrared absorption can be used with deuterated compounds because the large mass difference between ^{1}H and ^{2}H shifts the vibrational spectrum considerably. Atomic spectroscopy has been used to measure isotopes of the very light (e.g., Li) and very heavy (e.g. Hg, Pb, U) elements, based on electronic energy level differences caused by nuclear mass and nuclear volume, respectively. Mössbauer spectrosocopy has been used with isotopes like ^{57}Fe, and is capable of measuring several other nuclei of interest.

For the bulk of the metallic elements in the middle of the periodic table, the two most widely used measurement techniques for stable isotopes are neutron activation analysis and mass spectrometry. Of all available techniques for measuring stable isotopes, these two are the most versatile and satisfactory.

Neutron activation analysis (NAA) involves the production of a radioactive isotope by the capture of neutrons by the nuclei of a stable isotope of an element and subsequent measurement of the radioactivity. Not all nuclei (stable isotopes) give rise to a radioactive isotope and some produce species with a half-life too short for practical use. NAA is applicable to stable isotope dilution methods in many cases (as well as stable isotope tracer studies), and has some important advantages over other trace element analysis procedures. Generally, sensitivity and specificity are high. Interference from other elements during the analysis is minimized because after irradiation the samples' integrity can be threatened only by contamination with radioactive material. Another important advantage of NAA is that it is capable of multielement analysis. Some limitations of NAA include: access to a nuclear reactor and counting facilities is needed; many isotopes cannot be measured (neutron absorption cross-section too small, nonradioactive product, half-life too long or too short, emitted radiation difficult to measure); and positron emitters which produce 0.511 MeV annihilation radiation.

The most popular and perhaps most versatile way to measure stable isotopes is with a mass spectrometer. These are usually of the magnetic sector or quadrupole types (although time-of-flight instrumentation has also been used), and differ mostly in the ionization method or system used. The more popular ionization methods or systems have been (are): spark source, thermal ionization, electron impact, laser enhanced, and inductively-coupled plasma.

The high-voltage spark source is capable of ionizing all elements with approximately the same sensitivity. Theoretically then, all elements could be determined at the same time. These two factors – multielement capability and uniform sensitivity – are significant advantages over other techniques.

The thermal ionization instruments have demonstrated the highest precision and accuracy of all of the stable isotope methods. It is also one of the most time consuming, requiring considerable operator skill, and, until recently, quite expensive instrumentation. One manufacturer now produces a thermal ionization instrument based on a quadrupole (rather than magnetic) instrument, at significantly reduced cost.

The relatively simple electron impact ion source has been used recently for stable isotope measurements on combined gas chromatograph/mass spectrometer (GC/MS) instruments. With quadrupole instruments, the cost is kept down, and sample throughput with these combined systems is very high. The analyte in the sample is converted to a thermally stable, volatile chelate which is introduced into the mass spectrometer via the gas chromatograph. These systems are fast, reasonably precise, and have been applied to a number of elements, such as Mg, Cr, Fe, Cu, Zn and Se, limited only by the availability of a suitable chelating agent.

Laser enhanced ionization is a relatively new technique still being developed and not yet readily available commercially. Powerful dye lasers are used to ionize the analyte element of choice by 2-photon absorption, rendering the technique highly specific and capable of overcoming isobaric interferences (2 isotopes having the same nominal mass, e.g., ^{58}Fe and ^{58}Ni).

Inductively coupled plasmas (ICP's) are widely used as sources in atomic emission spectrometry. Two mass spectrometer manufacturers have recently interfaced these to their instruments as ionization sources, ions being sampled into the instrument through a small orifice. These instruments look very promising for multielement analysis as well as being able to measure stable isotopes. They have the advantages of speed, and ionization external to the instrument. Few biological applications have appeared thus far, but these instruments bear watching since development will be rapid.

Isotope dilution

The concept of elemental determinations by stable isotope dilution is truly elegant, and provides us with an extremely powerful and accurate analytical tool. It is based on the addition of a known quantity of an *enriched* stable isotope of the analyte element (spike) to a sample, chemical equilibration of the spike and analyte, and measurement of the isotope *ratio* of the spike and another isotope of the analyte.

Let us look at an example of how this works. Magnesium consists of 3 stable isotopes, ^{24}Mg, ^{25}Mg and ^{26}Mg. The natural abundance of each is almost exactly 79, 10 and 11 percent, respectively. Thus, if we looked at the mass spectrum of Mg in an ideal mass spectrometer, we would see 3 peaks having relative peak heights of 79, 10 and 11 for 3 isotopes. The measured isotope ratio of, for example, ^{24}Mg/^{26}Mg would be 7.18 (79/11). Let us suppose that we have *enriched* Mg stable isotopes as well, say, ^{25}Mg (96 %, containing 3 % ^{24}Mg and 1 % ^{26}Mg) and ^{26}Mg (90 %, containing 6 % ^{24}Mg and 4 % ^{25}Mg). If we add a known amount of enriched ^{26}Mg as a spike to a sample containing an unknown amount of natural (unenriched) Mg, we will have altered the measured isotope ratio (R) of ^{24}Mg/^{26}Mg. It will now be less than 7.18. We can express the isotope ratio, R, of ^{24}Mg/^{26}Mg as follows:

$$R = \frac{^{24}\mathrm{Mg}}{^{26}\mathrm{Mg}} = \frac{0.79\,\mathrm{Mg_n} + 0.06\,\mathrm{Mg_{sp}}}{0.11\,\mathrm{Mg_n} + 0.90\,\mathrm{Mg_{sp}}}$$

where Mg_n is the amount of natural (unenriched) analyte, and Mg_{sp} is the amount of ^{26}Mg spike added to the sample. The numerator represents all sources of ^{24}Mg, i.e., natural (at 79 % abundance) plus the 6 % abundance contributed by the added spike. The denominator is ^{26}Mg from natural (11 % abundance) plus the added spike (90 % abundance). Since R is measured and Mg_{sp} is known, the amount of Mg present in the original sample (Mg_n) can be calculated.

From this we can see that stable isotope dilution has some important advantages over other analytical techniques. First, an "ideal" internal standard (the spike) is used, because it is chemically identical to the analyte. Second, this means that quantitative recovery of the analyte and equilibrated spike during the analysis procedure is not necessary, since the measured ratio will not be affected. Third, this can be an "absolute" method (like osmotic pressure is an absolute measurement, for example) because one makes only a *ratio* measurement. One does not have to calibrate instrument response with standards, as with most other techniques.

However, there is an assumption made that is critical to isotope dilution, namely, that at some point the analyte and spike equilibrate chemically prior to analysis. For inorganic samples this is not much of a problem but it becomes much more important for biological samples. In these samples, the analyte is often part of an organic compound and could behave quite differently from an added inorganic spike. In practice, samples of this type are usually ashed or wet digested to destroy the organic matter and in the process this destroys the chemical history of the analyte. At that point, both analyte and spike are inorganic and can equilibrate. Once equilibrated, further manipulations will affect both equally and the measured isotope ratio will be unaffected.

Tracer studies

Since many elements are composed of several stable isotopes, one can use one or more enriched isotopes as tracers, and another as the internal standard spike. In the previous example with Mg, the enriched ^{25}Mg could be used (e.g., fed to human subjects) as a tracer. Samples would now contain natural Mg and some enriched ^{25}Mg tracer. They would be spiked with ^{26}Mg as before, and two isotope ratios measured, as follows:

$$R_1 = \frac{^{24}Mg}{^{26}Mg} = \frac{0.79\,Mg_n + 0.06\,Mg_{sp} + 0.03\,Mg_t}{0.11\,Mg_n + 0.90\,Mg_{sp} + 0.01\,Mg_t}$$

$$R_2 = \frac{^{25}Mg}{^{26}Mg} = \frac{0.10\,Mg_n + 0.04\,Mg_{sp} + 0.96\,Mg_t}{0.11\,Mg_n + 0.90\,Mg_{sp} + 0.01\,Mg_t}$$

where Mg_t is the amount of enriched ^{25}Mg present in the sample, and the other terms are as before. The two ratios, R_1 and R_2, are measured, and the amount of enriched ^{26}Mg spike added to the sample, Mg_{sp}, is known. Thus, we have 2 equations with 2 unknowns, Mg_n and Mg_t, i.e., the amount of natural (unenriched) Mg in the sample and the amount of enriched ^{25}Mg tracer in the sample. These can be solved simultaneously, giving us a quantitative estimate of the ^{25}Mg tracer and the natural, unenriched Mg present in the sample. As a bonus, we also get the total Mg content of the sample, which is simply the sum of these.

Application in metabolism

As indicated above, stable isotopes can readily be used in tracer studies. For the trace elements of biological interest, most of these studies have concentrated on metabolism of the element of interest. It would not be practical to discuss all of the excellent stable isotope work of this nature that has taken place in the last few years, but it is evident that interest in this area is growing. The activities of some of these groups have been described by Veillon and Alvarez (1983).

I would like to use as an example of an application of stable isotopes to tracer studies in humans, one of our first studies. This involved stable isotopes of selenium to study the utilization of this element in pregnant and nonpregnant women (Swanson, et al. 1983). But first, the ^{76}Se tracer was incorporated into a food source (Swanson, et al. 1983).

Laying hens were fed a low Se (0.06 μg/g) basal diet for 39 days to deplete their selenium stores. They were then fed the basal diet supplemented with enriched ^{76}Se (0.3 μg/g) for 35 days. The incorporation of the metabolic tag into edible tissues was followed with a double isotope dilution measurement of chelated Se with a combined GC/MS instrument. The ^{76}Se content of the basal diet was increased by a factor of 9.7 by the addition of the enriched stable isotope. This high level of enrichment was approached in egg yolk (9.5-fold) and liver (9.0-fold) during the 35-day repletion/labeling period. Interestingly, the enrichment was lower in egg white (7.2-fold). The breast meat enrichment reached 5.0-fold. The level of enrichment in a given tissue reflected both the turnover rate of the tissue and its natural selenium content. Considering the degree of enrichment, enrichment level and Se level, egg yolks were found to be the most efficient food source of the ^{76}Se tracer.

Thus, egg yolk from this study was incorporated into the human study mentioned earlier. The objective of the study was to compare selenium utilization of women in early and late pregnancy to that of nonpregnant controls. The subjects were fed a defined, constant diet providing

150 μg Se/day, for 20 days while confined to a metabolic ward. On day eight, the ^{76}Se intrinsically labeled egg yolk was substituted for that in the diet, such that the total daily intake remained at 150 μg, but for day eight, this included 40 μg of enriched ^{76}Se and 110 μg of natural Se.

The amount of selenium absorbed was constant at about 80 % for all three groups. Pregnant women conserved selenium by decreasing urinary

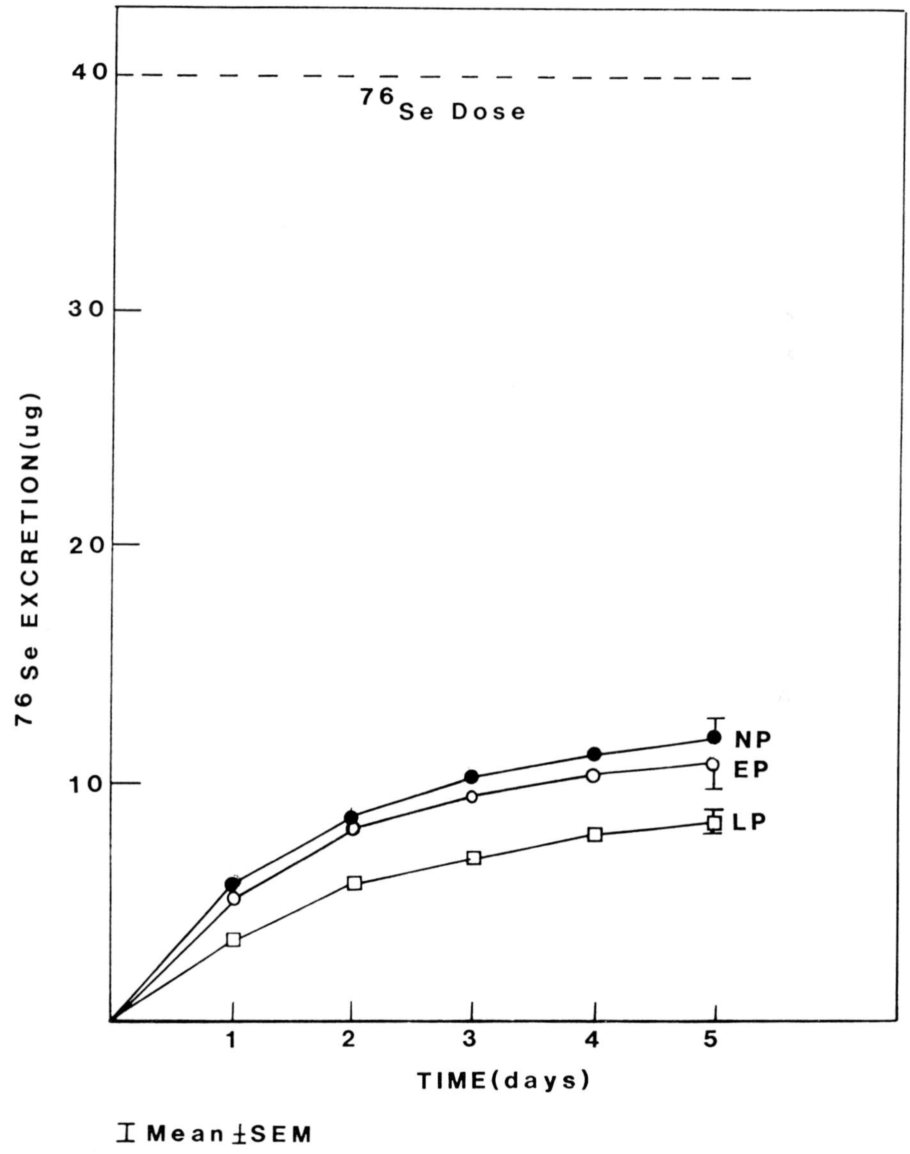

Figure 1. Cumulative urinary excretion of ^{76}Se.

excretion. Figure 1 illustrates the cumulative urinary excretion of the ^{76}Se dose. Here it can be seen that the women in late pregnancy (LP) excreted considerably less of the tracer than the nonpregnant women (NP), with the early pregnant (EP) women falling between these two. It is interesting to note that only 20–30 % of the single 40 μg dose of ^{76}Se tracer had been excreted by the urine 5 days following the tracer dose. In addition to greater retention during pregnancy (presumably due to the fetus), Fig. 1 illustrates that most of the recently absorbed Se goes into long-term or slower-turnover pools within the body. It is definitely not a case of "first in, first out".

Since these early studies, designed in part to demonstrate that foods can be intrinsically labeled with stable isotopes, and that metabolism of these can be followed in subjects that could never be studied using radioactive tracers, and that far more information can be obtained than by analysis for unenriched analyte, we have completed or are working on several other studies taking further advantage of the tremendous potential of this analytical technique. Hopefully, these will form the basis for future discussions at symposia such as the present one. Most assuredly this methodology will be with us and serve us well, well into the 21st century.

References

1. Veillon, C. and R. Alvarez. 1983. *Determination of trace metals in biological materials by stable isotope dilution.* Chapter 5 in "Metal Ions in Biological Systems", H. Sigel, ed. Marcel Dekker, Inc. New York, pp. 103–122.
2. Swanson, C.A., D.C. Reamer, C. Veillon, J.C. King and O.A. Levander. 1983. *Quantitative and qualitative aspects of selenium utilization in pregnant and nonpregnant women: an application of stable isotope methodology.* Am. J. Clin. Nutr. 38: 169–180.
3. Swanson, C.A., D.C. Reamer, C. Veillon and O.A. Levander. 1983. *Intrinsic labeling of chicken products with a stable isotope of selenium (^{76}Se).* J. Nutr. 113: 793–799.

Flow sorting of cells and organelles

DANIEL PINKEL
Lawrence Livermore National Laboratory, Biomedical Sciences Division, University of California, P.O. Box 5507, L-452, Livermore, CA 94550, U.S.A.

Abstract Flow techniques offer the possibility of rapidly and precisely determining many characteristics of each member of a population, and of efficiently separating desired subpopulations for detailed study or culture. Measurements on objects can be made based on their electrical and acoustical properties, light scatter, uptake of specific fluorescent dyes, affinity for antibodies, and binding of nucleic acid probes. Asymmetrical objects can be oriented in flow and scanned or imaged to obtain morphological information. Spatial resolution below 1 μm has been achieved. Measurement rates are limited only by problems of coincidence between objects; operation above 10^5/sec has been achieved. High efficiency, high purity sorting at rates above 2×10^4/sec has been demonstrated. The detection threshold for fluorochromes is determined primarily by autofluorescence of the objects being measured.

In the future we are likely to see improved handling of multiparameter data (measurements of 4 or more characteristics of each object are not unusual), increased application to the measurement of functional states of cells, greater emphasis on kinetic measurements on time scales ranging down to milliseconds, increased use of antibodies and nucleic acid sequences for specific labeling, and development of rugged, inexpensive instrumentation for clinical and field use.

Introduction

Flow cytometry and sorting offer efficient methods to rapidly obtain information on each member of a heterogeneous population of objects, and to separate desired subpopulations. The particles to be measured are suspended in a fluid which carries them through the instrument where they pass one or more measurement stations. While imaging and scanning of particles to obtain morphologic information is sometimes done, most applications rely on light scatter, electrical volume determinations (Coulter), and measurement of fluorescence from staining reagents specific for particular properties of the particles. The past several years have seen the dramatic growth of monoclonal antibodies as the vehicle to achieve fluorescent labeling of cellular components, and we now stand at the very beginning of what may be a similarly explosive use of nucleic acid probes for specific labeling of DNA and RNA sequences. In this article I will describe in more detail flow cytometry and sorting as currently practiced and illustrate these principles with various applications.

Basic concepts

A representative cell sorter is illustrated in Fig. 1. It consists of a flow chamber which establishes and controls the flow of the sample particles,

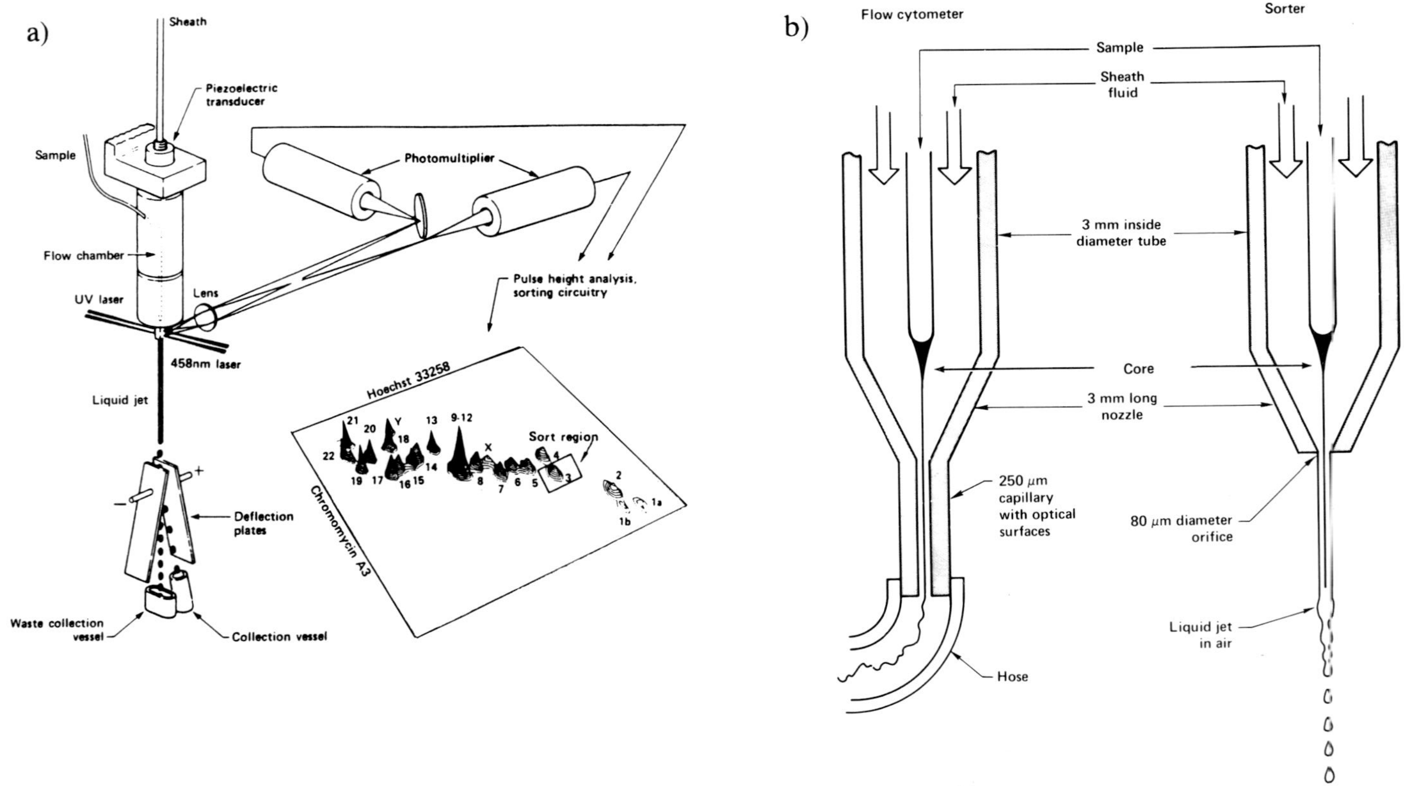

Figure 1. (a) Representative Flow Instrument. Particles are measured as they pass sequentially through two laser beams. The light emitted at each beam intersection, fluorescence and/or scatter, is imaged onto a separate photomultiplier tube. Spectral filters (not shown) in the light paths select the wavelength range that is detected. Data are recorded and displayed as a multivariate distribution. Illustrated is the measurement of human chromosomes stained with two fluorescent dyes. The chromosome type contained in each peak of the distribution are indicated. Particles can be separated by vibrating the flow chamber with a piezoelectric transducer to cause the liquid jet to break up into droplets. Droplets containing the selected particles, in this case chromosome 3, are charged and deflected by an electric field. (b) Details of a Flow Cytometer and Sorter Flow Chamber. In both chambers the sample is released from a small tube into the center of a flow of sheath fluid. The flow accelerates as it passes through the nozzle, which decreases the diameter of the core. In a flow cytometer the flow is usually contained for its whole path. In most sorters the liquid flow exits the chamber as a jet. Measurement may occur either in the jet or in an optical capillary just before the exit. Vibration of the nozzle causes the jet to break up into regular droplets.

laser beams to excite the fluorescent dyes, detectors to measure the fluorescence and scattered light, electronics to quantitate the signal intensities and display the results, and a mechanism to separate selected subpopulations of the sample. In this section I will discuss these components in more detail, concentrating on the underlying physical principles.

Flow cytometry. The flow chamber of a flow instrument, Fig. 1(b), typically consists of a tube which ends in a nozzle. As the sheath fluid passes through the chamber, its velocity increases by approximately the ratio of the crossectional areas of the entrance and exit. The diameters of these regions in biological instruments are roughly 3 mm and 0.1 mm respectively, yielding a velocity increase of a factor of 100 to 1000. Flow velocities at the nozzle exit are typically in the range of 1 to 20 m/sec. The sample particles, suspended in fluid, are introduced into the sheath flow through a small tube in the center of the chamber. The sample flow is carried along by the sheath in a confined stream, called the core, and accelerates through the nozzle with it. The diameter of the core is reduced as the velocity increases, causing all of the particles to follow very nearly the same path through the chamber. This process, called hydrodynamic focusing, is very important. Precise measurements require that all the particles pass through the region of uniform sensitivity of the measurement stations, for example the proper optical focal point. Without hydrodynamic focusing one would have to use very small nozzle orifices to achieve adequate confinement of the particle trajectories, which makes nozzles very susceptible to plugging. Even if plugging were not a problem, the spread of the particle trajectories across the entire nozzle would cause problems. Those particles near the center of the flow would be moving faster than the others since the fluid near the walls is slowed by drag. For most applications the variation in velocities would substantially reduce the precision of the measurements, (for example, the particles would spend different amounts of time in the illumination beam) and make it impossible to use time delays to predict particle positions subsequent to the measurement. The latter point is very important for particle sorting as discussed below. The principles of hydrodynamic focusing have been well understood for a century (Reynolds, 1883) and can be used with many types of fluids. The use of streams of smoke in wind tunnels to visualize flow around aircraft is one very familiar application. Indeed one of the first flow measurements reported involved light scattering from aerosols (Gucker et al., 1947).

To a first approximation the sample particles are simply carried with the sheath through the flow chamber. However this can't be completely true since the particles in general will not have exactly the same density as the fluid, and they are at least partially rigid. Thus, as the flow accelerates toward the nozzle orifice, the particles will feel substantial forces from the surrounding fluid. For example, if we follow an imaginary cylindrical fluid element through the nozzle we find that its diameter is reduced roughly by the ratio of the diameters of the nozzle entrance and exit, a factor of 10 to 30 or so in typical instruments. To conserve volume its length must increase by the square of this factor, approximately a factor of 100 to 1000. This hydrodynamic environment will attempt to impose the same deformation on particles. Thus a cell will experience stretching forces on passage through the flow chamber. These forces can be used to stretch and orient particles for measurement as will be discussed below, but they may also damage them. In standard instruments particle accelerations are in the range of 10^4 times gravity near the exit of the nozzle (Pinkel and Stovel, 1985).

The rate at which particles can be measured depends on the fluid velocity, which in turn is dependent on the pressure difference between the nozzle entrance and exit. If the flow does not travel very far in a constricted channel after exiting the nozzle, the exit velocity is given by the Bernoulli equation. In practical units $v = 3.7P^{0.5}$, where v is the velocity in meters per second and P is the pressure in pounds per square inch. The square root dependence of velocity on pressure has some striking consequences. To achieve a velocity of 1 m/sec at the nozzle exit all that is needed is 0.07 psi, which can be achieved with a syphon with a height difference of 0.05 m between the reservoir and nozzle. Velocities in the standard range of 10 m/sec require about 7 PSI, about half an atmosphere or 5 m height difference. To make another factor of 10 increase in velocity requires a 700 PSI system. This latter pressure is not extreme for laboratory instruments, but the complexity of dealing with flow systems rises rapidly with pressure. Only a few instruments operating at pressure much above 20 PSI have been built (Steen and Lindmo, 1979; Peters et al., 1985).

For a given flow velocity, the maximum particle measurement rate depends on the time required to make a measurement on a single particle and the degree of coincidence between particles that can be tolerated in a particular application. These in turn depend on the intensity of the signals, the speed of the signal processing system, and the size of the particles. For example, 10 μm diameter particles moving at 10 m/sec require 2 μsec to traverse a 10 μm focal spot of a laser beam. These particles can potentially be measured at a rate of 5×10^4/sec if a 10 % chance of having 2 particles in the laser beam at the same time is acceptable. Some instruments contain electrical circuitry that allows recognition and discarding of signals from some of the coincident particles. With these systems, one gains in measurement rate at the expense of throwing away information from a fraction of the population.

Sorting. Sorting requires making a decision on each particle on the basis of the measurements. The liquid containing a particle that meets the criteria set by the experimenter is then diverted into a collection vessel. If the particles are moving at 10 m/sec they will travel 1 cm in the millisecond or so which is needed to process the signals and make a sorting decision. A number of methods of diverting the desired particles have been developed. They fall into two classes. The first are fluidic switching systems in which the flow is always confined within a closed channel network and can be redirected rapidly from one branch to another (Kamentsky and Melamed, 1967). These systems have tended to have slower sorting rates than the droplet sorters discussed next, but they have the advantage that the particles remain confined at all times. This is especially attractive for dealing with hazardous materials. While many designs of this type have been built, the first commercial instrument of this type is just becoming available (Partec ag, Switzerland).

The most prevalent sorter design involves breaking the liquid flow into droplets and deflecting those that contain the desired particles (Fulwyler, 1965). Droplet generation and deflection was originally developed for ink jet printing (Sweet, 1965). The principle is illustrated in Fig. 1. The liquid carrying the particles forms a jet in air after exiting the nozzle. Such a jet is unstable and will break up due to random perturbations. However, if the nozzle is purposely vibrated (accomplished in practice with a piezoelectric crystal) the jet will break up into very regular droplets. Stroboscopic examination of the droplet generation process demonstrates that the location of the droplet break off point, Fig. 2, can be made to be stable to within a small fraction of a drop diameter over many hours. Consequently the time between measurement of a particle and its confinement in a drop is the same for each particle. To sort, one puts a voltage pulse on the sheath fluid (which needs to be a conductor, for example a salt solution) just before the droplet containing the desired particle breaks off from the jet. Charge is carried down the jet to its tip, and is captured in the droplet as it breaks off. The stream of droplets passes through an electric field between the deflection plates, and those with a charge are deflected. The direction and magnitude of deflection depends on the sign of the charge put on the droplet. In most cases, more than one droplet is deflected for each sorting event to be sure that the particle is separated. Most sorters allow simultaneous separation of two

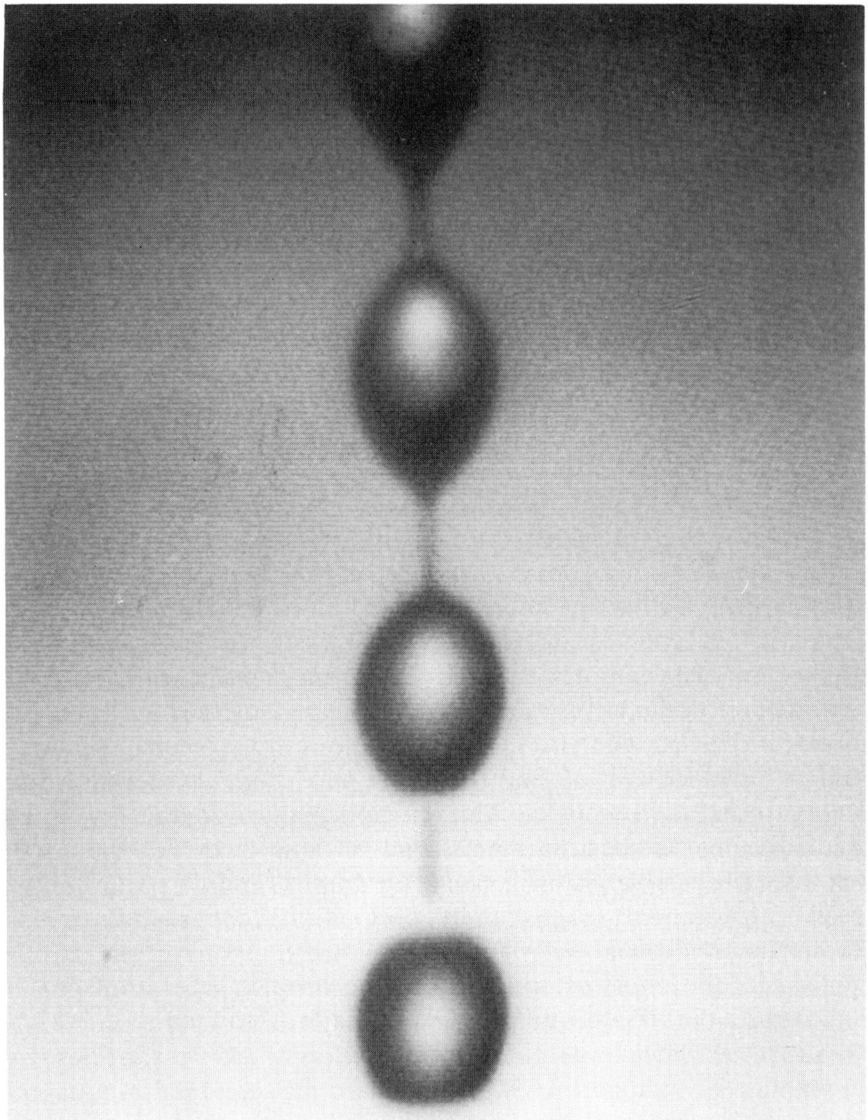

Figure 2. Droplet Formation in a Sorter. The oscillations put on the liquid jet by the piezoelectric transducer grow, eventually causing the jet to break into droplets. In this picture the lowest drop is unconnected to the ones above it. If the jet were charged, this droplet would capture some of the charge as it broke off. In some cases the thin necks between the droplets break free at both ends and form small satellite droplets. This is a stroboscopic image, indicating that the droplet formation process is stable. Droplets are being produced at 2×10^5/sec.

classes of particles using left and right deflections. Only limited use has been made of variable deflection, which can be accomplished by controlling the magnitude of the charge on each droplet. If care is taken the droplets can be placed within several tens of micrometers of the desired location (Stovel and Sweet, 1979). An important application of this ability is the sorting of single viable cells into individual wells of a culture tray for analysis of the resulting colonies. This is normally done by sorting a particle into a well, inhibiting sorting while the tray is moved to bring a new well into position, and then sorting the next particle. Approximately 1 particle per second can be sorted in this way. Attachments to commercial instruments are available that use standard 96 well plates, but custom systems using 10 cm by 10 cm plates containing 1000 wells have been made (Pallavicini et al., 1987).

To maintain purity of the sorted populations it is necessary to assure that unwanted particles are not contained in the droplets that are deflected to get the desired object. Most sorters contain electronic circuitry that can detect such a coincidence and the experimenter can elect to cancel sorting in these instances. The probability of having more than one particle present depends on the ratio of the measurement rate to the droplet generation rate, and increases as the measurement rate increases. If sorting is canceled for these coincident events, the achieved sorting rate will actually decrease if the measurement rate is increased beyond a certain point (Pinkel and Stovel, 1985). Thus, the higher the droplet generation rate the higher the potential sorting rate.

The droplet generation rate has an upper limit which depends on the details of the flow. The oscillations imposed on the jet by the piezoelectric crystal will grow and lead to droplet formation if their wavelength is longer than πd (Raleigh, 1879), where d is the diameter of the jet. The wavelength is the jet velocity divided by the driving frequency. Thus the maximum droplet frequency at a given operating pressure is given by $f = 3.7P^{0.5} \times 10^6/\pi d$, where P is in psi and d is in μm. Jet diameters are typically in the range of 50 to 100 μm, with the mid to higher ends favored for routine work to reduce the problems due to plugging of the nozzle by clumps. Thus at the 10–20 psi operating pressures of most sorters, droplets are produced at a rate of 30–50 kHz. Droplet generation up to 500 kHz has been demonstrated and a practical instrument operating at 200 kHz has been built (Peters et al., 1985). This instrument routinely sorts chromosomes, making high resolution fluorescence measurements on 3×10^4 particles per second.

Particle measurement. Measurements of the optical and electrical properties of particles have dominated flow research, although acoustical meas-

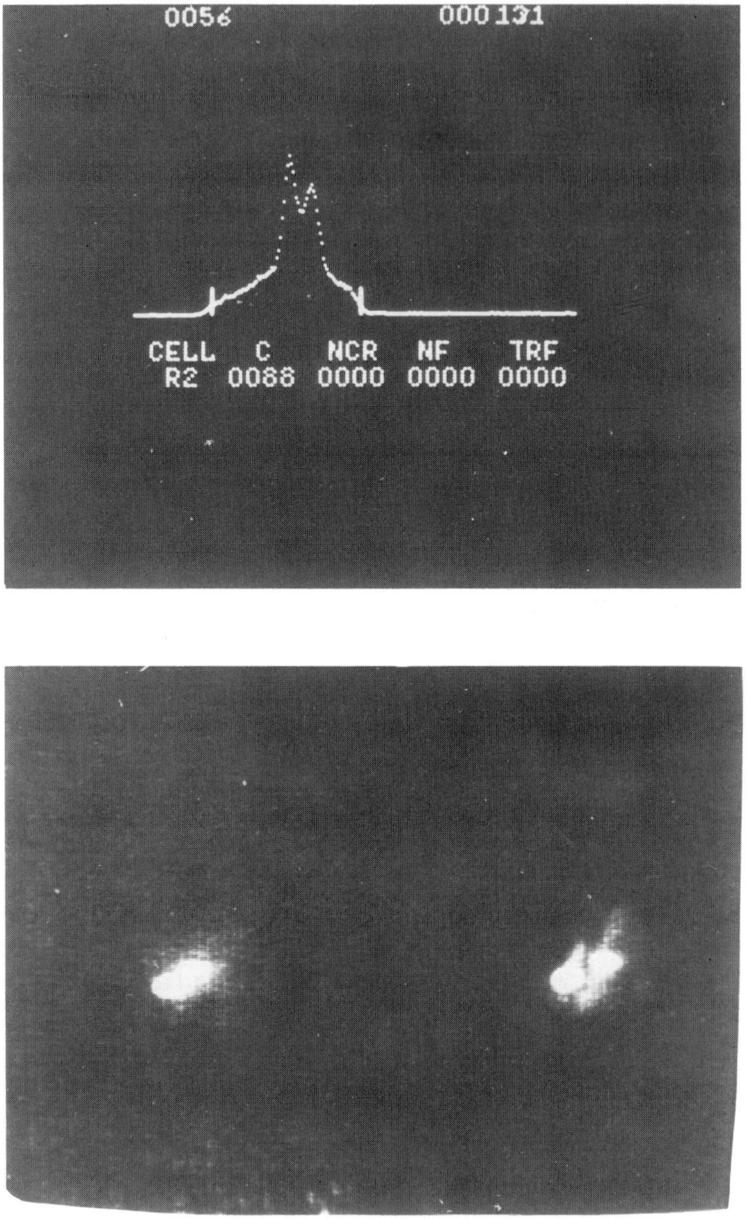

Figure 3. Imaging in Flow. The lower panel shows two images of a cell doublet in flow. The angles of view are separated by 90 degrees. Fluorescence from the cells was measured with an intensified vidicon detector. The upper panel shows profile of the fluorescence signal as this object passed through a tightly focused laser beam.

urements have also been explored (Sweet, unpublished). Particle counting and volume determinations were the first widespread application of flow technology. These sensed the change in electrical impedance of a short orifice as a particle flowed through. This technique, developed by Coulter, has been combined with optical measurements in some flow cytometers and sorters. It has been extended to allow determination of the dielectric breakdown properties of the cell membrane (Rieman et al., 1975).

Optical measurements may be relatively simple, such as measuring the extinction of light on passage of a particle, or complex, such as performing temporal (Cram et al., 1985) or spectral analysis of the fluorescence emission. Various types of imaging of particles in flow have been accomplished, Fig. 3 (Cram et al., 1985; Wheeless and Kay, 1985). Arc lamps and lasers have been used as light sources for most applications. Light from these sources is focused onto the sample stream and scattered or emitted light is collected by detectors. Arc lamp illuminated systems are usually similar to fluorescence microscopes, with the flow chamber taking the place of the microscope slide (Steen and Lindmo, 1979; Dittrich and Gohde, 1969). In these systems the wavelength of the excitation light can be selected using band pass filters. These instruments tend to be relatively inexpensive and simple to operate, and are very useful for measurements where the signals are sufficiently bright.

Lasers are used for illumination in most cytometers and sorters at the current time. They have the advantages of monochromatic, highly collimated output. The collimation allows easy transport of the light from the source to the flow chamber, and permits diffraction limited focusing of the beam on the sample stream. The monochromatic nature of laser emission has advantages and disadvantages. On the positive side, all of the light energy is in a very narrow wavelength band, allowing efficient excitation of fluorochromes if their absorbtion bands match the laser line. On the other hand, laser output is not continuously tunable so the match to the fluorochromes is not always very good. This has led to the use of multiple laser systems using lasers of different types, for example ion lasers filled with argon or krypton, or inclusion of a dye laser, to get adequate spectral coverage. Although the emission of dye lasers is continuously variable, large wavelength changes require changing of the dye. Thus in practice it may be cumbersome to reconfigure an instrument to measure a new set of fluorochromes.

Usually it is desired that the illumination be uniform across the core so that all particles are equally excited. To achieve this, the laws governing light propagation require that the illumination extend substantially beyond the core boundary; the intensity can not be made to drop

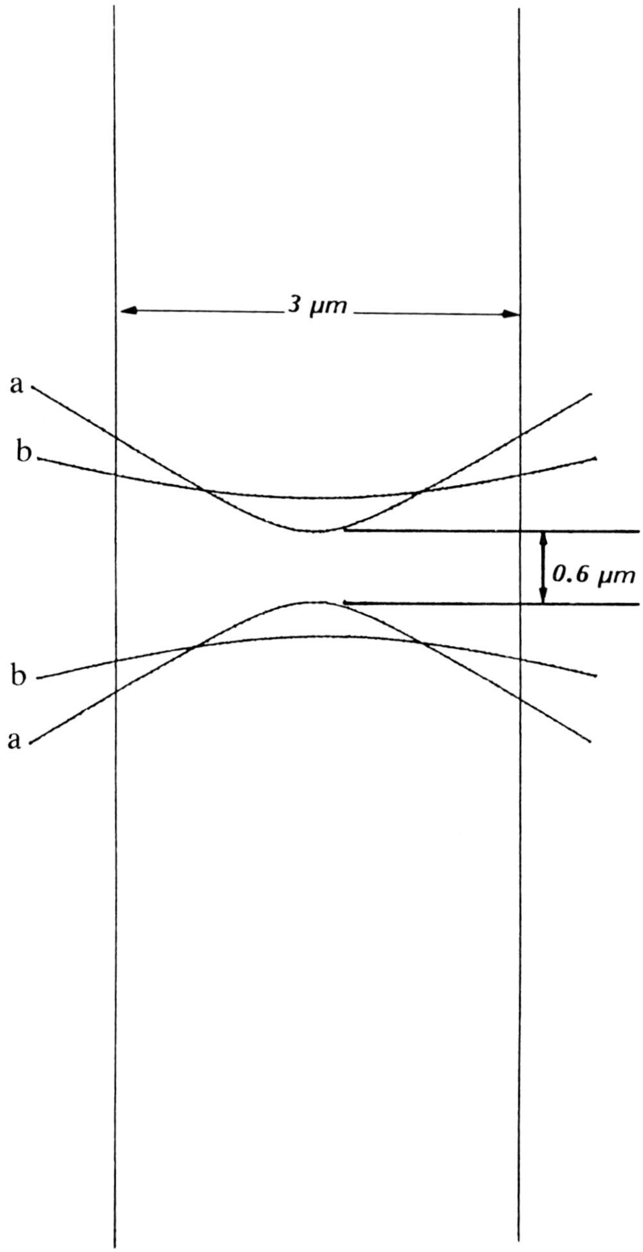

Figure 4. Effect of Focal Spot Dimension on Depth of focus. The more highly convergent beam, (a) is focused to a smaller spot than beam (b), but the minimum dimension is maintained for a shorter distance.

abruptly to zero outside the core. Thus a substantial portion of the excitation beam is not useful for exciting the particles. Similarly, it is often desirable to have the beam dimension along the direction of the sample flow be as small as possible. This reduces the probability of having multiple particles in the beam at the same time, and allows scanning of the fluorochrome distribution along the length of the particles. However the beam dimension along the flow cannot be made arbitrarily small, again due to the properties of light propagation. The minimum dimension of the beam on the core depends on the wavelength of the light and the *f* number (or numerical aperture) of the lenses doing the focusing, Fig. 4 (Wheeless and Kay, 1985). As the angle of convergence of the focused beam increases, the size of the focal spot and the depth of field for which this spot size is maintained, decrease. Thus as the beam is focused more tightly, the core diameter has to be correspondingly decreased to assure that all the particles are illuminated within the depth of field of the beam focus. The goals of spreading the focal spot transverse to the particle flow, and compressing it along the flow results in the use of beam focusing optics designed to produce elliptical illumination patterns in many instruments.

Optical detectors have ranged from arrays of photodiodes used to measure light scatter at multiple angles around the beam-core intersection point, to systems for imaging the scatter or fluorescence. A typical system for fluorescence intensity measurements is illustrated in Fig. 1(a). A lens collects light emitted and scattered from the particles at each intersection with a laser beam and brings it to a focus. Since the excitation points are spacially separated, their focal spots are separated. Small apertures (not shown) at each focus allow only light that originates from the corresponding intersection point to pass. Thus light originating from each intersection point can be independently directed to different detectors. The collected light is converted to an electrical signal by a photomultiplier tube. The wavelength range that is measured is selected by spectral filters placed into the light path. Frequently multiple detectors, each sensitive to a different wavelength range (for example fluorescence and scatter) are used for the light from each beam.

The amount of light collected from each particle depends on the number of fluorochromes it contains, the intensity and duration of the fluorescence excitation, and on the characteristics of the detection optics. In typical laser-based flow systems (1 watt beam focused to a 10 μm by 100 μm spot with a flow velocity of 10 m/sec) an efficient fluorochrome is excited 5 to 10 times during the 1 μsec that it is in the beam. The amount of this light that is collected depends on the design of the detection optics (Wheeless and Kay, 1985). Although systems using ellipsoidal reflectors which allow collection of the emission over almost all of the 4π solid angle around the particles have been built, the difficulties associated with their use have not in general made them worthwhile. High numerical aperture lenses such as microscope objectives, which can collect up to 25 % of the emitted light are more practical. The lower limit on the number of fluorochromes necessary to give detectable staining of a particle is typically set by such factors as autofluorescence of the particle, background light due to the fluorescence of unbound dye in the core, or Raman scatter from the sheath and core liquids. The unbound dye in the core is sometimes required to maintain stain equilibrium of the particles during their pass through the instrument. Measurements have indicated that as few as 1000 fluorescein molecules on a lymphocyte can be detected, the limit being autofluorescence of the cells. The orthogonality of the excitation and detection optics of most laser-based flow systems is ideal for sensitive detection of unbound fluorescent molecules in liquid suspension, such as at the output of HPLC systems. Studies are underway to determine the minimum detection limits (Dovichi et al. 1984).

Applications of flow cytometry and sorting

To illustrate the potential of flow measurements I have chosen a number of applications that are of current or potential interest to work of the U.S. Department of Agriculture. These include work in reproductive biology, bacteriology, and genetics of plants and animals.

Reproductive biology

Sperm separation. Accurate measurements of the DNA content of mammalian sperm are important for a number of applications, including detection of abnormal sperm and quantitation and sorting of the sex-determining sperm populations. However, such measurements have proven more difficult than for other cell types. The basis of the difficulty is that the sperm nuclei are roughly planar and the chromatin in them is highly condensed. The condensation causes difficulty in getting stoichiometric penetration of the stains and leads to a nuclear index of refraction that is much higher than that of water. The index of refraction coupled with the non-spherical nuclear shape leads to asymmetric emission of fluorescence. In general more light is emitted in the plane of the cell than normal to the flat faces. Unless care is taken to control these optical effects, the fluorescence distribution measured for a sperm population does not reflect the distribution of DNA content, Fig. 5(a).

Accurate staining was originally accomplished by slightly swelling the nucleus by breaking the disulfide crosslinks in the protamines surround-

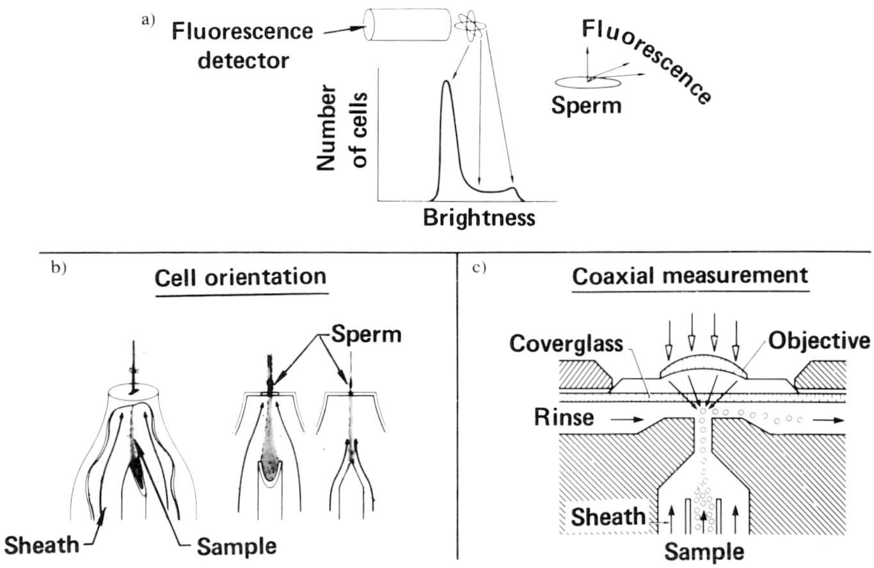

Figure 5. (a) Fluorescence from a DNA stain in a sperm nucleus is refracted so that it is most intense in the plane of the cell. If the orientation of the cell relative to the fluorescence detector is not controlled, the fluorescence distribution from a sperm sample will cover a broad range of intensities. This obscures accurate measurement of stain content. (b) Use of a tapered sample injection tube and/or a noncircular nozzle orifice produce a ribbon-like core. This results in predominant orientation of the nuclei in the plane of the ribbon. Having the optical axis of the excitation beam and fluorescence detector perpendicular to plane of the sperm gives the highest measurement precision. (c) Use of an optical system whose axis is coincident with the core excites and measures all nuclei with their thin edge pointing towards the objective lens. Sperm to sperm variation in the orientations about the flow axis do not affect the measurement.

ing the DNA, followed by mild proteolytic digestion (Otto et al., 1979). More recently it has been found that slight sonication followed by incubation in the DNA dye Hoechst 33342 gives quantitative staining (Johnson et al., 1987a). The optical difficulties have been overcome with two types of flow cytometers. In one, Fig. 5(b), the orientation of the plane of the nuclei about the flow direction is controlled by shaping the tip of the sample injection tube, and sometimes in addition using a rectangular nozzle orifice to give a ribbon-like core flow (Pinkel et al., 1982; Johnson and Pinkel 1986). The sperm nuclei are illuminated by a laser beam incident on one of their flat faces and fluorescence emitted normal to the opposite face is measured by a detector looking directly into the beam. The other type of flow cytometer that has been found useful for sperm DNA content measurement is illustrated in Fig. 5(c) (Otto et al., 1979). In this instrument the sperm nuclei are measured as they flow along the optical axis of the instrument, which is basically an epi-illumination fluorescence microscope with the microscope slide replaced by the flow chamber. The same lens is used to focus the excitation light and collect the fluorescence. As mentioned above, hydrodynamic forces orient the nuclei with their long axis parallel to the flow direction. There is no control over cell to cell variations in the orientation of the plane of the nuclei about the flow axis or over which end of the nucleus goes through first. However since the flow is parallel to the optical axis of the instrument, the signals are insensitive to the rotational distribution of the sperm. In addition sperm nuclei from many species are sufficiently similar at the front and back so that this variation does not interfere unduly with accurate fluorescence measurements. Both of these instruments are capable of achieving coefficients of variation below 1.5 % for DNA content measurements, which is sufficient to discriminate the X and Y populations in most domestic animals. These techniques do not reliably resolve the X and Y populations of human sperm, perhaps due to the greater sperm to sperm variability in morphology and chromatin condensation in human ejaculates as compared to those of other animals.

Accurate DNA content measurements can be performed on fresh and cryopreserved semen. They have been used to detect breed differences among Y chromosomes in cattle (Garner et al., 1983) and to test claims of techniques purported to separate the X and Y populations for sex-determined artificial insemination (Pinkel et al., 1985). None of the sperm separation methods so far tested has been found to be effective. The shaped sample tube alone is sufficiently effective at orienting sperm so that it can be used with conventional sorters to obtain small quantities of X or Y bearing nuclei, Fig. 6 (Johnson et al., 1987b). While these separated sperm are not functional in the normal sense, they have demonstrated some ability to begin development after micro-injection into hamster eggs (Libbus et al., in preparation). The availability of sorted sperm populations may lead to the discovery of characteristics other than DNA content that will allow efficient bulk separation of viable X and Y sperm populations.

Fertility assessment. Particular aspects of the functional state of cells can be determined by the use of dyes whose fluorescence properties change depending on the intracellular state. For example some dyes such as carboxyfluorescein diacetate (CFDA) are not fluorescent until cleaved

Figure 6. Sorting of X and Y sperm. Fluorescence staining resolves the X and Y sperm populations of Chinchilla, which differ in DNA content by 7.5%. The coefficient of variation of the measurement is 1.3%, which would permit resolution of the two populations in most domestic animals. The shaded regions of the two peaks were sorted. Reanalysis of the sorted fractions is shown in the middle and lower panels.

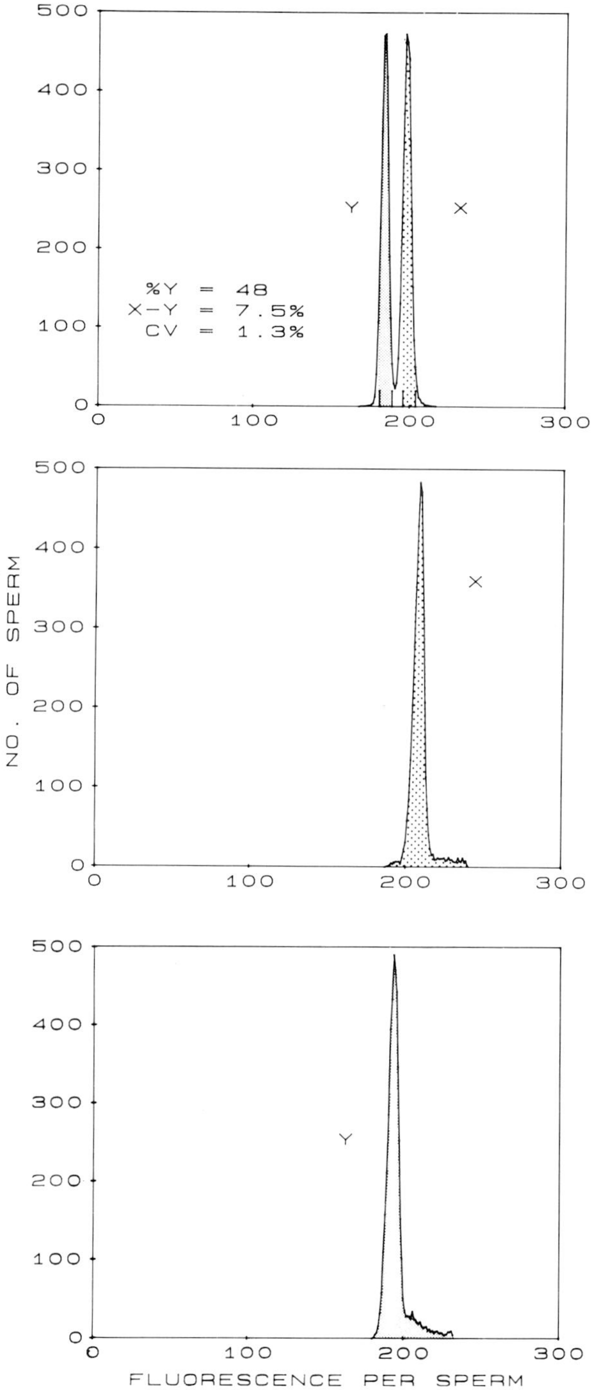
%Y = 48
X−Y = 7.5%
CV = 1.3%
Y
X
NO. OF SPERM
FLUORESCENCE PER SPERM
0
100
200
300
400
500

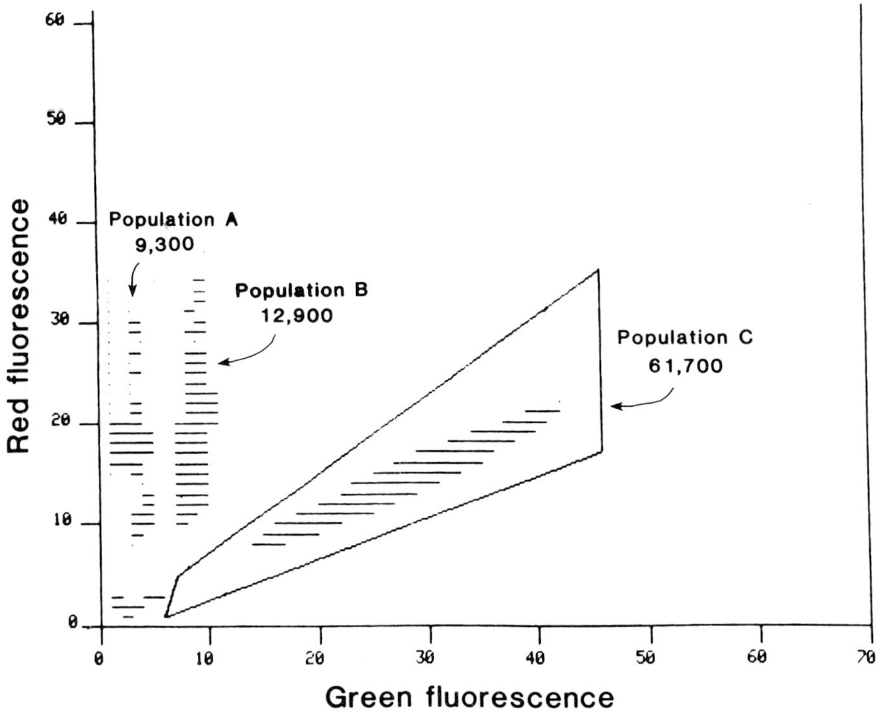

Figure 7. Measurement of CFDA and Propidium Iodide Stained Sperm. Three populations are seen when sperm stained with propidium iodide (red) and CFDA (green) are measured. Populations A and B represent dead cells and moribund cells respectively, while population C contains the live cells. The number of cells in each population are indicated.

by cellular enzyme activity, releasing the green fluorescent dye fluorescein. While CFDA readily penetrates cell membranes, the fluorescent product does not. It is therefore retained in viable cells. Thus the buildup of fluorescence intensity can be used as a measure cellular activity (Garner et al., 1986). Rhodamine 123 (R123) is specific for functioning mitochondria because of the high electronegativity of this organelle (Evenson et al., 1982). These dyes have been used along with DNA stains such as ethidium bromide and propidium iodide to explore fertility assessment of semen samples. Intact sperm membranes are relatively impermeable to the DNA stains. Thus viable sperm are expected to have a combination of low DNA stain and high CFDA fluorescences. Figure 7 shows measurements on a cryopreserved bovine semen sample stained with CFDA (green emission) and propidium iodide (red emission). Three distinct populations are detected. Population A, characterized by high red and low green fluorescence are dead cells. The intermediate population, B, contains moribund cells, while population C contains the viable sperm. The proportions of cells in each of these groups, and the changes in those proportions with time after thawing of the sample, are cor-

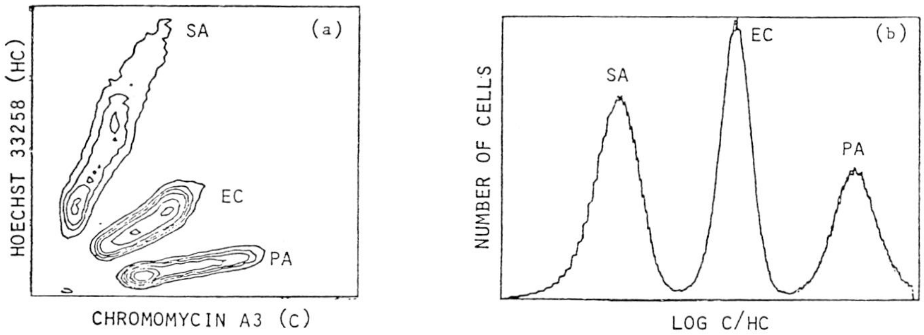

Figure 8. Flow Cytometric Measurement of a Mixture of Three Types of Bacteria. (a) Use of two DNA stains allows discrimination of bacteria based on GC content. The Hoechst stain (HC) preferentially binds to AT base pairs, while chromomycin (C) binds to GC pairs. S. aureus (SA) has the lowest proportion of GC base pairs, while P. aeruginosa (PA) has the highest of the three types shown. E. coli (EC) is intermediate. The extension of each population along a radial line from the origin the range of DNA content due to proliferation and perhaps clumpings of the cells. (b) The ratio of the two fluorescence signals, here plotted as log C/HC, separates the bacteria according to DNA base composition.

related with fertility. Other techniques for fertility prediction based on the susceptibility of sperm chromatin to denaturation are being developed (Ballachey, 1987).

Bacteriology. Light scatter, DNA stains, and antibodies have been used for bacterial analysis (Betz et al., 1984; Steen and Boye, 1980; Van Dilla et al., 1983). Such studies have demonstrated the potential for identification of bacteria, determination of their growth kinetics, and rapid assessment of their susceptibility to various antibiotics. Measurements can be performed on laser or mercury lamp-illuminated flow systems. The simultaneous determination of the proportions of multiple bacterial strains in a sample is illustrated in Fig. 8 (Van Dilla et al., 1983). The bacteria in this sample were stained with two DNA specific dyes, Hoechst 33258 and chromomycin A3 and measured in a dual laser flow cytometer so that both dyes can be efficiently excited. An ultraviolet beam is used to excite the Hoechst and a beam at 457 nm excites the chromomycin. While both of these dyes bind DNA, Hoechst binds relatively more intensely to AT rich regions while chromomycin prefers GC base pairs. Bacterial types differ greatly in their relative proportions of AT and GC, ranging from about 30 % to 70 % GC. Thus the type with the highest percentage of GC base pairs, P. aeruginosa in Fig. 8, has the brightest relative chromomycin fluorescence. The relative proportions of each can be easily determined by adding up the number of particles in each ridge in the distribution. Analyses like this may be of clinical interest and may allow improved understanding of the interactions in cultures containing multiple bacterial types.

Genetics. One of the major efforts of modern biology is the genetic manipulation of cells and organisms. This ranges from selection of naturally occurring genetic variants to purposeful alteration of the genetic constitution. Flow cytometry and sorting have been applied to many aspects of this effort, ranging from selection of desired cell phenotypes to the cloning of genes. In this section I will briefly describe several examples of this type of work: selection of plant heterokaryons, detection of gene transfections, production of chromosome-specific DNA libraries, and finally the use of nucleic acid probes to fluorescently label cells for flow cytometric analysis.

Protoplast fusion. Animal cells can be analyzed and separated based on a variety of cellular properties. Extension of these applications to plant cells is straightforward in principle, but requires attention to certain details. Among the particular problems are the fragility and large size of protoplasts relative to animal cells. This has required the use of sorter nozzles of larger diameter than customary to prevent plugging and to assure that the large particle diameter does not interfere with the breakup of the sorter jet into droplets. If care is taken viability of the sorted protoplasts can be maintained (Gailbraith et al., 1984).

One important application of sorting is the selection of heterokaryons, which result from the fusion of protoplasts of two types (Alexander et al., 1985). Production of heterokaryons allows investigation of the properties of combined genomes that cannot be achieved by sexual reproductive techniques. The difficulty is that the fusion frequency is usually low, on the order of 0.1 %, so that it is important to have an efficient method of selecting the successful fusions. Cell sorting can be used for purification if labels which differentiate the parents and fusion products are employed. This process is illustrated in Fig. 9, for barley leaf tips and Datura innoxia. The parent protoplasts were identified by the natural fluorescence (red) of the chlorophyll in the barley and the staining of the Datura using fluorescein diacetate (green), much as in the sperm viability test described above. In Fig. 9(a) the measurements of the parent populations prior to fusion are shown. Most of the cells have either red or green fluorescence and thus appear along the axes of the distribution. The few particles indicating both red and green fluorescence probably are due to two protoplasts sticking together. After fusion, Fig. 9(b), many more particles contain combined red and green signals, indicating complete or partial fusion of some of the cells. Sorting of these particles finds that approximately half of them are true fusions, the other half possibly representing partially fused cells that came apart due to the hydrody-

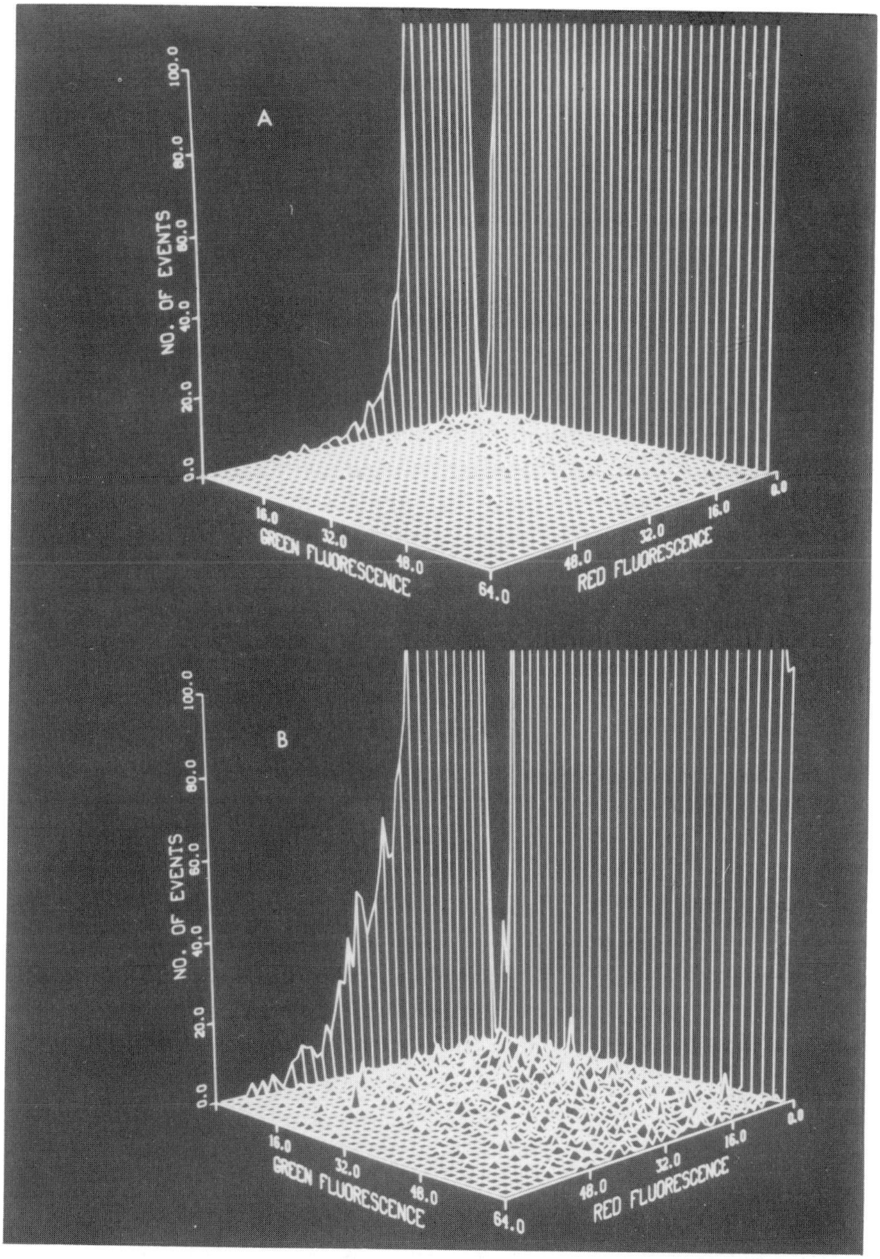

Figure 9. Bivariate Measurement of Plant Protoplast Fusion. (a) Measurement of a mixture of barley (red chlorophyll fluorescence) and Datura (green FDA fluorescence). Most objects show either a red or green signal and thus appear along the axes of the distribution. (b) After fusion, the proportion of objects showing both red and green fluorescence is increased. Sorting of these dual colored objects results in a highly enriched population of heterokaryons.

namic forces in the sorter. Sorting results in a more than a 200 fold increase in the concentration of the heterokaryons.

Gene transfection. Cell sorting is a powerful method for selection of cells that are expressing surface markers resulting from transfected genes. The rarely occurring positive cells can be recognized by staining with an antibody for the surface component in question, separated by sorting, and cultured. The source of the transfecting DNA can be a cloned gene or genomic DNA from almost any tissue of the donor organism (Hsu et al., 1984). One application of this approach is the cloning of the genes that code for the surface markers.

When genomic DNA is used for a transfection it is found that about one cell in a million expresses the desired gene product. This frequency of positive events is difficult to reliably detect with a flow cytometer since any fluorescent debris or clumps of particles may give a false positive signal. In addition, true positive events come only rarely, once every 5 minutes or so when processing several thousand cells per second. This situation can be substantially improved by adding an additional selective step. The procedure is illustrated in Fig. 10. It turns out that only a subset of the target cells are competent to accept the transfecting DNA. This subset can be recognized by a cotransfection technique. Target cells

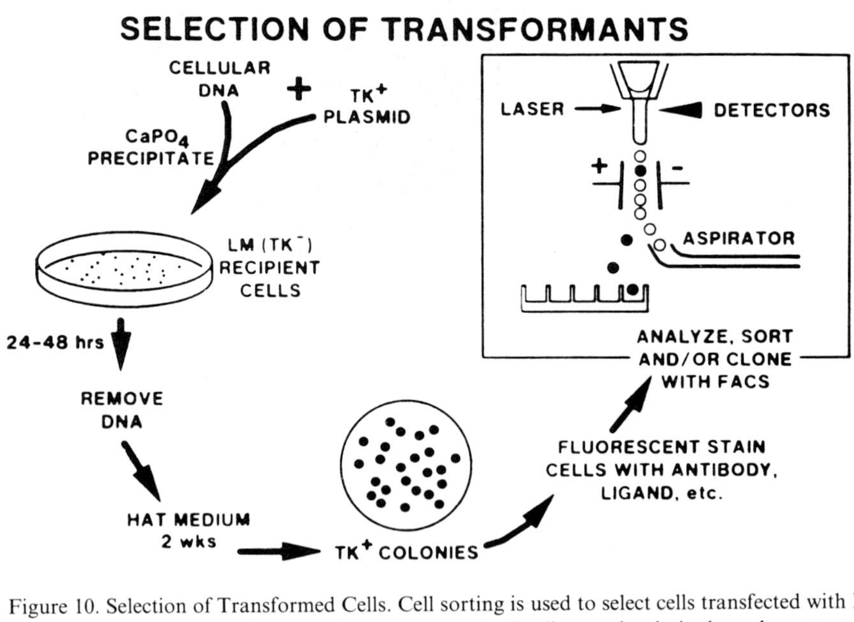

Figure 10. Selection of Transformed Cells. Cell sorting is used to select cells transfected with DNA which are expressing particular surface markers. Antibodies to the desired marker are used to identify the positive cells. Multiple selective steps sometimes produce cells in which the transfected genes are amplified.

that are deficient in thymidine kinase, TK, are simultaneously exposed to a cloned TK gene and genomic DNA. Cells which have accepted DNA contain both the TK gene and some of the transfecting genomic DNA. This subset, approximately one cell in a thousand, can be recognized by growth on selective media. These can then be stained and cells positive for the particular surface marker are found at a frequency of 10^{-3} to 10^{-4}. The sorted cells can be cultured, and those expressing the highest amounts of the surface marker selected. This has resulted in selection of cells that have spontaneously amplified the desired gene, making its eventual cloning easier (Kavathas and Herzenberg, 1983).

Chromosome-specific DNA libraries. One of the major applications of flow cytometry has been the analysis of chromosomes. Techniques have been developed for isolating metaphase chromosomes from cells, staining them with fluorescent dyes, and quantifying the normal and abnormal chromosomes that are present (Gray and Langlois, 1986). The most effective current technique for analysis of human chromosomes involves the simultaneous use of two DNA stains, Hoechst 33258 and chromomycin A3, as described previously in the section on bacteria. The result is a bivariate fluorescence histogram, such as the one illustrated in Fig. 1(a). The larger chromosomes appear farther from the origin of the histogram than the smaller ones because of their larger DNA content. However the various chromosome types are scattered off of the diagonal of the histogram because of the differences in the ratio of the uptake of the two dyes due to differences in their average AT:GC base pair ratio. Measurements on human chromosomes from many sources show that the positions of the peaks are very similar in normal individuals. However there are slight variations, and these measurements are capable of resolving differences in the homologues of a chromosome type within an individual. The stability of the measurements among normal individuals permits detection of cytogenetic abnormalities. The relative number of copies of a particular chromosome can be determined by the volume of the peaks, permitting detection of aneuploidy. Structural aberrations such as translocations can be identified by the presence of peaks in abnormal locations. Frequently the chromosomes involved in the translocation can be identified by quantitative analysis of the change in peak volumes of the normal chromosomes and the position of the newly appearing derivative chromosome peaks.

Once the chromosomes have been resolved they can be sorted, as indicated by the box around the chromosome 3 peak in Fig. 1(a). This has provided source material for the cloning of specific chromosomes, and has now led to production of a complete set of libraries for all the

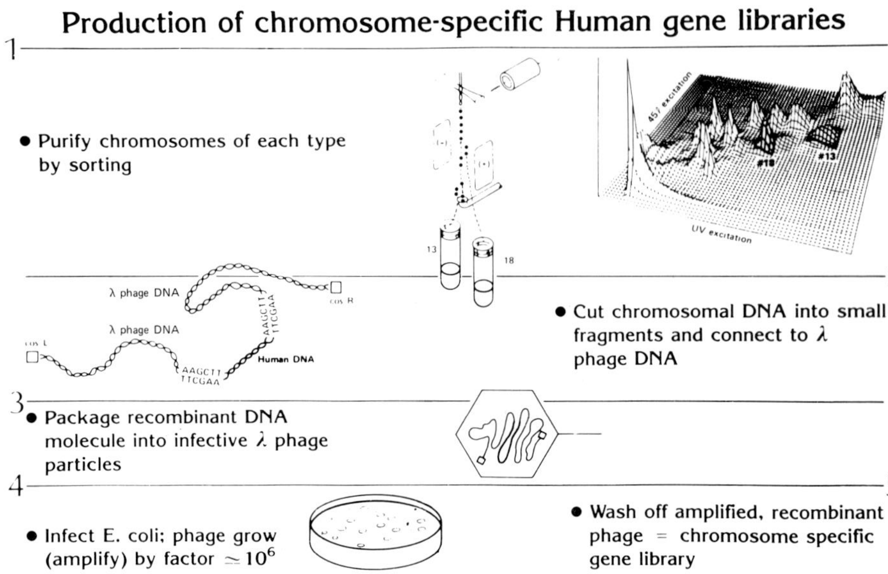

Figure 11. Production of chromosome-specific DNA libraries from sorted chromosomes. Simultaneous sorting of chromosomes 13 and 18 is illustrated.

human chromosomes (Van Dilla and Deaven, 1986). An outline of the cloning procedure is illustrated in Fig. 11. After sorting, DNA is isolated from the chromosomes and cut into many small pieces with a restriction enzyme. This enzyme cuts at a specific short DNA sequence so that the fragments produced from each chromosome are identical. For mammals, approximately 10^6 fragments are needed to cover the entire genome. These fragments are then inserted into a vector such as a plasmid or phage, which can be propagated in bacteria, permitting production of large amounts of specific genetic sequences for any desired purpose. On the order of a million chromosomes are needed to produce a library in lambda phage. In the human there are roughly 20 different chromosome types, thus during sorting only 5 % of the objects are the desired chromosome. Using a standard sorter measuring at about 1000 chromosomes per second implies that a maximum of 50 of the desired chromosomes can be sorted per second. It takes a day of sorting at this rate to separate a million, and given aggravating difficulties that are frequently encountered during long sorts where the highest measurement resolution must be maintained, this often must be spread over several days. The high speed sorter discussed previously can separate this much material in a fraction of a day (Peters et al., 1985).

One of the exciting new uses of recombinant DNA libraries is as a source of probes to label specific nucleic acid sequences in cells and

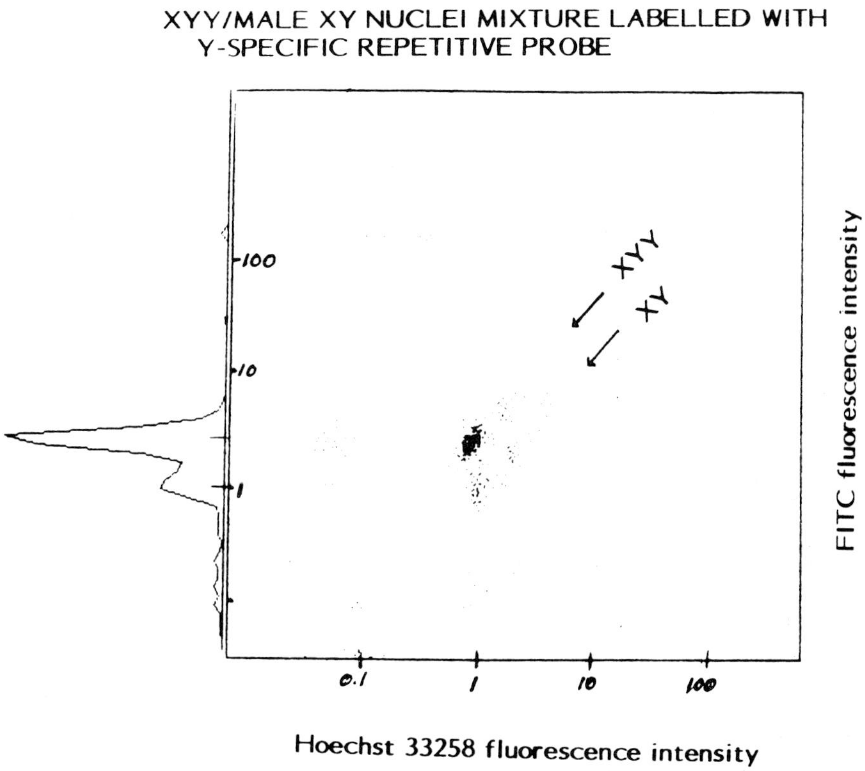

Figure 12. Fluorescence *in situ* Hybridization and Flow Cytometry. Normal XY and abnormal XYY human cell nuclei were hybridized with a probe specific for the Y chromosome. The probe DNA was biotinylated and the bound probe detected with fluorescein-avidin after the hybridization. All of the nuclear DNA was stained with Hoechst. The bivariate distribution shows the doubling of Y-specific fluorescence in each cell population as the cells traverse the cell cycle (double in Hoechst fluorescence). The XYY population has approximately twice the probe fluorescence of the XY.

chromosomes by the procedure of *in situ* hybridization. The binding specificity of the probes comes from matching complementary base sequences in the target cell or chromosome to the probe. Traditionally the probes have been radioactively labeled and the probe binding detected by autoradiography. Recently it has become possible to use chemically modified probes for hybridization, followed by reagents such as antibodies that are specific for the modification in the probe (Langer et al., 1981; Landegent et al., 1984). These reagents can carry fluorochromes, permitting fluorescent detection of *in situ* hybridization. *In situ* hybridization has been restricted to cells and chromosomes on slides or in tissue sections, but techniques have now been developed that allow use of nucleic acid probes for fluorescent labeling of cells in suspension, thus permitting flow cytometric analysis (Trask et al., 1985). One can begin to think of use of nucleic acid probes in the same way as one now does for monoclonal antibodies. Figure 12 demonstrates hybridization flow cytometric measurement of a probe specific for the human Y chromosome. The probe has been hybridized to normal (XY) and to abnormal (XYY) human cells. The precision of the measurements is such that one can detect the doubling in binding of the Y specific probe between the XY and XYY cells as they traverse the cell cycle. The sensitivity of this technique is approaching that which will be useful for detecting the presence of amplified genes in a cell population. Techniques for measuring the presence of specific mRNA sequences in flow are under development. Thus the circle is closing, flow cytometry can produce cloned DNA libraries which then become the source of probes for subsequent flow analysis of the genetic constitution of other cells and chromosomes.

Summary

Flow cytometry and sorting are very powerful methods of rapidly analyzing individual cells and organelles and separating desired subpopulations. At the beginning of development of these techniques attention was focused on instrumentation designed to explore different measurement principles and to establish their physical limits and ranges of applicability. As the field has matured emphasis has shifted to using the instruments as integral parts of biological experiments. The development of monoclonal antibodies, and now nucleic acid probes, as methods to label specific cellular or chromosomal constituents has dramatically increased the range of applicability of flow techniques, and permits biologically sophisticated measurements with simple instruments. One of the major directions of development in the future is likely to be in the area of very compact field instruments, for example using solid state lasers for illumination.

Acknowledgements

It is a pleasure to acknowledge D. Garner, J. Gray, L. Hertzenberg, J. Jett, L. Johnson, J. Mullikin, D. Peters, B. Trask and L. Wheeless, for supplying information and figures used in the preparation of this manuscript. Work performed under the auspices of the U.S. Department of Energy, Office of Health and Environmental Research by the Lawrence Livermore National Laboratory under contract number W-7405-ENG-48.

References

1. Alexander, R.G., E.C. Cocking, P.J. Jackson and J.H. Jett. 1985. *The characterization and isolation of plant heterokaryons by flow cytometry.* Protoplasma 128: 52–58.
2. Ballachey, D.E., W.D. Hohenboken and D.P. Evenson. 1987. *Heterogeneity of sperm nuclear chromatin structure and its relationship to fertility in bulls.* Bio. of Reprod. in press.
3. Betz, J.W., W. Aretz and W. Hortel. 1984. *Use of flow cytometry in industrial microbiology for strain improvement programs.* Cytometry 5: 145–150.
4. Cram, L.S., M.F. Bartholdi, L.L. Wheeless and J.W. Gray. 1985. *Morphological analysis by scanning flow cytometry.* Pages 164–193 in M. Van Dilla, P. Dean, O. Laerum and M. Melamed, eds., *Flow Cytometry: Instrumentation and Data Analysis.* Academic Press, London.
5. Dittrich, W. and W. Gohde. 1969. *Impulsfluorometric bei einzelzellen in suspensionen.* Z. Naturforsch. 24b: 360–361.
6. Dovichi, N.J., J.C. Martin, J.H. Jett, M. Trkula and R.A. Keller. 1984. *Laser induced fluorescence of flowing samples as an approach to single molecule detection in liquids.* Analytical Chem. 56: 348–354.
7. Evenson, D.P., Z. Darzynkiewicz and M.R. Melamed. 1982. *Simultaneous measurement by flow cytometry of sperm cell viability and mitochondrial membrane potential related to cell motility.* J. Histochem. Cytochem. 30: 279–280.
8. Fulwyler, M.J. 1965. *Electronic separation of biological cells by volume.* Science 150: 910–911.
9. Gailbraith, D.W., C.L. Alfonso and K.R. Harkins. 1984. *Flow sorting and culture of protoplasts: Conditions for high frequency recovery, growth, and morphogenesis from sorted protoplasts of suspension cultures of nicotiana.* Plant Cell Reports 3: 151–155.
10. Garner, D.L., B.L. Gledhill, D. Pinkel, S. Lake, D. Stephenson, M.A. Van Dilla and L.A. Johnson. 1983. *Quantification of the X- and Y-chromosome bearing spermatozoa of domestic animals by flow cytometry.* Biol. Reprod. 28: 312–321.
11. Garner, D.L., D. Pinkel, L.A. Johnson and M.M. Pace. 1986. *Assessment of spermatozoal function using dual fluorescent staining and flow cytometric analysis.* Biol. Reprod. 34: 127–138.
12. Gray, J.W. and R.G. Langlois. 1986. *Chromosome classification and purification using flow cytometry and sorting.* Ann. Rev. Biophys. Biophys. Chem. 15: 195–235.
13. Gucker, Jr, F.T., C.T. O'Konski, H.B. Pickard and J.N. Pitts Jr. 1947. *A photo electronic counter for colloidal particles.* J. Am. Chem. Soc. 69: 2442.
14. Hsu, C., P. Kavathas and L.A. Herzenberg. 1984. *Cell surface antigens expressed in L-cells transfected with whole DNA from non-expressing and expressing cells.* Nature 312: 68–69.
15. Johnson, L.A., J. Flook and M. Look. 1987a. *Flow cytometry of X- and Y-Chromosome bearing sperm for DNA using an improved preparation method and staining with Hoechst 33342.* Gamet Res., submitted.
16. Johnson, L.A., J. Flook, M. Look and D. Pinkel. 1987b. *Flow sorting of X and Y chromosome bearing spermatozoa into two populations.* Gamet Res. In Press.
17. Johnson, L.A. and D. Pinkel. 1986. *Modification of a laser-based flow cytometer for high resolution DNA analysis of mammalian spermatozoe.* Cytometry 7: 268–273.
18. Kamentsky, L.A. and M.R. Melamed. 1967. *Spectrophotometer cell sorter.* Science 156: 1364–1365.
19. Kavathas, P. and L.A. Herzenberg. 1983. *Amplification of a gene coding for human T-cell differentiation antigen.* Nature 306: 385–387.
20. Landegent, J., N. Jansen In De Wal, R. Baan, J. Hoeijmakers and M. Van der Ploeg. 1984. *2-acetylaminofluorene-modified probes for the indirect hybridocytochemical detection of specific nucleic acid sequences.* Exp. Cell. Res. 153: 61–72.
21. Langer, P., A. Waldrop and D. Ward. 1981. *Enzymatic synthesis of biotin labeled polynucleotides: Novel nucleic acid affinity probes.* Proc. Natl. Acad. Sci. 78: 6633–6637.
22. Otto, F.J., U. Hacker, J. Zante, J. Schman, W. Gohde and M.L. Meistrich. 1979. *Flow cytometry of human spermatozoa.* Histochemistry 61: 249–254.

23. Pallavicini, M.G., P.N. Dean, J. Mullikin, P.W. Matsson, L.J. Summers, D. Peters and R. Barlett. In preparation.
24. Peters, D., E. Branscomb, P. Dean, T. Merrill, D. Pinkel, M. Van Dilla and J.W. Gray. 1985. *The LLNL high speed sorter:design features, operational characteristics, and biological utility.* Cytometry 6: 290–301.
25. Pinkel, D., S. Lake, B.L. Gledhill, M.A. Van Dilla, D. Stephenson and G. Watchmaker. 1982. *High resolution DNA content measurements of mammalian sperm.* Cytometry: 1–9.
26. Pinkel, D., D.L. Garner, B.L. Gledhill, S. Lake, D. Stephenson and L.A. Johnson. 1985. *Flow cytometric determination of X- and Y-chromosome-bearing sperm in samples of purportedly separated bull sperm.* J. Anim. Sci. 60: 1303–1307.
27. Pinkel, D. and R. Stovel. 1985. *Flow chambers and sample handling.* Pages 77–128 in M. Van Dilla, P. Dean, O. Laerum and M. Melamed eds., *Flow Cytometry: Instrumentation and Data Analysis.* Academic Press, London.
28. Raleigh Lord J.W.S. 1879. *On the capillary phenomena of jets.* Proc. Roy. Soc. London. 29: 71.
29. Reiman, F., U. Zimmermann and G. Pilwat. 1975. *Release and uptake of haemoglobin and ions in red blood cells induced by dielectric breakdown.* Biochimica and Biophysica Acts 394: 449–462.
30. Reynolds, O. 1883. *An Experimental investigation of the circumstances which determine whether the motion of water shall be direct or sinuous, and the law of resistance of parallel channels.* Philo. Trans. Roy. Soc. London 174: 935–982.
31. Steen, H.B. and T. Lindmo. 1979. *Flow cytometry: A high resolution instrument for everyone.* Science 204: 403–404.
32. Steen, H.B. and E. Boye. 1980. *Escherichia coli growth studied by dual parameter flow cytophotometry.* J. Bacteriol. 145: 1091–1094.
33. Stovel, R.T. and R.G. Sweet. 1979. *Individual cell sorting.* J. Histochem. Cytochem. 27: 285–288.
34. Sweet, R.G. 1965. *High frequency recording with electrostatically deflected ink jets.* Rev. Sci. Instr. 36: 131–136.
35. Trask, B., G. van den Engh, J. Landegent, N. Jansen In De Wal and M. van der Ploeg. 1985. *Detection of DNA sequences in nuclei in suspension by in situ hybridization and dual beam flow cytometry.* Science 230: 1401–1403.
36. Van Dilla, M.A., R.G. Langlois, D. Pinkel, D. Yajko and W.K. Hadley. 1983. *Bacterial characterization by flow cytometry.* Science 220: 620–622.
37. Van Dilla, M.A. and L.L. Deaven. 1986. *Human chromosome-specific DNA libraries: construction and availability.* Biotechnology 4: 537–552.
38. Wheeless Jr. L.L. and J.B. Kay. 1985. *Optics, light sources, filters and optical systems.* Pages 22–77 in M. Van Dilla, P. Dean, O. Laerum and M. Melamed, eds., *Flow Cytometry: Instrumentation and Data Analysis.* Academic Press, London.

Field-flow fractionation of biological materials

KARIN D. CALDWELL
Department of Bioengineering, and Center for Biopolymers at Interfaces, University of Utah, Salt Lake City, UT 84112, U.S.A.

Abstract Most biological fractionation problems require techniques which perform rapid separations of delicate components without inflicting denaturation or loss of activity. The larger the components, the fewer are the methods available for this task.

Field-Flow Fractionation (FFF) is an analytical technique, which has proven ideally suited for a great number of separation problems involving fragile macromolecules and particles. This review article summarizes existing theoretical models for retention and zone broadening in FFF. Based on these models, equations are developed to optimize the resolution of given pairs of components in time. It is shown that by slightly increasing the selected fields, onc can significantly reduce the times required for resolution of given pairs of components.

The performance of several FFF subtechniques is illustrated by examples of biological interest, and qualitative demonstrations of the optimization strategy are discussed.

Introduction

Our understanding of biological structure-function relationships is intimately linked to our ability to isolate compounds of interest in pure form and with intact biological activity. Numerous high resolution methods exist for the purification of small metabolites. Proteins, nucleic acids, and other polymers are generally separated in high yield using ion-exchange or reverse phase chromatography, which base their selectivity on chemical differences among the samples, or by gel permeation chromatography which resolves samples due to differences in molecular size.

Separation of large molecules or particles such as viruses, subcellular organelles, and whole cells have in general been performed in one-phase systems where discrimination results from differential interactions with an applied field. Ultracentrifugation, carried out either as a rate-zonal or as an equilibrium separation in a medium of graded density, is by far the most common method used for such separations, although various forms of free zone electrophoresis have proven well suited for similar purification tasks.

Characteristic for these methods is a differential sample migration in the direction of the field, where higher resolution is associated with longer migration paths which, in turn, requires long separation times or high field strengths.

Field-flow fractionation (FFF) is a class of separation techniques where the field induced migration occurs in a direction perpendicular to the direction of separation. Although the same types of fields are used as in the direct-field methods, there is no longer a requirement of resolution in the direction of the field; the separation times are therefore shorter and the field strength requirements much less stringent. In addition to these advantages, FFF is an elution method, much like ordinary chromatography, and permits the collection of fractions during the course of a separation. Thanks to the well developed theory for FFF retention (Giddings, 1973 a, b; Giddings 1976; Giddings and Caldwell 1987), a given elution volume can be directly interpreted in terms of some physical characteristic of the sample, e.g. molecular weight, diffusion coefficient, or electrophoretic mobility.

By combining the advantages of speed and elution which characterize chromatography, with the gentleness towards particulate samples and the exact mathematical relationships which govern migration in the direct-field methods, FFF opens up a new dimension in the fractionation of macromolecules and particles. In the following text we will briefly summarize the theoretical underpinnings of the technique, and will then attempt to illustrate its performance with examples of biochemical relevance.

Theory

The general FFF channel, see Fig. 1, is a thin duct with rectangular cross section whose breadth b exceeds its thickness w by a factor of about one hundred (Giddings and Schure, 1987). Except in very narrow regions near the side walls, flow through such channels proceeds as it would between infinite parallel plates. In the absence of any orienting influences, a sample injected into the channel will be evenly distributed across all flow lines present between the plates, and will therefore travel with the average velocity of the fluid $\langle v \rangle$; its emergence at the end of the channel

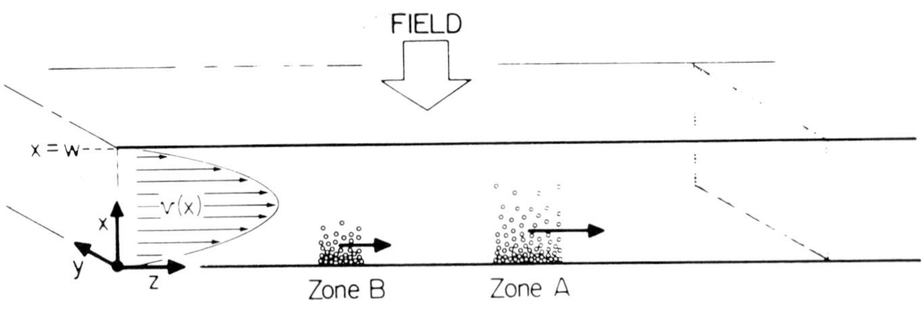

Figure 1. Side view of the FFF channel, identifying the directions of field (x-coordinate) and flow (z-coordinate). Zones move along the z-coordinate with velocities determined by the average layer thickness, l (zone A, fast; zone B, slow). Typical channel dimensions are: $10\,\mu m < w < 540\,\mu m$, $1\,cm < b < 2.5\,cm$, $40\,cm < L < 100\,cm$ (L being channel length).

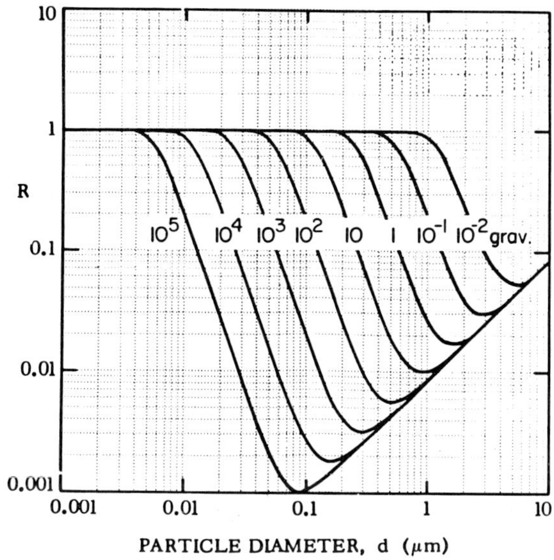

Figure 2. The relationship between R and particle diameter d ($\Delta\varrho = 1.0\,\text{g/cm}^3$, $\gamma = 0.7$) for sedimentation FFF/steric FFF. The minimum in R marks the equivalence between layer thickness l and particle radius $r = d/2$. Small d values show "normal" FFF behavior (eqn. 3 of Table 1), while R for particles of large diameters is entirely dominated by steric effects (eqn. 7). The position of the minimum varies with l, i.e. with strength of the applied field. (Reproduced from Myers and Giddings, 1982, with permission).

will mark the void-time, t^0, which is the time needed to sweep out one void-volume V^0.

Sample layer. The application of a field across the channel perturbs the sample and forces its components to move and equilibrate in exponential distributions near one of the walls. These distributions will be more or less compact depending both on the degree of interaction between field and sample and on the diffusivity of the sample. In terms of the coordinate system defined in Fig. 1, the concentration distribution $c(x)$ of any one component can be expressed as

$$c(x) = c_o \exp(-x/l) \tag{1}$$

Here, parameter l is a measure of the thickness of the equilibrated zone, as illustrated in the figure. Both l and its dimensionless analog λ reflect the balance between a dispersive flux due to diffusion and a concentrating flux which results from the field induced transport of sample into the wall region (Giddings, 1973a)

$$\lambda = l/w = D/Uw \tag{2}$$

In this equation, w has the usual meaning of channel thickness, D is the diffusion coefficient of the sample, and U represents its drift velocity. At steady state, U is directly proportional to the strength of the applied field S' and inversely proportional to the sample's friction coefficient f

$$U = S'\phi/f \tag{3}$$

The field susceptibility ϕ is a sample specific parameter, which takes different forms depending on the nature of the applied field. To date, sedimentation, flow, electrical and thermal fields have been successfully used to achieve separation in the FFF mode, and the specific expressions for λ which apply to these fields are summarized in Table 1 (Giddings and Caldwell, 1987).

Retention. The laminar flow of carrier through the channel will exert a drag on the various sample zones, and propel them forward in proportion to their spatial extension, so that less compact zones (larger λ values) will move ahead of those confined to the near-wall region where flow is sluggish. These differences in migration velocity are reflected in the retention parameter R, characteristic for each zone (Giddings, 1973a)

$$R = 6\lambda\,[\coth(1/2\lambda) - 2\lambda] \tag{4}$$

For small values of λ, eqn. 4 assumes the simple limiting form

$$R = 6\lambda \tag{5}$$

This relationship is accurate to within 5% for a sample retained five column volumes ($R = 0.2$), while its accuracy is better than 2% at a twenty column volume retention ($R = 0.05$).

Experimentally, R is determined as the ratio of void volume V^0 to the

Table 1. Relationship between field strength S' and reduced layer thickness λ for different types of field

	Type of field	Field strength	Field susceptibility (sample specific)	Expression for λ
1.	General	S'	ϕ	$\lambda = kT/S'\phi w$
2.	Sedimentation	G	$m(\Delta\varrho/\varrho_s) = m'$	$\lambda = kT/Gm'w$
			$M(\Delta\varrho/\varrho_s) = M'$	$\lambda = RT/GM'w$
3.	Sedimentation	G	$d^3\pi\Delta\varrho/6$	$\lambda = 6kT/Gd^3\Delta\varrho\pi w$
4.	Flow	$U = \dot{V}_c/A$	f	$\lambda = DA/\dot{V}_c w = DV^0/\dot{V}_c w^2$
5.	Electrical	E	μf	$\lambda = D/E\mu w$
6.	Thermal	$dT/dx \approx \Delta T/w$	$D_T f$	$\lambda \approx D/D_T\Delta T$

The symbols in this table have the following meaning: k Boltzmann's const, R gas constant, T temperature, w channel thickness, G gravitational accel., m particle mass, M molar mass, d particle diameter, ϱ_s particle density, $\Delta\varrho$ density difference, ΔT temperature drop, D_T thermal diffusivity, E electrical field, μ electrophor. mobility, $\dot{V}_c$ volumetric cross-flow, A area of accum. wall, f coefficient of friction, $D = kT/f =$ diffusion coeff., V^0 void volume.

observed elution volume V_r or, equivalently, as the ratio of the time analogs t^0 and t_r

$$R = V^0/V_r = t^0/t_r \qquad (6)$$

By a combination of eqns. 4 and 6 one can convert an experimentally observed value for V_r or t_r into the corresponding value for parameter λ, which gives access to information about e.g. molecular weight or diffusivity for the eluting species.

Steric effects. The relationship between retention and zonal thickness described above will hold for ideal systems, where sample molecules are behaving as point masses, moving freely without interactions with one another or with the accumulation wall. Large particles will be sterically hindered from sampling flow lines near the wall, and will thus cause departures from the retention model outlined above (Giddings, 1978).

For particles who radii exceed layer thickness l, the layer concept loses meaning, and retention will instead become a function of particle size (Giddings and Myers, 1978)

$$R = 3\gamma d/w \qquad (7)$$

In this relationship, d represents the particle diameter. The magnitude of factor γ is unity, but its exact value will vary with carrier velocity, settling field, and particle symmetry in a manner not yet fully understood (Caldwell et al., 1979; Koch, 1986).

The steric effect causes a reversal of the elution order observed in "normal" FFF, where more massive particles in general trail those of lesser mass. The change in retention mechanism is illustrated in Fig. 2, which is a graphic representation of the variation in R with particle diameter during sedimentation FFF. An increase in field strength has the effect of shifting the minimum in R towards smaller diameters, as seen in the figure. Similar shifts will result from changes in the density difference $\Delta\varrho$ between the particle and its surrounding medium.

Although these graphs are developed for sedimentation FFF, the same type of behavior is expected for all subtechniques (Myers and Giddings, 1982).

Selectivity. The relative ability of an elution based separation method to resolve compounds with a given difference in molecular weight M, or in particle diameter d, can be described by a selectivity index S defined as (Giddings, 1979)

$$S_M = |d \ln V_r / d \ln M| \qquad (8)$$

or

$$S_d = |d \ln V_r / d \ln d| \quad (9)$$

For FFF, the appropriate expressions for selectivity are obtained through differentiation of eqn. 4 (Gao et al., 1985)

$$S = |(d \ln V_r / d \ln \lambda) S_{max}| = 3(R/36\lambda^2 + 1 - 1/R) S_{max} \quad (10)$$

where S_{max} is the maximum selectivity of the system. This maximum, which is reached under conditions of high retention, assumes different values depending on whether the selectivity is mass based

$$S_{max\,M} = |d \ln \lambda / d \ln M| \quad (11)$$

or size based

$$S_{max\,d} = |d \ln \lambda / d \ln d| \quad (12)$$

The bracketed function in eqn. 10 varies between zero (at zero retention) and 1/3 (at high retention, where eqn. 5 becomes applicable). S_{max}, in turn, will depend on the choice of subtechnique, as seen in Table 2 (Giddings, 1979). Given the ability of two subtechniques to handle a particular sample, the method of choice will be the one which offers the highest selectivity.

Zone broadening. Aside from being selective, a separation system must also be efficient, i.e. it must generate differences in retention without undue spreading of the component zones. In case of "normal" FFF, the zone broadening can be accurately modeled, thanks to a clear understanding of the retention mechanism, and to the simple geometry of the FFF channel (Giddings et al., 1975).

Traditionally, zone broadening is discussed in terms of plate height H (Giddings, 1965a) which is a measure of the generation of variance σ^2 in a zone under migration along the separation coordinate (z-direction).

$$H = d\sigma^2/dz \approx \sigma^2/L \quad (13)$$

Table 2. Mass and size based selectivities for different FFF subtechniques

Subtechnique	$S_{max,M}$	$S_{max,d}$
Sedimentation FFF	1	3
Flow FFF	0.33–0.8[a]	1
Electrical FFF	N/A[b]	N/A[b]
Thermal FFF	0.33–0.8[a]	1

[a] The exact value is system dependent, and reflects the interaction between sample and carrier.
[b] The charge on sample molecules or particles vary e.g. with pH and ionic strength, and have no direct relationship to M or d, unless special arrangements are made.

If column and retention processes are uniform throughout, H is well approximated by the variance (in length units) of the eluting peak divided by the length L of the column, as shown in the right hand expression above. The total variance of the peak is the sum of contributions due to different zone broadening effects. Some of these arise from the instrumental design (surface roughness, wall effects etc.), the extra-column dispersion in tubing and detector, and the width of the injected sample slug. These contributions, collectively labelled ΣH_i below, can be minimized with good operating procedures, and the major influence on H will be the nonequilibrium broadening H_n. This effect, present in all forms of chromatography, is due to the distribution of the sample over volume elements which move with different velocities, and to the sluggish transport of sample between such elements. Longitudinal diffusion effects, which are important causes of zone broadening in gas chromatography, play a negligible role in liquid systems, and the experimentally observed FFF plate height H can therefore be summarized as (Smith, Myers and Giddings, 1977; Karaiskakis et al., 1981))

$$H = H_n + \Sigma H_i + (H_p) \tag{14}$$

The nonideal term H_p, added in parenthesis, is generated by polydisperse samples whose various constituents will be retained to differing degrees and collectively emerge as a broadened zone. This term, which is sample- rather than process-induced, is irrelevant for the optimization of a system's performance. However, since it is easily separated from the other plate height contributions it does add important analytical information to the fractionation of polydisperse samples, as will be discussed below.

With increased retention, the FFF nonequilibrium zone broadening becomes progressively smaller. This effect, which is the direct opposite of what is observed in chromatography, is due to increased compression of the sample zone, and therefore to shorter diffusion distances and lesser velocity differences between sample particles. Accounting for these effects, H_n has been expressed as (Giddings, 1968; Giddings et al., 1975)

$$H_n = \chi(\lambda) w^2 \langle v \rangle / D \tag{15}$$

where w and $\langle v \rangle$ are the channel thickness and average velocity, respectively. D is the sample's diffusion coefficient, and $\chi(\lambda)$ rapidly approaches $24\lambda^3$ as retention increases. (The approximation carries a 37% error for $R = 0.2$, but is in error of only 6% at $R = 0.05$.)

Polydispersity. The fractionation of a polydisperse sample will yield a peak which is broadened by differential migration of its components, in addition to the standard instrumental and nonequilibrium effects, as expressed by eqn. 14. The relationship between H_p and the standard deviation σ_d in particle diameter, or the polydispersity index μ (i.e. the ratio of weight to number average molecular weight M_w/M_n), can be expressed in analytical form (Giddings et al., 1983; Gao et al., 1985)

$$H_p = LS_M^2(1 - 1/\mu) = LS_d^2(\sigma_d/d)^2 \tag{16}$$

The more selective the technique, the more will differences in the molecular weight or particle size of the sample's components manifest themselves as differences in retention, and therefore in width of the eluting peak. The mass and size based selectivities S_M and S_d will differ with choice of subtechnique as well as with experimental conditions, in accordance with eqns. 10–12 (see Table 2).

Through eqn. 16, one has the means for experimental evaluations of a sample's polydispersity, provided H_p can be accurately determined. If the sample is retained by the normal FFF mechanism there are no velocity effects on either retention or selectivity, and H_p is therefore independent of the selected flow of carrier.

Of the three plate height contributions listed in eqn. 15, H_n is the only term which shows an explicit dependence on $\langle v \rangle$. Under constant field, variations in carrier flow will linearly affect the nonequilibrium term, and by plotting the measured values for plate height H against velocity $\langle v \rangle$, it is possible to evaluate the sum of H_p and ΣH_i from the intercept of this curve. Since, for the most part, ΣH_i is either negligible or can be determined from studies of monodisperse standards, the intercept method provides an excellent tool for assessing sample polydispersities (Martin, Myers and Giddings, 1979; Giddings et al., 1983).

Resolution and optimization. The most important characteristic of any separation method is its ability to resolve complex mixtures into their components. For the purpose of comparing resolution between methods, or between different operational conditions applied to one method, it is convenient to introduce the resolution index *Rs*, commonly defined as (Giddings, 1965b)

$$Rs = \Delta V_r/4\sigma_v \tag{17}$$

where ΔV_r is the difference in retention volume between a given pair of components, and σ_v represents the average standard deviation (volume units) in peak width. *Rs* will vary from zero (for completely overlapping peaks) to infinity (for infinite spacing between peaks), with the value of unity representing baseline resolution of the two components.

The spacing between two peaks (ΔV_r) increases with increased selectivity S of the separation method while the peak width, measured by σ_v and its analog the plate height H, decreases with increased efficiency of the method. The resolution of pairs of components with a relative difference $\Delta M/M$ in molecular weight, or $\Delta d/d$ in particle size, can therefore be expressed in terms of S, H, and column length L (Giddings, 1979)

$$Rs = (\sqrt{L}/4)(\Delta M/M)S_M/\sqrt{H} = (\sqrt{L}/4)(\Delta d/d)S_d/\sqrt{H} \qquad (18)$$

One very important use of eqn. 18 is to help select experimental conditions which will accomplish a desired separation in the shortest amount of time. Although the general nature of the equation does permit comparison between such different techniques as e.g. gel permeation chromatography and thermal FFF (Giddings, Yoon and Myers, 1975; Gunderson and Giddings, 1986), it is particularly helpful in providing insight into the effects associated with variations in field strength and flow rate for a given FFF technique. In order to examine these effects, one needs to recall the expression for zone broadening given in eqn. 14. Since ΣH_i is generally either negligible, or finite but invariant with changes in the parameters to be examined, the nonequilibrium term H_n becomes the only plate height contribution of importance in this context.

The discussion is simplified by assuming operation under conditions of high retention where S is constant (S_{max}), R is given by eqn. 5, and χ in the expression for H_n (eqn. 15) equals $24\lambda^3$. With these assumptions, eqn. 18 can be rearranged to give

$$Rs = A/\sqrt{\lambda^3\langle v\rangle} \qquad (19)$$

where A is a constant for a given pair of components. Since λ bears an inverse relationship to field strength S' (Table 1), it is obvious that an increase in S' by a factor α can sustain an increase in channel velocity $\langle v\rangle$ by a factor α^3 without impairing resolution. How this affects the time required to resolve the pair is seen by writing eqn. 19 in a slightly different form. From eqn. 6 we know t_r to equal t^0/R. Since t^0 is the column passage time for an unretained compound, moving with average velocity $\langle v\rangle$, t^0 may be replaced by $L/\langle v\rangle$. Using the high retention approximation for R (eqn. 5), we can now rearrange eqn. 19 in a form which better illustrates the effect of field strength S' on the time of resolution

$$Rs = BS'\sqrt{t_r} = B\alpha S'\sqrt{t_r/\alpha^2} \qquad (20)$$

where B is a constant. From eqn. 20 it is clear that an increase in field strength by a factor α leads to separations at constant resolution in times that are shorter by the factor α^2.

Applications

Sedimentation FFF. From the discussion of selectivity above it is evident that sedFFF is superior among those FFF methods which have shown retention at the present time. The molecular weight range accessible with this high selectivity depends primarily on the design of the seal which connects spinning and stationary parts in the system (Giddings et al., 1980). In the systems built at the University of Utah, the seals will permit spin rates of up to 3000 rpm which, with the rotor radius of 16 cm, implies a field of just above 1600 gravities. This field strength is generally enough to show retention of samples whose molecular weight exceeds 5×10^7 dalton.

The SF^3 Particle Fractionator (Kirkland, Dilks and Yau, 1983), which was introduced by DuPont in March of 1986, is capable of spin rates up to 20 000 rpm which, for a rotor radius of 9 cm, translates into a field strength of about 40 000 gravities. The use of this instrument, therefore, allows retention of molecular weights ranging from just above 10^6 dalton to the about 10^{18} characteristic for the high range of steric FFF (ca. 100 microns).

The mass range accessible by the sedFFF subtechnique makes it a uniquely useful tool for colloid fractionation and characterization, tasks which are difficult to accomplish with conventional methods. The array of samples studied by this technique includes suspensions of latex particles, both monodisperse and polydisperse (Giddings, Yang and Myers, 1974; Giddings et al., 1983; Yang, Caldwell and Giddings, 1983; Kirkland, Rementer and Yau, 1981; Giddings, Caldwell and Jones, 1986), particles in stream water (Karaiskakis et al., 1982; Beckett et al., 1986), viruses (Giddings, Yang and Myers, 1975; Caldwell et al., 1980; Kirkland et al., 1980; Caldwell, Karaiskakis and Giddings, 1981; Yonker, Caldwell and Giddings, 1985), subcellular particles (Fox, Schallinger and Kirkland, 1985; Caldwell et al, 1986, Mozersky et al., manuscript in preparation), protein aggregates (Caldwell et al., 1981a; Caldwell et al., 1984a), nucleic acids (Schallinger, Yau and Kirkland, 1984; Schallinger et al., 1984), emulsions and liposomes (Caldwell et al., 1981b; Kirkland, Yau and Szoka, 1982; Yang et al., 1983; Yang et al., 1984). Below, we will give a few illustrations of the practical workings of the technique.

(a) Virus characterization. Gypsy moth larvae, infected with the host specific nuclear polyhedrosis virus (NPV), will encase newly formed viruses in inclusion bodies. Electron micrographs of these particles clearly indicate that in their interior the rod-like viruses are found either singly, or arranged in bundles of up to a dozen rods. The protein matrix,

which embeds these structures, is easily dissolved at moderately alkaline pH, permitting the released viruses to be separated into monomers, dimers, and higher order aggregates. This is traditionally done through ultracentrifugation in sucrose gradients (Summers and Volkman, 1976). Adequate separation does not require the very time consuming complete equilibration of the various sample components at their respective isopychnic densities. Instead it can be accomplished using a rate-zonal protocol based on a 4 hour, 24 000 rpm spin in an SW 27 swinging bucket rotor. The collection of fractions has to be carried out with great care, in order not to disturb the zonal banding pattern, and elution times are therefore about an hour in length. In order to prepare the various fractions for electron microscopy, the sucrose must be eliminated from the viral suspension medium. This should ideally be done in a gradual manner, e.g. through dialysis, in order to protect the particles from potentially disruptive osmotic shocks.

All in all, the traditional method for preparing pure Gypsy Moth NPV is both time consuming and labor intensive, which pointed to the sedimentation FFF as a convenient alternate route to obtaining virus aggregates of discrete size. As an analytical technique, FFF is performing most accurately with small loads of sample, and preparative scale fractionations are still best performed by density gradient sedimentation. However, for morphological studies the typical FFF sample loads of around 1–2 mg are fully adequate.

Figure 3(a) illustrates a fractionation of the alkali liberated virus aggregates, which took place under a field of 173 gravities. The cross-hatched areas in the fractogram indicate cuts taken for morphological identification of the eluting particles, and the first emerging retained peak was identified as containing exclusively monomers, with the second containing dimers, and so on. After completed elution of the trimer, the run was aborted by stopping the centrifuge. As a result, all higher order aggregates were swept out of the column to emerge as one peak.

Elution of the monomer took place in just over 50 minutes at the selected flow rate of 24 mL/h. The previous optimization discussion indicates that it is possible to substantially decrease the elution time without compromising resolution (eqn. 20). This is done through a simultaneous increase in both field strength and flow rate, as indicated in eqn. 19. Indeed, by increasing the field to 280 gravities, and the flow rate to 60 mL/h, it was possible to collect the monomer fraction in 27 minutes (Fig. 3(b)), as compared to the 53 minutes required under the conditions used in Fig. 3(a) (Caldwell et al., 1980).

From the elution patterns in Fig. 3 it is possible to determine an "effective mass", i.e. the product of mass and buoyancy ($m\Delta\varrho/\varrho_s$ in eqn.

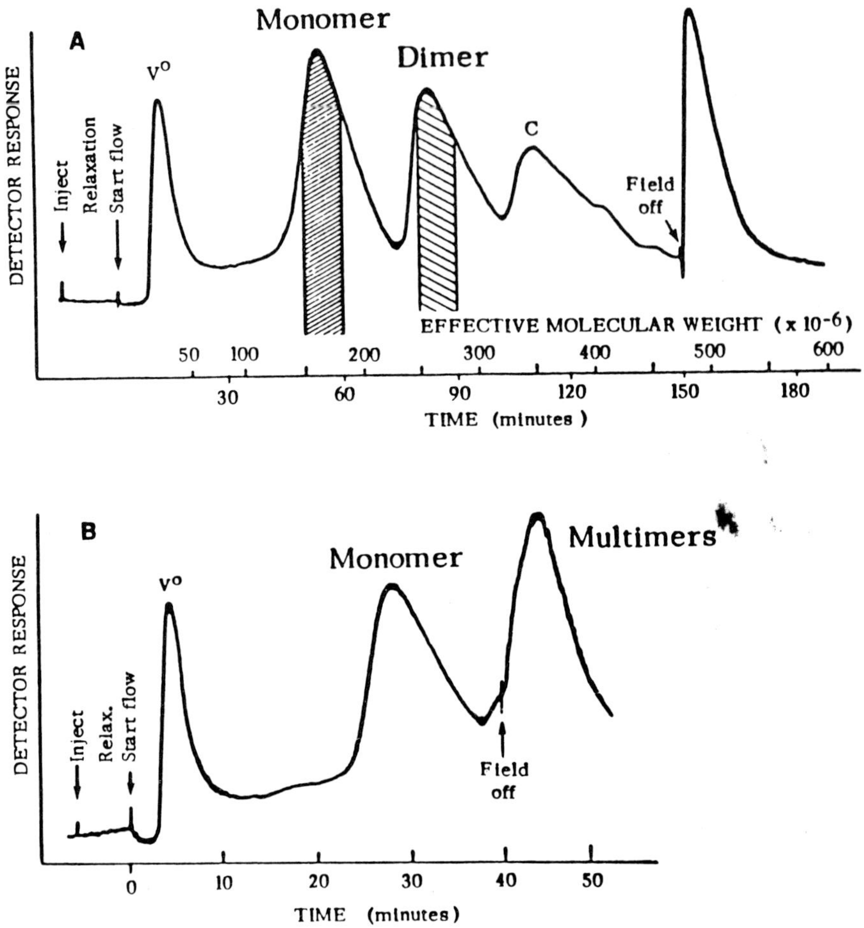

Figure 3. Sedimentation FFF of Gypsy Moth NPV. (a) Under a field strength of 173 gravities and a flow rate of 24 mL/h the monomer emerges after 53 min. The elution volume scale is transformed into a scale in effective molecular weight M' through use of eqns. 4 and 6, and eqn. 2 of Table 1. (b) Under a field strength of 280 gravities and a flow rate of 60 mL/h the monomer elutes after 27 min. (Reproduced from Caldwell et al., 1980, with permission).

2 of Table 1), for each type of virus aggregate. These effective masses should not be expected to be even multiples of the value for the monomer, since each type of particle is surrounded by a lipid envelope which will change both the mass and density, calculated on a *per monomer* basis, with changes in aggregate size.

The determination of effective mass begins by identification of the elution time t_r or elution volume V_r associated with the peak maximum for the desired component. Through eqn. 6 either of these quantities is converted to a value for R, which is then transformed into the important retention parameter λ via eqns. 4 or 5. Table 1 lists the expressions for

λ which applies to sedimentation FFF. From knowledge of the applied field strength G and the temperature of the experiment, the particle's effective mass m' or effective molecular weight M' (the product of m' and Avogadro's number N_A) is easily determined. For the monomeric form of the Gypsy Moth NPV, M' was found to be $(160 +/- 6.8) \times 10^6$ dalton in six different determinations on different preparations, analyzed under comparable experimental conditions.

A problem arises when M' is to be converted into an actual molecular weight M. This process requires accurate values for the densities of both sample particle and suspension medium ($\Delta\varrho/\varrho_s$ in eqn. 2 of Table 1). The latter is easily found e.g. by pychnometry. However, density determinations of solids are generally based on dry-weight determinations of particle suspensions of known weight and volume. In the case of viruses and other hydrated particles such measurements are quite prone to errors, and the accuracy in M'-derived molecular weights suffers accordingly. One way around this problem is to run the FFF analysis in media of different densities (Giddings, Karaiskakis and Caldwell, 1981). The different R-values observed under the selected conditions are then used to yield numerical values for λ, and from knowledge of the solvent densities ϱ, associated with given λ:s, one calculates the particle parameters M and ϱ_s. Although, in principle, only two runs are needed to yield the two unknowns, it is preferable for high precision work to overdetermine the system.

Through a slight rearrangement of eqn. 2 of Table 1 one can organize the experimental data in a form well suited for graphic representation, as seen in eqn. 21

$$\varrho = \varrho_s - \varrho_s(RT/MGw)\lambda^{-1} \tag{21}$$

By plotting the carrier density ϱ versus λ^{-1} one obtains a linear relationship that, when extrapolated to zero (no retention), gives an intercept equal to the particle density ϱ_s. From the slope of the line, in combination with its intercept, one finally determines molecular weight M. This technique was first established for latex particles (Giddings, Karaiskakis and Caldwell, 1981), and was subsequently applied both in the analysis of T4D virus (Caldwell, Karaiskakis and Giddings, 1981) and the algal virus PBCV (Yonker, Caldwell, Giddings and Van Etten, 1985). It should be noted that the above procedure is only valid for cases in which the physical characteristics of the particle remain invariant with the composition of the medium.

(b) Cross-linked BSA spheres. Recently, pharmaceutical chemists have shown an interest in using colloidal particles as intravenous carriers for

drugs with prolonged action, and as probes of obstructions in the reticuloendothelial system (Scheffel et al., 1972). Such particles must be nonimmunogenic, and their size must be large enough to make them perform as desired, but not so large as to cause obstructions in the vascular system. Cross-linked serum albumin appears well suited for this purpose, but any preparation of such protein particles must be carefully sized prior to injection into the human body. A test of the applicability of sedFFF to sizing problems of this nature was performed using particles of bead-polymerized bovine serum albumin (BSA) (Caldwell et al., 1981a). Under the selected experimental conditions, the fractogram of Fig. 4 was recorded in less than one hour. The retention volume associated with the maximum amplitude of the retained peak was used to calculate an average particle diameter d of 0.349 μm for the BSA beads. This was readily done with the aid of eqns. 4 and 6, and eqn. 3 of Table 1, assuming a protein density of 1.36 g/cm^3.

Through application of eqns. 14–16, the measured peak width was subsequently translated into a polydispersity for the sample. This calculation involved an estimation of the nonequilibrium term H_n in eqn. 15, by inserting the value for λ at the peak maximum (previously used to determine d), the average linear velocity of the carrier $\langle v \rangle$, and the average diffusion coefficient D obtained from d (by means of the Stokes–Einstein relationship (Tanford, 1961)). By further assuming that the instrumental band broadening was negligible, the subtraction of H_n from the measured H gave a value for the polydispersity contribution H_p which, in turn, was converted to a standard deviation in d (0.040 μm) using eqn. 16. All in all, this characterization procedure took little more than an hour to perform, and the results agreed remarkably well with those obtained by the considerably more laborious electron microscopic procedure which was used as a reference.

(c) Emulsion characterization. The fragile vesicles or droplets of an emulsion are difficult to characterize by most methods used for solid particles. For narrow size distributions, the quasi-elastic light scattering technique gives a good estimate of the average droplet size, but if the distribution is broad, this method becomes less reliable. SedFFF has been shown to reproducibly give detailed information on the droplet size distribution in several emulsions (Yang et al., 1983; Yang et al., 1984). Such information is often useful as a diagnostic of product stability, in which case one examines the fractogram for variations in time. A stability, or quality control, evaluation of this kind does not require a fullfledged transformation of the elution volume/time axis into a diameter axis;

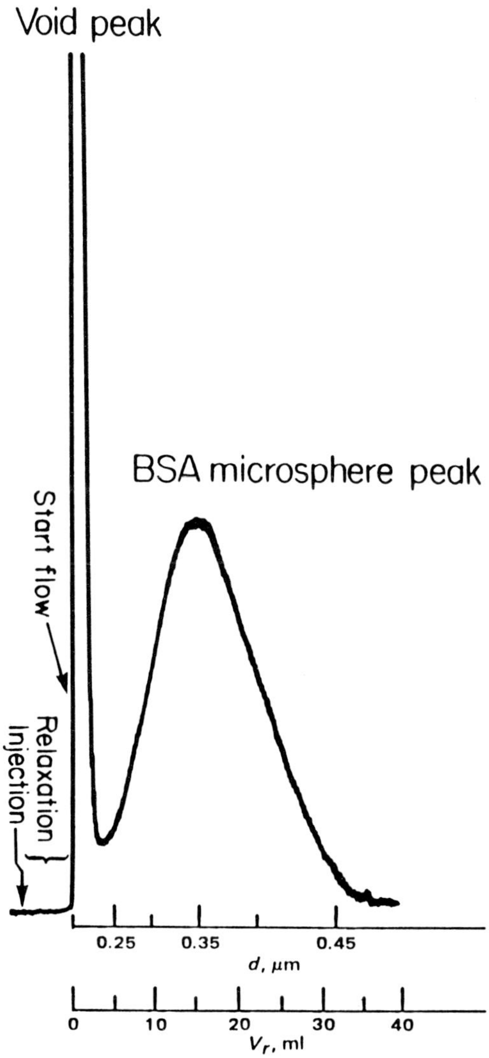

Figure 4. Sedimentation FFF of BSA microspheres. The field strength was 13.8 gravities and the linear flow rate $\langle v \rangle$ was 0.33 cm/s. The elution volume scale is converted into a scale in particle diameter by use of eqns. 4 and 6 together with eqn. 3 of Table 1. $V^0 = 0.90$ mL and $w = 0.0254$ cm. (Reproduced from Caldwell et al., 1981a, with permission).

instead, the fractogram is merely used as a "fingerprint" of the sample, as seen in Fig. 5. The two traces in this figure represent droplet size distributions for two batches of a lipid emulsion used in parenteral nutrition. These and similar samples were analyzed repeatedly over several months without showing signs of change. However, the manufacturer's continuous efforts to produce an emulsion with smaller droplets

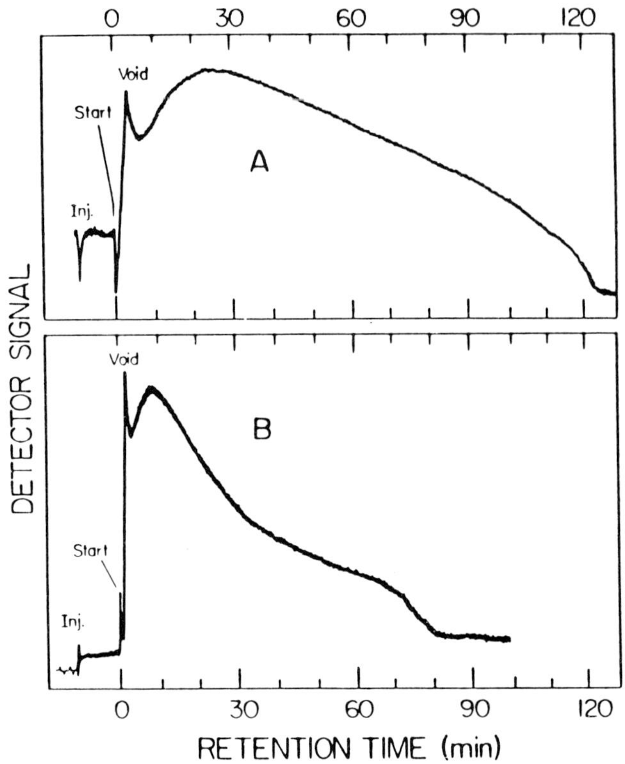

Figure 5. Comparison of fractograms for two commercial samples of soybean oil emulsion, identified by their expiration dates: (a) 3/79, and (b) 6/80. Both runs were made under a field strength of 194 gravities and a flow rate of 60 mL/h. Channel parameters were: $V^0 = 2.0$ mL, and $w = 0.0127$ cm. (Reproduced from Yang et al., 1983, with permission).

(presumably to meet the challenge of a competitor) was evidenced by a shift in the peak maximum towards lesser retention (smaller diameters) for samples with later production dates.

Flow FFF. The flow FFF subtechnique is based on the application of a cross-flow of carrier, perpendicular to the axial flow which transports the different zones along the separation axis. In order to permit the "field-flow" to occur, the channel is constructed out of some porous material (Giddings et al., 1980), and a semipermeable membrane is inserted to cover the accumulation wall.

The permeability of this membrane defines the lower molecular weight limit to the range of samples which can be processed by the channel. In recent years, the most extensively used membranes include the PTGC product from Millipore, which has a nominal molecular weight cut-off of 10 000 dalton, and the YM5 from Amicon, whose cut-off limit is quoted as 5,000 dalton. With a separation range that extends from molecular weights of around 10^4 dalton to the onset of steric effects, which occurs for particles with "molecular weights" in the vicinity of 10^{11} dalton, the flow FFF offers some unique advantages in work with highly polydisperse materials. However, its selectivity (see Table 2) is inferior to that offered by its sedimentation analog, and whenever the sample permits (i.e. for molecular weights in excess of 10^6 dalton), the sedimentation system is in general the method of choice.

Evaluation of a physical property of the sample molecule from its retention volume is more straightforward for flow FFF than for any of the other FFF subtechniques. This can be realized from an examination of Table 1, which summarizes expressions for λ applicable to the different methods. In all but the flow FFF case (eqn. 4 of the table) the retention parameter depends on a combination of properties, e.g. molecular weight and density in sedimentation FFF, diffusivity and thermal diffusivity in thermal FFF, or diffusivity and electrophoretic mobility in electrical FFF. In flow FFF, by contrast, the retention volume is uniquely related to the diffusivity of the sample, which permits direct conversion of the axis representing V_r, or t_r, into an axis representing component diffusivity D.

The flow FFF technique has been used to separate synthetic polymers in both aqueous (Giddings, Lin and Myers, 1978a; Wahlund et al., 1985) and nonaqueous systems (Brimhall et al., 1984), as well as colloidal particles of both polystyrene (Giddings, Yang and Myers, 1976) and silica (Giddings, Lin and Myers, 1978b). Among samples of biological interest, the method has been applied to several proteins (Giddings, Yang and Myers, 1977a), carbohydrates, nucleic acids, viruses (Giddings Yang and Myers, 1977b), and subcellular particles (Giddings et al., 1980). A typical fractogram is seen in Fig. 6. Here the two bacteriophages Qβ and P22 are baseline resolved from each other and from bovine serum albumin, which emerges virtually unretained under the selected cross-flow (Giddings, Yang and Myers, 1977b). The conditions chosen for the separation were not optimized in time, and the massive P22 phage required about four hours to elute.

The optimization of an FFF separation according to eqn. 20 requires that the system performs ideally even under high fields, when the zones are compacted in the immediate vicinity of the wall. Since the layer thickness l is no more than 2 microns for a zone retained by 20 column volumes in a channel with a thickness w of 250 microns, it is evident that any surface roughness, even of the order of fractions of a micron, will have an effect on both R and H at this level of retention. The influence of surface roughness on these two parameters has been studied in some

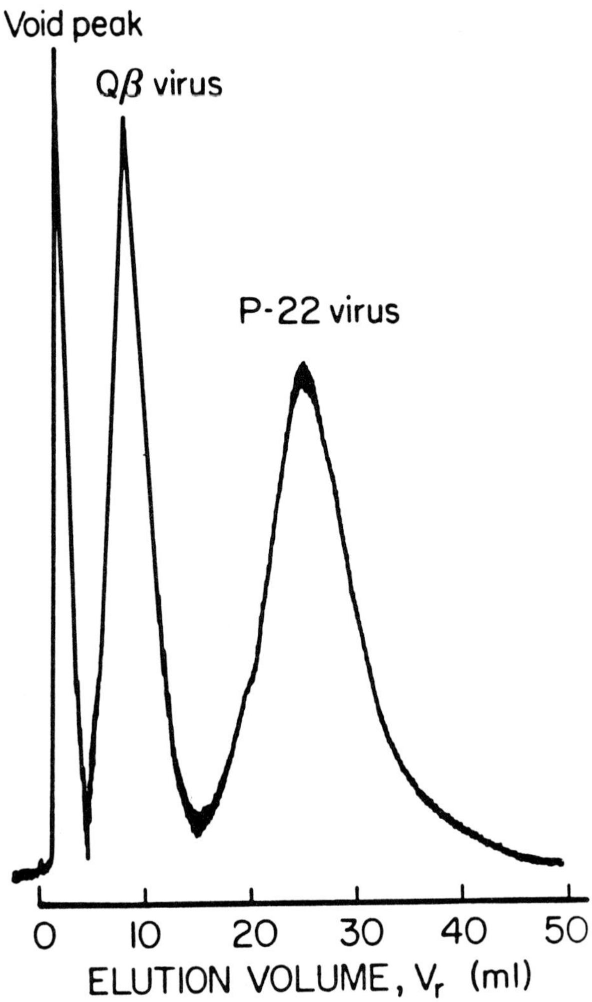

Figure 6. Flow FFF separation of bovine serum albumin (void peak) and the two bacteriophages Qβ and P-22. The cross-flow was 22 mL/h and the channel flow was 8 mL/h. $V^0 = 1.84$ mL; $w = 0.038$ cm. (Reproduced from Giddings, Yang and Myers, 1977b, with permission).

detail for thermal FFF (Giddings, Smith and Myers, 1978). So far, none of the membranes used for flow FFF comes close to the smoothness of the highly polished metal surfaces which serve as accumulation walls in the sedimentation and thermal FFF units. As a result, one often observes excessive retention under high fields using the flow subtechnique. Such was the case in the virus study which produced Fig. 6. Plots of λ versus $1/\dot{V}_c$, which ideally should be straight lines with zero intercept, all crossed the abscissa at a point significantly removed from the origin, which for practical purposes set an upper limit to useful cross-flows for that system

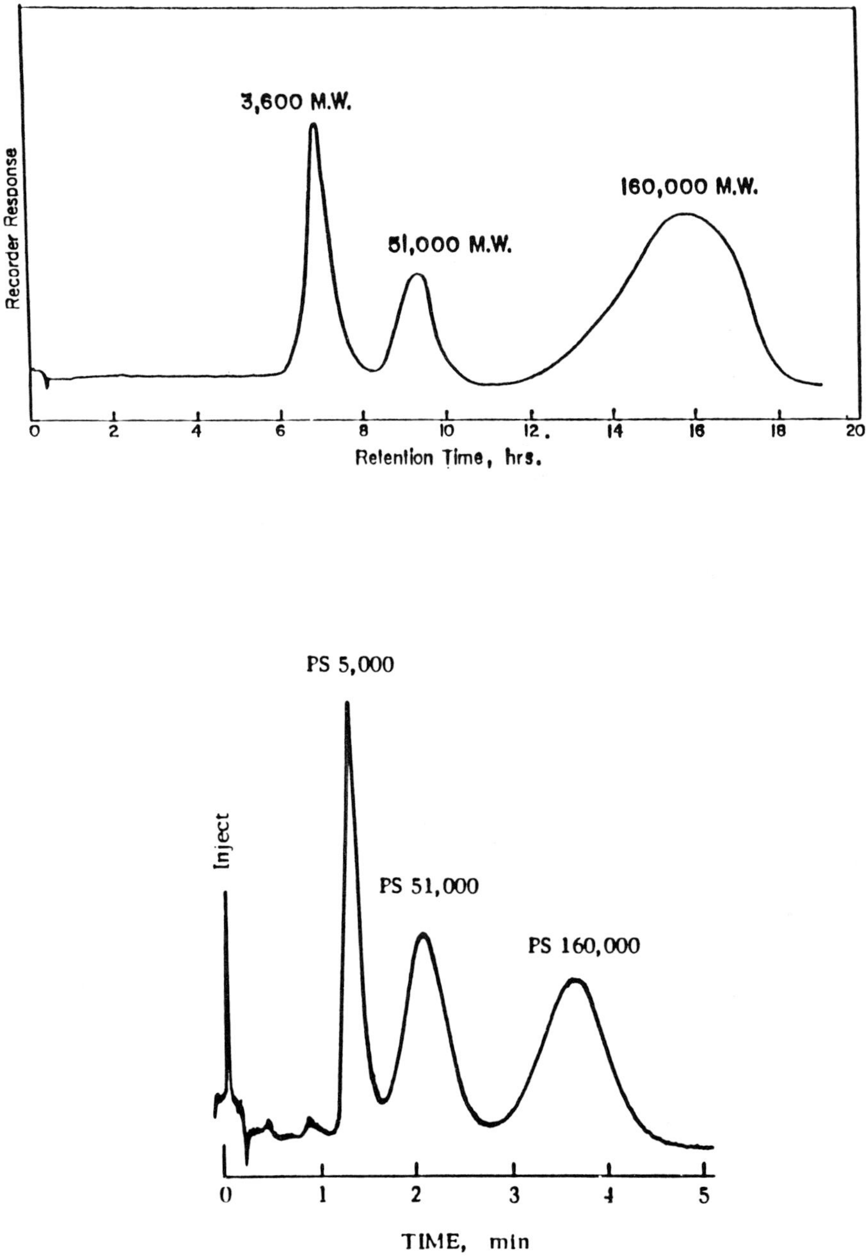

Figure 7. Thermal FFF of linear polystyrene standards in toluene. (a) Separation performed in an early channel (1969), whose approximate dimensions were: $L = 3.05$ m, $b = 1.2$ cm, $w = 0.025$ cm. ΔT was 72°C, and the carrier flow was 1.32 mL/h (0.012 cm/s). (b) Separation in a later, more refined unit (1978) whose dimensions were: $L = 42$ cm, $b = 2.0$ cm, $w = 0.0051$ cm. ΔT was 60°C and the carrier flow was 21 mL/h (0.56 cm/s). (Reproduced from Thompson, Myers and Giddings, 1969; and Giddings, Martin and Myers, 1978, with permission).

at 50 mL/h ($U = 2.9 \times 10^{-4}$ cm/s). Similar trends have been observed for other systems.

For reasons not yet fully understood, the nonideal effects of high cross-flow are reduced at higher channel flows (Wahlund et al., 1985, which permits the separation of standard test mixtures of proteins in times comparable to those recorded with high performance GPC (Wahlund K.G., unpublished results). This is exemplified by the baseline separation of bovine serum albumin from its dimer, which took place in 8 minutes under a cross-flow of 330 mL/h ($U = 1.85 \times 10^{-3}$ cm/s) and a channel flow of 468 mL/h.

It has proven difficult to obtain a uniform cross-flow throughout the channel, and in a recent design the flow FFF unit (Wahlund and Giddings, 1987) has been equipped with an impermeable upper wall. The entire flow now enters at the head of the channel, and the relative amounts which are to exit through the accumulation wall and through the channel outlet respectively, are controlled by flow restrictors at each exit. With this system one has the option of focusing the zone during its relaxation by forcing carrier to enter also through the channel outlet. The two opposing flows force the sample into a narrow band at some position determined by their relative magnitude. During the actual run, all carrier enters at the head of the channel, and the focused zones elute as quite sharp peaks.

Electrical FFF. Although electrical fields are used routinely in today's many powerful electrophoretic separation methods, their use in the FFF mode has been fraught with problems (Caldwell et al, 1972; Giddings, Lin and Myers, 1976; Kesner et al., 1976). Early experiments were able to show a qualitative agreement between experimentally measured retentions for a series of proteins, and those predicted from theory. Shifts in pH from 4.5 to 8.0 were shown to have the expected influence on the elution order of BSA (pI 4.5) and IgG (pI 7.4). However, the range of field strengths E for which a plot λ versus $1/E$ remains linear has proven very limited indeed (0 to about 10 V/cm) for the cellulose membranes which we have used as channel walls so far. The reason for this is not entirely clear, but it is expected that both electroosmotic effects and membrane polarization are influencing the retention process. This fact, together with the significant zone broadening caused by our channels to date, has prevented the achievement of resolutions similar to those seen with the other FFF subtechniques.

Thermal FFF. Thermal FFF was the first subtechnique to show measurable retention of injected polymers (Thompson Myers and Giddings,

1967). Due to the simplicity of the column assembly, and the ease with which the channel walls can be polished to a high finish, this technique also permitted the application of theoretical principles of optimization in order to generate rapid, high resolution separations of relatively complex samples. The crucial importance of a controlled channel geometry, represented by absolutely smooth and parallel channel walls, is illustrated by Fig. 7. The early fractogram (Fig. 7(a)) was recorded in a channel far from ideal in this respect (Thompson, Myers and Giddings, 1969). The apparatus induced zone broadening was significant, and the separation had to be performed under a very slow channel flow in order for the differential retention to overcome the low efficiency and result in resolution. While it took nearly eighteen hours to accomplish baseline resolution of a three component mixture of linear polystyrenes (Mw 3,600, 51 000, and 160 000 dalton) with the early apparatus, a later and more refined thermal FFF system, operating at four times higher field, performed a similar task in about four minutes (Fig. 7(b)) (Giddings, Martin and Myers, 1978).

There are two principal facts which limit the usefulness of thermal FFF for biological fractionations. The first and least problematic is the inherent thermal instability associated with a great number of biological macromolecules. Intuitively, one hesitates to expose e.g. large protein

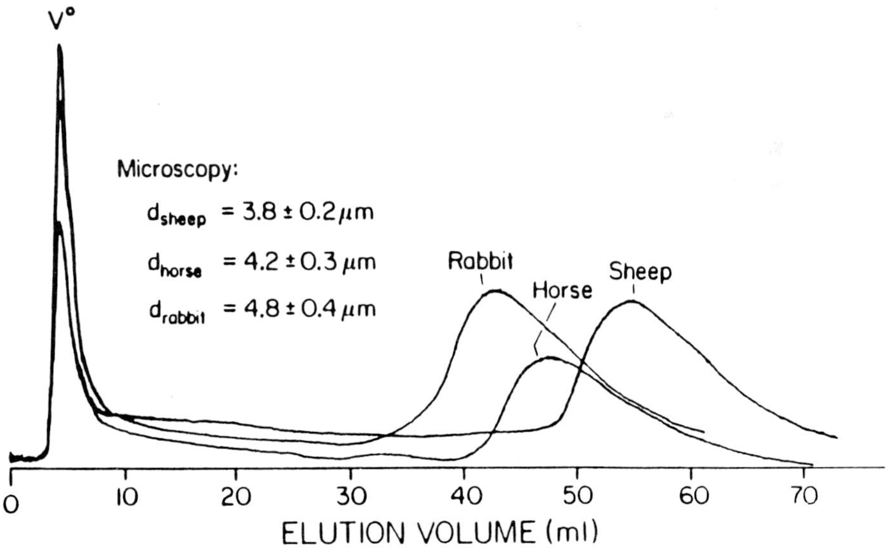

Figure 8. Steric FFF of whole blood from different animals (three runs superimposed). The main retained peak represents the red cells. The elution order is that predicted from their microscopically determined sizes by eqn. 7. Settling field: 3.6 gravities; flow rate 36 mL/h; V^0 = 4.7 mL; w = 0.0254 cm. (Reproduced from Caldwell et al., 1984b, with permission).

molecules to the temperature drops of 50–100 degrees C, typically used for polymer fractionation. This problem is, however, easily circumvented by constructing a system with a split inlet (Giddings, 1985), and introducing the sample through the cold wall, where it eventually is to accumulate. When carefully executed, this mode of introduction prevents the sample from ever contacting the hot wall of the channel.

The second problem is potentially much more severe, and arises from the fact that very few materials appear to experience thermal diffusion (the transport mechanism in thermal FFF) in aqueous medium. Several workers have failed to observe migration of e.g. proteins, nucleic acids, or polysaccharides (Bonner, 1973), even under steep temperature gradients, when these compounds have been present in aqueous solution. Kirkland and Yau (1986) were able to retain a small number of water soluble synthetic polymers, but the majority of samples examined in their study appeared unaffected by the field. By switching carriers from water to DMSO, we have managed to significantly retain a dextran sample that, in aqueous solution, passed the channel entirely unaffected by the field (Giddings, Caldwell and Myers, 1976). For biological samples which are stable in nonaqueous solvents, the thermal FFF technique could therefore still prove to be a useful complement to existing separation techniques.

Steric FFF. In the extreme case of particles whose size significantly exceeds the theoretical layer thickness l in eqn. 2, the retention will directly reflect their particle diameter d, as expressed in eqn. 7. This sterically controlled behavior implies a reversal of the normal FFF elution order, since larger particles will migrate ahead of smaller ones. Although the effect is general in nature, and occurs for all types of applied fields, our initial steric FFF separations were all the result of a gravitational or sedimentation settling force (Giddings and Myers, 1978; Giddings et al., 1979; Caldwell et al., 1979).

Figure 8 shows a superposition of three fractograms, representing heparinized whole blood from rabbit, horse, and sheep (Caldwell et al., 1984b). Under the applied field of 3.6 gravities, the serum proteins elute unretained in the void, and the major retained peak in each of the three traces represents the red cells. These cells were sized by microscopy, and the elution order of the three types of red cells is that predicted from eqn. 7: rabbit (4.8 μm), horse (4.2 μm), and sheep (3.8 μm).

High channel velocities tend to decrease particle retention, i.e. to increase R. This effect is thought to arise from lift forces generated by the gradient in velocity which is spanned by the particle (Caldwell et al, 1979). Since these lift forces can be overcome by an increase in the strength of the settling field, it is possible to simultaneously increase both field and flow while keeping retention at a constant level. Figure 9 illustrates this principle for a sample of human red cells. Under a field of 3 gravities, an R of 0.15 can be accomplished with a flow rate of 25 mL/h, whereas the same retention is recorded at 150 mL/h under a settling field

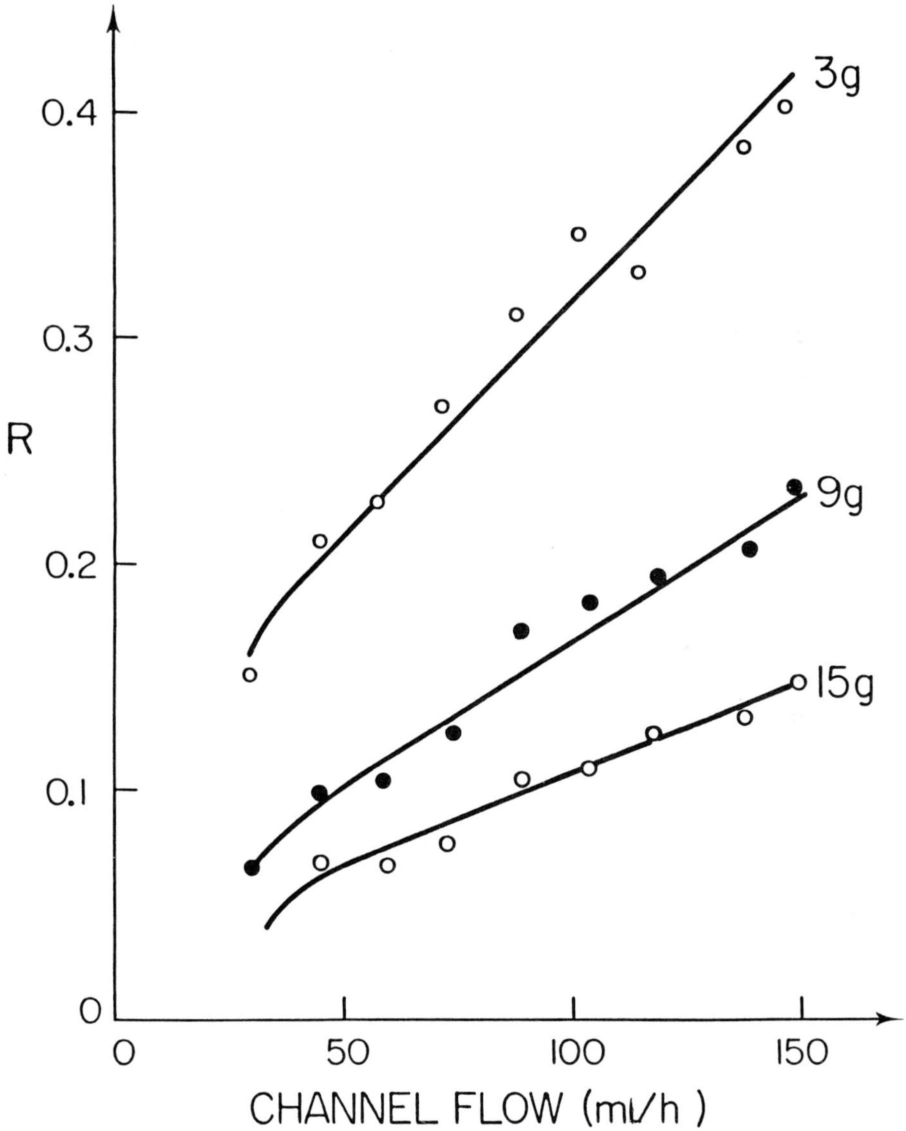

Figure 9. Variation in retention of human red cells with variations in flow rate and settling field (gravities). (Reproduced from Caldwell et al. 1984b, with permission).

of 15 gravities. Through a five-fold increase in field strength it was therefore possible to reduce the elution time by a factor of six.

Since a velocity increase does not produce a significant increase in zone broadening for this mode of separation, one has the ability to perform extraordinarily rapid and selective fractionations using this technique. An early example showed baseline separation of fixed avian and human red cells (with diameters of 9.5 and 6.0 μm, respectively) in 7.25 minutes following relaxation (Caldwell et al., 1984b). With better pumping equipment it was possible to increase the carrier flow to the 2280 mL/h used to develop the fractogram in Fig. 10 (Koch and Giddings, 1986). This figure, which demonstrates the baseline separation of a seven component mixture of latex particles performed in less than four minutes, gives an indication of the capability for speed and selectivity which is offered by the steric FFF approach. The potential use of the technique as a cell

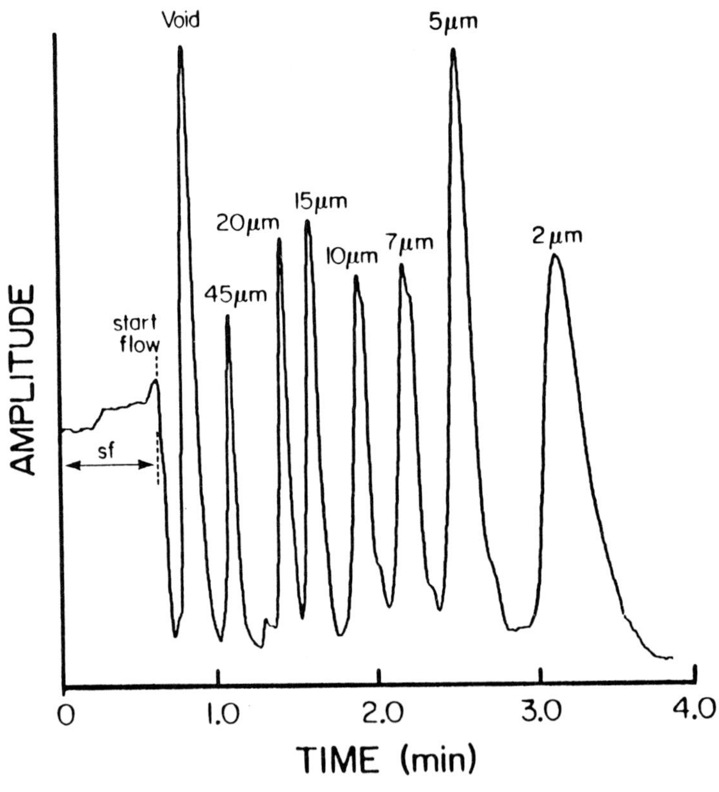

Figure 10. Steric FFF of standard latex particles using a sedimentation field of 345 gravities and a channel flow of 2280 mL/h (12.5 cm/s). V^0 was 4.2 mL and w was 0.0254 cm. (Reproduced from Koch and Giddings, 1986, with permission).

separation tool is supported by the demonstrated viability of several cell samples following passage through the separator (Caldwell et al., 1984b).

Conclusion

The thin FFF channels are ideally suited for rapid fractionations, as transit times across the channel are short even at very modest applied fields. This can often prove valuable in work with delicate and short-lived tissues. The open, unobstructed FFF channel allows separation of even very large particles with a minimum of destructive shear, leaving virus particles infectious and cells in a viable condition following separation. In general, this process is allowed to take place in a carrier medium that is chosen for maximum compatibility with the sample.

Since the mathematical foundation for the separation process is well understood, it is possible to determine specific physical properties for the separated components from their elution characteristics. Another significant advantage afforded by a well-developed theory is the ability to optimize the system for a given separation problem. With field strengths and flow rates under external control, such optimizations are in theory easy to implement in the FFF systems. Their success hinges on the possibility to construct the crucial accumulation walls out of materials that are both very smooth and inert towards biological samples. As we gain more knowledge of the surface interactions between macromolecules and such materials which can be used to manufacture future FFF systems, we should be able to enhance the speed of resolution far beyond what has been demonstrated today.

Acknowledgement

This review has been prepared with support by Public Health Service grant nr. GM 10851-29 from the National Institutes of Health.

References

1. Beckett, R., G. Nicholson, B.T. Hart, M. Hansen and J.C. Giddings, *Separation and Size Characterization of Colloidal Particles in River Water by Sedimentation Field-Flow Fractionation*, Environ. Sci. Technol. (accepted).
2. Bonner, F.J. 1973. *Thermal Diffusion of Dextran and Polystyrene*, Chem. Scr. 3: 149–152.
3. Brimhall, S.L., M.N. Myers, K.D. Caldwell and J.C. Giddings. 1984. *Separation of Polymers by Flow Field-Flow Fractionation*, J. Polymer Sci.: Polymer Letters Edition 22: 339–345.
4. Caldwell, K.D., L.F. Kesner, M.N. Myers and J.C. Giddings. 1972. *Electrical Field-Flow Fractionation of Proteins*, Science 176: 296–298.
5. Caldwell, K.D., T.T. Nguyen, M.N. Myers and J.C. Giddings. 1979. *Observations of Anomalous Retention in Steric Field-Flow Fractionation*, Sep. Sci. Technol. 14: 935–946.

6. Caldwell, K.D., T.T. Nguyen, J.C. Giddings and H.M. Mazzone. 1980. *Field-Flow Fractionation of Alkali Liberated Polyhedrosis Virus from Gypsy Moth (Lymantria Dispar, Linnaeus)*, J. Virol. Methods 1: 241–256.
7. Caldwell, K.D., G. Karaiskakis and J.C. Giddings. 1981. *Characterization of T4D Virus by Sedimentation Field-Flow Fractionation*, 215: 323–332.
8. Caldwell, K.D., G. Karaiskakis, M.N. Myers and J.C. Giddings. 1981a. *Characterization of BSA Microspheres by Sedimentation Field-Flow Fractionation*, J. Pharmaceutical Sci. 70: 1350–1352.
9. Caldwell, K.D., G. Karaiskakis, M.N. Myers and J.C. Giddings. 1981b. *Characterization of Liposomes by Sedimentation Field-Flow Fractionation*, Colloids and Surfaces 3: 233–238.
10. Caldwell, K.D., B.J. Compton, J.C. Giddings and R.J. Olson. 1984a. *Sedimentation Field-Flow Fractionation: A Method for Studying Particulates in Cataractous Lens*, Invest. Ophthalmol. Vis. Sci. 25: 153–159.
11. Caldwell, K.D., Z.-Q. Cheng, P. Hradecky and J.C. Giddings. 1984b. *Separation of Human and Animal Cells by Sedimentation Field-Flow Fractionation*, Cell Biophysics 6: 233–251.
12. Caldwell, K.D., S.M. Mozersky, J. Nagahashi and R. Barford. 1986. *Sedimentation Field-Flow Fractionation of Corn Root Membranes*, in *Chemical Separations*, Vol. 1, J.D. Navratil and C.J. King, eds., Litarvan, Denver pp. 59–74.
13. Fox, A., L.E. Schallinger and J.J. Kirkland. 1985. *Sedimentation Field-Flow Fractionation of Bacterial Cell Wall Fragments*, J. Microbiol. Methods 3: 273–281.
14. Gao, Y.S., K.D. Caldwell, M.N. Myers and J.C. Giddings. 1985. *Extension of Thermal Field-Flow Fractionation to Ultra High (20×10^6) Molecular Weight Polystyrenes*, Macromolecules 18: 1272–1277.
15. Giddings, J.C. 1965a. *Dynamics of Chromatography*, Vol. 1, Dekker, New York, Chapter 2.
16. Giddings, J.C. 1965b. *Dynamics of Chromatography*, Vol. 1, Dekker, New York, Chapter 7.
17. Giddings, J.C. 1968. *Nonequilibrium Theory of Field-Flow Fractionation*, J. Chem. Phys. 49: 81–85.
18. Giddings, J.C. 1973a. *The Conceptual Basis of Field-Flow Fractionation*, J. Chem. Ed. 50: 667–669.
19. Giddings, J.C. 1973b. *Parameters for Optimum Separations in Field-Flow Fractionation*, Sep. Sci. 8: 567–575.
20. Giddings, J.C., F.J.F. Yang and M.N. Myers. 1974. *Sedimentation Field-Flow Fractionation*, Anal. Chem. 46: 1917–1924.
21. Giddings, J.C., F.J.F. Yang and M.N. Myers. 1975. *Application of Sedimentation Field-Flow Fractionation to Biological Particles: Molecular Weights and Separation*, Sep. Sci. 10: 133–149.
22. Giddings, J.C., Y.H. Yoon and M.N. Myers. 1975. *Evaluation and Comparison of Gel Permeation Chromatography and Thermal Field-Flow Fractionation for Polymer Separations*, Anal. Chem. 47: 126–131.
23. Giddings, J.C., Y.H. Yoon, K.D. Caldwell and M.E. Hovingh. 1975. *Nonequilibrium Plate Height for Field-Flow Fractionation in Ideal Parallel Plate Columns*, Sep. Sci. 10: 447–460.
24. Giddings, J.C. 1976. *Field-Flow Fractionation: Extending the Molecular Weight Range of LC to One Trillion*, J. Chromatogr. 125: 3–16.
25. Giddings, J.C., F.J. Yang and M.N. Myers. 1976. *Theoretical and Experimental Characterization of Flow Field-Flow Fractionation*, Anal. Chem. 48: 1126–1132.
26. Giddings, J.C., K.D. Caldwell and M.N. Myers. 1976. *Thermal Diffusion of Polystyrene in Eight Solvents by an Improved Thermal Field-Flow Fractionation Methodology*, Macromolecules 9: 106–112.
27. Giddings, J.C., G.C. Lin and M.N. Myers. 1976. *Electrical Field-Flow Fractionation in a Rigid Membrane Channel*, Sep. Sci. 11: 553–568.
28. Giddings, J.C., F.J. Yang and M.N. Myers. 1977a. *Flow Field-Flow Fractionation as a Methodology for Protein Separation and Characterization*, Anal. Biochem. 81: 395–407.
29. Giddings, J.C., F.J. Yang and M.N. Myers. 1977b. *Flow Field-Flow Fractionation: A New Method for Separating, Purifying, and Characterizing the Diffusivity of Viruses*, J. Virol. 21: 131–138.

30. Giddings, J.C. 1978. *Displacement and Dispersion of Particles of Finite Size in Flow Channels with Lateral Forces. Field-Flow Fractionation and Hydrodynamic Chromatography*, Sep. Sci. Technol. 13: 241–254.
31. Giddings, J.C. and M.N. Myers. 1978. *Steric Field-Flow Fractionation: A New Method for Separating 1–100 μm Particles*, Sep. Sci. Technol. 13: 637–645.
32. Giddings, J.C., G.C. Lin and M.N. Myers. 1978a. *Fractionation and Size Distribution of Water Soluble Polymers by Flow Field-Flow Fractionation*, J. Liq. Chromatogr. 1: 1–20.
33. Giddings, J.C., G.C. Lin and M.N. Myers. 1978b. *Fractionation and Size Analysis of Colloidal Silica by Flow Field-Flow Fractionation*, J. Colloid Interface Sci. 65: 67–78.
34. Giddings, J.C., M. Martin and M.N. Myers. 1978. *High Speed Polymer Separations by Thermal Field-Flow Fractionation*, J. Chromatogr. 158: 419–435.
35. Giddings, J.C., L.K. Smith and M.N. Myers. 1978. *Surface Barriers for Retention Enhancement in Field-Flow Fractionation*, Sep. Sci. Technol. 13: 367–385.
36. Giddings, J.C. 1979. *Field-Flow Fractionation of Polymers: One-Phase Chromatography*, Pure & Appl. Chem. 51: 1459–1471.
37. Giddings, J.C., M.N. Myers, K.D. Caldwell and J.W. Pav. 1979. *Steric FFF as a Tool for the Size Characterization of Chromatographic Supports*, J. Chromatogr. 185: 261–271.
38. Giddings, J.C., M.N. Myers, K.D. Caldwell and S.R. Fisher. 1980. *Analysis of Biological Macromolecules and Particles by Field-Flow Fractionation*, in *Methods of Biochemical Analysis*, Vol. 26, D. Glick, ed., Wiley, New York, pp. 79–136.
39. Giddings, J.C., G. Karaiskakis and K.D. Caldwell. 1981. *Density and Particle Size of Colloidal Materials Measured by Carrier Density Variations in Sedimentation Field-Flow Fractionation*, Sep. Sci. Technol. 16: 607–618.
40. Giddings, J.C., G. Karaiskakis, K.D. Caldwell and M.N. Myers. 1983. *Colloid Characterization by Sedimentation Field-Flow Fractionation. I. Monodisperse Populations*, J. Colloid Interface Sci. 92: 66–80.
41. Giddings, J.C. 1985. *Optimized Field-Flow Fractionation System Based on Dual Stream Splitters*, Anal. Chem. 57: 945–947.
42. Giddings, J.C. and M.R. Schure. 1987. *Theoretical Analysis of Edge Effects in Field-Flow Fractionation*, Chem. Eng. Sci. (in press).
43. Giddings, J.C., K.D. Caldwell and H.K. Jones. 1987. *Particle Size Distribution of Simple and Complex Colloids by Sedimentation Field-Flow Fractionation*, in *Particle Size Distribution: Assessment and Characterization*, ACS Symposium Series No. 332, T. Provder, ed., (in press).
44. Giddings, J.C. and K.D. Caldwell. 1987. *Field-Flow Fractionation*, in *Physical Methods of Chemistry*, Vol. IIIB, B.W. Rossiter, ed., Wiley, New York (in press).
45. Gunderson J.J. and J.C. Giddings. 1987. *Comparison of Polymer Resolution in Thermal Field-Flow Fractionation and Size Exclusion Chromatography*, Anal. Chim. Acta, (in press).
46. Karaiskakis, G., M.N. Myers, K.D. Caldwell and J.C. Giddings. 1981. *Verification of Retention and Zone Spreading Equations in Sedimentation Field-Flow Fractionation*, Anal. Chem. 53: 1314–1317.
47. Karaiskakis, G., K.A. Graff, K.D. Caldwell and J.C. Giddings. 1982. *Sedimentation Field-Flow Fractionation of Colloidal Particles in River Water*, Intern. J. Environ. Anal. Chem. 12: 1–15.
48. Kesner, L.F., K.D. Caldwell, M.N. Myers and J.C. Giddings. 1976. *Performance Characteristics of Electrical Field-Flow Fractionation in a Flexible Membrane Channel*, Anal. Chem. 48: 1834–1839.
49. Kirkland J.J., W.W. Yau, W.A. Doerner and J.W. Grant. 1980. *Sedimentation Field-Flow Fractionation of Macromolecules and Colloids*, Anal. Chem. 52: 1944–1954.
50. Kirkland, J.J., S.W. Rementer and W.W. Yau. 1981. *Time-Delayed Exponential Field-Programmed Sedimentation Field-Flow Fractionation for Particle Size Distribution Analyses*, Anal. Chem. 53: 1730–1736.
51. Kirkland, J.J., W.W. Yau and F.C. Szoka. 1982. *Sedimentation Field-Flow Fractionation of Liposomes*, Science 215: 296–298.
52. Kirkland, J.J., C.H. Dilks and W.W. Yau. 1983. *Sedimentation Field-Flow Fractionation at High Force Fields*, J. Chromatogr. 255: 255–265.

53. Kirkland J.J. and W.W. Yau. 1986. *Thermal Field-Flow Fractionation of Water Soluble Macromolecules*, J. Chromatogr. 353: 95–107.
54. Koch, T. and J.C. Giddings. 1986. *High-Speed Separation of Large (> 1 μm) Particles by Steric Field-Flow Fractionation*, Anal. Chem. 58: 994–997.
55. Koch, T. 1987. *Steric Field-Flow Fractionation*, Ph.D. Thesis, Univ. of Utah.
56. Martin, M., M.N. Myers and J.C. Giddings. 1979. *Nonequilibrium and Polydispersity Peak Broadening in Thermal Field-Flow Fractionation*, J. Liq. Chromatogr. 2: 147–164.
57. Myers, M.N. and J.C. Giddings. 1982. *Properties of the Transition from Normal to Steric Field-Flow Fractionation*, Anal. Chem. 54: 2284–2289.
58. Schallinger, L.E., W.W. Yau and J.J. Kirkland. 1984. *Sedimentation Field-Flow Fractionation of DNA*, Science 225: 434–436.
59. Schallinger, L.E., J.E. Gray, L.W. Wagner, S. Knowlton and J.J. Kirkland. 1985. *Preparative Isolation of Plasmid DNA with Sedimentation Field-Flow Fractionation*, J. Chromatogr. 342: 67–77.
60. Scheffel, U., B.A. Rhodes, T.K. Natarjan and H.N. Wagner. 1972. *Albumin Microspheres for the Study of the Reticuloendothelial System*, J. Nucl. Med. 13: 498–503.
61. Smith, L.K., M.N. Myers and J.C. Giddings. 1977. *Peak Broadening Factors in Thermal Field-Flow Fractionation*, Anal. Chem. 49: 1750–1756.
62. Summers, M.D. and L.E. Volkman. 1976. *Comparison of Biophysical and Morphological Properties of Occluded and Extracellular Nonoccluded Baculovirus from In Vivo and In Vitro Host Systems*, J. Virol. 17: 962–972.
63. Tanford, C. 1961. *Physical Chemistry of Macromolecules*, Wiley, New York, Chapter 6.
64. Thompson, G.H., M.N. Myers and J.C. Giddings. 1967. *An Observation of a Field-Flow Fractionation Effect with Polystyrene Samples*, Sep. Sci. 2: 797–800.
65. Thompson, G.H., M.N. Myers and J.C. Giddings. 1969. *Thermal Field-Flow Fractionation of Polystyrene Samples*, Anal. Chem. 41: 1219–1222.
66. Wahlund, K.-G., H.S. Winegarner, K.D. Caldwell and J.C. Giddings. 1986. *Improved Flow Field-Flow Fractionation System Applied to Water-Soluble Polymers*, Anal. Chem. 58: 573–578.
67. Wahlund, K.-G. and J.C. Giddings. 1987. *Properties of an Asymmetric Flow Field-Flow Fractionation Channel Having one Permeable Wall*, Anal. Chem. (submitted).
68. Yang, F.-S., K.D. Caldwell and J.C. Giddings. 1983. *Colloid Characterization by Sedimentation Field-Flow Fractionation II. Particle Size Distribution*, J. Colloid Interface Sci. 92: 81–91.
69. Yang, F.-S., K.D. Caldwell, M.N. Myers and J.C. Giddings. 1983. *Colloid Characterization by Sedimentation Field-Flow Fractionation. III. Emulsions*, J. Colloid Interface Sci. 93: 115–125.
70. Yang, F.-S., K.D. Caldwell, J.C. Giddings and L. Astle. 1984. *Stability of Perfluorocarbon Blood Substitutes Determined by Sedimentation Field-Flow Fractionation*, Anal. Biochem. 138: 488–494.
71. Yonker, C.R., K.D. Caldwell, J.C. Giddings and J.L. van Etten. 1985. *Physical Characterization of PBCV Virus by Sedimentation FFF*, J. Virol. Methods 11: 145–160.

Research instrumentation for the 21st century: progress toward complete genomic maps and sequence data bases, and indexes of protein gene products

NORMAN G. ANDERSON, N. LEIGH ANDERSON and J.-P. HOFMANN
Large Scale Biology Corporation, 7503 Standish Place, Rockville, MD 20855, U.S.A.

Abstract The instrumentation requirements for agricultural research in the 21st century hinge on whether or not complete analyses of animal and plant cells will be required, on the cost of performing the analyses, and on the cost and complexity of the analytical instrumentation needed. It is emphasized that nature's response to modern agriculture is to evolve pests and diseases better fitted to attack new genetically uniform plants, or to escape pesticides or herbicides. For all future history, new strains of crop plants, new herbicides and pesticides, and new animal strains will be required periodically. Existing native plant varieties now provide genes for resistance which are empirically discovered, and bred into high-yield strains by conventional methods. It is suggested that this process will not be endlessly successful, and that at some future time the required genes will not be found. Then, true genetic engineering with rational design and optimization will be required. This means the complete sequencing of plant and animal genomes, the identification and characterization of all gene products, computer modelling of cell function, and ultimately the introduction of sets of modified or totally synthetic genes. We review the status of some of the instruments required, and suggest future directions for their development.

Introduction

In 1986 a large fraction of the scientific community concluded that it was technically feasible to map and completely sequence the entire human genome (3.4 billion base pairs) (Anderson and Anderson, 1985; Palca, 1986; Lewin, 1986; Martin and Davies, 1986; Wada and Soeda, 1987) and a smaller fraction believed that the project should be started now. It has been previously proposed that nearly all human protein gene products could be separated and indexed (Anderson and Anderson, 1982a, b and 1984; Taylor, Anderson, Scandora et al., 1982), and methods have gradually developed which allow nearly all known metabolites to be chromatographically separated and identified. The medical importance of such efforts is obvious. The questions we ask here are these: Will similar analytical efforts be required in agriculture, what will motivate such projects, what instruments or systems will be required, and will the results justify the costs?

The discovery of new elements, of new subnuclear particles, or of new planets has turned out not to be endless. Biological systems, in contrast, have been found to be so complex that few have wished to consider, even briefly, the question of whether living cells could, even in theory, be so completely understood that predictable and successful modifications of them could be made. Fortunately the organizers of this symposium have provided both a license and a forum for discussing these questions – at least once.

Will a complete understanding of living cells be necessary?

At present, new enzymes, new biochemical pathways, new structural proteins, and new chemicals which affect plants and animals are being discovered at an increasing rate. Genes for plant proteins are being isolated and sequenced, and transferred from one strain or species of plant to others. The astonishing rate of acceleration of research, however, obscures the fact that only a very tiny fraction of any plant genome has been sequenced, and the structure and function of an almost equally small fraction of the protein gene products of plant genes are known. It is unpopular to attempt to measure or indicate how much potentially useful knowledge we lack. However we cannot escape the fact that we have hardly begun to amass the bodies of knowledge which a continued human food supply will, we believe, ultimately require.

The first biotechnology revolution gave rise to agriculture and animal domestication, made possible civilization, and ultimately facilitated a rapid expansion of the human population. Most of the advances involved understanding and manipulation at the whole organism level, including the modification of human behavior to match agricultural requirements.

The second biotechnology revolution has started within the memory of most practising scientists who, incidentally, also constitute the majority of all scientists who have ever lived (De Sola Price, 1986). Nature itself will provide us with the motivation to keep this revolution going. Herbicides provide means for selectively killing undesirable plants in crop areas, and nature has and will continue to respond by providing new mutations for herbicide resistance in weed and non-crop plant populations. Pesticides are and will continue to stimulate selection of new resistant strains of insects, and antifungal agents will give us resistant fungi. As antibiotics become more widely used, antibiotic resistant bacteria have emerged and spread, holding the promise that in time all presently used antibiotics may become useless. As genetically homogeneous crop plants are cultivated, nature provides, often within as little as ten years, "improved" pests and diseases to attack them. In man and animals immune systems have provided marvelously versatile responses to viral and bacterial infections. Nature's latest response is the AIDS virus which attacks the immune system itself.

These reponses may be viewed as nature's attempts to prevent any one species from reaching its full reproductive potential, and overgrowing others. Once we have started, through modern technology, to break free of nature's control mechanisms, which include starvation, disease, and predation, then we must attempt to counter nature's responses *in perpetuo*. For the rest of human life on earth, barring a catastrophic decline in the human population, we will require new crop varieties, improved agricultural animals, new herbicides and pesticides, and new antibiotics. We are barely into this new era, but we have already passed the point of no return.

While the green revolution now provides sufficient world food, it would appear that cycles of new pests and diseases, followed by development of new resistant strains will occur for ensuing centuries, possibly on shorter and shorter cycles. We must ask, can our response be as varied and resourceful as nature's attacks?

Evolution in our time

Nature's responses to modern agriculture, animal breeding and husbandry, and to medicine may be described as *evolution in our time*, based on the random generation and selection of new or previously unnoticed mutations. Our response thus far is almost equally random, involving the search for existing genes for resistance or adaptation, with the work carried out by individuals or groups which by accident (possibly involving social mutations?) happened to be interested in a particular problem.

Even with the most comprehensive and systematic collection of strains and species related to present crop plants obtained from all the Vavilov centers of the world, there is no guarantee that resistance genes exist and will be found to fill all future needs, or that useful new mutations will occur by chance when needed. Thus the time required to breed new resistant strains may be expected to increase, while the chances of success may decrease.

Note that nature's responses are not fine-tuned. More species are extinct than now survive on earth, and there is no guarantee of long-term survival of any species, including man. While these considerations may appear pessimistic, prudence suggests that now is the time to begin to consider how the role of chance can be reduced in crop breeding, in herbicide and pesticide development, and in the development of antibiotics and antiviral agents. The only answer appears to be massive acquisition of new knowledge in the biological sciences, and serious attempts to understand and to model living systems in such detail that they may be engineered, that is, modified and improved in a rational and predictable manner.

Barriers to progress in biotechnology

In the past plant and animal breeders have systematically altered the genetic makeup of plants and animals by every means that they possessed, and have proliferated new varieties freely. Recombinant DNA techniques, however, have been viewed as allowing new and unique combinations of genes to be made which would never and could never exist in the wild, and hence might constitute new, unique, and dangerous threats to existing balances of nature.

If the genetic makeup of plants is fixed or shielded in such a way that gene transfer between non-interbreeding species *never* occurs in nature, and if many (or any) new combinations of genes can be made artificially which are better adapted to survive than are native ones, then the popularized dangers of genetic engineering are potentially real.

Gene transfer, however, appears to occur widely in nature. Many viruses are known which infect not only across species barriers, but between different classes and phyla. In addition many viruses readily transfer bits of DNA between hosts, providing a system for widely transfecting genes. While cancer virology has long emphasized the importance and universality of transfection, the concept appears to have escaped popular notice. If indeed miscellaneous bits of genetic material freely float around the plant and animal kingdoms, transported by a viral mailing system, one might expect that a significant amount of excess DNA would accumulate in host cells which is not expressed, as is indeed the case.

If there are few barriers in nature to gene transfer, and if it occurs widely and naturally, then it would appear that very few transferred genes confer advantage, and nearly all the advantageous (in nature) collections of genes already exist (Anderson, 1970).

The following simple analogy, for which I apologize in advance, makes the point. Some aircraft have almost as many parts as there are different structural genes in a higher plant or animal. If one learned how to take a few parts from one type of plane and put them into another, then one might (provided one was sufficiently ignorant of the details) propose that there was a danger that a few fighter parts could convert an airliner into a super fighter, better than either original craft. This, of course, is ludicrous because modern aircraft are designed from scratch and are carefully designed, highly-integrated systems of parts that fit and match. We believe that the same will be found to be true of living systems. In practice, cultivation and selection have always produced new varieties which were *less well* adapted to survival in the wild than were the native varieties. Otherwise we may well have overgrown the earth with some plant or other by now. Effective biological warfare – the limiting case against genetic engineering – turns out to be very difficult to do. Given tons of natural vectors spread over the earth's surface, mega-trillions of plants, and eons of time, nature can and does do experiments on a scale and of such a range of complexity as to relegate to total insignificance work done in all genetic engineering laboratories in the world. We are not trespassing on a pristine nature, rather we are under assault by a nature whose technical versatility and repertoire of tricks dwarfs our imagination. We do not believe that all the plant breeders or genetic engineers together could or will produce a better weed. These are not merely theoretical or philosophical arguments. Much of the future of agricultural research rests on the answers.

Progress in genome sequencing

New DNA and protein sequences are being recorded at a logarithmically increasing rate (Table 1 and Fig. 1). The strategy and tactics for mapping and sequencing the human genome are now subjects of intense discussion in the biomedical community. Sequencing the human genome at a cost of approximately $1 per base pair would cost over three billion dollars. However, with robotic sequencing systems now planned, costs could drop by one or two orders of magnitude, making it feasible to contemplate completion of plant and animal genomes. Given genomic libraries widely available on optical discs, and expert systems running on

Table 1. Genetic sequence data bank database summary GENBANK (R) Release 48.0 16 February 1987

Major groups of entries	Reports	Entries	Bases
1. PRIMATE Sequences	1,891	1,337	1,602,436
2. RODENT Sequences	2,307	1,612	1,460,441
3. Other MAMMAlian Sequences	361	310	325,323
4. Other VERTebrate Sequences	612	513	440,085
5. INVERTebrate Sequences	793	685	542,525
6. PLANT Sequences	828	689	819,740
7. ORGANELLE Sequences	498	423	840,194
8. BACTerial Sequences	1,321	916	1,268,642
9. Structural RNA Sequences	756	659	72,838
10. VIRAL Sequences	1,848	1,160	1,684,316
11. BacterioPHAGE Sequences	350	169	283,756
12. SYNTHETIC Sequences	287	251	78,943
13. UNANNOTATED Sequences	2,192	2,189	1,542,126
Summary:	13,774	10,913	10,961,365

Plant sequence data is approximately 0.75% of the total. 6.8% of plant sequences are for maize, and 4.5% for wheat. Less than 1/500 000th of the maize genome has been sequenced, while approximately 0.05% of the human genome has been sequenced. (Data kindly provided by Dr. Howard Bilofsky of Bolt, Beranek and Newman, Inc., Cambridge, Mass.).

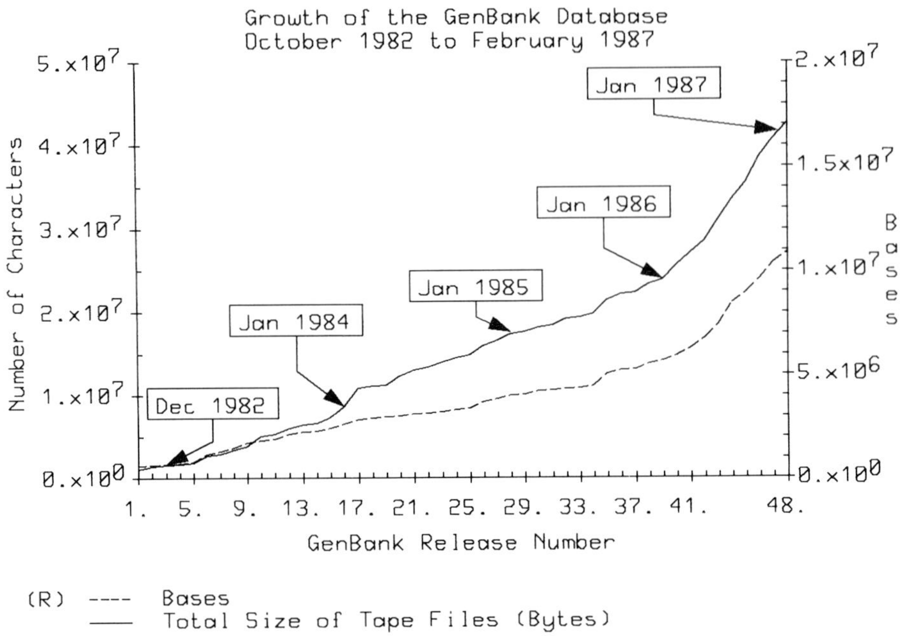

Figure 1. Rate of acquisition of DNA sequence data in GENBANK (Data courtesy Dr. Howard S. Bilofsky, Bolt, Beranek and Newman, Inc., Cambridge, Mass.).

currently available work stations, graduate students will be able to roam through and intercompare entire genomes to find interesting and useful differences, and to deduce functions of individual genes or effects produced by sets of genes. Courses in genomic architecture, on integration of gene systems, and on theoretical genomic biology will be commonplace. The instrumental requirements include the development of routine large-scale chromosome sorting, systems and methods for producing and separating large DNA fragments, robotic systems for routinely producing and cloning small fragments from larger ones, completely robotic and computerized systems for sequencing DNA and for integrating results back into complete chromosome structures, and computerized methods for performing model experiments and for keeping track of experimental results. All of these are in some stage of planning, development, or use now.

Information explosions

The scientific literature is now unmanageably large, and its sheer size severely inhibits interdisciplinary research. Radically new methods for evaluation, summarizing, transmitting, presenting and manipulating information will be required based on agreed-upon architectures and matching hardware. The key appears to be sophisticated work stations which provide real time staged access to an ever enlarging data base.

A DNA-based index to biology

It is technically feasible to develop complete data bases of plant or animal DNA with masses of useful data electronically linked to each coding region. On a computer-generated sequence map, one would merely point to a gene of interest, and the computer would provide menus listing the types of information available which relate to that gene, i.e., all known properties of the gene product including its name, function, known genetic variants, and keys to literature citations. The difficulty with DNA-based indexing of biological knowledge, however, is that long sequences are not easily remembered. This suggests that genes be named for their products, which are nearly all proteins. While at present many genes are "named" for a phenotypic effect or a genetic disease, this is most likely a temporary situation which will last until the relevant gene product (protein, or in a few instances an RNA) has been isolated, identified, characterized, and named. It appears that the index to any genome will ultimately be based on a list of proteins.

Protein-based indexing of biochemical knowledge

While it may ultimately be possible to deduce the three-dimensional structure and function of a protein from its primary sequence, which in turn may be deduced from the nucleotide sequence of the gene, this cannot be done now. The unifying concept is of a protein data base which is the index to a centralized DNA data base such as GENBANK.

Hence some way of separating and cataloguing proteins is required which also allows each protein to be linked to its DNA sequence. The protein data base should contain all available pertinent information on each protein organized in such a manner that one may retrieve specific information starting with a name, sequence, function, E.C. number, disease, or activity. The problem has been to evolve a protein analysis system capable of resolving thousands of proteins, and of developing image analysis and data base systems to match the complexity of living cells.

High-resolution two-dimensional electrophoresis resolves thousands of proteins (O'Farrell, 1975; Anderson and Anderson, 1978a and b, 1979, 1982, a and b, 1984) and provides the results in the form of maps containing large numbers of spots (Young, 1984). These maps may be scanned, processed, and displayed (Anderson, Taylor, Scandora et al., 1981; Anderson, Taylor, and Anderson, 1983) with both spacial (Taylor,

Anderson, and Anderson, 1983) and quantitative precision (Anderson, Nance, Tollaksen et al., 1985), and individual spots used as entry points into large data bases (Taylor, Anderson, Scandora et al., 1982). Linked data may include the location of the corresponding gene, the name of the protein (or subunit), the protein's function, structure, a description of genetically variant forms of the protein, the intracellular location, the cell types in which found, association with disease, and literature references. The development of a human protein index based on these concepts has been described in detail (Anderson and Anderson, 1982a). With the development of methods for partially sequencing proteins recovered from 2-D gels (Aebersold, Teplow, Hood et al., 1985), a systematic technique for linking large numbers of proteins to their respective genes by reverse translation has become available.

Gene product (protein) data bases

It is well to describe in some detail how image-linked data bases work, and how they lend themselves to research and analysis. The data bases are built around image spot files which include all of the proteins resolved from a particular class of sample. These files can be used to construct a "master" map such as the one we have prepared for corn (Fig. 2). Individual samples are analyzed with reference to this master pattern. One may ask whether a new sample contains any protein not previously seen, whether some proteins usually present are absent, and whether quantitative differences are detected relative to one standard reference pattern or to a reference set of patterns. The key point is that the results (differences) are displayed on a video screen either as changes in grey levels, pseudocolor differences, or by flickering. One sees directly spots related to the answers to experimental questions. When a cursor is placed on a spot identified as interesting, the computer provides its identifying number, and any additional information required from a separate menu (Anderson and Anderson, 1982a),.

Thus the proteins which differ between different parental strains of hybrid corn may appear on the screen in different and identifying colors. Or wheat endosperm proteins which correlate with superior baking quality may be similarly identified. One may then ask specific questions relative to either individual proteins found to be interesting, or to groups of proteins. For example the set of all mitochondrial proteins may be highlighted (from information in the data base), and compared with a set of proteins whose expression may be affected by a herbicide or plant hormone. One may ask for the intersection of these two sets to be displayed (i.e., all proteins which are both mitochondrial and are affected by the agent studied). Examples of such studies on lymphocytes have been previously published (Anderson and Anderson, 1982b). A fundamental objective is to arrange to manage complex data quantitatively and rapidly, to do statistical studies in real time, and to very rapidly relate the results to what is already known. Representative of agriculture-related studies which have been done include those on wheat endosperm proteins (Anderson, Tollaksen, Pascoe et al., 1985; Anderson and Anderson, 1987), wheat plant tissue (Zivy, de Vienne, and Hofmann, 1983), Douglas fir (Barhman, de Vienne, Thiellement et al., 1985), milk (Anderson, Powers, and Tollaksen, 1982) and muscle tissue (Giometti, Anderson, and Anderson, 1979).

We have concluded that high resolution computer-processed and CRT displayed color protein maps provide a unique solution to the problem of presenting and intercomparing complex data, as a point of entry into the rapidly growing data base of protein-related information, and as a link to DNA sequence information.

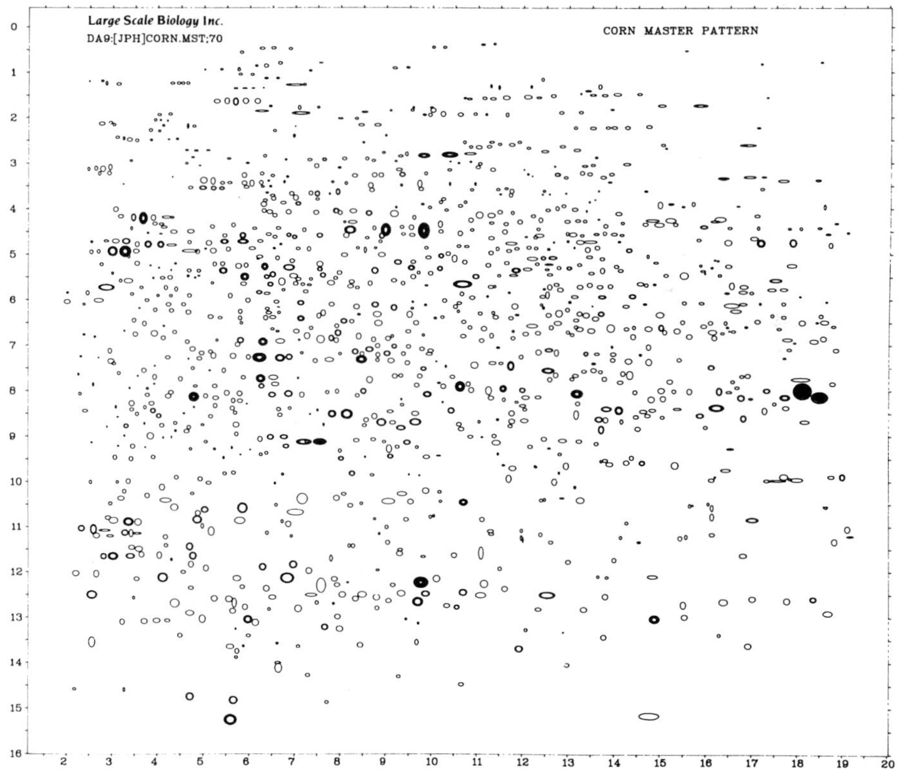

Figure 2. Master plot of corn seedling proteins. (Prepared by Dr. Jean-Paul Hofmann, LSB Corporation in collaboration with the Argonne National Laboratory).

The use of 2-D mapping in plant breeding is now under intensive investigation in several laboratories, and will become routinely feasible as the cost of the analysis drops due to automation and the use of robots.

Instruments for the 21st century

The basic instrumentation for resolving living cells includes systems for cell separation or cell sorting, for high-resolution subcellular fractionation, for separation of cellular metabolites by HPLC, for cloning, amplifying and sequencing DNA, for sequencing proteins, for peptide and oligonucleotide synthesis, and for resolving very complex protein mixtures. High resolution two-dimensional electrophoresis, as originally developed by O'Farrell (1975) and others, combined with sophisticated image analysis and data base systems such as the TYCHO system, allow the data provided by all of the instruments mentioned to be stored, organized, managed, and accessed. With the development of methods for partially sequencing proteins recovered from 2-D gels spots, efficient methods for linking proteins in 2-D maps with DNA data bases have become a reality. Since protein recovered from 2-D gels can also be used to produce antibodies, it is feasible to develop libraries of antisera against specific proteins to be used to isolate undenatured proteins, to identify associated subunits, and to cross-identify proteins between different laboratories (Anderson and Pearson, 1980).

All of these systems will doubtless be vastly improved before the end of this century.

The structure of research in the 21st century

While centralized instrumentation facilities are emerging in many universities (Hunkapiller, Kent, Caruthers et al., 1984) their reduplication and enlargement may not solve the massive data acquisition requirements we have discussed. Many scientists are now disturbed over the possibility that much of biological research will become "big science", with central organization and loss of individual initiative.

We propose that there will not be enough scientists in government, private, or university laboratories to acquire data on the scale required. Hence large centralized automated or robotic facilities for analytical data acquisition will be essential. They will perform an analytical function analogous to the synthetic and distribution function performed by large biochemistry supply houses. It is less than a generation since biochemists generally prepared most of their own reagents such as ATP.

Two trends follow from the views we have expressed. The first is enrichment of biological research by disciplines and technologies now somewhat foreign including robotics and sophisticated electronics.

The second is an increase in the importance of individual investigator-initiated research. Biology is now becoming so complex that many central problems cannot now be attacked by individual investigators in even well equipped laboratories because of the high cost and complexity of analytical instrumentation. With masses of centrally developed data available on optical discs, specific reagents available from large banks, and access to sophisticated analytical systems by overnight mail, the individual investigator at his computer work station *is* the man to explore arcane aspects of genomic architecture, to find genes for resistance to specific diseases, to pinpoint factors controlling yield, and ultimately to begin the design of new and unique genes. The challenge is to provide talented individuals who have questions with organized and useful data in real time, and with sophisticated tools that carry out instructions. He need not do manually tasks robots and computers do better.

It has been said that agricultural research in the United States has been too successful, and has worked itself out of a job. We propose that in the 21st Century the situation will be reversed, and agricultural leadership and productivity will be closely tied to very high research technology.

References

1. Aebersold, R.H., D.B. Teplow, L.E. Hood and S.B.H. Kent. 1985. *Electroblotting onto activated glass: High efficiency preparation of proteins from analytical SDS-polyacrylamide gels for direct sequence analysis.* J. Biol. Chem. 261: 4229–38
2. Anderson, N.G. 1970. *The evolutionary significance of virus infection.* Nature 227: 1346–7.
3. Anderson, N.G. and N.L. Anderson. 1978a. *Analytical techniques for cell fractions. XXI. Two-dimensional analysis of serum and tissue proteins: Multiple isoelectric focusing.* Anal. Biochem. 85: 331–340.
4. Anderson, N.G. and N.L. Anderson. 1979. *Molecular anatomy.* Behring Inst. Mitt. 63: 169–210.
5. Anderson, N.G. and N.L. Anderson. 1982a. *The human protein index.* Clin. Chem. 28: 739–748.
6. Anderson, N.G. and N.L. Anderson. 1982b. *The human protein index project and the molecular pathology data base.* Medical Laboratory 11: 75–94.
7. Anderson, N.G. and N.L. Anderson. 1984. *Introduction: Some perspectives on two-dimensional protein mapping.* Clin. Chem. 30: 1898–1905.
8. Anderson, N.G. and N.L. Anderson. 1985. *A program and plan for biotechnology.* (Editorial) Amer. Biotechn. Laboratory, Sept. issue.
9. Anderson, N.G. and N.L. Anderson. 1987. *The application of high resolution two-dimensional electrophoresis to the analysis of wheat, milk, and other agricultural products. In* New Directions in Electrophoretic Methods, Jorgenson, J.W., and M. Phillips, eds. American Chemical Society Symposium 335: 132–142.
10. Anderson, N.G., M.T. Powers and S.L. Tollaksen. 1982. *Proteins of human milk. I. Identification of major components.* Clin. Chem. 28: 1045–1055.

11. Anderson, N.G., S.L. Tollaksen, F. Pascoe and N.L. Anderson. 1985. *Two-dimensional electrophoretic analysis of wheat seed proteins.* Crop Sciences 25: 667–674.
12. Anderson, N.L., and N.G. Anderson. 1978. *Analytical techniques for cell fractions. XXII. Two-dimensional analysis of serum and tissue proteins: Multiple gradient slab gel electrophoresis.* Anal. Biochem. 85: 341–354.
13. Anderson, N.L., S.L. Nance, S.L. Tollaksen, F.A. Giere and N.G. Anderson. 1985. *Quantitative reproducibility of measurements from Coomassie-Blue-stained two-dimensional electrophoresis gels. Analysis of mouse liver protein patterns and a comparison of BALB/c and C57 strains.* Electrophoresis 6: 592–599.
14. Anderson, N.L., and T.W. Pearson. 1983. *Use of high-resolution two-dimensional electrophoresis for analysis of monoclonal antibodies and their specific antigens.* Methods in Enzymology 92: 196–200.
15. Anderson, N.L., J. Taylor and N.G. Anderson. 1983. *Image analysis and interactive data base manipulation applied to two-dimensional maps of human proteins.* Proc. National Computer Graphics Association, June 26–30, 1983, pp 69–76.
16. Anderson, N.L., J. Taylor, A.E. Scandora Jr., B.P. Coulter and N.G. Anderson. 1981. *The TYCHO system for computerized analysis of two-dimensional gel protein mapping data.* Clin. Chem. 27: 1807–1820.
17. Barhman, N., D. de Vienne, H. Thiellement and J.-P. Hoffman. 1985. *Two-dimensional gel electrophoresis of proteins for genetic studies in Douglas Fir (Pseudotsuga menziesii).* Theoret. Appl. Genet. 23: 247–255.
18. De Solla Price, D.J. 1986. *Little Science, Big Science . . . and Beyond.* Columbia University Press. N.Y. pp 30.
19. Giometti, C.S., N.G. Anderson and N.L. Anderson. 1979. *Muscle protein analysis. I. Development of high resolution two-dimensional electrophoresis of skeletal muscle proteins for analysis of microbiopsy samples.* Clin. Chem. 25: 1877–1884.
20. Hunkapiller, M., S. Kent, M. Caruthers, W. Dreyer, J. Firca, C. Giffin, S. Horvath, T. Hunkapiller, P. Tempst and L. Hood. 1984. *A microchemical facility for the analysis and synthesis of genes and proteins.* Nature 310: 105–111.
21. Lewin, R. 1986. *Proposal to sequence the human genome stirs debate.* Science 232: 1598–1600.
22. Martin, W.J. and R.W. Davies. 1986. *Automated DNA sequencing: Progress and prospects.* Bio/Technology 4: 890–895.
23. O'Farrell, P.H. 1975. *High resolution two-dimensional electrophoresis.* J. Biol. Chem. 250: 4007–4021.
24. Palca, J. 1986. *Department of Energy on the map.* Nature 321: 371.
25. Pearson, T. and N.L. Anderson. 1980. *Analytical techniques for cell fractions. XXVIII. Dissection of complex antigenic mixtures using monoclonal antibodies and two-dimensional gel electrophoresis.* Anal. Biochem. 101: 377–386.
26. Taylor, J., N.L. Anderson, A.E. Scandora Jr., K.E. Willard and N.G. Anderson. 1982. *Design and implementation of a prototype human protein index.* Clin. Chem. 28: 861–866.
27. Taylor, J., N.L. Anderson and N.G. Anderson. 1983. *Numerical measures of 2-D gel resolution and positional reproducibility.* Electrophoresis 4; 338–346.
28. Wada, H. and E. Soeda. 1987. *Strategies for building an automatic and high speed DNA sequencing system.* Proc. 4th Congr. Fed. Asina and Oceanian Biochemistr, in press.
29. Young, D.A. 1984. *Advantages of separations on "giant" two-dimensional gels for detection of physiologically relevant changes in the expression of protein-gene products.* Clin Chem. 30: 2104–2108.
30. Zivy, M., D. de Vienne and J.-P. Hofmann. 1983. *Studies on nuclear and cytoplasmic genome expression in wheat by two-dimensional gel electrophoresis. I. First results in 18 alloplasmic lines.* Theoret. Appl. Genet. 66: 1–7.

Current developments and future trends in electron microscopy

WILLIAM P. WERGIN
U.S. Department of Agriculture, ARS Plant Stress Laboratory, Plant Physiology Institute, Beltsville, MD. 20705, U.S.A.

Abstract This chapter will attempt to describe some of the diverse instrumentation that collectively comprises the field referred to as electron microscopy. The discussion will begin with a brief history of the development of instruments, trace the evolution of biological applications and finally speculate about a relatively new instrument, the scanning tunneling microscope (STM) which, in the near future, may allow the biologist to visualize routinely structure at the atomic level.

Development of instrumentation

The term "microscope" has been repeatedly modified during the past 25 years because of rapid developments in the field of electron microscopy (EM) (Table 1). Although the first EM was built in the 1930's, biologists continued to use the term "microscope" to refer to an optical instrument, namely the light microscope, in which image formation resulted from passing light through a specimen and a series of glass

Table 1. Diagram illustrating the development of terminology and instrumentation that biologists have come to use during the last 25 years. HVEM — High Voltage Electron Microscopy, CTEM — Conventional Transmission EM, STEM — Scanning Transmission EM, CSEM — Conventional Scanning EM, STE — Scanning Tunneling Microscopy

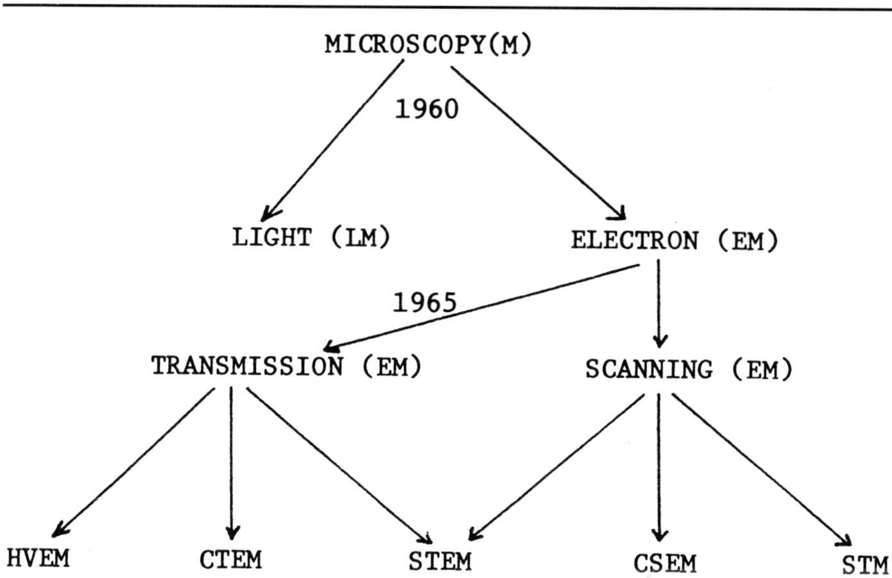

lenses. Not until the mid-1960s, which brought the development of the diamond knife, ultramicrotome, chemical fixation and embedding resins, did biologists express wide spread interest in this "new" instrument where image formation resulted from passing electrons through a specimen and a series of electromagnetic fields (lenses). At this time biologists had to qualify their use of the term "microscopy" by referring to the techniques as either light or electron microscopy. Use of the term "electron microscopy" referred to a simple transmission EM in which electrons were randomly scattered by heavy metal stains in a thin biological section or they were differentially transmitted through a shadowed replica.

A few years later, a second type of electron microscope became commercially available. In this instrument image formation consisted of collecting electrons from the surface of the specimen, i.e., secondary electrons that were excited as the primary electron beam systematically scanned across the tissue. This second type of EM necessitated additional distinction in terminology. As a result, reference to the technique, electron microscopy, was further qualified by using scanning EM or SEM to refer to the instrument in which image formation resulted from the collection and display of information that occurred when the primary electron beam excited surface or secondary electrons from the specimen and by using transmission EM or TEM to refer to the instrument in which image formation resulted from transmission of primary electrons through a specimen.

During the past 15 years, the field of electron microscopy has mushroomed. Numerous developments have resulted in a long list of acronyms that refer to the various analytical and spectroscopic techniques that fall under the general heading of EM (Table 2). The technique that was used initially to gain raw structural data about a biological sample can now be used to gain information about chemical composition, elemental and enzymatic localizations, three-dimensional reconstructions, as well as to visually enhance and satistically analyze images.

These diverse developments in EM have paralleled changes in the training and discipline of the biological microscopist. In the mid-1960's, the microscopist was typically trained as a taxonomist, virologist or morphologist who was attempting to extend LM observations by using the greater resolution available in the EM. Today the typical microscopist must not only be familiar with biological tissues but with computer science and biophysics as well in order to maximize the information retrieved from an EM study. In addition, the complexity of this field as well as limitations in personal time and mental faculties will only allow

Table 2. List of acronyms and their corresponding terms which represent some of the analytical and spectroscopic techniques that are included in the field of electron microscopy

EDX	ENERGY DISPERSIVE X-RAY ANALYSIS
WDX	WAVELENGTH DISPERSIVE X-RAY ANALYSIS
ETEAM	ELECTRON THERMOELASTIC ACOUSTIC MICROSCOPY
EPMA	ELECTRON PROBE MICROANALYSIS
SAM	SCANNING ACOUSTIC MICROSCOPY
SEAM	SCANNING ELECTRON ACOUSTIC MICROSCOPY
AES	AUGER ELECTRON SPECTROSCOPY
PIXE	PROTON-INDUCED X-RAY EMISSION
LAMMA	LASAR MICROPROBE MASS ANALYZER
SIMS	SECONDARY ION MASS SPECTROMETRY
XPS or ESCA	X-RAY PHOTOELECTRON SPECTROSCOPY
XAES	X-RAY INDUCED AES
SESD	SCANNING ELECTRON-STIMULATED DESORPTION
SLAM	SCANNING LASAR ACOUSTIC MICROSCOPY
BEIA	BACKSCATTERED ELECTRON IMAGE ANALYSIS
EELS	ELECTRON ENERGY LOSS SPECTROSCOPY

today's microscopist to specialize in one or two of the many techniques associated with EM.

The rapid and diverse development in the field of electron microscopy has resulted in the emergence of numerous societies and publications that concentrate on one or more areas of EM. Worldwide, more than a dozen societies pertaining to either TEM, SEM or analytical techniques exist; similarly, results are published not only in dozens of general biological journals but also in specialized journals that deal exclusively with technical or biological EM studies.

Finally, the cost of pursuing instrumentation has grown as dramatically as the field itself. In 1965, a "state of the art" TEM could be purchased for approximately $35 000. Today a full-performance conventional TEM with analytical accessories will cost 10 times that figure whereas a high voltage TEM can easily be in excess of $1 000 000.

Evolvement of EM applications

Transmission electron microscopy

As the instrumentation developed and the preparation techniques evolved the applications of EM became more refined. The initial phase of TEM was concerned with descriptions of tissues and their characteristic organelles. The cell could now be seen as a unit that was highly compartmentalized by an elaborate membrane system (Fig. 1). Not only could the organelles familiar to the light microscopist be readily observed, such as the nucleus, chloroplasts, mitochondria and vacuoles but the TEM allowed biologists to describe in detail the Golgi, microtubules, plasmodesmata and even the ribosomes. Each of these structural or-

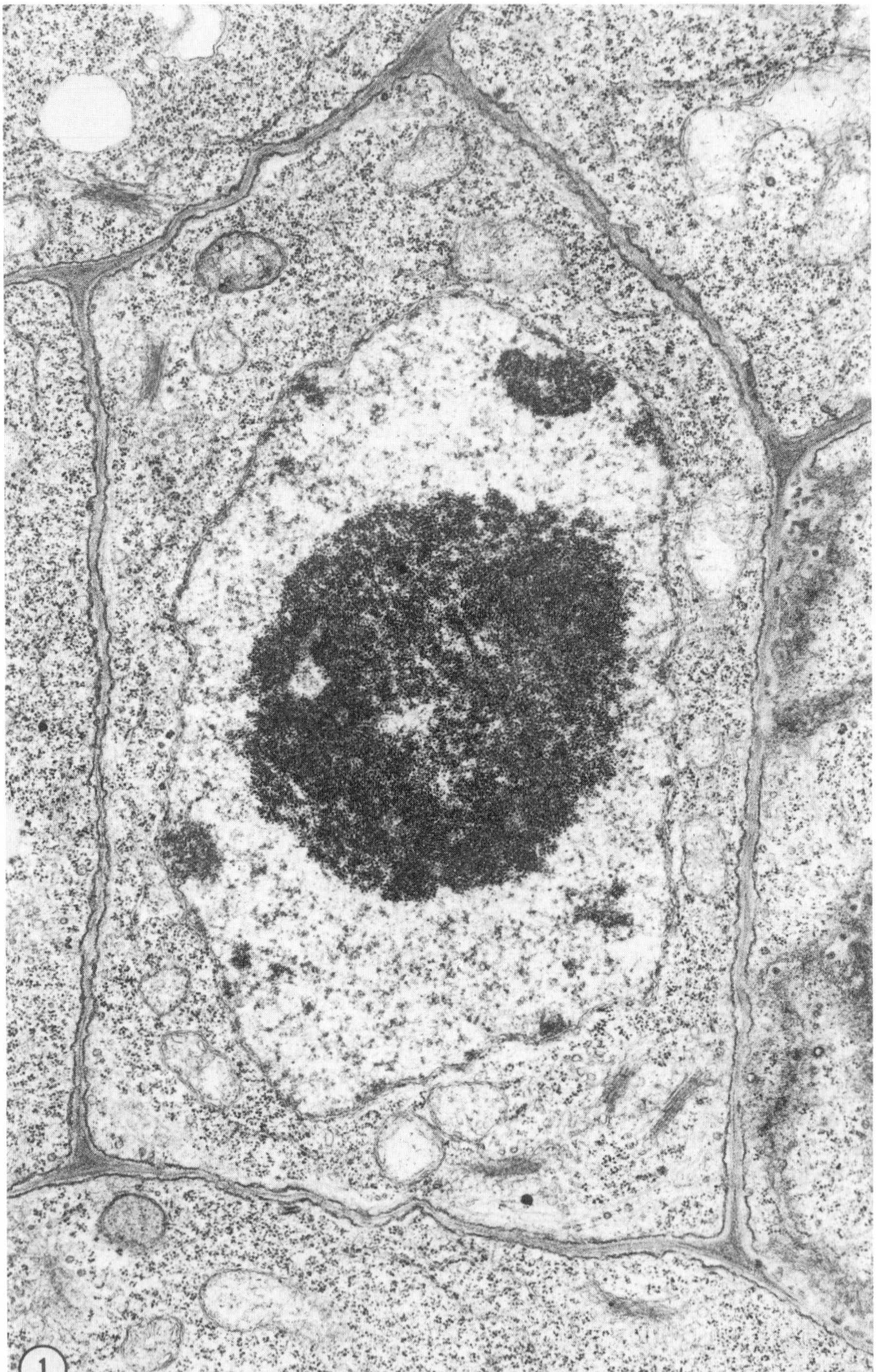
1

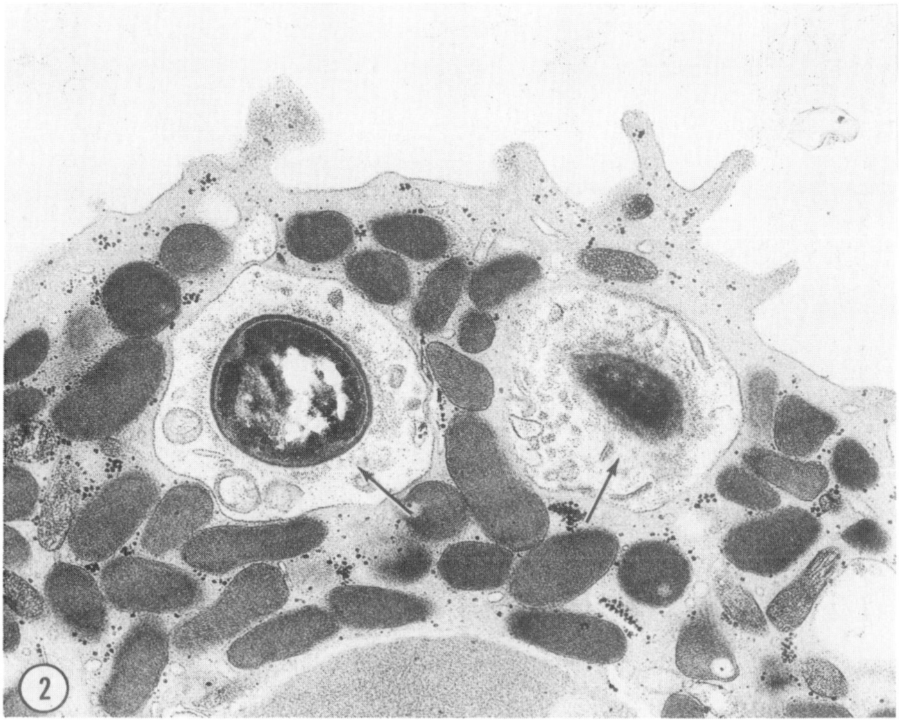

Figure 2. Portion of a polymorphonuclear leukocyte (PMN) isolated from bovine serum. The PMN has engulfed two bacterial cells (arrows) which are being enzymatically digested.

ganelles compartmentalized the cell into subunits with unique physiological functions.

Descriptions of tissues and their normal organelle complements laid a basis for evaluating how cellular ultrastructure was affected by adverse pathological, chemical or environmental conditions. For example, defensive cells called polymorphonuclear leukocytes, which circulate in blood of mammals, could be seen engulfing and digesting foreign bacteria (Fig. 2); or the primary structural effects of herbicides that inhibit photosynthesis could be seen affecting specific membrane components of the chloroplast (Fig. 3).

These types of observations brought two separate disciplines, namely cytology and physiology, closer together because structural observations could be related to organelles or membrane systems that were responsible for specific metabolic functions in the cell. The result of this close association between the two disciplines gave rise to cell biologists who,

---

Figure 1. Meristematic plant cell from bean root tip. TEM observations resolve cellular ultrastructure into a series of unique organelles which correspond to specific physiological functions.

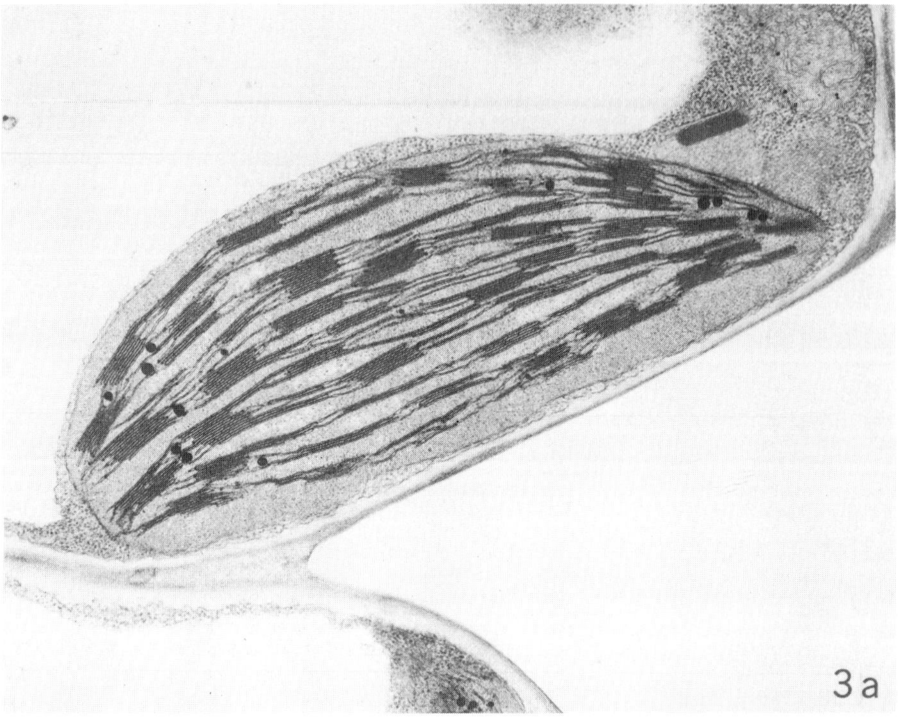
3a

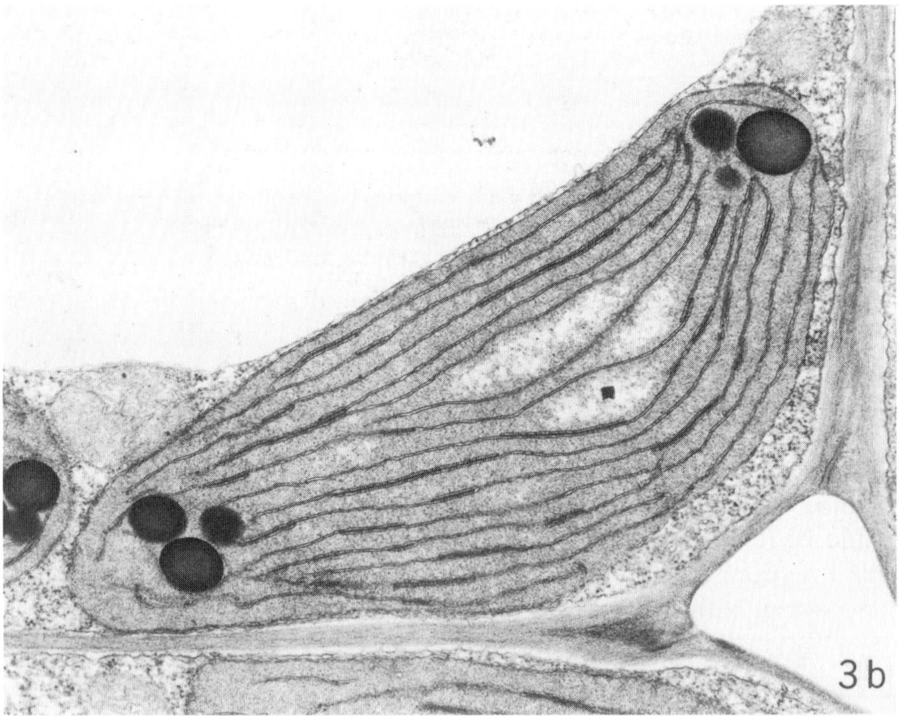
3b

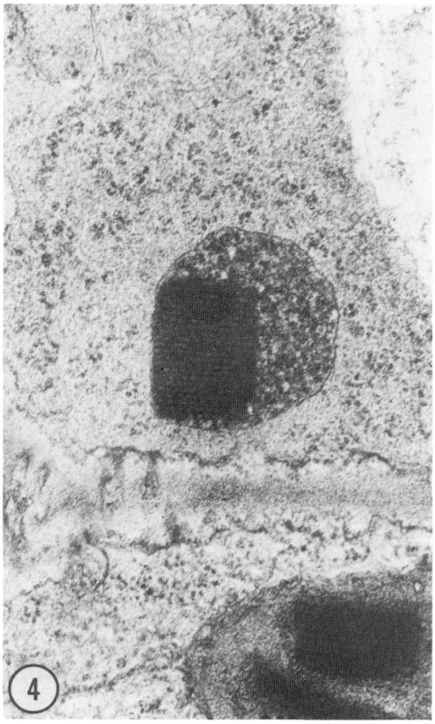

Figure 4. Portion of a leaf cell from tomato. The tissue has been selectively stained to illustrate the presence of catalase which appears as an electron dense deposit in an organelle known as a peroxisome.

in their attempts to study structure-function relationships, began to stain selectively specific enzymes or metabolic systems that could then be visualized and localized in the cell. Enzymes such as catalase could be observed restricted to a unique leaf organelle which corresponded to what the biochemist called a peroxisome (Fig. 4). Cells challenged by pathogens could be seen to produce peroxidase which initially appeared on vesicles of the ER, and was subsequently packaged by the Golgi, transported and deposited in the cell wall (Fig. 5) and PSI activity could be localized on specific membranes in the chloroplast (Fig. 6).

At the same time that thin sectioning techniques were being pushed to the enzyme level, freeze fracture and freeze etch development allowed individuals to observe membrane associated particles believed to be macromolecular proteins (Fig. 7) and to visualize nucleic acids (Fig. 8).

Figure 3. Normal chloroplast containing stacked grana lamellae which are interconnected by single stroma lamella (Fig. 3(a)). In the presence of sublethal concentrations of the herbicide fluometuron, normal grana stacking is inhibited (Fig. 3(b)).

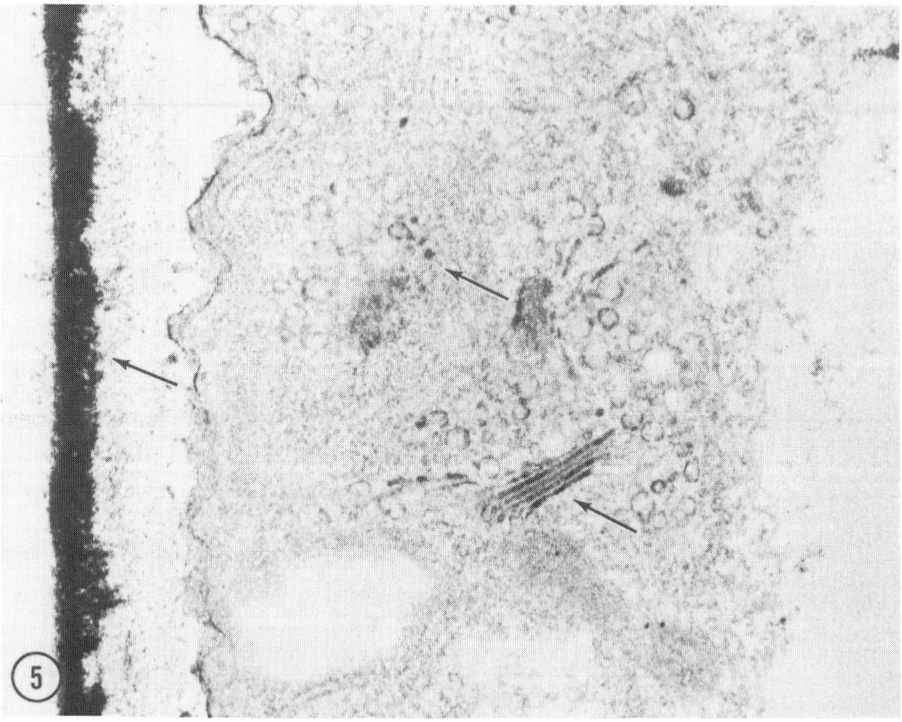

Figure 5. Portion of an epidermal cell from tomato root that was challenged by a fungal pathogen. The apparent response of the cell is the synthesis, packaging, transport and deposition of peroxidatic activity which appears as an electron dense product (arrows) in the ER vesicles, Golgi, plasma membrane and cell wall.

As a result, the TEMist today is equipped with the techniques to visualize not only cellular organelles but enzymatic activity, and macromolecules as well. In only thirty years, the biologist has achieved resolutions of less than 20A with the TEM. Future areas of development in TEM will undoubtedly come from image analysis and enhancement. These topics are briefly mentioned in the next section but detailed in a later chapter.

Scanning electron microscopy

The application of SEM has undergone a transition similar to that of TEM. The first commercial SEMs were introduced in 1965; within 10 years, chemical fixation, sputter coating and critical point drying, the preparation techniques widely used by biologists today, had been developed. The information gained from SEM observation found immediate application in the fields of taxonomy and morphology because of the resolution (50–100 A) and depth of field that was attainable with this instrument. These advantages were greatly enhanced by a specimen stage

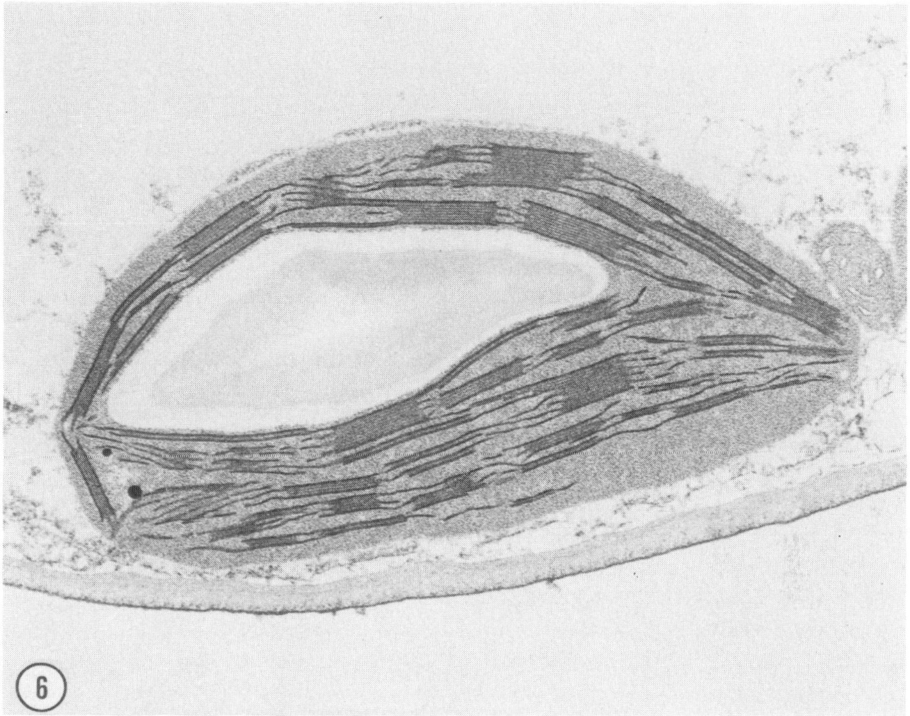

Figure 6. Chloroplast from duckweed. The tissue was selectively stained to localize PSI activity which appears as an electron dense product on the stroma lamellae and the outer lamellae of the grana.

which easily accepted tissue samples as large as 1 to 2 cm^3 and the diverse stage manipulations, which not only allowed specimen movement in the X and Y directions, but also allowed for Z movement, 360° rotation and tilt. The tilt and Z movements allowed microscopists to record two micrographs with parallax information that could be used to view specimens in true three-dimension. Three-dimensional analysis allowed investigators not only to clarify shapes and contours of objects (Fig. 9) but also to resolve spatial relationships (Fig. 10) and enhance resolution. In addition, systematic recording of three-dimensional micrographs allowed the investigator to perform stereometry, i.e., obtain accurate quantitative measurements from the three-dimensional images [1]. This marked a new beginning for the application of SEM; an instrument that initially was used to record structural details on high resolution micrographs could now be used to obtain quantitative data on specimens as well.

The quantitative aspects of the SEM can be greatly enhanced by

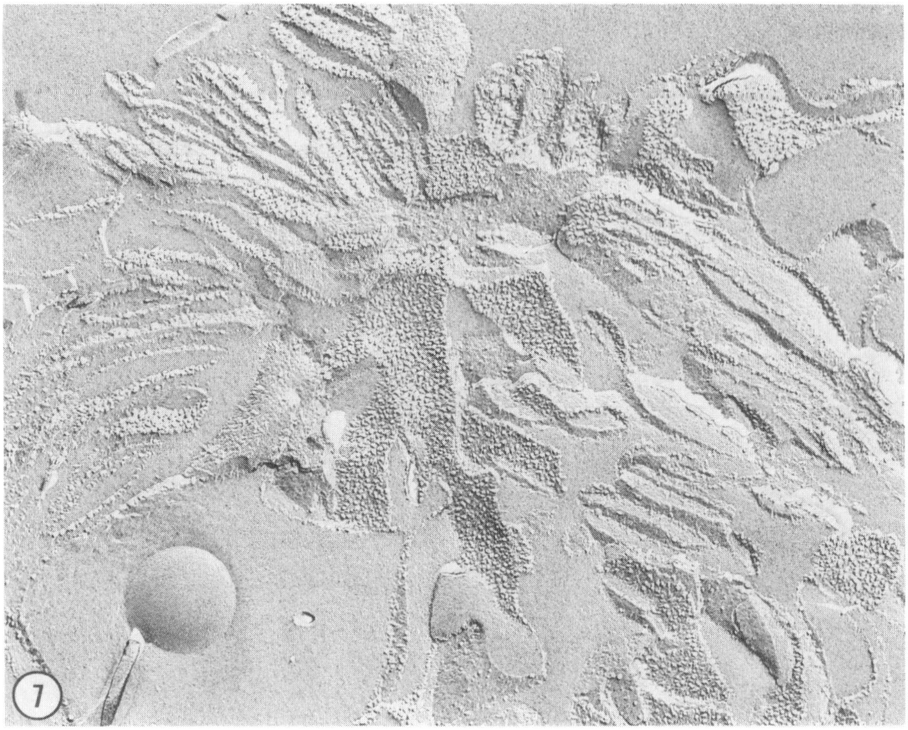

Figure 7. Freeze-fracture image of a portion of a chloroplast from Solanum. Membrane fractures reveal particles believed to represent protein macromolecules.

collecting and analyzing additional information that is generated in a microscope. When a primary electron beam interacts with a specimen, rearrangement of electrons in atomic orbits releases energy in the form of X-rays. The energy values of the X-rays are characteristic of the elements (and in fact the specific electron orbits) that are affected by this process. Therefore, if the X-rays are collected and analyzed with a multichannel analyzer, one can correlate the energy values of the X-rays with those that are unique for specific elements. This process produces qualitative and semi-quantitative data on the elemental composition of the sample. This type of system, which is referred to as energy dispersive X-ray analysis (EDX), is available for both SEMs and TEMs but appears to be more commonly associated with the former. For TEM, electron energy loss spectroscopy (EELS), a technique that produces similar information, is more adaptable [2].

Two aspects of EDX appear to have promise for the future; X-ray mapping and image analysis. Mapping allows one to localize the area(s) from which X-rays with a given energy value are being emitted. Since the

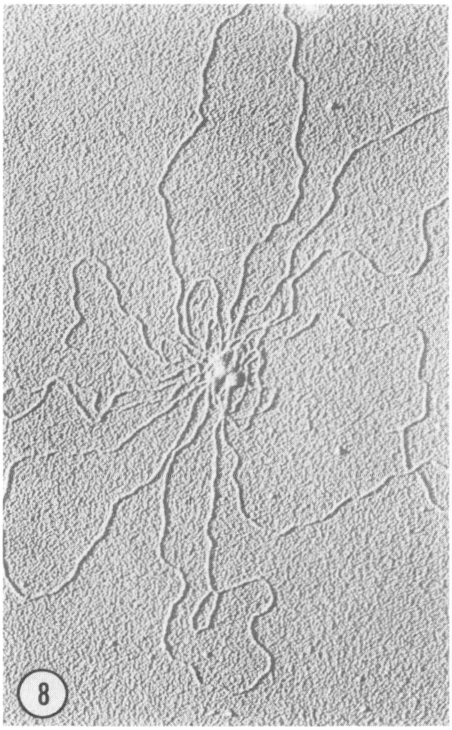

Figure 8. Shadowed preparation of bacterial double stranded DNA.

energy value is related to a specific element, the map can be compared to a normal SEM or secondary electron image to correlate the source of X-rays with structural detail in the micrograph. (Fig. 11).

Image enhancement and analysis are general techniques that have application in many areas of science. Because these subjects are covered in another chapter, they are only briefly mentioned here in reference to micrographs of images that are generated by EMs. In a micrograph, the resolution and contrast are limited by characteristics inherent in the specimen, the instrument and the photographic materials. Once a negative is obtained, any changes in resolution and contrast are dependent on the enlarger and the photographic papers that are used to process the final print. With the general availability of computers and specifically the minicomputers used to process EDX data, micrographs can be digitized, stored, processed, enhanced and displayed with improved image characteristics. In addition, the computer can be used to recognize unique shapes and gray levels in a micrograph. These features, which can be made to correspond to specific organelles or selectively stained enzymes, can then be used to generate statistical data. As a result, the future

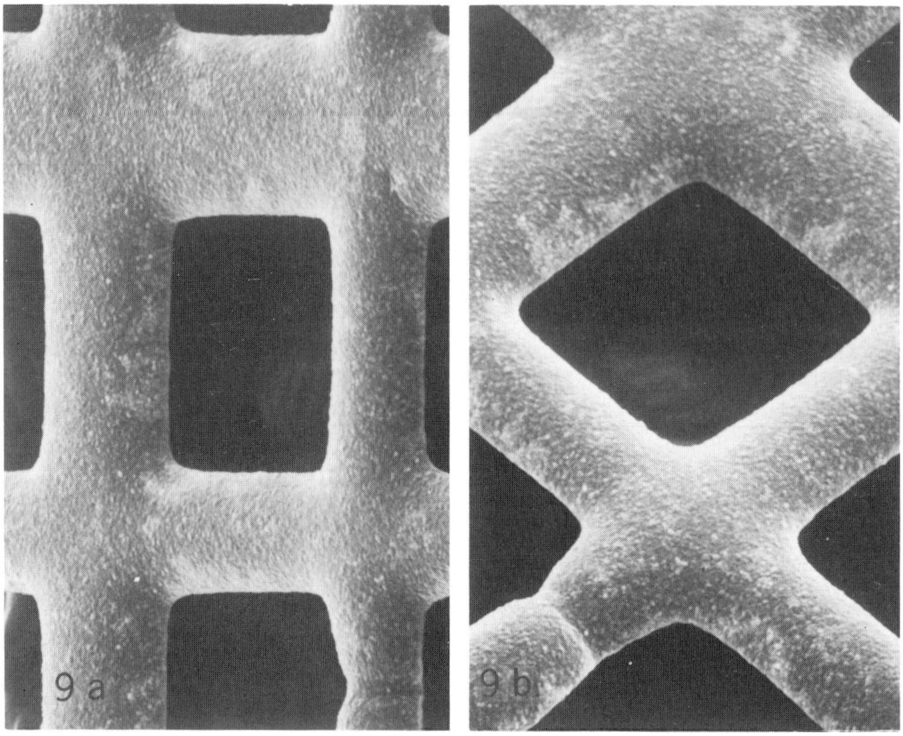

Figure 9. Copper mesh grid photographed with an SEM. Specimens may be tilted, rotated or moved along X, Y and Z axes in an SEM. Because these images are photographed on a flat two-dimensional film, the apparent shapes in the resulting micrograph can be misleading. The central opening in the copper mesh grid is identical in Figs (a) and (b); Fig (b) has merely been rotated 45° to produce the rhombic shape. Three-dimensional analysis is needed to determine the true shape, which is actually square.

analysis and interpretation of biological fine structure will become more quantitative and less subjective than it has been in the past.

Scanning tunneling microscopy (STM)

Because this symposium is concentrating on instrumentation of the future, we cannot overlook a new type of instrument, called the scanning tunneling microscope (STM) which is attracting considerable attention from scientists as well as the media. In fact, Drs. Gerd Binnig and Heinrich Rohrer, who perfected this instrument, were recently awarded the Nobel Prize in Physics for their achievement. The scanning tunneling microscope, which resolves surface topography at the atomic scale, has not been applied to agricultural problems; however its advantages and potential warrant a discussion in this chapter. Because the technique is

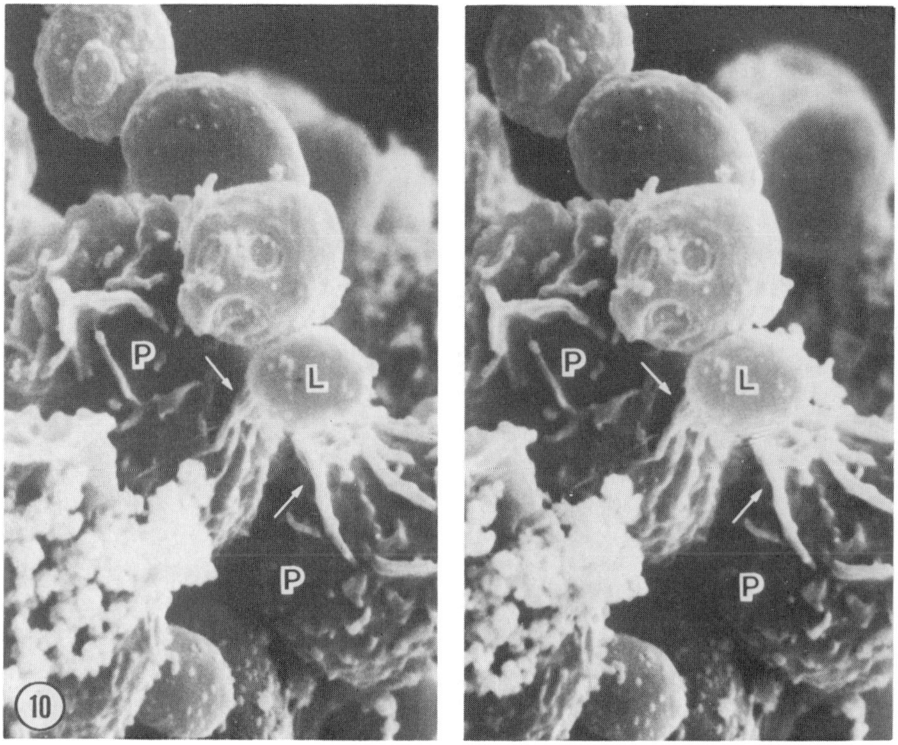

Figure 10. Example of how three-dimensional analysis resolves spatial relationships. In a single micrograph, pseudopods (arrows) from a polymorphonuclear leukocyte or PMN (P) appear to contact a yeast cell (L). If this were true, membrane contact would represent the initial stage of phagocytosis. In reality, when the stereo-pair is viewed, no contact has occurred between the PMN and the yeast cell.

only two years old and relatively new to most biologists, the discussion will briefly elucidate its advantages and describe the process of image formation. Finally, I will speculate about how the instrument may be applied to agricultural problems in the future.

Advantages

As we have seen, the SEM and TEM impose serious limitations on the biologists. First, both instruments operate under high vacuum; consequently, nearly all biological tissues must be chemically fixed and dehydrated. Second, good image formation with either instrument results from an interaction between the primary electrons that comprise the beam and high atomic number elements in the sample. Because biological tissue normally consists of relatively light elements, such as carbon, hydrogen, oxygen, etc., the specimen being examined must be either coated with gold or palladium for SEM observation or stained with

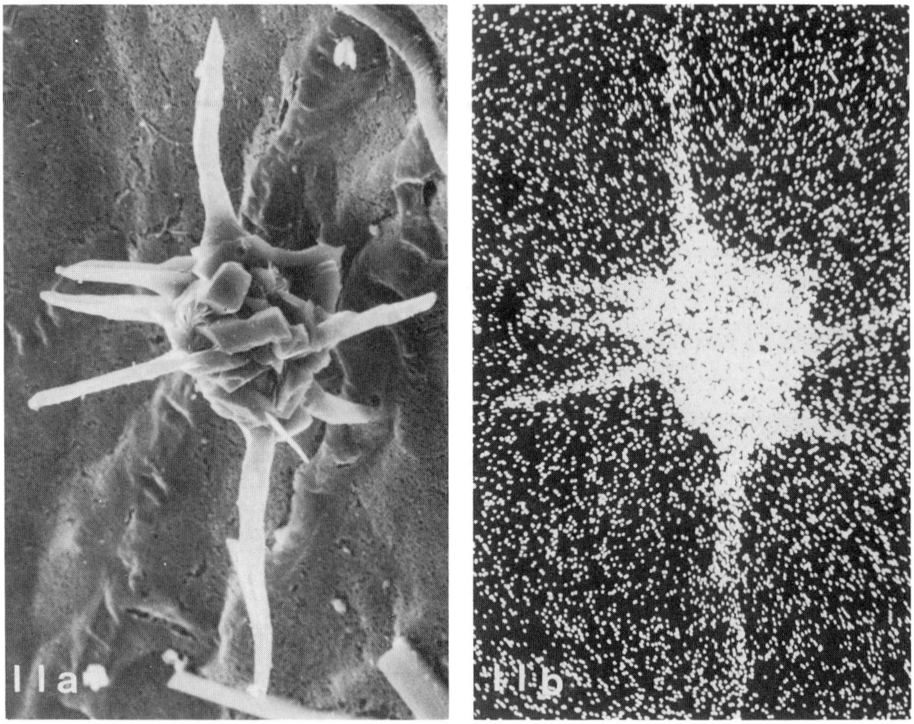

Figure 11. Secondary electron image (Fig. 11(a)) and X-ray map for calcium (Fig. 11(b)). Comparison of Figs. (a) and (b) shows that the calcium is restricted to a crystalline structure that is localized in a specific area of the fungal mycelium.

osmium, lead and uranyl salts for TEM imaging. Finally, biological specimens are normally subjected to extremely high radiation levels during examination in electron microscopes. For example, the radiation required to expose adequately a negative at 100 000 times, has been estimated to be 10^9 rads [3].

In the STM these serious disadvantages, which are common to SEM and TEM, may not occur. Even though STM is at its early stages of development, the instrument is currently capable of operating not only at atmospheric pressures but also at temperatures that extend from those of the ambient environments to that of liquid nitrogen. The information obtained with STM is not based on the size of atomic nuclei but rather on electron clouds; therefore, coating and staining techniques are not currently employed. Finally, this technique appears far less destructive; therefore, the biologist may be able to image living tissues. At this stage of development, even though the specific problems and needs of the biologist have been largely ignored by the physicists who are designing and perfecting the instrument, our most serious problems appear to be alleviated.

Image formation

The basis for STM, i.e., electron tunneling, is based on the principle of quantum mechanics established nearly 50 years ago by Fowler and Nordheim [4]. The idea that electrons could tunnel between two metal electrodes separated by a vacuum gap was introduced by Frenkel [5]. However, this theory was not experimentally verified until 1976 by Thompson and Hanrahan [6].

To appreciate how the instrument works we must first understand what the physicist means by tunneling. As we have heard earlier, a basic principle of quantum mechanics states that light and other forms of energy exhibit characteristics of waves as well as particles. In the case of electrons, waves or clouds representing electrons surround the surface nuclei of atoms. The probability of finding the electron beyond the surface of a conductor decreases exponentially with distance. However, if two conductors are placed sufficiently close, say within a few Angstroms of one another, the exponentially decaying electron clouds will overlap sufficiently so that some electrons have a probability of moving from one conductor to the other. This process is known as tunneling (Fig. 12). To enhance this tunneling effect a potential can be applied between two conductors; as a result the electron flow or a tunneling current can be measured. This current is exponentially dependent on the distance between the two conductors.

In practical use, the two conductors consist of a tip or probe and a conductive sample which are separated by a gap. The tip generally consists of a tungsten wire that is prepared by chemical etching, sputtering or annealing. Direct measurements of tip size on a scale relevant for tunneling are not yet available although it is assumed that the tunnel current depends so strongly on gap distance that effectively only the atom(s) closest to the sample contributes to the current.

Ideally, then the tip would terminate in a single atom which is believed to be the case when a scope is functioning optimally. However, atoms may be transferred from the sample to the tip or simply diffuse from the shank to the tip. Therefore, the nature of the tip is constantly undergoing changes. The size constrictions of the tip would indicate that it flows over the electron clouds of the sample rather than atomic nuclei. As a result the STM scans actually give the shape of the electronic wave functions near the Fermi level, roughly in the middle of the gap, rather than the positions of the atomic cores of elements in the surface of the specimen.

The insulating layer or gap between the tip and probe can consist of

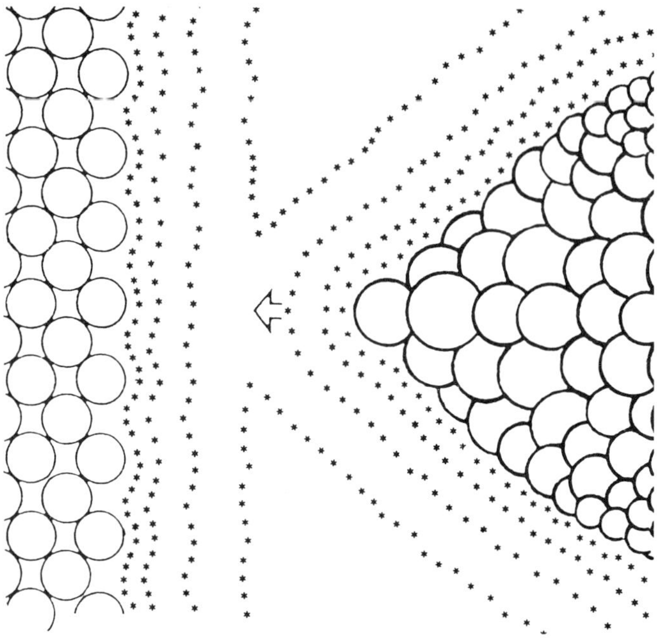

Figure 12. Diagram illustrating the tip or probe (right) and the sample (left). Electron clouds surrounding the atoms of the two conductors are represented by dotted lines while the gap between the overlapping clouds is shown by the arrow.

a liquid, gas or vacuum. To gain topographic information about the sample, the tip is brought to within a few A (5–10) of the sample whereby the electron clouds overlap. At this time, a voltage applied between the tip and sample causes electrons to flow (tunneling) which can be measured as a tunneling current. The density of the electron cloud falls exponentially with distance; for example, a change in distance equal to the diameter of a single atom, which is only a few A, may cause the tunneling current to change by a factor of 1000.

From this discussion, we can see that if the tip can be systematically moved or scanned across the specimen at a constant height by changing the gap voltage (from a few mV to a few V) and maintaining a constant tunneling current, a systematic plot of the gap voltage would reveal some topographic information about the surface atoms, i.e., the tip follows the atomic contour of the electron clouds. This can be accomplished by mounting the tip on piezoceramics whose motion can be monitored and processed by a computer and plotted or displayed on a cathode ray tube (CRT). By systematically sweeping the tip across the specimen a scanning image of the surface is obtained. This process, which has been successfully performed on surface topography of several samples, shows a vertical resolution of 0.1A and lateral resolution of 2 to 5 A.

As one can see, this type of movement is dependent on keeping a constant distance, usually 5 to 10 A, between the tip and the sample. In addition the tip must be stable with respect to any given point on the sample and must be capable of being systematically moved or scanned in X and Y directions. To accomplish these objectives, the sample must first be moved within 1 μm of the probe. No single mechanical or electrical system predominates to perform this manipulation. Binnig and Rohrer [7], who built the first successful STM, use a system referred to as a "louse" which consists of piezoelectric sticks or tubes that will elongate or contract when a voltage is applied. Other individuals are using frictional or magnetic clamping devices.

The precise movement of the tip is achieved by rectangular piezodrive sticks arranged in 3 normal directions (X, Y and Z). The Z piezo provides motion vertical to the surface whereas X and Y provide for lateral movements. Very simply this is accomplished by high voltage amplifiers whose output determines the expansion of the piezo arms (X, Y and Z) and thus the position of the tip with respect to the sample. The Z piezo is part of a feedback loop that senses tunnel current so that the constant gap width is maintained. The motion of the tip is monitored by a computer and either plotted or displayed on a CRT.

Results

Since the development of the STM was accomplished by physicists, the first samples examined were those of interest to the area of surface physics. For example, one of the initial observations, which involved silicon crystals, revealed bumps believed to correspond to single atoms. Twelve of these bumps were arranged in diamond shaped cells that measured 26 × 26 A units [8]. Subsequently, a group led by Quate at Stanford built a STM that would operate at cold temperatures. By comparing surface structure at room temperature with that at cold temperature, they were able to document growth of regions on surfaces believed to be associated with the phenomenon of superconductivity. Baro et al. [9] have confirmed data from scattering experiments that showed how oxygen was absorbed by nickel.

The initial biological applications have involved phage particles. Baro et al. [10] used STM at atmospheric pressure and room temperature during which they estimated the stability and resolution of their instrument to be 1 A vertical resolution and 5–10 A lateral resolution. This instrument was used to examine the oligomeric protein structure that forms the upper collar of a bacteriophage. After comparing their images to negatively stained TEM micrographs, they concluded that at this time the STM can be used to examine surface topography of air-dried untreated biological specimens at atmospheric pressure and room temperature without altering the structure.

Conclusions

In conclusion, one cannot be anything but optimistic when reviewing the early applications of the STM which have occurred during the past three years. Even though the design at the current time has essentially ignored the needs of the biologist, the instrument that has been developed by the physicist is capable of operating at atmospheric pressures and at temperatures which extend from ambient to those of liquid nitrogen. In addition, samples to be examined can be air dried or in aqueous saline solutions. Finally the technique appears to be less destructive as compared to TEM and SEM. Biological investigators who have worked in the field of EM realize that these features overcome the main problems that have existed for 30 years; namely that we work in a vacuum with material that must generally be chemically fixed, dehydrated, replicated, sectioned, stained and irradiated to obtain an image that hopefully resembles the living tissue. The potential biological applications afforded by STM has prompted Sonnenfeld and Hansma [11] to speculate that "further goals could include direct observation of biological processes such as DNA replication, enzymatic catalysis and membrane transport".

References

1. Wergin, W.P. 1984. *Importance of incorporating stereopsis and stereometry into a scanning electron microscopy course*. Scanning Electron Micros. III: 1225–1235.
2. Leapman, R. *Electron energy loss microspectroscopy and the characterization of solids*. Pages 217–233 in B.F. Kyser, M. Niedrig, D.E. Newbury and R. Schimizer, eds. in *Electron Beam Interactions with Solids*. 1984 SEM, Inc., AMF O'Hare (Chicago, IL).
3. Joy, R.T. 1973. *The electron microscopical observation of aqueous biological specimens*. Vol. V, pp 297–352 in R. Barer and V.E. Cosslett, eds., *Advances in Optical and Electron Microscopy*. Academic Press, New York.
4. Fowler, R.H. and L. Nordheim. 1928. *Electron emission in intense electric fields*. Proc. Roy. Soc. Lond. A119: 173.
5. Frenkel, J. 1930. *On the electrical resistance of contacts between solid conductors*. Phys. Rev. 36: 1604–1618.
6. Thompson, W. and S.F. Hanrahan. 1976. *Thermal drive apparatus for direct vacuum tunneling experiments*. Rev. Sci. Instr. 47: 1303–1304.
7. Binnig, G. and H. Rohrer. 1985. *The scanning tunneling microscope*. Scientific American 253: 50–56.
8. Binnig, G., H. Rohrer, C. Gerber, E. Weibel. 1983. *7 × 7 reconstruction on Si (111) resolved in real space*. Phys. Rev. Lett. 50: 120–123.
9. Baro, A.M., G. Binnig, H. Rohrer, Ch. Gerber, E. Stoll, A. Baratoff and F. Salvan. 1984.

Real-space observation of 2 × 1 structure of chemisorbed oxygen on Ni(110) by scanning tunneling microscopy. Phys. Rev. Lett. 52: 1304–1307.
10. Baro, A.M., R. Miranda, J. Alaman, N. Garcia, G. Binning, H. Rohrer, Ch. Gerber and J.L. Carrascosa. 1985. *Determination of surface topography of biological specimens at high resolution by scanning tunneling microscopy*. Nature 315: 253–254.
11. Sonnenfeld, R. and P.K. Hansma. 1986. *Atomic resolution in water*. Science 232: 211–213.

Electron probe x-ray microanalysis and digital elemental x-ray imaging and their application to biomedical histological sections and whole cells

GREGORY HOOK, Ph.D.
Surgical Neurology Branch, National Institute of Neurological and Communicative Disorders and Stroke, National Institutes of Health, Bethesda, MD 20892, U.S.A.

Abstract This paper will introduce electron probe x-ray microanalysis (EPMA) using both energy dispersive (EDS) and wavelength dispersive (WDS) spectrometry and digital EPMA elemental x-ray imaging to the biological researcher who has had little or no experience with these methods. The instrumentation, physics, quantitation and artefacts of these methods will be discussed, compared and contrasted. Biomedical applications of digital WDS-EPMA x-ray imaging of histological sections and quantitative EDS-EPMA of whole cells will be presented. Specifically, Pt distribution in abdominal wall of rats given the drug cisplatin, Ca and Al distributions in human pathological brain tissue, and Mg quantitation of individual peripheral blood mononuclear cells will demonstrate biomedical applications of these methods.

Introduction

Inorganic biomedical research is concerned with the essential, toxic and/or mutagenic effects of elements. Currently, there are 21 inorganic elements generally considered to be essential for human nutrition (K, P, Ca, S, Na, Cl, Mg, Fe, Cr, Co, Cu, F, I, Mn, Mo, Ni, Si, Se, Sn, V and Zn) and 4 elements are under consideration as possibly essential (Ti, Pb, As and Cd) [1]. At least 21 elements are known to be toxic and/or mutagenic (As, Ba, Be, Bi, Cd, Co, Cr, Cu, Fe, Hg, Li, Mn, Ni, Pb, Sb, Se, Sn, Te, Tl, V and Zn) [1] and other elements may also be toxic and/or mutagenic. Many elements are both essential and toxic or mutagenic with the biological effect of an element being determined by the elemental concentration in a dose dependent manner.

As is well known, human beings are composed of individual cells having a wide variety of structures and functions. The elemental concentrations and hence the biomedical effect can vary significantly among individual cells. The presence of a mutagenic element at high concentration in a few individual cells may cause the transformation of normal cells into cancer cells, the development of a tumor and the death of the individual. Thus, it may be necessary to determine the elemental concentration of individual cells in order to understand inorganic elemental effects.

Send correspondence to: Gregory Hook, Ph.D., Surgical Neurology Branch, NINCDS, National Institutes of Health, Bld. 10A. Rm. 3E68, Bethesda, MD 20892, U.S.A.

The cellular organization of tissue naturally divides inorganic analytical methods into two groups: microanalytical methods which analyze elements within individual cells and bulk analytical methods which determine the average elemental concentrations of many cells. Microanalytical methods include electron probe x-ray microanalysis (EPMA) using an energy dispersive x-ray spectrometer (EDS) or wavelength dispersive x-ray spectrometer (WDS), analytical electron microscopy (AEM) using an EDS spectrometer or electron energy loss spectrometer (EELS), energy filtering transmission electron microscopy (EFTEM), secondary ion mass spectrometry (SIMS), proton microprobe (PMPA), ion selective microelectrodes (ISM) and laser beam mass analysis (LAMMA). Bulk analytical methods include flame atomic absorption spectrometry (FAAS), inductively coupled plasma spectrometry (ICP), emission spectrometry (ES), nuclear magnetic resonance spectrometry (NMR), polaragraphic spectrometry (PS), electro-thermal atomic absorption spectrometry (ETAAS), x-ray fluorescence spectrometry (XRF), neutron activation analysis (NAA) and spark source mass spectrometry (SSMS). Several microanalytical methods such as EPMA, AEM, EELS, EFTEM, SIMS and PMPA can provide images of elemental distributions within cells as well as spectral information. Most bulk analytical methods detect analytes present at trace concentrations (1 ppm or less) whereas most microanalytical methods require higher analyte concentrations (1 ppm or more) [2]. For example, WDS and EDS-EPMA require analyte concentrations of approximately 10 and 100 ppm, respectively, for detection and thus only elements present in relatively high concentrations can be analyzed for within individual cells.

There are several organizations in the United States including the Microbeam Analysis Society (MAS), the Electron Microscopy Society of America (EMSA) and Scanning Electron Microscopy Inc. (SEM, Inc) which conduct scientific meetings devoted to microanalysis methodology and applications. This paper will discuss EPMA using WDS and EDS spectrometers and digital EPMA elemental x-ray imaging as these methods apply to the analysis of biomedical histological sections and whole cells. Pt images of the chemotherapeutic drug cisplatin in rat abdominal wall and Al and Ca images of human pathological brain tissue will demonstrate digital WDS-EPMA elemental imaging. Quantitative Mg analysis of individual, whole peripheral blood mononuclear cells will demonstrate quantitative EDS-EPMA spectral analysis.

EPMA instrumentation and physics

Electron probe and other electron column instruments

The electron probe and EDS and WDS spectrometers and their application to biological specimens are the subject of several excellent texts [3–5]. An electron probe is similar to a conventional scanning electron microscope (SEM) but is designed primarily for microanalysis rather than morphological imaging. As a result, a complete electron probe has WDS spectrometers, an EDS spectrometer, coaxial light/electron optics, and a very stable electron beam capable of high current. An SEM equipped with an EDS spectrometer also can do EDS microanalysis but not WDS-EPMA. Electron probes or WDS spectrometers are manufactured by Cameca (Stamford, CT), JEOL (Peabody, MA) and Microspec (Sunnyvale, CA). An electron probe equipped with one WDS and light optics costs approximately $350 000 and requires a service contract of approximately $15 000/year. The EDS systems are manufactured by many companies including Tracor Northern (Middleton, WI), KEVEX (Foster City, CA), LINK (Palo Alto, CA), EG&G ORTEC (Oak Ridge, TN), Princeton Gamma Tech (Princeton, NJ) and EDAX (Prarrie View, IL). An EDS system capable of spectral analysis and digital elemental imaging costs approximately $100 000.

The electron probe and SEM have electron optical systems which are very different from the transmission electron microscope (TEM), the most common electron microscope. At the heart of the electron microprobe is a finely focused, high energy (up to 50 kV) electron beam that systematically scans a microscopic field in a raster pattern in a vacuum environment. The interaction of the beam with the sample produces signals which are collected by a detector and produce an image on a display cathode ray tube. The electron beam striking the sample and the electron beam in the display cathode ray tube scan in synchrony and the detected signal modulates the display cathode ray tube intensity. The image is created in time and is analogous to a television image. In contrast, a conventional TEM, has a stationary, high energy (up to 100 or 200 kV in most microscopes) beam which is focused on an ultra-thin sample and a projected image is seen on a viewing screen below the sample. A TEM image is created in space and is analogous to a conventional transmission light microscope image. The basic electron optical systems of the electron probe or SEM and the TEM are diagramed in Fig. 1.

A TEM equipped with a scanning option (STEM) and an EDS spectrometer is called an analytical electron microscope (AEM). The AEM is capable of both morphological and elemental analysis. The AEM can

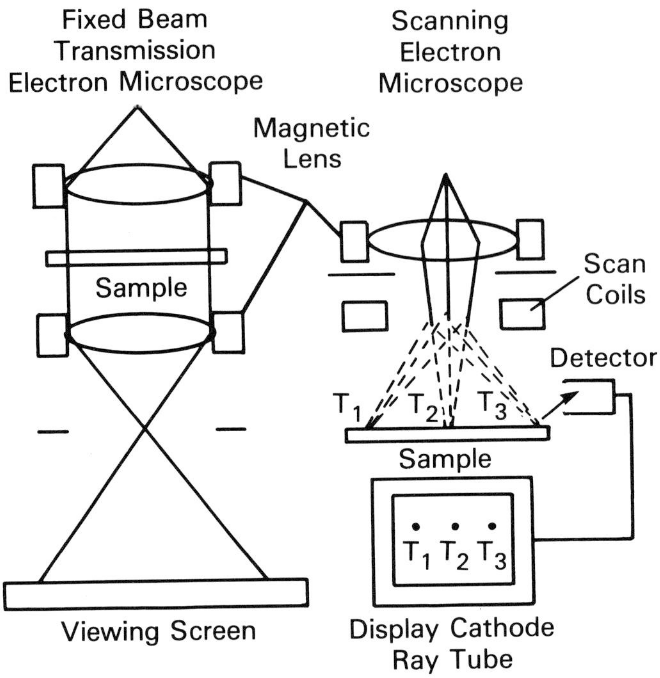

Figure 1. The electron optical systems of the scanning and transmission electron microscopes are compared. The electron probe has a finely focused beam that scans the sample and creates an image in time on a display cathode ray tube. The transmission electron microscope has a fixed position beam and projects an image on a viewing screen below the sample.

analyze for elements at high spatial resolution but requires ultra-thin (approximately 0.08 μm or less) specimens. The electron probe has a lower spatial resolution but can analyze for elements in much larger samples.

Electron beam-sample interactions

The interaction of the electron beam with the sample produces a variety of signals which can be collected by different detectors and provide morphological and elemental information (Fig. 2). Low energy secondary electrons (SE) are emitted from the surface of the sample and provide reflective, topographical information. High energy backscattered electrons (BE) make either topographical or average atomic number images. Transmitted electrons (STEM) through samples mounted on thin film supports make projection images. Cathodoluminescence (CL) is visible light which can provide chemical information. The x-ray signal is used in EPMA to identify, quantitate and map the elements within the sample.

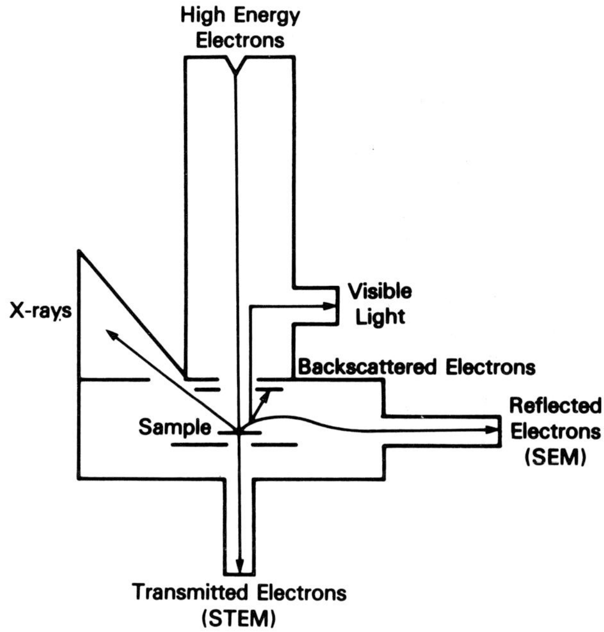

Figure 2. Many signals are produced by the electron beam-sample interaction which provide morphological and elemental information. The x-ray signal is used in EPMA to identify and quantitate the elements.

There are two types of x-rays produced by two different physical processes; characteristic energy x-rays which are specific for each element and continuum energy x-rays (also called background) which are not elemental specific (Fig. 3). Characteristic energy x-rays can be produced when a high energy beam electron ejects an inner shell orbital electron of an atom and an outer shell orbital electron falls into the vacant inner shell orbital. Due to the quantum nature of atomic orbitals, the orbitals have discrete energies for each element. The difference in energy between the two orbitals can be released as a high energy x-ray photon having a specific or characteristic energy. Since several transitions are possible for a given element, a group of x-rays having characteristic energies "fingerprints" an element.

Continuum x-rays, on the other hand, result from the coulombic interaction between the beam electron and the nucleus of an atom in the sample. The emitted x-ray can have any energy up to that of the incident energy but the probability of continuum x-ray production decreases with increasing x-ray energy. The continuum x-ray intensity increases with increasing atomic number of the sample because of the increased coulombic field in the core of heavy atoms compared to light atoms. Al-

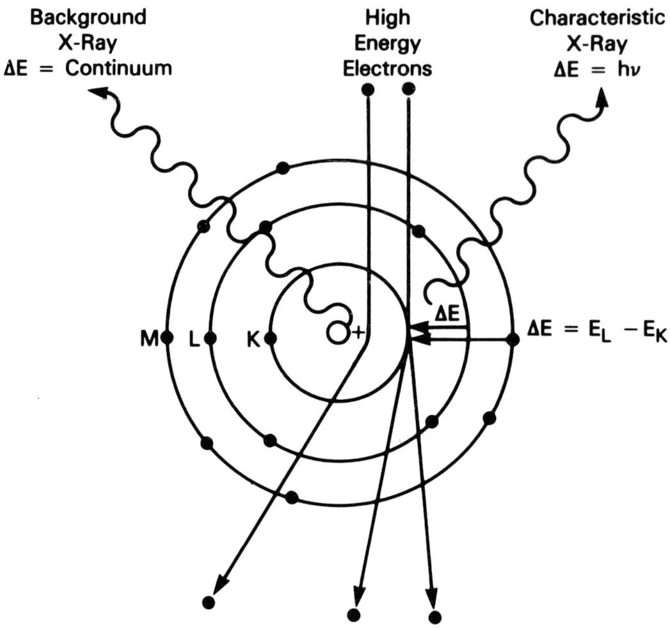

Figure 3. Two types of x-rays, characteristic and continuum energy x-rays, are produced by two different physical processes. Characteristic energy x-rays are elemental specific whereas continuum x-rays are not. Under appropriate conditions, the characteristic x-ray intensity is proportional to the elemental content and the continuum x-ray intensity is proportional to the mass of the sample.

though not elemental specific, continuum x-rays can, under appropriate conditions, be used to quantitate the mass of the sample in which the analyte is contained.

Since x-ray production is an atomic event, EPMA provides information on the total amount of element present in the sample without respect to the valence state of the element. Thus, free and bound elemental forms are detected. The bioavailability and activity of the free vs. bound element may be quite different.

An interaction volume is formed by the electrons penetrating and scattering in the sample and x-rays are generated through out much of this volume (Fig. 4). X-rays are detected from a sampling volume and the analysis is of a microvolume rather than the surface of the sample. The interaction volume increases with increasing accelerating voltage of the electron beam and decreasing atomic number of the sample. X-rays produced within the sampling volume can undergo absorption or fluorescence before leaving the sample but, for biological samples, x-ray absorption is the more significant problem. For histological sections, elements can be detected in cells that are within the thickness of the section but not observed on the surface of the section by SE topography. The spatial resolution of a x-ray image of a thick sample is always substantially lower than the corresponding topography image because the sampling volume from which x-rays are detected is larger than the sampling volume of SE.

WDS

The x-rays emitted from the electron beam-sample interaction must be detected, dispersed and analyzed to determine the elements present. WDS spectrometry is the classical, physical method of dispersing the x-ray spectrum based on crystal diffraction. The basic WDS-EPMA system is diagramed in Fig. 5. A x-ray is diffracted by an analyzing crystal if Bragg's law

$$n\lambda = 2d \sin\theta$$

is satisfied; where λ is the wavelength (energy) of the x-ray, d is the distance between atomic planes of the crystal and θ is the angle of incidence between the x-ray and the crystal. The diffracted x-ray is detected by a proportional detector. A x-ray is absorbed by an atom of

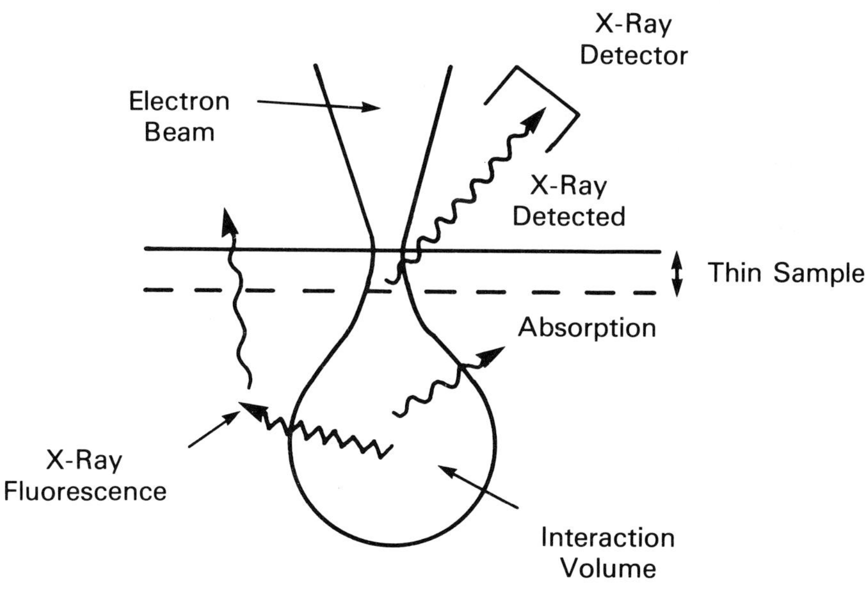

Figure 4. The electron beam penetrates the sample and x-rays are detected which are produced within a microvolume of the sample. X-rays can suffer absorption or fluorescence events before leaving the sample.

gas which ionizes many other atoms and gas electrons are attracted to a positively charged wire which creates an electrical pulse (Fig. 6). The positive potential on the charge wire is maintained within the proportional voltage region so that the amplified pulse is directly proportional to the x-ray energy. The pulse is passed to a single channel analyzer which has a discriminator to select pulses of a predetermined voltage and to produce rectangular waveform pulses of a fixed voltage. The selected pulses are passed to the scalar and counted. This method can determine all elements with an atomic number greater than or equal to 4 (Be). The WDS analyzes for x-ray energies in series and thus analyzes for one element at a time.

EDS

The EDS spectrometer system (with Be window) can detect all elements with an atomic number greater than or equal to 11 (sodium). The EDS has a solid state, liquid nitrogen cooled detector and multichannel analyzer which electronically disperses the x-ray spectrum and analyzes for all detectable elements in parallel. The fundamental components of a EDS-EPMA system are diagramed in Fig. 7. The x-ray signal from the sample goes through a thin Be window and into a lithium-drifted silicon detector, (the window maintains a vacuum environment around the crystal which stops atmospheric water from condensing on the crystal). Absorption of a x-ray by the crystal causes the ejection of a photoelectron which creates electron-hole pairs. The electron-hole pairs are swept

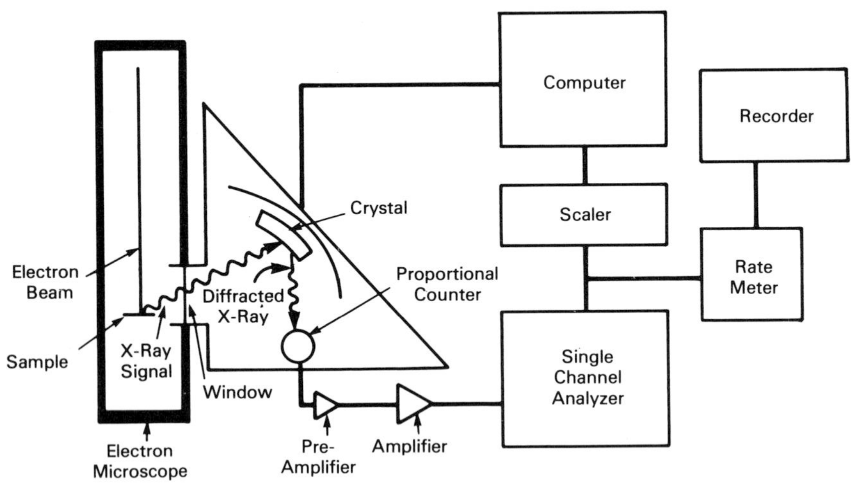

Figure 5. A WDS-EPMA system is diagramed (see text).

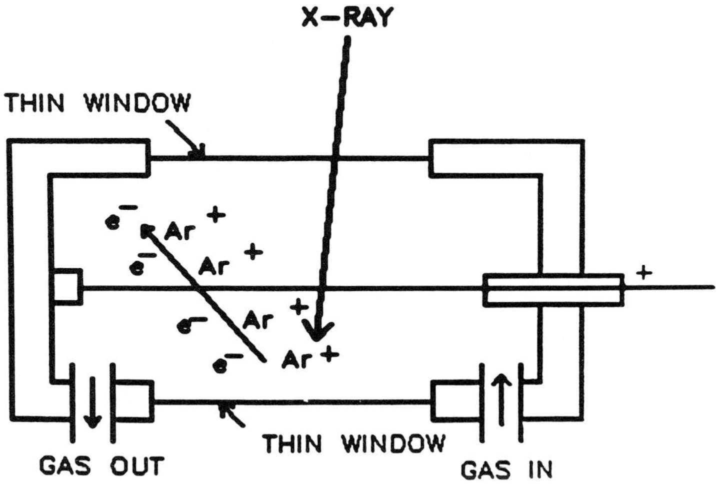

Figure 6. The proportional detector detects diffracted x-rays in the WDS-EPMA. The x-ray is absorbed by gas atoms which are ionized and electrons are attracted to a positive potential wire and an electrical pulse is formed proportional to the energy of the x-ray.

away by the bias on the detector and a voltage pulse is made by the charge-sensitive preamplifier. The number of electron-hole pairs and subsequent voltage pulse formed are proportional to the energy of the incident x-ray (Fig. 8). The signal is amplified by a main amplifier and sorted by voltage in a multichannel analyzer. The output of the multichannel analyzer is stored in computer memory and/or displayed on a cathode ray tube. The spectral information in the computer can be processed for peak identification or quantitation. To achieve the best energy resolution possible, the main amplifier needs enough time to process each pulse received. If the pulse rate is more rapid than the amplifier processing time, the pulses "pileup" and are placed in the wrong channel of the multichannel analyzer. This problem is eliminated by a pulse pileup rejection electronic circuit which eliminates counts that occur in too short a time span by temporarily shutting off the main amplifier. This processing results in "real time" and "live time" or "dead time" for the analysis. The actual time elapsed for the analysis is the real time whereas the live time is the real time minus the time the amplifier was turned off by the pulse pileup rejector. Often the dead time, the time the amplifier is turned off, is expressed and displayed as a percentage of the real time (e.g. 20% dead time).

EDS vs. WDS

There are several significant differences between the EDS and WDS

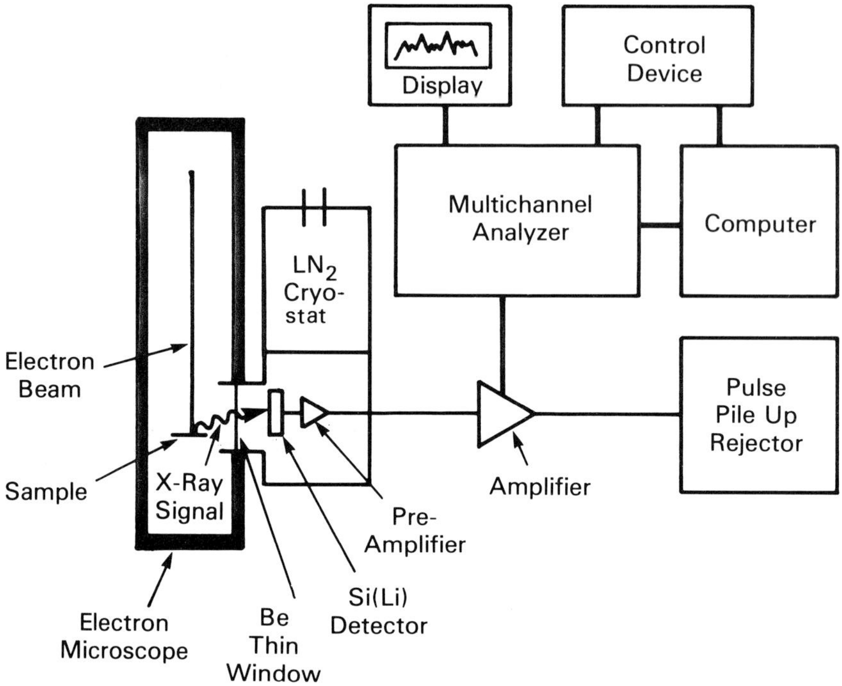

Figure 7. The EDS-EPMA system is diagramed (see text).

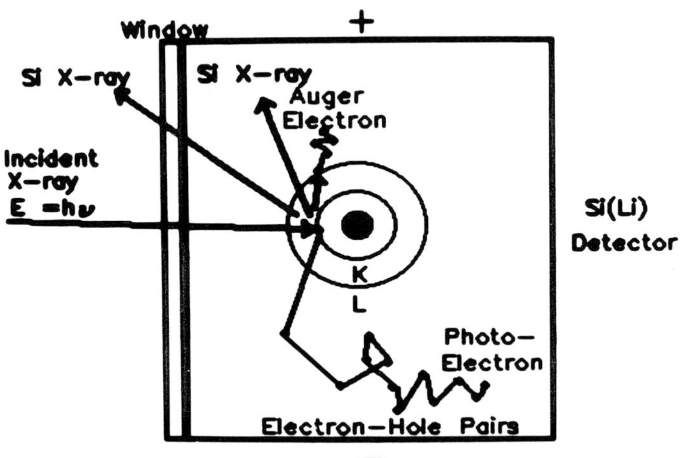

$$E = n\varepsilon$$

E = Energy of Incident X-ray

n = Number of Electron-Hole Pairs

ε = 3.8 eV for silicon

Figure 8. The EDS solid state detector detects x-rays on the EDS-EPMA. The incident x-ray creates electron-hole pairs in the crystal and the number of electron-hole pairs is proportional to the energy of the x-ray.

spectrometers which effect biomedical analyses some of which are summarized in Table 1. For example, the parallel multielemental analysis provided by the EDS spectrometer is a distinct advantage over the serial single elemental analysis of the WDS spectrometer for rapidly screening the elemental composition of the sample. It is important that the analyst be aware of the capabilities and limitations of each spectrometer inorder to choose the correct spectrometer for each research application. Further, the two types of spectrometers have different analytical artefacts which can effect the analysis.

Quantitation

The characteristic peak-to-continuum x-ray intensity ratio or Hall method is the most widely used and well established quantitation method for biological EDS-EPMA research. For thin specimens (4 μm or less) mounted on thin film supports, the characteristic x-ray intensity is proportional to the elemental content and the continuum x-ray intensity is proportional to the mass in the analytical volume (after correcting for x-ray signals originating in the sample support and microscope). The characteristic peak-to-continuum x-ray intensity ratio is thus proportional to the elemental concentration. The elemental concentration is determined by direct proportion between the characteristic peak-to-continuum x-ray intensity ratios obtained from the sample and a standard using the following relationship:

$$P_u/P_k \quad = \quad C_u/C_k$$

where P_u is the peak-to-continuum x-ray intensity ratio of the unknown

Table 1. Comparison of WDS and EDS spectrometers

Characteristic	WDS	EDS
X-ray dispersion	Physical	Electronic
Elements detected	$Z \geqslant 4$	$Z \geqslant 11$ (Be window) $Z \geqslant 6$ (windowless)
Elements detected per analysis	1	All
Energy resolution	5 eV	150 ev (5.9 keV)
Overall detection efficiency	0.05%	5%
Maximum count rate	50k counts/s (1 x-ray line)	2k counts/s (entire spectrum)
Artefacts	Geometrical focusing	Dead time Electron scatter Peak overlaps Window absorption Escape peaks

(sample). P_k is the peak-to-continuum x-ray intensity ratio of the known (standard), C_u is the weight fraction of the unknown (sample) and C_k is the weight fraction of the known (standard). The elemental concentration is usually in terms of dry weight because the samples are usually analyzed in a dry state but the wet weight can be determined if the sample is preserved in the frozen-hydrated state[6]. Biological EPMA generally has an analytical precision of approximately 10% although the precision depends on the counting statistics.

In the Hall method, the continuum x-ray intensity is measured at different x-ray energy from the characteristic x-ray energy free of characteristic x-ray peaks. For thick, irregularly shaped specimens, x-ray absorption by the sample can cause significant errors because x-ray absorption depends on the energy of the x-ray. The characteristic peak-to-local continuum x-ray intensity ratio method accounts for sample x-ray absorption because the characteristic and continuum x-rays have the same energy and hence suffer the same absorption fate. The characteristic peak-to-local continuum x-ray intensity ratio should be used to quantitate thick sections, whole cells or bulk biological samples.

A suitable biological standard should have a homogenous, known concentration of analyte and the same average atomic number as the unknown. The atomic number should be the same as the sample because the continuum x-ray intensity is proportional to atomic number. The standard should have a similar geometry and matrix as the unknown to account for absorption effects and radiation damage effects. Various standards have been used including frozen-hydrated cryosections of gelatin with elements, frozen-hydrated salt solutions, freeze-dried cryosections of gelatin with elements, and low atomic number glass beads or fibers.

Digital EPMA elemental x-ray imaging

Computer-control of electron microprobes and on-line image acquisition is a recent development which offers great advantages in biological research. A major advantage of EPMA elemental x-ray imaging is that elemental distributions can be observed within complex tissue. Different elements can be correlated or anticorrelated with each other or with specific cell types. Since the images are stored in computer memory, computer processing techniques can be applied to correct analytical artefacts and extract numerical information from the images. Computer-control allows automated data aquisition, greatly decreases the manual labor of data collection and increases analytical throughput.

In digital EPMA elemental x-ray imaging, the computer stops the focused electron beam at discrete points in a two-dimensional array on the specimen surface. Each point in the array is called a pixel and the array of numbers forms a digital image (Fig. 9). The array size typically ranges from 512 by 512 downwards and the greater the array size the better the spatial resolution of the image. A detector collects a signal emitted from the sample at each point and the number of counts collected is stored in computer memory (no photography is required). The image is displayed on a color monitor where the intensity of the display at each point is scaled to the number stored in memory. Different detectors can collect different signals and multiple images can be made simultaneously from one scan.

Computer-controlled electron microscope systems were initially only available at major research centers [7, 8] but commercial systems are now marketed (primarily by EDS and microscope manufacturers) and, as a result, digital EPMA elemental x-ray imaging will become more common in the near future. The physical size of the computer hardware can range from a small desk to a full room and the capability of the system can range from a dedicated system which controls one microscope to a general image processing facility which controls several microscopes and processes images simultaneously. A schematic diagram of the hardware for the system developed at the National Institutes of Health, Bethesda, MD. [7] is shown in Fig. 10. This system was used to make the elemental images described below. The microprobe (Cameca) is interfaced via a dedicated satellite computer (LSI 11/23) to a larger host computer (PDP 11/60). The acquired data is transferred to hard disk storage devices (2, 64 megabytes each) and loaded into an image display system (DeAnza) with four 512 × 512 × 8 bit planes (3 color, 1 black and white). For

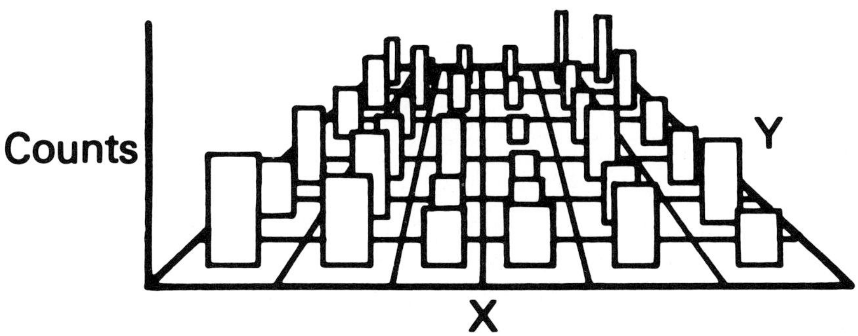

Figure 9. A digital image is a numerical array in which the number of counts at each point (pixel) in the image is stored in computer memory.

long term storage the images currently are stored on magnetic tape but will be stored in the future on optical (compact) disks. The software for the data aquisition, display and processing is organized using a branching tree logic system. To operate the system, the analyst does not need to know a computer language because he simply points to desired operations from a menu on the display monitor using an interactive mouse/pointer system. It is very important that the computer system be very "user friendly" so that analysts who are not computer specialists can be operators. The system supports basic image processing such as addition, division, smooth and contrast enhancement, etc.

Elemental EPMA x-ray images suffer from the same analytical artefacts associated with spectral analysis but, because people tend to interpret images more readily than numbers, elemental images are more likely to result in erroneous conclusions unless the analyst is aware of EPMA artefacts. The following is a discussion of some of the more important artefacts of EPMA elemental x-ray imaging and how to correct them.

A WDS-EPMA elemental x-ray image is made by tuning a WDS spectrometer to the characteristic x-ray peak and collecting the detected x-rays from all the pixels in the array. The EDS elemental x-ray images

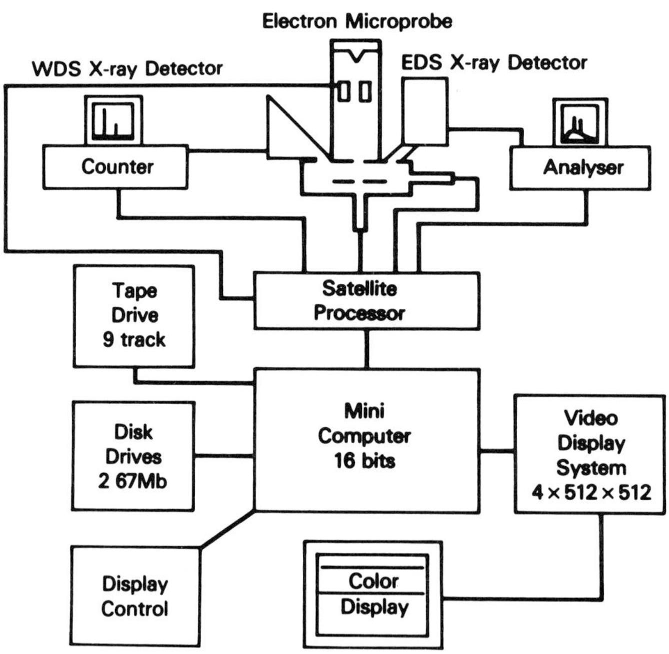

Figure 10. A computer-controlled EDS and WDS-EPMA system is diagramed (see text).

are made in a similar way by windowing one or several energy regions around characteristic x-ray peaks on the EDS spectrometer and collecting the x-rays in each window in separate images. The resulting WDS or EDS-EPMA digital elemental x-ray images are composites of the characteristic and the continuum x-ray signals. Since the continuum x-ray signal varies with the thickness, density and average atomic number of the sample, it is important to subtract the continuum x-ray component from the characteristic-plus-continuum x-ray signal. The continuum x-ray contribution to the image can be determined by making a continuum x-ray image of the same field by tuning the spectrometer to the continuum x-ray signal. A net elemental x-ray image can be made by subtracting the continuum x-ray image from the characteristic plus-continuum x-ray image at each pixel. The net elemental x-ray image is free of continuum artefacts caused by variations in thc thickness, density or average atomic number of the sample and is directly proportional to the elemental content of the sample [9]. To obtain an image proportional to the elemental concentration, a characteristic peak-to-continuum x-ray image must be made by dividing the net elemental x-ray image by the continuum x-ray image at each pixel.

An EDS-EPMA elemental x-ray image must be corrected for spectrometer dead time fluctuations. An increase in dead time causes a decrease in the number of x-rays that are processed and counted. If the dwell time at each pixel is of a fixed duration but the dead time among pixels varies, the resulting image will reflect the variation in dead time. This artefact can be compensated for by setting the dwell time at each pixel to a fixed live time on the EDS spectrometer rather than a fixed real time. This will result in the electron beam remaining at each pixel for a variable of amount of real time and a constant amount of spectrometer live time. A WDS elemental image does not have to be corrected for dead time because the WDS has a high maximum x-ray count rate which is usually not exceeded.

The low collection efficiency of the WDS spectrometer and low levels of elements usually present in tissue require that a high beam current (approximately 0.2 μA) and long image collection times (4 to 18 hours) be used to obtain sufficient x-ray counts to make an image. These conditions require a very stable electron beam current and specimen stage. The high beam current causes significant amounts of mass to be lost from the sample and experiments must be made to determine if the analyte of interest is volatile or translocates under high beam current. If the analyte is stable, the loss of other elements (C, N, O, H, etc.) in the sample causes the concentration of the analyte which facilitates detection. The EDS has a much higher x-ray collection efficiency and much lower beam currents (approximately 5–20 nA) and shorter image acquisition times are generally required (1–4 hours) although mass loss still occurs.

The electronic processing in the EDS spectrometer system causes elemental characteristic peaks to broaden and some characteristic peak overlaps occur (e.g. K and Ca). When elemental characteristic peaks overlap, the relative contributions of each element to the resulting peak must be deconvolved inorder to determine the individual elemental distributions. Deconvolution methods have been developed for spectral analysis which must also be used in EDS-EPMA elemental x-ray imaging when elemental overlaps occur. Using a WDS spectrometer, all characteristic x-ray peaks are resolved and no deconvolution is necessary.

A WDS-EPMA elemental x-ray image must be corrected for geometrical focus artefacts. To maximize the signal at the detector, WDS use a curved-crystal so that all x-rays originating from a point source will have the same incident angle to the crystal and will be brought to focus at the same point on the detector. There is a finite volume of the sample which is in focus to the spectrometer and the maximum sensitivity to spectrometer defocusing occurs along the z direction of the sample (i.e. the optical axis). The coaxial light microscope of the electron probe is prealigned so that the optical focus of the sample is the focus of the WDS spectrometer and the sample must be in optical focus inorder to obtain a WDS-EPMA x-ray image. Further, low magnification (less than 500 ×) WDS-EPMA x-ray images must be corrected for defocus artefacts in the x, y dimensions because the collected x-rays from different regions of the image make different angles with the diffracting crystal. To compensate for the x, y defocus, the crystal must be dynamically rocked during the image aquisition to maintain the same angle for all positions in the image. The EDS spectrometer, on the other hand, is a defocused system and does not have this problem but can have artefacts due to the collection of x-rays emanating from places other than at the site of the electron beam-sample interaction.

Digital EPMA x-ray images must be displayed on a monitor and the scale used to display the image can have important effects on interpretation of the results. Most monitors have 256 levels of intensity ranging from black to maximum brightness. Usually, the minimum and maximum values of the x-ray image are scaled to the black and maximum intensity, respectively. The minimum/maximum scaling method uses the maximum intensity range of the monitor and gives the best image detail when the image values fill the entire range. However, there are occasions when the minimum/maximum scaling method is not the best method. For example, if an image has a few pixels with extremely high counts, the minimum/maximum scaling method will cause the vast majority of the image to be compressed into low intensities and the resulting image will be dim and without detail. This situation can be corrected by ignoring the extreme counts and scaling the brightest intensity to a lower value. The maximum/minimum scaling method may not be the best method for visually comparing two images which have very different values. For example, if a net elemental x-ray image has values which are much greater than the corresponding continuum x-ray image, it may be better to scale both images to the minimum value of the continuum image and the maximum value of the elemental image inorder to directly compare the two images.

There are several image processing methods which can be applied to extract more information from digital x-ray images. Spatial averaging or smoothing can be very useful for x-ray images because x-ray images are often very noisy which may hide elemental information. Smoothing averages the values from several adjacent pixels in both dimensions for all pixels in the image. The number of adjacent pixels in each dimension is the kernel size of the smoothing routine. There are two common types of smoothing routines; a block smooth which uses the value of each pixel in the kernel for the averaging or a tent smooth which weights the value of each pixel in the kernel based on the distance from the central pixel. There are many other image processing methods which can be applied but all image processing must be done judiciously and with extreme caution because numerous image artifacts can be created.

Biomedical applications

The sample preparation is probably the most important aspect of biomedical microanalysis because it determines the biological information obtained from the microanalysis data. The sample preparation potentially can cause analyte extraction, analyte redistribution or elemental contamination prior to analysis. There are many different ways of preparing the sample for microanalysis [10] but this paper can only introduce the reader to a few general concepts for preparing histological sections or whole cells. In general, samples prepared for microanalysis have very poor morphology and it is essential that conventional light or electron microscopy morphological preparations are made in parallel in order to correlate the elemental distributions with structure.

Probably the optimum sample preparation method currently available for microanalysis of diffusable elements is to rapidly freeze the sample without chemical fixation and examine the sample in the frozen-hydrated state [6]. This method preserves endogenous elements, minimizes elemental redistribution within the sample and avoids elemental contamination. Further, frozen-hydrated sample preservation allows quantitating the water in the sample by measuring the continuum x-ray intensity before and after freeze-drying the sample. Low temperature analysis of samples (either frozen-hydrated or freeze dried) reduces sample mass loss due to radiation damage compared to room temperature analysis. This is particularly important for S analysis which moves during room temperature but not low temperature microanalysis. However, frozen-hydrated sample preservation requires specialized equipment and analysts must learn new cryoelectron microscopy techniques. Cryosample preparation equipment is now commercially available and easier to use and therefore frozen-hydrated sample preservation will become more common in the near future.

A technologically simpler sample preparation method is to freeze the sample without chemical fixation, cut histological cryosections on a cryostat, melt the sections on supports, carbon coat and analyze in the electron probe at room temperature. This method avoids elemental contamination of the sample and preserves endogenous elements. However, elements may redistribute within the sample during melting especially if the sample contains a large amount of water and very little cellular material (e.g. plant vacuole).

Conventional chemical fixation sample preparation methods used in morphological studies cause loss of diffusable endogenous elements (e.g. K) and elemental contamination of the sample prior to analysis. Osmium tetroxide, uranyl acetate, lead citrate and cacodylate contain heavy elements which cause characteristic peak overlaps using EDS and increased continuum x-ray production. For these reasons, heavy element fixation and stains must be avoided for most microanalysis studies. Samples fixed in gluteraldehyde or formalin in sodium phosphate buffer also lose endogenous elements and are contaminated with Na and P but, nonetheless, tightly bound endogenous elements other than Na and P can be analyzed for.

Several methods have been developed for whole cell microanalysis [11] but the fundamental problem with all methods is removing the extracellular medium while preserving the analyte in the cell. The extracellular medium must be removed for several reasons. A significant amount of the analyte may be in the extracellular medium and the analysis would be of extracellular and cellular analyte. The dried proteins and salts in the medium cover the cells so that they cannot be seen by SE or STEM imaging. The dried medium adds a significant amount of mass to the cell which introduces a large analytical error. Robust cells can be washed in distilled water to remove extracellular medium but

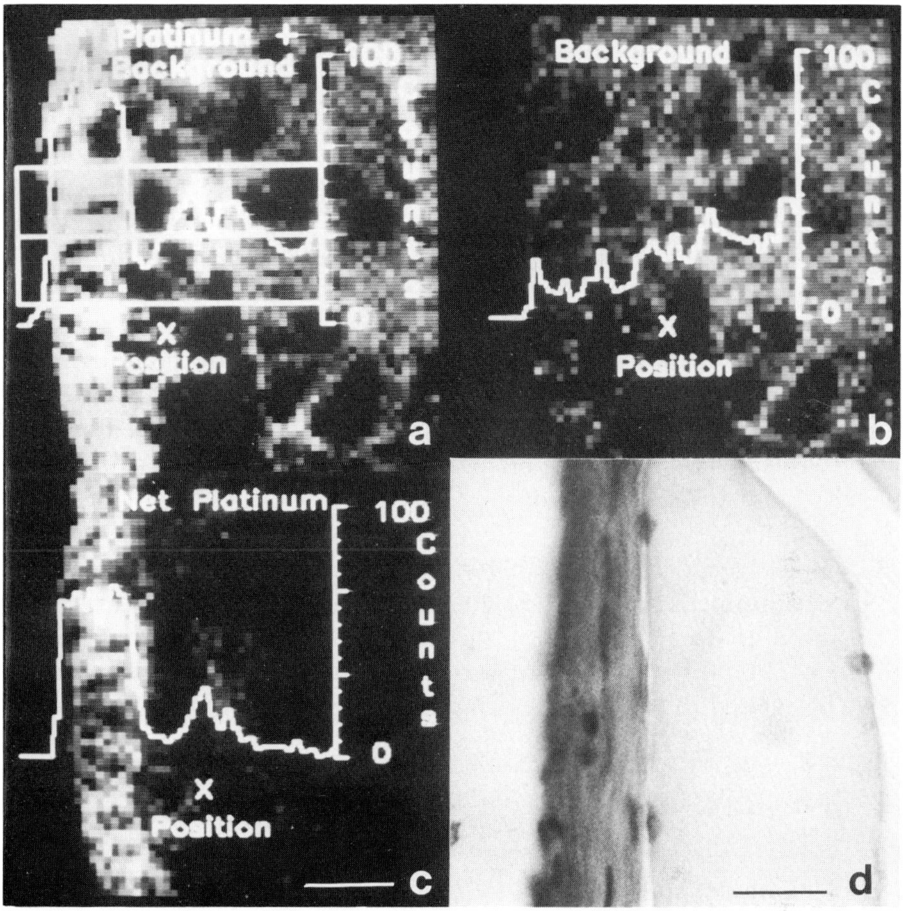

Figure 11. The Pt in the drug cisplatin is localized within a transverse histological section of rat abdominal wall using digital WDS-EPMA elemental x-ray imaging. (a) A Pt characteristic-plus-continuum x-ray image is shown. The Pt-plus-continuum x-ray counts are plotted as a function of position in the section. Position is the *x* coordinate of the rectangular region and x-ray counts are for all pixels from the bottom to the top of the rectangle for a given x-coordinate. (b) The continuum x-ray image of the same region as in Fig. 1 is shown. Continuum x-ray counts are plotted in the same rectangular region as in Fig. 1. (c) The net Pt x-ray image is obtained by subtracting the continuum x-ray image from the Pt-plus-continuum x-ray image at each pixel. The net Pt x-ray counts are plotted as a function of position in the rectangular region seen in Fig. 1 and shows a large Pt content in a band along the peritoneal surface of the section. Bar = 20 μm. (d) A staine light microscopy sister section shows the large Pt band is in the parietal peritoneum of the abdominal walla [12]. Bar = 20 μm.

many osmotically sensitive cells will lyse when exposed to distilled water. Alternatively, cells can be washed in isoosmotic, pH 7.4 solutions which form volatile salts upon drying (i.e. ammonium acetate or ammonium nitrate). Washing cells in any solution may affect the analysis because analytes may leak out of intact cells. Cells should be washed for the minimum amount of time and analytes which are bound within the cell are best for this type of analysis.

Samples can be mounted on either thin film supports or low atomic number solid supports (usually spectroscopically pure carbon stubs). Thin film supports minimize the amount of continuum x-rays produced and thus allow detection of analytes present in low concentration. Solid supports provide a very stable platform which is very important for making very long elemental WDS images. Samples must be coated (usually with carbon) to make the sample electrically conductive to reduce electrostatic charge build up on the sample and the distortion of SE images. In general, microanalysis samples should not be coated with metals as is used in SE morphological studies for the same reasons heavy metal stains must be avoided.

Histological localization of cisplatin in the rat abdominal wall

Cis-diamminedichloroplatinuum II (cisplatin) is a chemotheraputic drug used to treat cancers of the testicals, ovaries, head, neck and bladder. Digital Pt WDS-EPMA x-ray images of cisplatin in the tissue have been used to localize the drug in experimental rats given cisplatin

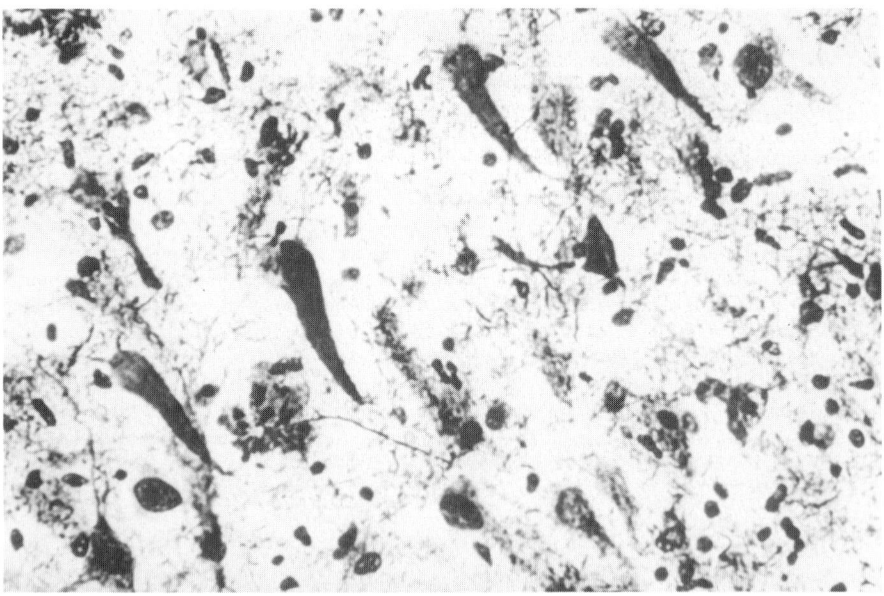

Figure 12. A stained light microscopy section of the hippocampus of a patient having parkinsonism-dementia of Guam shows pathological neurofibrillary-tangle bearing pyramidal neurons [13].

intraperitoneally [12]. Supratheraputic doses of cisplatin were administered inorder to detect the Pt. The tissue was frozen without chemical fixation, cryosectioned and melted onto warm supports. Correlative sister sections for conventional light microscopy were required to identify the Pt in the perietal peritoneum (Fig. 11). Digital WDS-EPMA elemental x-ray images can localize drugs containing heavy metals in histological sections.

Histological localization of Ca and Al in pathological brain tissue

Abnormal Ca and Al accumulations have been associated with parkinsonism-dementia of Guam. A specific pathological cell type, neurofibrillary-tangle bearing pyramidal neurons, in the hippocampus are seen by conventional light microscopy from patients with parkinsonism-dementia of Guam (Fig. 12). Digital Ca and Al WDS-EPMA elemental x-ray images (Figs. 13 and 14) have localized these elements in neurofibrillary-tangle bearing pyramidal neurons from patients with this condition [13]. The tissue was fixed in formalin in sodium phosphate buffer, frozen, cryosectioned and melted onto warm carbon supports. The Ca

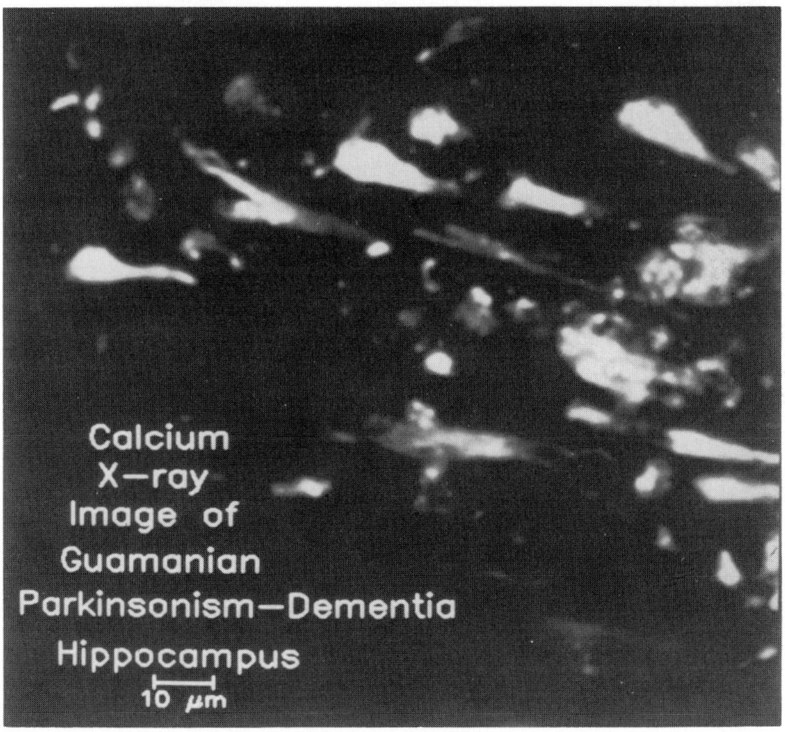

Figure 13. A digital WDS-EPMA net Ca x-ray image of a sister section of Fig. 12 shows large accumulations in the neurofibrillary-tangle bearing pyramidal neurons [13].

and Al were tightly bound in the cells. Correlative sister sections were used to identify the presence of neurofibrillary-tangle bearing neurons. Digital WDS-EPMA elemental x-ray imaging can be used to localize abnormal accumulations of elements in various other neuropathological conditions [14].

Quantitative Mg analysis of human peripheral blood mononuclear cells

Total body Mg status is inversely correlated with cardiovascular disease, hypertension, tetany and seizures. Mg status has been difficult to determine clinically because plasma Mg is a poor indicator but recently, Mg in peripheral mononuclear blood cells (MBC) has been shown to be a good indicator of Mg status. MBC Mg has been studied by quantitative EDS-EPMA and FAAS in normal adult volunteers [15]. The MBC were isolated from whole blood by density sedimentation, washed in ammonium nitrate, spray deposited on thin film supports, air dried

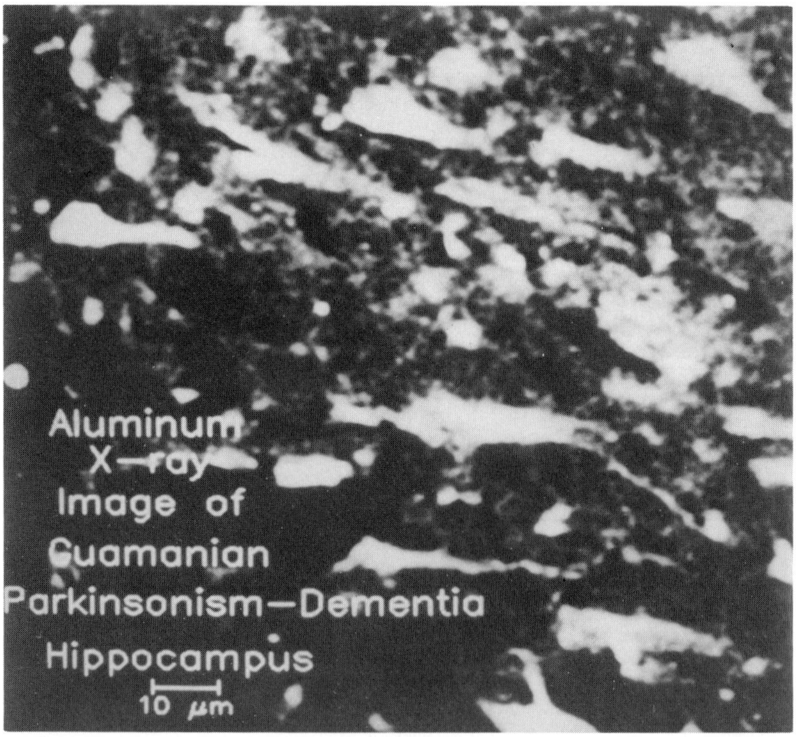

Figure 14. A digital WDS-EPMA net Al x-ray image of the same region seen in Fig. 13. The Al accumulation is in the neurofibrillary-tangle bearing neurons and correlates with the Ca distribution [13]. Both the Ca and Al are tightly bound in the cells because diffusable endogenous elements were lost when the tissue was fixed in Formalin prior to analysis.

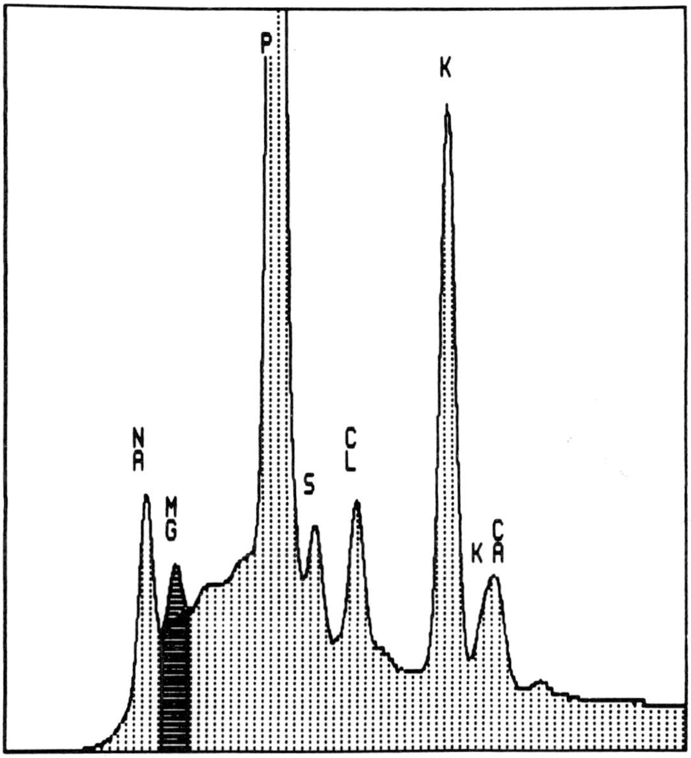

Figure 15. An EDS-EPMA spectrum obtained from analysis of an individual whole mononuclear blood cell mounted on a thin film support. The spectrum demonstrates the parallel multielemental information that can be obtained from one analysis by the EDS spectrometer system. The hatched region shows the Mg peak and local continuum x-rays below the peak. The Mg peak-to-local continuum x-ray intensity ratio is used to quantitate the Mg concentration (see text) [15].

Table 2. Human peripheral blood mononuclear cell Mg concentrations determined by EDS-EPMA and FAAS (mmol/kg dry weight)

Donor	EDS-EPMA	FAAS
RE	51.3 +/− 8.2	36.7
GH	46.1 +/− 16.5	34.2
JR	50.2 +/− 16.5	63.6
MR	68.9 +/− 19.9	63.2
JL	65.3 +/− 8.9	49.3
SS	64.0 +/− 20.6	68.4
DB	74.0 +/− 12.2	54.0
DG	95.0 +/− 34.7	32.3
PB	56.3 +/− 13.8	58.7
RS	62.0 +/− 13.7	63.0
Mean	64.0	52.2
Standard Deviation	14.3	13.8

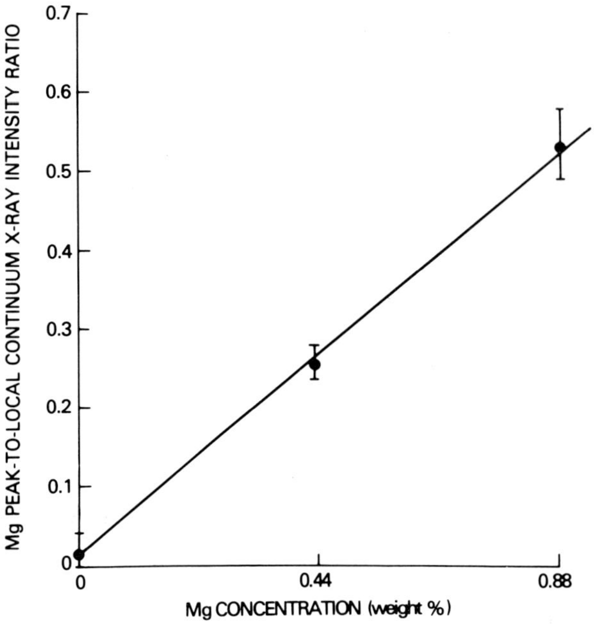

Figure 16. A standard curve is shown of the Mg concentration vs. the Mg peak-to-local continuum x-ray intensity ratio for low atomic number glass microsphere standards having known Mg concentrations. The plot shows EDS-EPMA particle analysis can be quantitative and has an analytical precision of 10%.

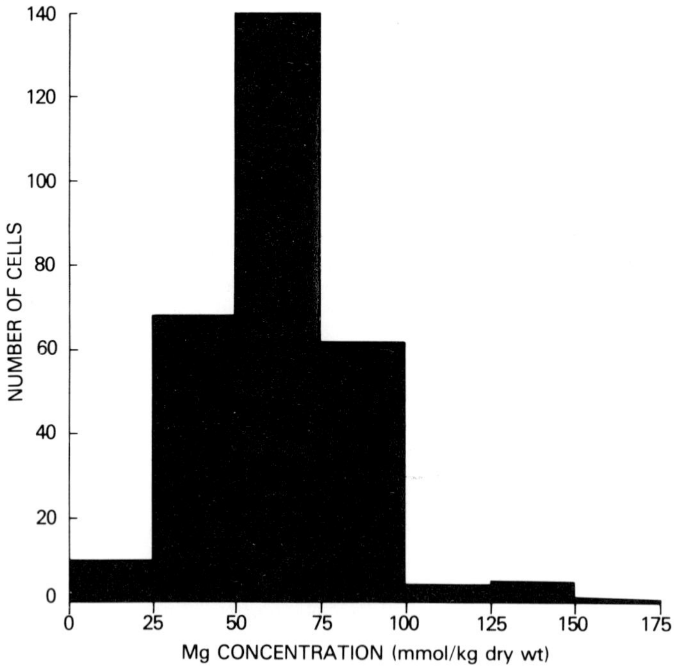

Figure 17. The Mg concentration (mmol/kg dry weight) distribution of individual mononuclear blood cells is plotted and appears to be normally distributed [15]. EDS-EPMA can quantitate elemental concentrations among individual cells.

and carbon coated. Individual cells were analyzed by adjusting the electron beam raster so that individual cells were circumscribed by the scanned area. The Mg peak-to-local continuum x-ray intensity ratio was used to measure concentration (Fig. 15). The EDS-EPMA was calibrated using lithium borate glass microspheres with known Mg concentrations which were analyzed in the same manner as the cells (Fig. 16). The mean Mg concentrations determined by EDS-EPMA and FAAS were not significantly different (Table 2). The Mg concentrations among individual MBC appeared to be normally distributed (Fig. 17) Thus, EDS-EPMA can be quantitative and used to study elemental variation among individual whole cells.

Conclusions

The fundamentals of EDS and WDS-EPMA have been presented and contrasted. The new method of digital EPMA elemental x-ray imaging has been discussed, especially in regards to the associated artefacts. Biomedical applications have shown that digital WDS-EPMA elemental x-ray images can be used to determine the elemental distribution within histological sections and quantitative EDS-EPMA can determine the amount of element in individual whole cells. These methods offer a way of studying the biomedical effects of inorganic elements on individual cells.

References

1. Morrision, G.H. 1984. *Elemental trace analysis of biological materials*. In *Critical Reviews in Analytical Chemistry Vol. 8, Issue 3*. CRC Press, Boca Raton, pp. 287–320.
2. Hook, G.R., Hosseini, J.M. and Elin, R.J. 1985. *Analytical approaches for biomedical elemental analysis*. J. Am. Col. Nutr. 4: 599–612.
3. Goldstein, J.I., Newbury, D.E., Echlin, P., Joy, D.C., Fiori, C.E. and Lifshin, E. 1984. *Scanning Electron Microscopy and X-ray Microanalysis*. Plenum Press, New York.
4. Echlin, P. 1984. *Analysis of Organic and Biological Surfaces*. John Wiley & Sons, New York.
5. Hren, J., Goldstein, J.I. and Joy, D.C. 1979. *Introduction to Analytical Electron Microscopy*. Plenum Press, Inc., New York.
6. Sauberman, A.J. and Scheid, V.L. 1985. *Elemental composition and water content of neuron and glial cells in the central nervous system of the North American medicinal leech* (Macrobdella decora). J. Neurochem. 44: 825–833.
7. Gorlen, K.E., Barden, L.K., Del Priori, J.S., Fiori, C.E., Gibson, C.C. and Leapman, R.D. 1984. *Computerized analytical electron microprobe for elemental imaging*. Rev. Sci. Instrum. 55: 912–921.
8. Nelson, A.C. 1986. *Computer-aided microtomograph with true 3-D display in electron microscopy*. J. Histo. Cytochem., *in press*.
9. Fiori, C.E., Swyt, C.R. and Gorlen, K.E. 1984. *Continuum correction of x-ray images in scanning electron column instruments*. In *Microbeam Analysis – 1984*. San Francisco Press, San Francisco, pp. 179–184.
10. Chandler, J.A. 1977. *X-ray microanalysis in the Electron Microscopy Vol 5*. North-Holland Pub. Co., Amsterdam, pp. 377–423.

11. Wroblewski, J. and Roomans, G.M. 1984. *X-ray microanalysis of single and cultured cells.* In *Scanning Electron Microscopy/1984/IV*, SEM Inc., AMF O'Hare (Chicago), pp. 1875–1882.
12. Hook, G.R. 1986. *A sample preparation and microprobe x-ray imaging method for histological localization of cisplatin.* In *Proc. Electron Microscopy Society of America.* San Francisco Press, San Francisco, pp. 370–371.
13. Garruto, R.M., Fukatsu, R., Yanagihara, R., Gajdusek, D.C., Hook, G. and Fiori, C.E. 1984. *Imaging calcium and aluminum in neurofibrillary tangle-bearing neurons in parkinsonism-dementia of Guam.* Proc. Natl. Acad. Sci. USA 81: 1875–1879.
14. Duckett, S., Galle, P. and Fiori, C.E. 1985. *Electron probe microanalysis of normal and pathological neural tissue with wavelength dispersive x-ray spectrometry.* In *Metal Ions in Neurology and Psychiatry.* Alan Liss, Inc., New York, pp. 367–396.
15. Hook, G.R., Hosseini, J.M., Elin, R.J. and Fiori, C.E. 1986. *Quantitative electron probe x-ray microanalysis and flame atomic absorption spectroscopy of human mononuclear blood cell magnesium.* In *Microbeam Analysis – 1986*, pp. 217–220.

Applications of microspectrophotometry

R.J. GRIESBACH
Florist and Nursery Crops Laboratory, USDA, Agricultural Research Service, BARC-West, Beltsville, MD 20705, U.S.A.

Introduction

Microspectrophotometry is the application of standard macrospectrophotometric techniques through a microscope. The basic instrumentation requires a microscope, controllable wavelength light source and a phototube for detection. Most microscope manufacturers now produce microspectrophotometers. The commercially produced units in addition to having all the necessary hardware have computers and the required software for many of the more common applications of microspectrophotometry. There are two different applications of microspectrophotometry – quantitative and qualitative (Berlyn & Miksche, 1976; Wied, 1966).

Quantitative microspectrophotometry involves the precise measurement of the absorption of some material at a specific wavelength of light. This measurement is then used to determine the mass or concentration of that substance. Qualitative microspectrophotometry involves measuring the absorption spectrum of some material at a range of wavelengths. This information can then be used to identify the material. Qualitative microspectrophotometry requires a sophisticated spectrophotometer, whereas quantitative microspectrophotometry only requires the use of filters.

The lower limits of detection using microspectrophotometry are about 10^{-13} liters and 10^{-6}M, while the lower limits for macrospectrophotometry are about 10^{-3} liters and 10^{-10}M. In general, macrospectrophotometry is more sensitive; however, with the use of multichannel analyzers and image intensifiers the sensitivity of microspectrophotometry can approach 10^{-14}M.

Qualitative measurements

A high quality spectrophotometer or monochromator is essential for measuring the absorption spectrum of any object or material. The light source, however, is the limiting factor in many instances. Because of optics of the microscope and the very small path lengths, the light source must have a high intensity over a wide spectral range. In addition, the light source should have a stable and constant output without sharp emission lines. Many light sources (e.g. tungsten, halogen or xenon) are suitable for visible wavelengths. For ultraviolet wavelengths, hydrogen,-deuterium or xenon must be used. Because of low intensities of light with wavelengths below 200 nm, the monochromator should be highly efficient at these wavelengths. A diffraction grating system instead of interference filters or a prism is the method of choice for dispersing light (Wied, 1966).

The optical system must be very precise. Spherical and chromatic aberrations, as well as general light scattering, can lead to different measurements for the same structure when located in different areas of the field since light travels through the center of a lens with a different transmittance from light which travels through the len's perimeter. In addition, the amount of light which is scattered by the specimen is not linearly proportional to the area illuminated. Another factor which affects scattering is the light source, for a fraction of the scattered light leaving the source is directed into the measuring diaphragm. This becomes important when the illumination diaphragm approaches the size of the measuring diaphragm. On the other hand if the illumination is too large, scattering will also affect measurements (Wied, 1966).

The method for obtaining a microspectrum is quite simple. The sample is measured and the instrument response is substracted from the sample measurement. Commercially available units have simple computer programs for substracting the instrument response and then plotting the sample's absorption over a range of wavelengths.

There are many applications of qualitative microspectrophotometry. DNA/protein identification is one such example. The spectra of unknown inclusions within a living cell can be obtained (Fig. 1). From a comparison of known pure protein, pure DNA or known protein/DNA complexes, one can determine the composition of the inclusion. In this way, it is possible to tell if cellular virus inclusions are protein, nucleic acid or a combination of both. One can determine to what extent virus

---

Figure 1. Absorption spectra of nucleic acids, protein and combinations of the two between 290 and 240 nm. (A) protein, (B) DNA, (C) 50% DNA, 50% protein, (D) 66% DNA, 33% protein, (E) 33% DNA, 66% protein. The ratio of the absorbance at 260 nm to 280 nm can be used to quantify both the amount of protein and nucleic acid in the sample (see quantitative measurements section). The amount of DNA equals:

$$X = \frac{A_{280} \cdot (Y - 1/2)}{18}$$

where: A_{280} = absorbance at 280 nm. $Y = A_{260}/A_{280}$. The amount of protein equals: $X = A_{280}(1.6 - 0.8Y)$.

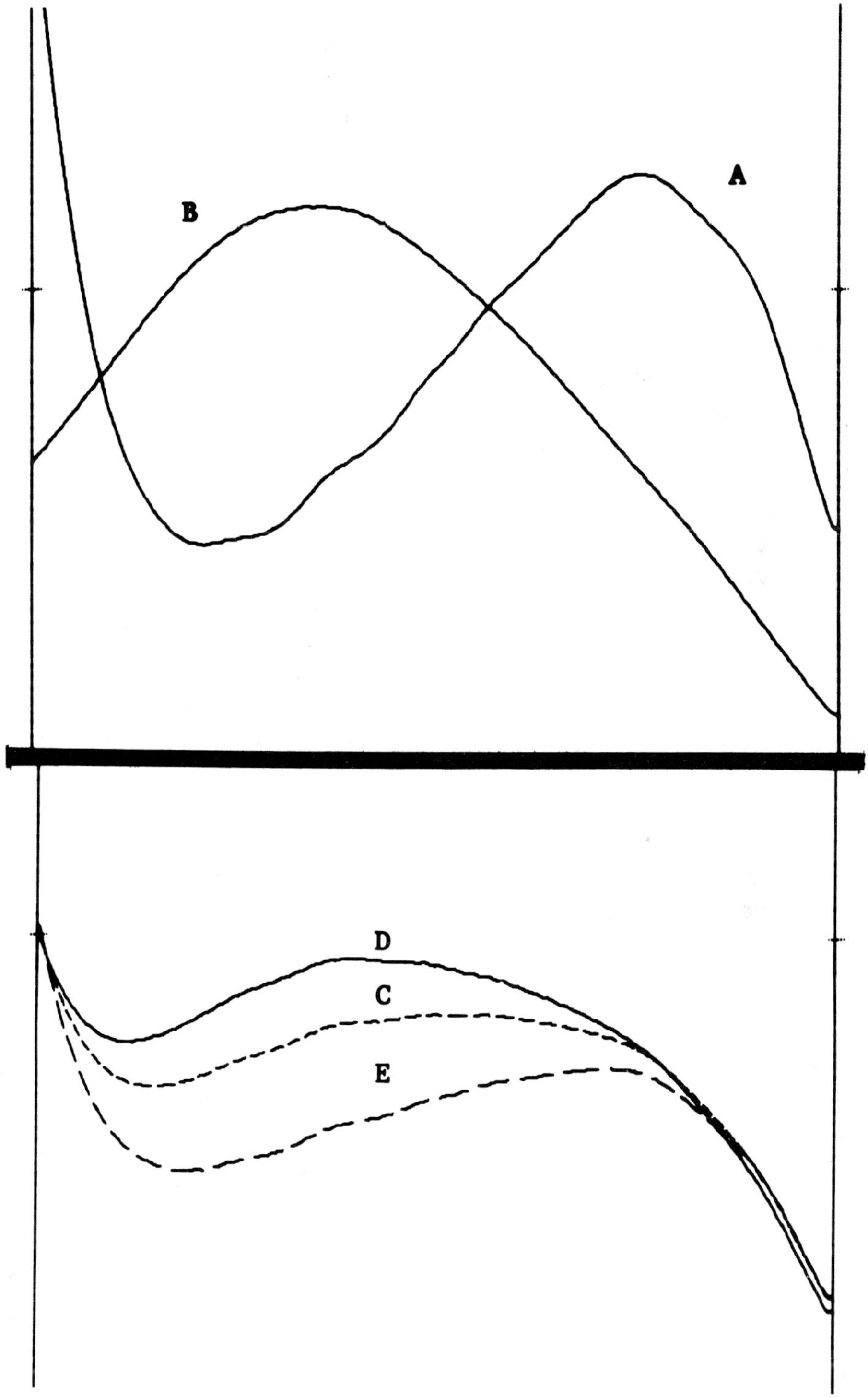
A
B
D
C
E

assembly has progressed by measuring the ratio of protein to nucleic acid. This information is very important when analyzing genetic mutants which are defective in virus assembly. Classical methods of analysis require isolating intact inclusions, a process which is very difficult and labor intensive. Microspectrophotometric analysis should be faster, easier and more accurate.

Another application is in the biochemistry of flower color. Anthocyanins are the principal pigments which determine flower color. In addition, carotenoids and flavonoid copigments can affect the color of flowers. For example, red flowers can be red in several ways. The flower could have a red anthocyanin and lack carotenoids or the flower could have a magenta anthocyanin and contain carotenoids. There are many other reasons behind red flower colors. Similarly, a flower could be blue in one of several ways. The flower could have a blue anthocyanin and lack copigments or could have a magenta anthocyanin and contain copigments. Once again, there are many other reasons for blue colors. Absorption spectra can help identify the different underlying basis for flower colors. At the microscopic level, the cells with a flower may not be uniformly pigmented (e.g. spotted flowers); therefore absorption spectra of individual cells (microspectrophotometry) is usually more meaningful than a spectra of an *in vitro* extract. Absorption spectra have been used to determine that an orange sport of the azalea cultivar "Red Wing" was the result of a reduction in the concentration of flavonoid copigments (Asen et al., 1971). In addition, microspectrophotometry helped determine that the red spots in certain orchids were the result of carotenoid/anthocyanin combinations and not solely due to anthocyanin (Griesbach, 1984). *In vivo*, microspectra can be obtained on living, unfixed material unlike the classical means of differential staining of fixed sections. Differential staining can however be quite useful for quantitative microspectrophotometry.

Shifts in absorption spectra have been used *in vitro* to study the conformation of enzymes (Radda, 1973). One example uses the fluorescent dye *N,N* dimethylnaphtheurhodine which ionically binds to the sulfhydyl group of polypeptides. As the secondary and tertiary structure of the protein changes, the absorption spectrum of the bound dye also changes. This technique was used to show that a conformation change occurs in the enzyme glycogen phosphorylase when its cofactor Mn^{++} binds. An additional conformation change occurs when AMP binds. Mathematical models predict that the spectra changes when Mn^{++} binds translates into an 8Å lengthening in the distance between the dye binding site and the Mn^{++} binding site. When the AMP binds, this distance is shortened by 1.5Å.

Quantitative measurements

The only physical requirement for quantitative microspectrophotometer is that the transmittance of the substance which is being measured must follow the Beer–Lambert law which is:

$$T = I/I_0 = 10^{-Ecd}$$

where

T = transmittance
I = energy transmitted by substance
I_0 = energy of incident radiation
E = molar absorptivity
c = concentration
d = path length.

From this equation the concentration of the substance is:

$$C = \frac{\log(1/T)}{dE} = \frac{A}{dE}$$

where the log $(1/T)$ equals the absorbance (A) or optical density.

There are 3 common methods for using the Beer–Lambert law in taking quantitative measurements through a microscope-Core Method, Two Area Method, or Wavelength Method. The simplest is the core method (Wied, 1966). This method however, requires that the substance or object being measured have a uniform thickness and even distribution of absorption. For spherical objects, the mass is:

$$M = \frac{Ar^2\pi}{FE}$$

where:

M = mass
A = absorbance
r = radius of test area
F = fraction of sphere's volume in test area
E = molar absorptivity.

The two area method is suitable for measurements for irregularly shaped objects which are flat (Wied, 1966). In this case the mass is:

$$M = \frac{A_{obj}a_{total}}{E} \cdot \frac{1 - T_{total}}{1 - T_{obj}}$$

where:

A = absorbance
a = area
E = molar absorptivity
T = transmittance.

The two wavelength method is suitable for most irregular shapes if there is little stray light and if measurements for large differences of energy are linear (Wied, 1966). The mass in this instance is:

$$M = \frac{a_{\text{total}} A_{1/2\lambda\,\text{max}} C}{\ln 10\, E_{1/2\lambda\,\text{max}}}$$

where:

A = area
$A_{1/2\lambda\,\text{max}}$ = absorbance at $1/2\lambda$ max
$E_{1/2\lambda\,\text{max}}$ = molar absorptivity at $1/2\lambda$ max

$$C = \frac{1}{2 - (A_{\lambda\,\text{max}}/A_{1/2\lambda\,\text{max}})} \cdot \frac{1}{\ln(1/(A_{\lambda\,\text{max}}/A_{1/2\lambda\text{max}}) - 1)}.$$

The cellular concentration of DNA can be quantitated using microspectrophotometry (Jacqmard et al., 1972; Kapuscinski & Skoczylas, 1977). The principal for measurement involves staining the DNA with Schiff's reagent. First, the DNA is chemically fixed with acid which breaks the purine-base glycoside linkages. When these bonds are broken the aldehyde groups on the sugars are free. Schiff's reagent then can react with the aldehyde groups producing a DNA-specific unsoluble magenta compound with an absorbance maximum at 569 m. Using the core method one can measure the amount of DNA within a cell.

Through the use of known internal standards (e.g. 2.62 μg DNA in chicken erythrocyte = 9.7 absorption units), one can determine the DNA concentration in an unknown sample.

Another dye, 4′-6-diamidino-2-phenylinodole (DAPI) can also be used for quantitative DNA measurements (Hammerschlag, 1983; James & Jape, 1978). DAPI preferentially binds to AT rich regions of double-stranded DNA. A DAPI-DNA complex will fluoresce at 450 nm when excited with light of 360 nm. DAPI-DNA fluorescence is much more sensitive than Schiffs-DNA staining and has been used to measure the amount of DNA within a chloroplast. DAPI can also be used on living, unfixed cells, unlike Schiff's reagent.

These techniques have been used in many areas of biology. For example, microspectrophotometry has been used to demonstrate that differential DNA replication is involved in the differentiation of certain root cells (Van't Hof & Bjerknes, 1982). Another example in the area of plant development involves the measurement of heterochromatin within a nucleus (Nagl, 1979). Results suggest that in plants, unlike in animals, differential DNA condensation does not play a significant role in development.

Combinations or slight modifications of these three methods for determining the mass of some absorbing object or chemical can be applied to almost any situation (Kohen et al., 1973). Intracellular proteins can be similarly quantitated using naphthal yellow-S. Carbohydrates can be assayed using the periodic acid-Schiff's reaction. Lipids can be measured with sudan III. Procedures have also been developed for measuring cholinesterase, lipase, glucose, alkaline phosphatase, $NADP_2$, and lactate dehydrogenase.

A little imagination and some basic knowledge about compounds which absorb differently under different conditions make quantitative microspectrophotometry a powerful technique. For example, these methods can be used to determine the pH of microliter samples (Stewart et al., 1975). The measurements are based upon the ability of certain dyes to have different colors at different pHs (Fig. 2). In the case of bromophenol blue, two absorption maxima are found at 590 and 435 nm. The relative height of these absorption maxima is dependent upon pH. If one plots the ratio of absorbance at 590 nm to the absorbance at 435 nm vs. pH, a sigmodal curve is obtained (Fig. 2(b)). An equation can be obtained for the linear portion of the curve between pHs 2.5 + 4.5: $\mathrm{pH} = 3.437 + 1.084 \log (A_{435}/A_{590})$. The standard error being 0.038 pH units.

This information has applied to microspectrophotometry. Interference filters (590 nm + 435 nm) were moved in and out of the light path ahead of the specimen. By placing into a light path a 590 nm interference filter, the A_{435} can be measured. The opposite is true for the 435 nm filter. In addition, a broad-pass filter was inserted after the specimen but before the phototube to illuminate the stray radiation. The absorbance of the tissue and water was substracted from that of the tissue and the dye to correct for instrument response and the absorption due to specimen. Using this system, very accurate pH measurements were made on volumes of approximately 0.5 nl.

Another technique for the microspectrophotometric measurement of pH relies upon the flurorochrome 9-aminoacridine (Manigualt et al., 1983). 9-aminoacridine is a lipophilic base which can enter living cells in its neutral form by diffusion. Once inside the cell, the quenched, protonated form is concentrated in the acidic vacuole. The ratio between the intracellular and extracellular concentration of the dye can be used to measure pH:

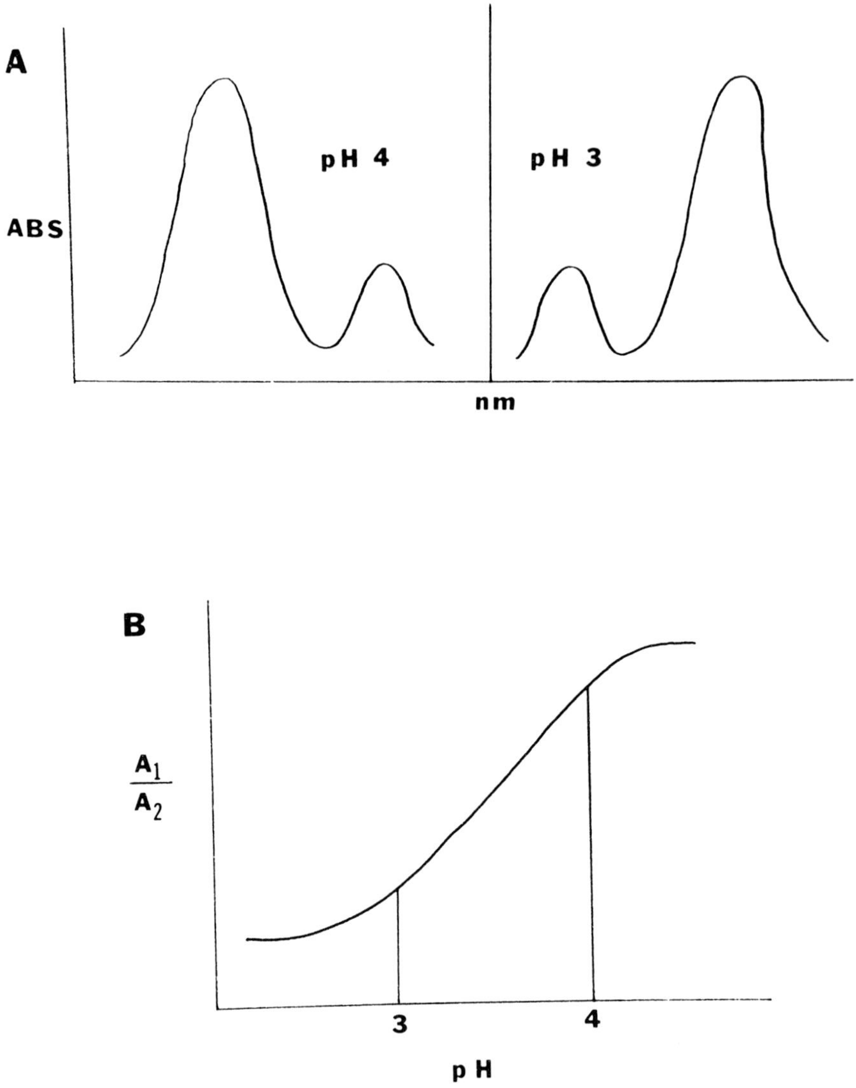

Figure 2. Use of indicator dyes in the measurement of pH. (a) Absorption spectra of bromphenol blue at pH 3 and 4; absorption maxima are at 435 + 590 nm. (b) Plot of the ratio of the absorbance at 435 to 590 nm vs. the pH. The curve is linear from pH 2.5 to 4.5. The equation pH = 3.437 + 1.084 log A453/A590 can be derived from the slope.

$$C_i/C_e = \frac{1 + 10^{pK - pH_i}}{1 + 10^{pK - pH_e}}$$

where:

C_i = internal concentration of dye

C_e = external concentration of dye.

The accumulation ratio is:

$$C_i/C_e = \frac{V(A_1/A_2 - 1)}{v} + 1$$

where:

V = total volume of measured area
v = volume of cell
A_1 = absorbance of measured area
A_2 = absorbance of extracellular area.

Conclusion

There are many *in vitro* and *in vivo* applications of microspectrophotometry in Biology. Microspectrophotometry has two advantages over macrospectrophotometry. First, much smaller samples can be measured. Second, microspectrophotometry can be used to take measurements within or on a single living cell. With these advantages come a single disadvantage. Microspectrophotometry requires a strong signal for detection. New technology (e.g. image intensifiers) may eliminate this requirement in the future. Used properly, microspectrophotometry can be a very powerful tool.

References

1. Asen, S., R.N. Stewart and K.H. Norris. 1971. *Co-pigmentation effect of quercetin glycosides on absorption characteristics of cyanin glycosides and color of Red Wing Azalea.* Phytochemistry 10: 171–175.
2. Berlyn, G.P. and J. Miksche. 1976. *Botanical microtechniques and cytochemistry.* pp. 245–262. Iowa State Univ. Press, Ames, Iowa.
3. Griesbach, R.J. 1984. *Effects of carotenoidanthocyanin combinations on flower color.* J. Heredity 75: 145–147.
4. Hammerschlag, F. 1983. *Factors influencing the frequency of callus formation among cultured peach anthers.* Hort. Sci. 18: 210–211.
5. Jacqmard. A., J. Miksche and G. Bernier. 1972. *Quantitative study of nucleic acids and proteins in shoot apex of sinapsis during transition from the vegetative to the reproductive condition.* Am. J. Bot. 59: 714–721.
6. James, T.W. and C. Jape. 1978. *Visualization by fluorescence of chloroplast DNA in higher plants by means of the DNA-specific probe DAPI.* J. Cell Biol. 79: 623–630.
7. Kapuscinski, J. and B. Skoczylas. 1977. *Simple and rapid fluorimetric method for DNA microassay.* Anal. Biochem. 83: 252–257.
8. Kohen, E., B. Thorell, C. Kohen and M. Michaelis. 1973. *Rapid microfluorometry for biochemistry of the living cell.* In: *Fluorescence Techniques in Cell Biology*, eds. A. Thaer and M. Sernetz, Springer-Verlag, New York. pp. 219–232.
9. Manigualt, P., J. Manigualt and A. Kurkdjian. 1983. *A microfluorometric method for vacuolar pH meaurements in plant cells using 9-aminoacridine.* Physiol. Veget. 21: 129–136.

10. Nagl, W. 1979. *Nuclear ultrastructure: Condensed chromatin in plants in species-specific; but not tissue-specific*. Protoplasma 100: 53–71.
11. Radda, G. 1973. *Probes for enzyme conformation*. In: *Fluorescence Technique in Cell Biology*, eds. A Thaer and M. Sernetz. Springer-Verlag, New York. pp. 261 272.
12. Stewart. R.N., K.H. Norris and S. Asen. 1975. Microspectrophotometric measurement of pH and pH effect on color of petal epidermal cells. Phytochem. 14: 937–942.
13. Van't Hof. J. and C.A. Bjerknes. 1982. Cells of pea that differentiate from G-2 phase have extrachromosomal DNA. Mole. and Cell. Biology 2: 339–345.
14. Wied, G.L. 1966. *Introduction to quantitative cytochemistry*. Academic Press, New York.

Satellite instrumentation for monitoring agricultural conditions

JOHN C. PRICE
Remote Sensing Research Laboratory, Beltsville Agricultural Research Center, Beltsville, MD 20705, U.S.A.

Abstract Remote sensing by satellites provides a speedy and efficient tool for assessing conditions of large agricultural areas. This data source is used by several USDA operational agencies to help fulfill their reporting requirements to the public and other government agencies. The evolution of satellite instrumentation is described, illustrating that many levels of cost and complexity are available. The government policy of commercialization of earth viewing satellites will force greater attention to cost recovery and providing those data users require. Rapid transmission of satellite data to individual farmers and county agents is now technically possible, but cost and utility considerations suggest that this will not occur in the near term.

Introduction

Although U.S. agriculture is a fragmented industry by virtue of the large number of producers, it is an important factor in both the national economy and the international marketplace. For many years the U.S. Department of Agriculture (USDA) has been charged with obtaining current information on foreign and domestic agricultural conditions, and reporting its findings to the public. The cost of a good information system is repaid to society through the lower costs resulting from efficient economic interactions.

In addition the Department of Agriculture operates a number of programs which require knowledge of the acreage planted in various crops, the status of soil conservation measures, and the extent of affected areas when weather conditions cause areas of general crop failure. The evolution of technology and the increasing cost of site visits by USDA agents point toward the use of remote sensing by aircraft or satellite to fulfill some of these requirements. Since the observing requirements may be addressed by many forms of instrumentation it is useful to summarize the existing status of remote sensing within the USDA:

(a) Major agricultural areas in the U.S. are photographed from aircraft with 35 mm film for the Agricultural Stabilization and Conservation Service (ASCS) to monitor federally legislated farm program acreage compliance. Typically the photography is flown so that each slide is centered on a section or square mile of crop or rangeland.

(b) The National High Altitude Photography (NHAP) program is a multi-federal agency program to acquire systematic coverage of the United States. The U.S. Geological Survey and the Soil Conservation

Service (SCS) are the major contributors to the program. In 1985 a pilot program by the Soil Conservation Service explored the use of such data for the congressionally mandated National Resources Inventory, which is carried out at 5 year intervals (USDA, 1982).

(c) Satellite data from the Landsat satellites, which are now operated by the EOSAT corporation, are used by the Statistical Reporting Service (SRS) of the USDA to refine estimates of the acreage devoted to major U.S. crops. Eight states in the central U.S. are currently analyzed in the SRS program, but the program is not yet considered operational, partly because of the uncertainties concerning the commercialization of the Landsat series of satellites, and partly because future data costs have not been established.

(d) The Foreign Agriculture Service (FAS) uses data from the meteorological satellites to assist in the evaluation of crop conditions in the major agricultural areas of the world. Information from these satellites is combined with that from attache reports, from popular news media, etc, in estimating the potential supply and demand from our important competitors.

(e) The Agricultural Research Service (ARS) carries out research in the use and analysis of satellite data to support other USDA agencies. At Weslaco, Texas an effort is currently underway to develop a low cost video camera based system which might be flown on aircraft for monitoring farms and rangeland, and the Hydrology Laboratory at Beltsville, MD is supporting NASA in the development of a microwave instrument for estimating soil moisture. With these exceptions the ARS is not involved in instrumentation per se for remote sensing, but only in understanding the data as they may be useful to the agricultural community.

Since the above factors justify maintaining a capability in remote sensing, it is appropriate to discuss the satellite instrumentation which is available and planned, relating the technical specifications to applications of the data. The new government policy (US and French) is to have the user pay a greater portion for cost recovery and eventually make satellite systems profitable. This represents a change from earlier policy which treated satellite data as a research tool which could be distributed without charge.

Basic properties of remotely sensed data

The most widely known type of remotely sensed data is aerial photography. A camera is placed in an aircraft or satellite and the film in the camera is exposed over the sites of interest, after which the film is returned to a processing center for chemical treatment and production of the desired products, typically prints or photographic transparencies. Aerial photography is straightforward and cost effective for some applications, but satellite photography is costly and inefficient, due to the high cost of accelerating the camera to orbital speed (7 kilometers per second), then later slowing down the film canister for recovery following descent through the atmosphere. This statement does not apply universally, as military surveillance requires extremely high quality imagery over very specific sites at infrequent time intervals. Photographic techniques are suitable if not preferable for such applications. On the other hand some large area agricultural and natural resource inventories require repetitive surveys, but not the very fine spatial detail of photographic imagery. In addition agricultural interpretation frequently requires information on subtle spectral properties of soils and vegetation, which must be obtained from analysis of numerical data. Photographic images lack the quantitative aspect required for such applications.

Generally satellite remote sensing instruments provide an electronic signal representing the region viewed by the satellite as it circles the earth. This signal is transmitted directly by radio to a ground receiving station, or else the signal is stored in the satellite for later transmission. After recording on the ground the electronic signal may be sent to a film device for creation of a "hardcopy" print, or to a TV display for assessment of quality and suitability for further analysis. Usually the signal is converted into a computer readible form after addition of ground supplied information such as time and date. Computer formatted data are suitable for refinements such as correction for geometric distortions associated with the sensor's scan mechanism, radiometric calibration through use of auxilliary information which may not be supplied in the sensor data stream, and for quantitative manipulation by a scientist or end user who wishes to infer some property about the imaged area.

Figure 1 illustrates the general form of satellite data, showing that the imaged area is partitioned into a number of brightness values which, displayed on a television monitor, produce an image resembling a photograph (Gonzalez and Wintz, 1977). The significant features of this data type are:

(1) The image is a set of numerical values stored in a computer readable medium, rather than a visual product such as a print or photographic negative. In a common computer format the numbers corresponding to the first line are stored as a computer record, those in the next line are stored as the next record, etc. Specialized devices may be used for converting the numerical values into a photographic product.

(2) Each numerical value represents the brightness of an area deter-

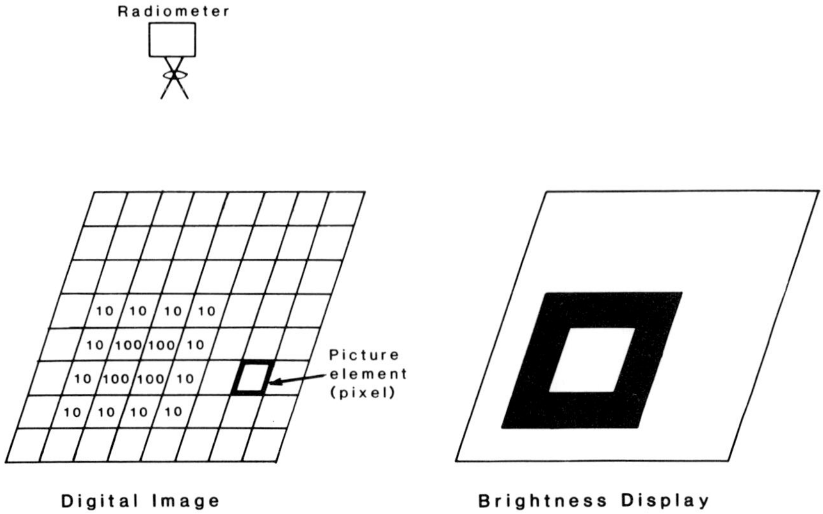

Figure 1. An electronically acquired image consists of a series of numerical values which form a picture when displayed as brightness elements.

mined by the characteristics of the sensor. The value is thus an average of the brightness elements of all the smaller features within this area. This smallest area of an image is called a pixel, standing for picture element. The expression spatial resolution is also used, with high spatial resolution implying small picture elements and low spatial resolution implying larger picture elements. While photographic products may be enlarged until the graininess of the film negates further expansion, the spatial resolution of the digital data from a given area cannot be improved, except to the degree that numerical interpolation techniques may be applied successfully. Interpolation may be appropriate for geometrically regular cases, such as at boundaries of large fields, but it adds little to the interpretation of data which are quite chopped up or "busy", as when a city is imaged with a ground resolution larger than features such as buildings, the width of roads, etc.

(3) Areas may be imaged in many spectral intervals. (Fig. 2) This multispectral aspect of digital remote sensing represents an advance over photography, in which the possible number of spectral intervals is limited by the number of photosensitive compounds in the film emulsion. While color photography consists of mixtures of 3 pigments, a sensor presently operating, the Thematic Mapper, acquires electronic imagery in 7 distinct spectral intervals. This number of spectral images will be much larger if current plans for satellite sensors are realized.

Having defined the geometric characteristics of electronic satellite data I next consider the properties of radiation as measured by a satellite

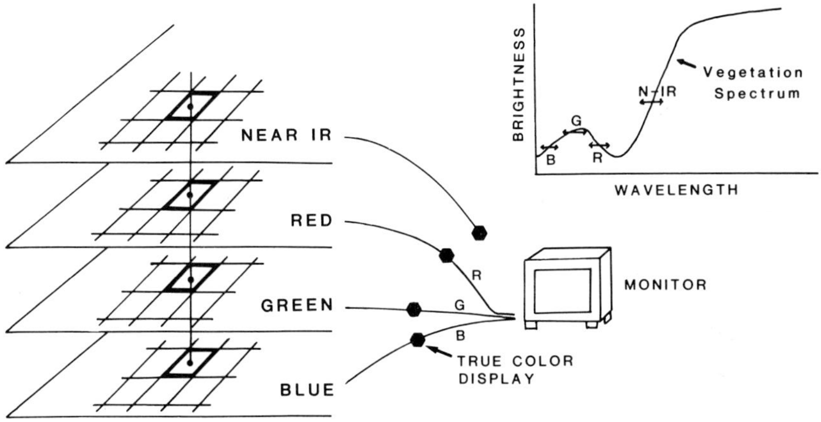

Figure 2. Several spectral intervals may be imaged at the same time. Any set of three spectral channels may be displayed as a color image, but many data applications do not require an image at all.

sensor. Since the atmosphere is not uniformly transparent to radiation of all wavelengths, images of the earth's surface can be obtained only at certain spectral intervals. These relatively transparent intervals are referred to as "windows".

Figure 3 illustrates atmospheric transparency in wavelength regions of interest. There are 4 relatively clear windows in the 0.4 micrometer to 10 centimeter range, corresponding to the visible, near infrared, thermal infrared, and microwave.

The visible portion of the spectrum, from 0.4 to 0.7 micrometers, is best understood. The fact that some objects are more reflective than others, e.g. yellow sand versus charcoal, and that colors are diagnostic for vegetation types and condition, is a basic property of the human visual system and may be refined to some degree through study of

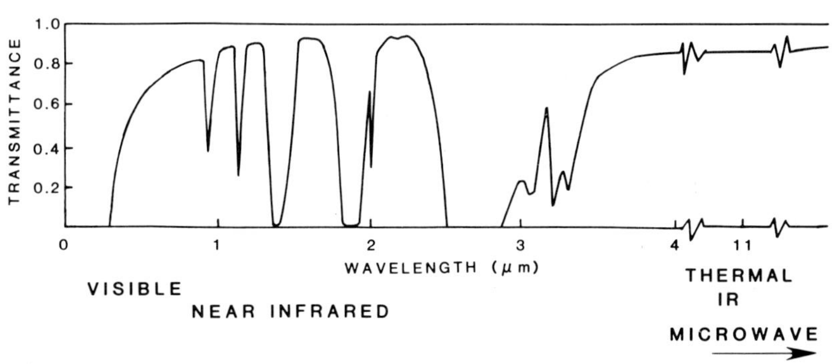

Figure 3. Instrumentation can provide measurements over a broad range of the electromagnetic spectrum, but satellite observations of the earth require spectral regions with high atmospheric transmittance.

photointerpretation techniques. Quantitative application of this knowledge when working with digital data is not common, and in fact one seldom uses visual color information at all when working with satellite data, instead relying on recognition of geometric properties such as field boundaries, houses, water bodies, etc.

Near infrared radiation, from 0.7 to about 3 micrometers, is invisible to the human eye, but is highly useful for monitoring vegetation when appropriately displayed. In this wavelength range healthy vegetation is more reflective than virtually all natural materials. This spectral interval is useful for discriminating vegetation from other materials when both appear green to the human eye, e.g. color infrared film was developed for the military for camouflage detection. The high reflectivity of live vegetation causes it to be a sensitive indicator of the extent and vigor of vegetation. This fact remains true even at the relatively low spatial resolution of the data from the meteorological satellites.

The thermal infrared, from 3 to about 100 micrometers, provides an indication of surface temperature. Thermal radiation is affected by the heating effect of the sun and the cooling effect of transpiration in healthy well watered crops. Although high quality satellite data in the 11 micrometer spectral range have been available since 1978, they have been investigated only in research programs and results are not yet suitable for USDA operational programs.

At longer wavelengths lies the microwave region of the spectrum in the 1 millimeter to 10 centimeter range. This spectral interval is potentially useful for monitoring soil moisture by passive sensors as well as for active (radar) measurements of surface roughness and soil moisture. However these sensors will not be flown in a civil satellite program before approximately 1995. Thus this portion of the spectrum will not be discussed.

Basic forms of instrumentation

Besides photography there are two principal procedures for imaging the earth from a satellite (Slater, 1980). These possibilities result from the existing technology for converting radiation into an electrical signal. The conversion device, called a detector, can be used in two different forms of imaging systems. The first, an example of which is a vidicon, converts an image at the detector surface into a charge variation. This charge may be read out by scanning an electron beam over the surface. The most familiar example is commercial television. Figure 4 illustrates this technology, which is referred to as image plane scanning because an image is formed by a mirror or lens on the detector plane before conversion to an electrical signal. The detector itself may be relatively large, e.g. 1 cm^2.

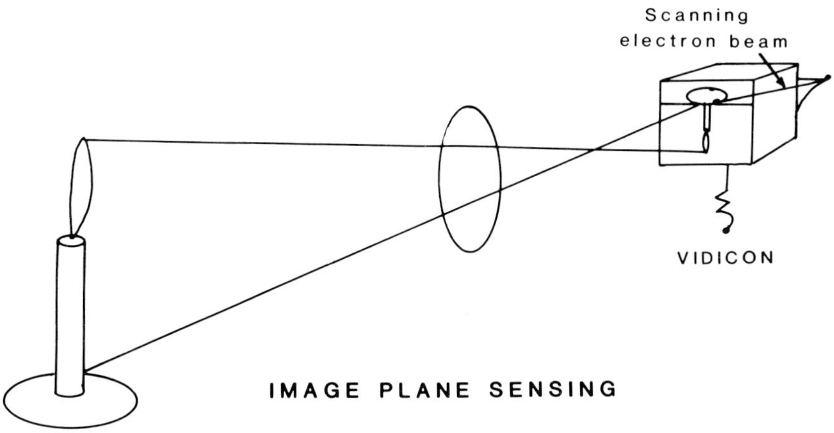

Figure 4. Image plane technology is conceptually similar to photography, as the entire scene is imaged on the detector surface, which is then scanned electronically.

The second sensor technique uses a moving optical component, various forms of which will be discussed in the next section, to scan the image over a single detector, as illustrated in Fig. 5. This approach is referred to as object plane scanning, because the detector in effect looks at the object, pixel by pixel, as the scan device operates. Detectors are very small, e.g. the detectors in a current satellite instrument, the Thematic Mapper, are about .003 cm across. There are two basic differences between object and image plane scanners:

1. Generally an image plane scanner is shuttered and creates an image with a single exposure. It is subject to blurring or geometric aberration only to the extent that the sensor moves during the brief exposure of the

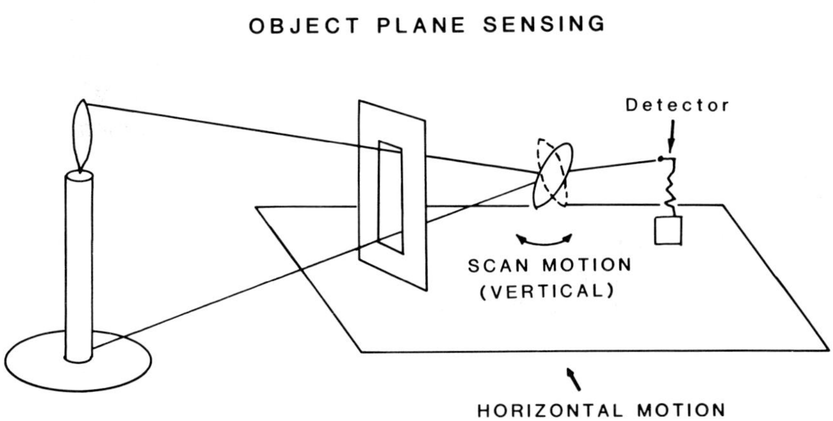

Figure 5. Object plane technology uses a moving optical element to cause the detector(s) to scan the image. For most satellites the orbital motion provides the scan in one direction.

vidicon. In contrast the object plane scanner creates an image over an extended period of time while the scan device traces out the image, pixel by pixel. The geometric fidelity of the object plane scan is thus susceptible to variations in attitude of the host vehicle (satellite) during the time while an area is imaged.

2. It is much more difficult to calibrate an image plane scanner, as the detector in effect consists of many small detectors which are read out as the electron beam passes over them. Variation of sensitivity across the detector surface is expected during the fabrication process, and changes during operation are common. Assuring a uniform response across this surface is a difficult task, and the precision of data from such a device is fundamentally limited. In contrast the object plane scanner generally uses a single small detector to scan the image, so that in principle the variability of the output signal is associated only with the variation of incident radiation.

Although each option has advantages and disadvantages, quantitative considerations argue strongly for use of object plane scanning in viewing the earth. It is possible to use a large number of detectors in an object plane scanner. This does indeed present calibration problems, but these problems are minor compared to those of the image plane scanner. The vidicon device may be preferable for satellite missions to other planets, since the precision of the data is relatively unimportant (pictures are of highest priority, and these require minimal precision data), while the reliability problems associated with the scan mechanism for the object plane scanner become more severe if the satellite becomes cold as it moves away from the sun.

Satellite considerations

Major attributes of a satellite instrument are defined from elementary principles of physics. A primary consideration is the data rate from the instrument. An earth satellite at 800 kilometers altitude passes over the earth at approximately 7 km per second. If the scene width imaged by the sensor is given by w, then the area (A) swept out per second is $A = w *$ (7 km/sec). If the detector signal represents a pixel size L by L meters, the measurement rate N is $N = A/(L * L)$ measurements per second. If each measurement is to a precision P, which is typically 8 bits, i.e. the detector output is sampled in the range 0–255, then the resulting data rate R is $N * 8$ bits per second.

$$R = A * P/(L * L) \quad \text{bits per second}$$

Let us consider a recent example, the Thematic Mapper (TM) which operates on Landsats 4 and 5. The scene width is 185 km, the data are sampled to 8 bits, and 7 spectral channels are imaged simultaneously in the visible, near IR, and thermal IR, with a pixel size of (30 × 30 meters) for 6 of the 7 channels. The data rate is thus

$$R = \frac{(185\,000) * (7000\,\text{m/sec}) * (7\ \text{channels}) * (8\ \text{bits})}{(30\,\text{m}) * (30\,\text{m})}$$

80 Megabits/second

By most any standard this is a high data rate, and the expression "drinking from a fire hose" has been applied to the ground receiving station. Basic parameters of instrument design affect directly the telemetry requirements of the satellite/ground system.

An additional major consideration results from operation in the space environment. The satellite must carry both the sensor and the subsystems which support the sensor's environmental requirements. These subsystems themselves represent complex engineering design problems. The major components are:

(1) A power system to maintain the operation of the sensor, as well as all other subsystems. This power system consists of solar panels, which acquire power during the sunlit portion of each orbit, a battery which stores energy for operations during the night half of the orbit, and control electronics which provide stable voltages to the other systems.

(2) An attitude control system to maintain the proper orientation of the satellite with respect to the sun and the area to be imaged. This control system should not generate abrupt satellite motions as these would cause distortions in the imagery. The better the ground resolution of the imaging sensor the higher the performance requirements for the attitude control system.

(3) An antenna and transmitter for relaying sensor and other subsystem data to earth, and a receiver for accepting ground instructions to control the satellite's operations. The transmitter power requirements and the antenna size increase with increasing data rate.

(4) An onboard computer system which maintains control of basic support functions including monitoring engineering parameters, controlling heating and cooling devices, storing command sequences for operations which occur out of reach of ground transmissions, regulating electronic circuitry, and formating sensor data for transmission to the ground.

(5) A tape device for storing data from the acquisition point to a location where the data may be transmitted to ground. This element is not required if a communications link is used for realtime transmission to a ground station from any satellite location. Such a link uses direct (line of sight) transmission from the observing satellite to a distant geosynchronous receive/transmit satellite and back to the ground sta-

tion. Several relay satellites are sufficient to provide sensor observations from any location on earth.

These factors illustrate that the sensor data acquisition rate is a dominant factor controlling the cost and complexity of an earth viewing satellite.

The evolution of satellite instrumentation

I consider the current and proposed instrumentation which is suitable for earth observations. I will discuss the sensors in order of increasing complexity, illustrating the relationship of design to the characteristics of the data. The significance of these attributes for data users will be discussed later.

The geosynchronous satellites

The geosynchronous satellites (GOES) illustrate a simple design. From orbital mechanics a satellite at an altitude of approximately 36000 kilometers will circle the earth at precisely the earth's rotation rate. Such a satellite in the earth's equatorial plane appears to be stationary in the sky, and, conversely, the satellite sees the earth as stationary beneath it. This feature is useful for obtaining frequent views of the same area on earth, as for viewing the evolution of storm systems and large scale weather patterns. It has the disadvantage that the distance to the earth's surface is 30–50 times greater than that of the 800 kilometer altitude polar orbiting satellites which obtain the data most suitable for agricultural assessment. This distance factor translates into a great cost differential in the design of a high spatial resolution instrument. As with a camera – if you want a good close up picture of something, get close to the object.

Since the main purpose of the geostationary satellites is to obtain images for meteorological interpretation a high quality image acquisition system is not required. In particular a sophisticated attitude control system is not needed to point the satellite. Instead the whole satellite spins, with it's axis of rotation parallel to that of the earth. This orientation is stable, i.e. no adjustment is required after the satellite has commenced operations from the desired location in space. In addition the satellite spinning motion obviates the need for an east-west scanning mechanism to sweep the earth's image across the sensor detectors. Only a mirror is required which steps through a small angle once per satellite revolution to produce the north-south scan (Hambrick, 1980). In practice the mirror steps through the north to south earth view in 20 minutes, after which the mirror returns to the north viewing position during a 10 minute retrace period. (Fig 6) The major problem with this arrangement

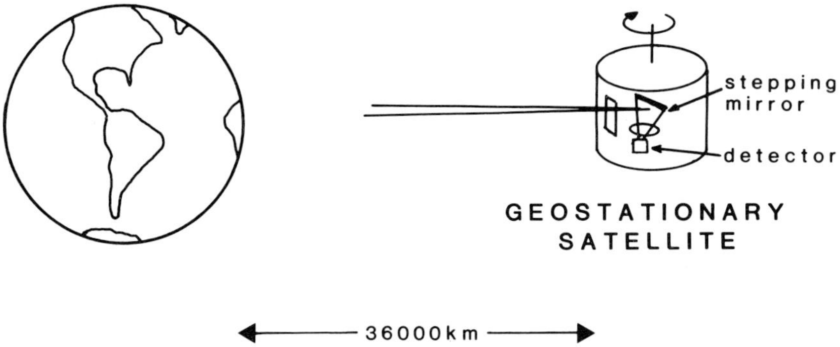

Figure 6. In the geosynchronous satellites the east-west scan is provided by satellite rotation, while a stepping mirror provides the north-south scan.

arises from the fact that the sensor looks at the earth during only 15 degrees of the 360 degree revolution. This requires a high data acquisition rate while the most of the scan, a view of space, is not useful. To simplify matters for users, NOAA has a sophisticated communications and data processing system on the ground. The satellite transmits data continuously to a receiving station in Wallops, Virginia. This station receives the data, strips out that portion representing the view of the earth, then sends this portion back to the satellite at a lower data rate. The satellite in turn retransmits these data to users. Whereas the data rate from the real time scan is 28 Mbits per second the earth view data are transmitted at 1.8 Mbits per second. The National Weather Service provides subsets of the GOES images by telephone transmission for display on a facsimile machine.

The spatial resolution of the GOES sensor is 0.9 kilometers in the visible, and 8 kilometers in the thermal infrared. Due to the earth curvature the effective ground resolution degrades near the edges of the satellite view. The thermal infrared data are useful for tracking cloud motions at night.

In principal these data are of interest to the USDA because they are acquired throughout the day and provide information on daily insolation. Variability of insolation has an effect on the yield of some crops. These data are not used operationally by any USDA agencies.

The polar orbiting satellites

For frequent coverage of large areas of the earth meteorological satellites of the National Oceanic and Atmospheric Administration (NOAA) carry the Advanced Very High Resolution Radiometer (AVHRR). This instrument acquires data in 5 spectral bands at 0.6–0.7, 0.7–1.1, 3.6–3.9, 10.3–11.3, and 11.5–12.5 micrometers (NOAA, 1981).

The image is scanned by a rotating mirror, with satellite motion providing the scan advance from image line to image line (Fig. 7). The sensor acquires data from earth horizon to horizon with a spatial resolution of 1100 meters at nadir, although the equivalent pixel size increases greatly at the edges of the image. Because the field of view of the instrument is large a relatively large energy flux passes through the optics to the detectors, and highly precise measurements of radiant flux are possible. The sensor delivers 10 bit data, i.e. numbers in the range 0 to 1023, despite the relatively low scan efficiency E, which is defined as the ratio

$$E = \frac{\text{Duration of observing interval}}{\text{Scan repetition interval}}.$$

Since the scan mirror rotates at constant angular velocity the efficiency is the ratio of the angle of observation, 114 degrees, divided by the full rotation angle of 360 degrees or about 33%. During the observing interval 2048 measurements corresponding to the number of pixels in a horizon to horizon scan are obtained in each of 5 spectral intervals. Thus each measurement is allotted a time interval T

$$T = \frac{(\text{scan efficiency}) \times (\text{rotation time} = 1/6\,\text{sec})}{2048}$$

$$\approx 25 \text{ microseconds}$$

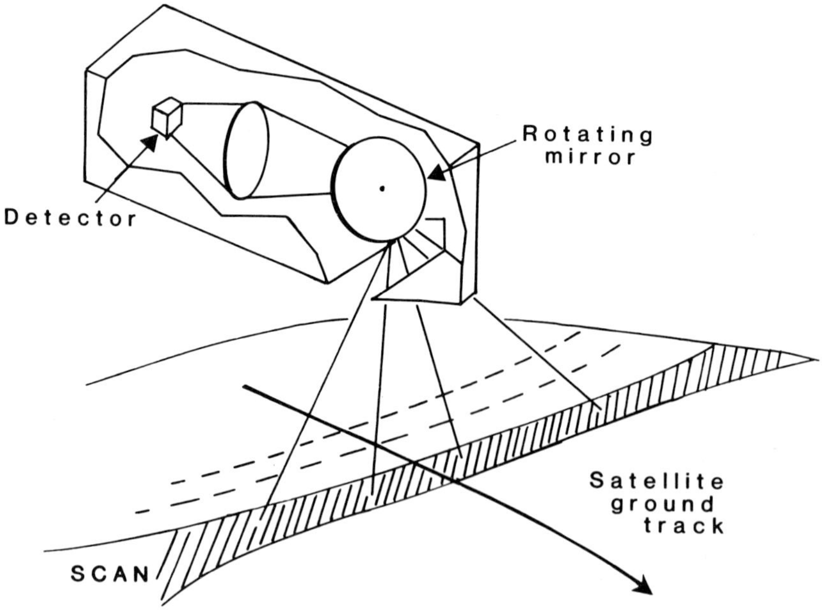

Figure 7. The NOAA meteorological satellites scan along track by virtue of satellite motion, while a rotating mirror provides cross-track scanning.

The sensor data rate from the AVHRR is 2 Megabits per second. For a given image quality higher performance detectors are required as the sampling time gets shorter and shorter.

One may wonder why the meteorological satellites do not rotate like the geosynchronous satellites, thereby eliminating the need for a scan mirror, and decreasing the performance requirements for the attitude control system. This constraint results from the need for other observing instruments to view the earth continuously, which would not be possible if the satellite rotated.

The visible and near infrared channels of the AVHRR are used for by the Foreign Agriculture Service for monitoring vegetation status. The frequent global coverage of this instrument is appropriate for this task, even though the higher resolution Landsat sensors provide much more detail for small areas. The high observing frequency of the meteorological satellites is unmatched by the higher resolution instruments.

The Landsat satellites

The first high quality earth viewing instrument, the Multispectral Scanner (MSS), was placed in orbit onboard Landsat 1 in 1972. Currently virtually identical sensors are operating on Landsat 4 and 5, which were launched in 1982 and 1984, respectively. These instruments image the earth in 4 spectral bands at 0.5–0.6, 0.6–0.7, 0.7–0.8 and 0.8–1.1 micrometers. The ground resolution is 80 × 80 meters for a swath 185 kilometers wide. From data rate considerations the improved ground resolution of the MSS compared to AVHRR has been offset by a reduced swath width. This translates into a 16 day repeat cycle of coverage as the satellite ground track moves to the west from one day to the next. As a second consideration the viewing angle subtended by the coverage swath is very small compared to the 360 degrees which would be scanned by a rotating mirror. In order to maintain a reasonable radiation level at the detectors it is necessary to increase the scan efficiency of the instrument. This is achieved by means of a vibrating mirror which rocks back and forth in a plane perpendicular to the satellite motion. (Fig. 8) Because of this sawtooth pattern of the ground track it is possible to acquire data on only one half (left to right) of the instrument scan. Even with this improvement of scan efficiency to 50% a higher radiation level is required to achieve an acceptable signal to noise level from the detectors. The radiation level is increased by using 6 detectors in each spectral band so that a 480 meter strip is covered with each left to right scan. This swath width corresponds to 14 back and forth mirror cycles per second as the spacecraft circles the earth at approximately 7 kilometers per second. The time interval for each of 3000 measurements along a scan line is about 25 microseconds, as for the

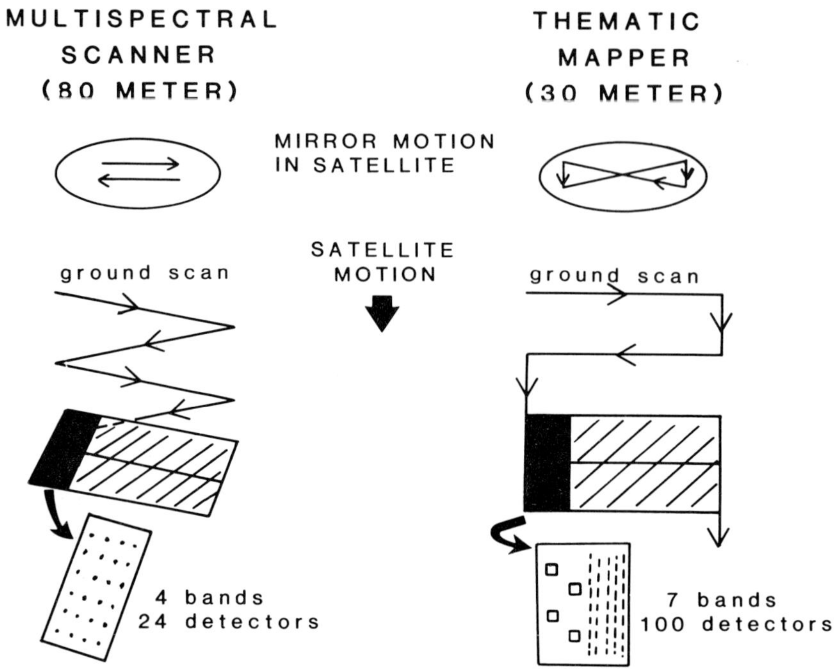

Figure 8. A rocking mirror is used to achieve a scan efficiency near 50% for the Landsat Multispectral Scanners. The far greater complexity of mirror motion of the Thematic Mapper provides only a factor of 2 improvement in scan efficiency, while adding greatly to system cost.

AVHRR, but the data are sampled to only 6 bit accuracy, i.e. 0–63. Lower precision results because the 80 meter field spatial resolution provides a much lower radiation level than is obtained from the 1100 meter spatial resolution of the AVHRR, and it is not feasible to compensate by using very large telescopic optics in the spacecraft. The MSS acquires data at the rate of 15 Mbits per second during the active portion of the scan.

The Thematic Mapper (TM), first launched in 1982 on Landsat 4, is more complicated in all respects (Engle, 1980, USGS and NOAA, 1984). Instead of having 6 detectors per band, it has 16 in each of the first 6 spectral bands, producing 30 meter spatial resolution compared to 80 meters of the MSS. Three spectral bands are in the visible, 0.45–0.52, 0.53–0.61, 0.62–0.69 micrometers, and three are in the near infrared at 0.77–0.91, 1.57–1.78 and 2.1–2.3 micrometers. The seventh band at 11 micrometers has 4 detectors corresponding to 120 meter resolution. The spatial resolution in the thermal infrared is poorer due to the lower radiant energy level associated with thermal emission from the earth as compared to reflected sunlight. Thus the TM has 100 detectors instead of the 24 of the Multispectral Scanner, and acquires 8 bit data instead of the 6 bits of the MSS. In order to achieve desired signal to noise figures an improved scan efficiency is required. This was achieved through a complicated two directional motion of the mirror within the TM optical train, as illustrated in Fig. 8. This motion creates the appropriate ground pattern so that near 100% scan efficiency is achieved. Besides the obvious complexity of the scan device a second factor was recognized well after the major design aspects of the instrument were completed. Because the TM scans the same width strip, 480 meters per scan, but in both directions, the mirror bounces at half the frequency of that for the MSS. As a result the two mirror vibrations interact with structural resonances in the spacecraft itself, and the spacecraft flexes. This realization lead to the addition of an additional component in the system: a highly precise attitude sensor which senses the small amplitude high frequency variations of the view direction of the TM. These attitude variation data are transmitted along with the sensor data to the ground, where geometric corrections are computed and applied to the imagery in order to remove the geometric distortions which are inherent in the data from the vibrating instruments. As mentioned previously the TM acquires data at 80 Mbits per second. Both the TM and MSS acquire data over a 185 kilometer wide swath, but the time for each measurement is essentially the same because of the higher scan efficiency, 100% versus 50%, and the use of more detectors in the TM.

SPOT

On February 21, 1986 a French satellite, the Satellite Pour l'Observation de la Terre (SPOT), was launched into an orbit similar to that of the Landsat satellites. The performance specifications of the instrument represent a significant improvement over other satellite systems, as 20 meter data are acquired in three spectral channels (0.50–0.59, 0.61–0.68, and 0.79–0.89 μm), and 10 meter data are obtained in a single visible channel (0.51–0.73 μm). A 10 meter pixel is 1/9 the area of the Thematic Mapper pixel. This spatial resolution is adequate to resolve many cultural features such as houses, small roads and bridges, etc, and opens possibilities for mapping which have previously been unavailable from civilian satellites (SPOT, 1984). This capability will be most valuable in the developing world, and in areas in this country where year to year changes of land use and construction must be monitored. Perhaps as significant for SPOT as the spatial resolution is the level of technology used within the instrument. The linear array technology (Thompson, 1979) used in SPOT allows nearly 100% scan efficiency without the need for a mechanical scan (Fig. 9). The detectors are "charge coupled", so

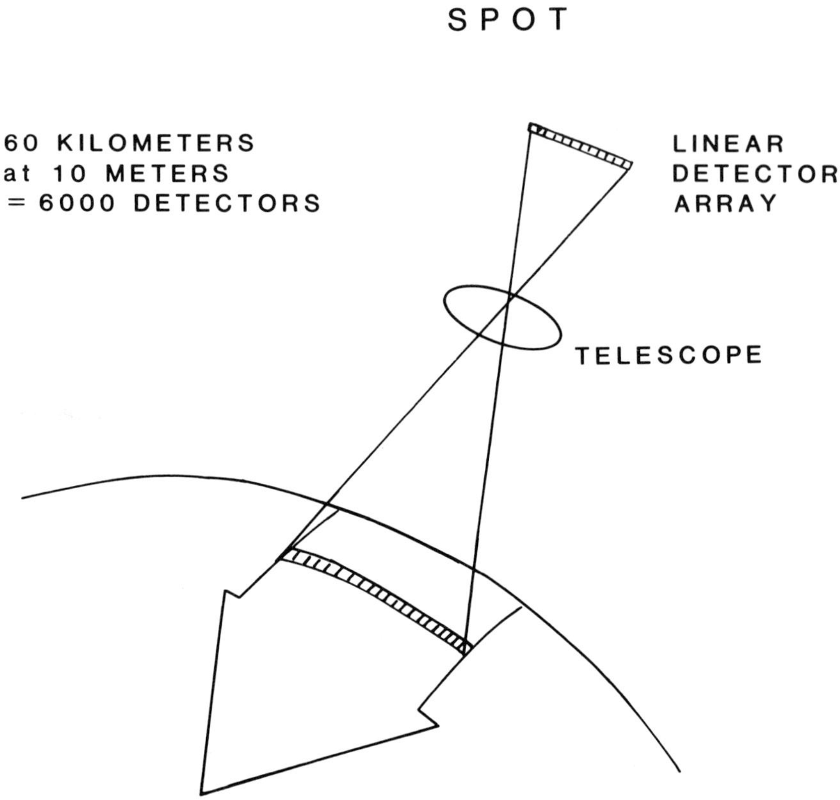

Figure 9. The French SPOT satellite employs linear arrays of detectors. The absence of internal moving parts greatly simplifies many aspects of design.

that they may be read out from one end of the array and complex detector wiring is not needed. The read interval decreases the effective scan efficiency slightly, but the absence of mechanical scan has many beneficial implications for attitude control of the satellite, geometric fidelity of the image data, and system reliability. The time interval for each measurement is the spatial resolution divided by the spacecraft velocity, i.e. more than a millisecond. This long sampling time permits accurate measurements even from the sensors 10 meter pixels. In SPOT the 10 meter channel data from the instrument are partially processed in the satellite to reduce the data rate, with resulting reduction of transmitter and data storage requirements. The telemetry data rate is 25 Megabits per second from each of two identical instruments, representing either 3 channels of 20 meter data or the single channel of 10 meter data over a swath width of 60 kilometers. The 10 meter data are decoded during ground processing. The on board coding technique is facilitated by the microchip technology developed in the past 10 years. The implications of this technology for data users have not yet been explored, but in principle they extend as far as compressing the satellite information to a compact form which could be transmitted by phone to a farmer within hours after the satellite overpass.

Proposed instrumentation

The trend of instrumentation should be evident from the previous discussions. The ultimate goal is extremely high spatial resolution data over a wide coverage swath, with many spectral bands, e.g. 100 bands in the visible, near IR, and thermal IR. Figure 10 illustrates a conceptual design for a proposed instrument, with a two dimensional array of detectors providing essentially 100% scan efficiency for a high spatial resolution, a broad coverage swath and a spectral resolution, which is better than our knowledge of the spectra of most types of soils, vegetation, rocks and water bodies. Such instruments will measure essentially all the spectral variability which is accessible from a satellite platform. Design studies are in progress (NASA, 1984) for such instruments with the goal of launch and operations in the 1995 time frame.

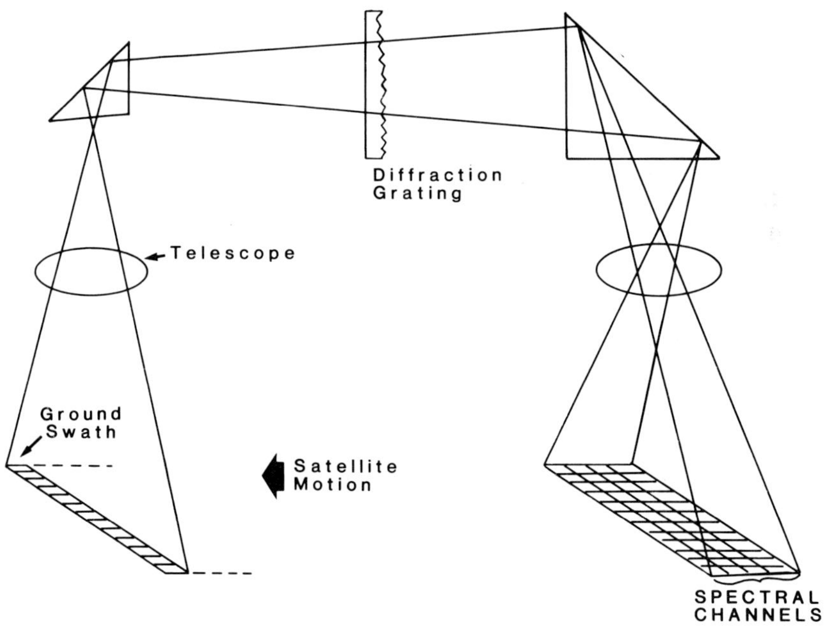

Figure 10. As array technology matures one may consider substantial expansion of the imaging capability.

Trading off instrumentation, data interpretation and applications

The previous discussion illustrates that the technical capability for better and better instrumentation continues to improve rapidly. The technology is basically limited by the availability of sensitive detectors which can convert very low energy fluxes into an electrical signal, and by the capabilities of specialized computers which may support the acquisition of encoded data at the ground. Realistically speaking the state of instrumentation has progressed so far that financial support is the major limiting factor. Proposed instruments for the future Earth Observing System (EOS) can be expected to cost roughly \$100 000 000 each. These costs exceed greatly the available funding of potential user agencies such as the USDA. Thus for the USDA the major consideration is optimizing existing satellite systems (Price, 1982). For global vegetation monitoring it is essential to obtain data on a regular basis, e.g. every 3–5 days at a minimum. As we have seen data transmission limitations imply an instrument which acquires a wide swath of imagery. Better spatial resolution than the AVHRR's 1 kilometer would be desirable, but frequent satellite coverage is most important because many regions on the earth are covered by clouds about 50% of the time. On the other hand the geologic community prefers very high spatial resolution data, since a single good image of an area is sufficient until an even better instrument is flown in space. In addition the geologic community can afford to pay a high price for a single satellite image, while the agricultural interests cannot support a large per scene charge for data which must be purchased day after day, year after year. The present earth viewing satellite instruments represent a compromise among the conflicting needs of many disciplines.

Ironically the main limitation to use of satellite data is the lack of an adequate research base to support the applications. Figure 10 illustrates

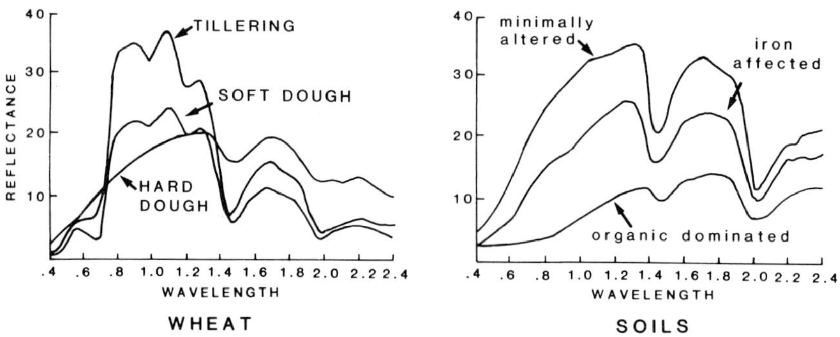

Figure 11. The similarity of spectra of from many macroscopic objects tends to hamper the interpretation of satellite observations, since only a few spectral intervals are measured by present instruments.

typical spectra from vegetation and soils (Stoner and Baumgardner, 1981), as would be measured in the field with a hand held radiometer. However, the satellites see a large area with a single measurement so each pixel is typically an average over soils and vegetation in various states of maturity, with minor spectral variations due to moisture differences, availability of fertilizer and trace elements, etc. This makes interpretation of crude spectra represented by 4–7 spectral bands a confusing task. Historically instrumentation has been developed by NASA mainly as an engineering challenge, and users have simply been provided with data and encouraged to use them. Combined satellite/field studies have been rare, and very little research has been devoted to spectroscopy as a discipline, as opposed to visual interpretation of data. Although it is feasible to optimize spectral band selection through statistical studies of representative data sets (Price, 1984), this concept has not been used in instrument design.

A reawakened interest in basic research has opened new questions which have been in the background since the days of the first Landsat. Radiometric calibration must be considered if data from one satellite are to be compared with those from another, or from ground measurements. The interaction of radiation with a vegetation canopy is a subject of field investigation and numerical modeling. The technical questions of optimizing a color display for visual interpretation are being addressed. With better theoretical understanding of the physics of radiation, we may expect the quantitative criteria to be developed for selecting the scientific parameters for the instruments which are now in the design phase. The recurring issues are: what spatial resolution and swath width, how many spectral channels and at what wavelengths, and what level of precision of measurement? Historically the applications community has been happy to supply to the instrument builders a "shopping list" requesting extremely high spatial resolution, many spectral bands, frequent coverage, and high precision. The concepts discussed in this paper suggest that selection of the rate of data acquisition, based on cost considerations, followed by user's choices for other parameters, would be a more effective approach.

Future

The U.S. program of satellite earth observations stands at a crossroads. On the one hand the potential for improved instrumentation is without precedent. The development to commercial scale of multiple detector arrays, and the coincident development of small sophisticated microprocessors for data manipulation, open the possibility of instrument specifications which far exceed all previous limitations. To the degree that onboard data processing (i.e. data compression) can reduce the need for a large transmitter on the satellite and a large data archive on the ground, it is feasible to consider data acquisition rates far beyond present satellite capabilities. Realistically speaking the limitations are now associated with cost rather than engineering and system complexity.

However cost recovery is now the government's standard for all except research satellite instrumentation, and the last major research satellites, Nimbus 7 and Seasat, were launched in 1978. The prolonged time lapse suggests that future research satellites will be launched sporadically at best, and that the tradeoffs of science and applications versus engineering costs will be evaluated by a new set of managers and congressional overseers. The next major research satellite, the Earth Observing Satellite, will be considered by Congress around 1990, with launch scheduled sometime in the mid 1990's. It is most unlikely that users' data purchase fees will support design and fabrication of improved sensors for perhaps a decade or more. Will the program of earth observations for practical purposes come to an end? Probably not. For at least the next 5–10 years a subsidized U.S. corporation, EOSAT, and a subsidized French corporation, SPOT Image, will provide data from high quality instrumentation. The near term answer will represent a political and scientific balance between cost recovery, the steady improvements of technology, and the desire for national prestige.

Acknowledgement

I am indebted to J. Ritchie, G. Hart and H. Ostrow for careful review of the draft manuscript.

References

1. Engle, J.L. 1980. *Thematic Mapper – An Interim Report on Anticipated Performance*, Proceedings of the AIAA Sensor Systems for the 80's Conference, Colorado Springs, CO, pp. 25–37.
2. Gonzalez, C.G. and Wintz, P. 1977. *Digital Image Processing*, Addison Wesley Publishing, Reading, Mass, pp. 5–11.
3. Hambrick, L.N. 1980. *Earth Locating Data of Spin-Stabilized Geosynchronous Satellites*, NOAA Technical Memorandum NESS 111, 49 pp.
4. *Landsat 4 Data User's Handbook*. 1984. U.S. Geological Survey and National Oceanic and Atmospheric Administration.
5. NASA. 1984. *Earth Observing System*, Technical Memorandum 86129, Goddard Space Flight Center, Greenbelt MD, 2 Volumes.
6. *NOAA Polar Orbiter Data User's Guide*. 1981. U.S. Department of Commerce.
7. Price, J.C. 1982. *Satellite Orbital Dynamics and Observation Strategies in Support of Agricultural Applications*, Photo. Eng. and Rem. Sens., 48: 1603–1611.
8. Price, J.C. 1984. *Comparison of the Information Content of Data from the Landsat-4 Thematic Mapper and the Multispectral Scanner*, IEEE Transactions on Geosci. and Rem. Sens., 22: 272–281.

9. Slater, P.N. 1980. *Remote Sensing, Optics and Optical Systems*, Addison Wesley Publishing, Reading, Mass, pp. 401–437.
10. *SPOT Simulation Applications Handbook*. 1984. American Society of Photogrammetry, Falls Church, VA 274 pp.
11. Stoner, E.R. and M.F. Baumgardner. 1981. *Characteristic Reflectance of Surface Soils*, Soil Science J. of Am. 45: 1161–1165.
12. Thompson, L.L. 1979. *Remote Sensing Using Solid-State Array Technology*, Photo. Eng. and Rem. Sens. 45: 47–50.
13. USDA, 1982, *Basic Statistics 1977 National Resources Inventory*, Soil Conservation Service and Iowa State University, Statistical Bulletin 686, 267 pp.

Near-infrared used in process chemistry

BRUCE R. BUCHANAN, JONATHAN PERKINS and DAVID E. HONIGS
Center for Process Analytical Chemistry, University of Washington, Seattle, WA 98195, U.S.A.

Introduction

Dr. Gary Heiftje has written a text book on Analytical Chemistry which some consider to be a classic. That textbook contains one sentence on near-infrared spectroscopy which states that "The NIR is generally a useless region for an analytical chemist." While this statement is indeed correct for classical studies for functional group assignment, laboratory technicians and process engineers out-perform conventional spectroscopy time and time again by using near-infrared (NIR) spectroscopy for routine analysis. NIR analysis, or NIRA as it is called by some, relates NIR diffuse reflectance or transmission spectra to analyte concentrations. It is, therefore, primarily a quantitative technique, although it can be used for qualitative purposes as well.

The NIR wavelength range extends from approximately 1000 nm up to 2600 nm, but a better way of saying that would be that the NIR region stretches roughly from about where the human eye gives out, up to the first fundamental vibrational bands, so it is entirely an overtone region. There are very few electronic transitions, or electronic absorbances in this region. There are also very few fundamental absorbances in this region. The electronic absorbances occur in semi-conductors. The fundamentals take place in metal hydrides. But with the exceptions of these two classes of compounds, it is overtones that are sought. Because the bands are overtones, they are formally not allowed by quantum mechanics. In actuality, the bands are the anharmonic portion of what is typically modelled as a harmonic oscillator in a mid-infrared vibration. The fact that the transitions are not allowed quantum mechanically does not mean that they do not happen, it just means that they happen with less intensity or at less if a probability. Thus, the typical NIR band intensities are down by a factor of a hundred to a thousand compared to the intensities of the corresponding fundamental ones.

Why then is the NIR chosen over other methods? It has three distinct characteristics. On the surface these characteristics do not appear as advantages. However, it can be shown that the characteristics that appear to work against this region are actually quite favorable. Of the characteristics, the first is the near-infrared has very small absorptivities, as has been previously stated. Second, it has very broad and overlapping features. It is affected extremely by hydrogen bonding, particle size and other physical properties. Finally, the band assignments are very complex. Although it is not simple to go through in the mid-infrared and calculate what bands should be present, it is a much more difficult problem to go through in the NIR and calculate what bands would not be present, as these are the ones which will be observed. All the possibilities of combinations, differences and overtones make NIR band assignments very complex. If these are the advantages, it obviously would be unwise to discuss the disadvantages of the technique. But is is necessary to explain the method in a little more detail. Also it will be shown how the reasons that kept people out of the NIR region are actually the precise reasons why NIR is going through a very large explosion of agricultural and industrial use.

NIR special features

First of all, the reason weak band intensities prove to be an advantage. Weak bands are a disadvantage if attempting the analysis of trace components. However, if one is analyzing bulk constituents, weakness of absorbance can actually be used to arrange the path length to be compatible with an industrial product. Another nice thing about the weak absorbances is that they make diffuse reflectance a very well behaved phenomenon. At a particular particle size an effective path length exists that the light passes through before it is reflected. As the particle size grows smaller this pathlength decreases and the absorbance decreases. This is why a finely powdered substance appears lighter than coarser particles of the same substance. In the visible region the bands are very intense because of electronic absorbances. This means that the light does not penetrate very far into the sample at all. This is why UV-visible monitoring of a diffusely reflecting compound is so difficult. Only the first monolayer, or couple of monolayers on the sample are being monitored. Only the dirt that has settled on the top of the product is being analyzed and not the product itself. Likewise, in the mid-infrared, the band intensities are still very strong and the same problem exists. In the NIR, the band absorbances are very weak. This means that the radiation not only hits the surface, but it penetrates to some extent. So when the diffuse reflectance signal, $-\log(1/R)$, is measured, the signal comes from well into that sample, perhaps a centimeter or further into the depth of that sample. Therefore, the dirt or oxidation layer that is on the surface is now only a small fraction of the measurement, a minor interference instead of being the major observable. Furthermore, much larger particle sizes can be monitored. The effect of particle size is clearly seen in the visible region. Visible light barely makes it through large particles and the diffuse reflectance, $-\log(1/R)$, is not very linear with concentration. However, even with very large particle sizes, the NIR light will traverse the particle, perhaps even several particles, before it gets scattered back. Thus, the signal remains much more linear with concentration. There are theories that explain how to linearize diffuse reflectance data. One could, for instance, use the Kubelka–Munk function, which linearizes diffuse reflectance with concentration. However, diffuse reflectance can be treated just like transmission to a first order. The minus the log of the reflectance is very much proportional to the absorbance of this material. This allows one to use the pseudo-absorbance term in Beer's law, analogously to conventional spectroscopy.

This ability to sample is foremost in the importance of NIR. NIR is a highly flexible technique, then, because it can measure the solids or the liquids, using reflectance, tranmission or transflectance techniques. This flexibility means that the sample needs little or no preparation. The sample can be scanned as it exists in the inventory, or as it is gathered off-line in a sampling operation, or, perhaps, the material can be used as it exists in the process. True on-line monitoring can be done and the NIR would be used as a feedback or a feedforward control to improve the quality and consistency of the product.

Finally, NIR can handle aqueous samples. This capability is very unique. Water, like most materials, has weak absorptivities. So transmittance through what would be considered rather large pathlengths of water can be done. This capability has advantages not only for aqueous process streams, but also for instrumentation in general. One does not have to worry about nitrogen purging or background water fluctuations in the air, because these fluctuations are so small that they do not interfere with the measurement. The carbon dioxide in the air also does not interfere.

The broadness of the bands has another effect. It allows for less stringent tolerances in instrument manufacture. A spectrometer which will perform very admirably in the NIR can be made from relatively simple equipment. The mechanical tolerances are not as high. The optics are much simpler because things like quartz, sapphire or even glass, can be used. These are materials that stand up very well to a process. Very good sources are available. As a matter of fact, the emissivity of tungsten lamps peaks in this region. Similarly, very nice reflective optics are available. Finally, the detectors do very well because the energy in the photons is much greater than the mid-infrared. Thus, a detector such as lead sulfide, which has been highly optimized for the military and night viewing scopes, or detectors like silicon, which have been optimized by the telephone company for telecommunications, all perform very well. They are all available from Radio Shack or other manufacturers, and they cost very little compared to more exotic mid-infrared detectors.

The broadness of the bands is not a major drawback. The fact that the information is present in such a broad form is only the outline of the hydrogen bonding shape. Very often, we are interested in changes in hydrogen bonding that affect physical properties. These changes can actually be rather narrow and they are simply hidden under a very broad background. So data treatments that enhance the narrow features can be used. One such treatment is to take the NIR spectrum, such as the one modeled in Fig. 1, and to take the derivative. If the first derivative is calculated it can be seen that even the very broad bands start to show a much more narrow character. As a matter of fact, that broad band might be made of two or three narrow components. Those components start to be resolved in the derivative spectra. The individual effects start to emerge, rather than one big lumpy potato-like band. If the second derivative is used, the information is enhanced further. The second derivative is often used because the band intensities have peaks where peaks were in the absorbance or pseudo absorbance spectrum. The reason derivatives enhance the information is that the data treatments act like filters. If one considers the band width of the information present, the first derivative is discriminating very highly against broad information or very broad bands. The second derivative is discriminating even further towards narrower and narrower features. If this type of discrimination is coupled with a smoothing filter, like a Savitsky–Golay algorithm, a band pass is effectively created. One region has broad band widths cut out by the derivative functions, and the narrow band widths cut out by the smoothing operation, and so it is kind of like electronic filtering. The portion of the spectrum that contains the chemical information is returned and the broadness of the bands doesn't happen to matter very much.

The final characteristic listed was complex band assignments. It is not the fact that the band assignments are complex that is really an advantage, it is what makes them complex. The complexity arises because of the overtone transitions, the anharmonicities, which are affected as much by molecular structure as they are by secondary and tertiary chemical

Figure 1. (a) Model of a near-infrared spectrum. (b) Second derivative of near-infrared spectrum modelled in part (a).

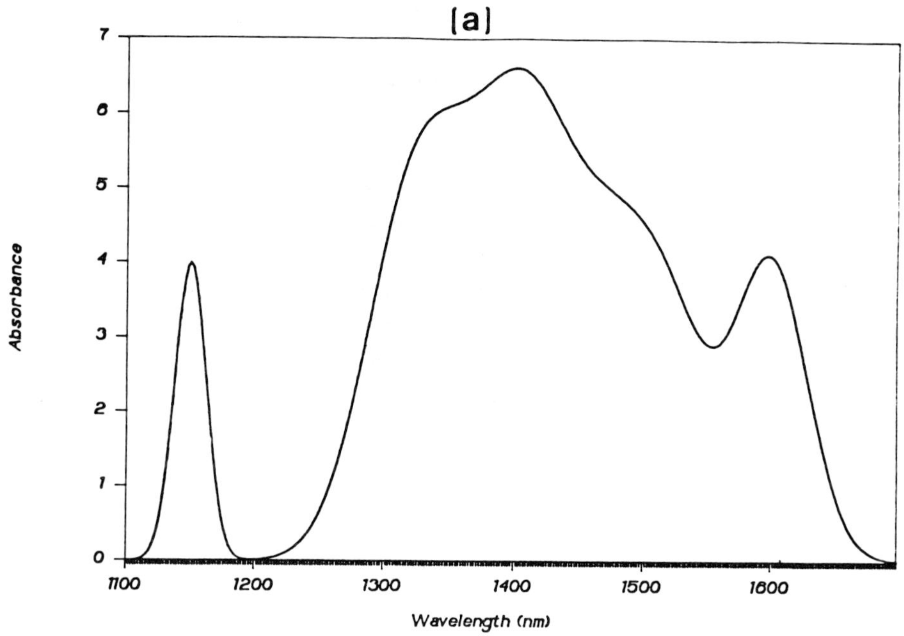
(a)
7
6
5
4
3
2
1
0
Absorbance
1100
1200
1300
1400
1500
1600
Wavelength (nm)

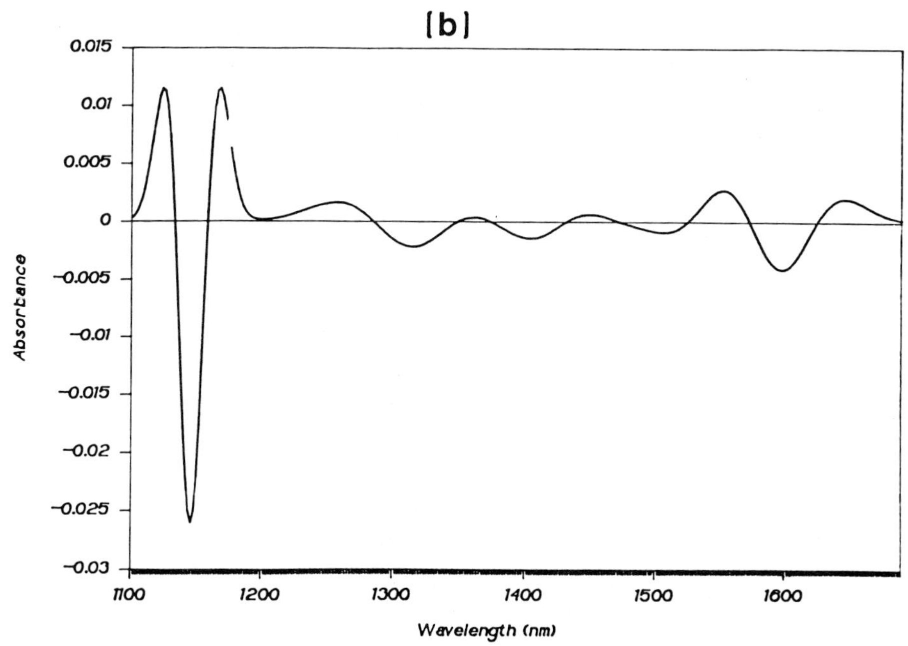
(b)
0.015
0.01
0.005
0
−0.005
−0.01
−0.015
−0.02
−0.025
−0.03
Absorbance
1100
1200
1300
1400
1500
1600
Wavelength (nm)

effects. These structural differences can be used to make calibrations for a wide number of things other than chemical species. The calibration technique of NIR is actually the real difference between NIR and mid-IR or UV-visible. This calibration procedure was made necessary originally because the bands were so complex. Just like computers in FTIR, what was originally a disadvantage rapidly got launched into an advantage. The presence of a computer in Fourier-Transform IR made possible all kinds of new techniques. Because we had to resort to mathematics as a first way to solve a problem in NIR, mathematical calibration techniques have advanced faster in this area than, perhaps, in the mid-infrared or UV region of the spectrum. These mathematical techniques can be summarized in a few steps. Room does not allow a detailed analysis of each step, but here is an overview. First the samples must be prepared and the spectra obtained. This is done by illuminating the samples with NIR energy at a series of wavelengths. A survey scan is obtained, because the mathematics determine which regions of the spectrum contain information that is important. Next, a calibration equation is found and the results are calculated. It is necessary to have a series of samples that the spectrometer learns to interpret. Finally, the calibration is used to determine the values for unknown samples. It will be placed in an industrial quality control or process control situation. The calibration that was learned by the mathematics will be used.

Automation of methods development

The NIR methods development is actually an automation of a portion of what the analytical chemist or analytical spectroscopist used to do. Previously many of the effects of automation have been seen, such as in the way instruments take data, or the way the readouts are handled. Automation is beginning to creep into the area of sample presentation/sample preparation, with the use of robots. NIR spectroscopy, in a way, takes automation one step further, as it automates the way that a method is developed. Before automation, things tend to be repetitious. Calibrations tend to be dependent on a particular operator, and there is tedium or boredom associated with the work. Or if the tedium or boredom do not exist, then the best operators do such good jobs that they get promoted and the Peter Principle takes over. Good people are promoted until one person who cannot do the job takes the position and that is the person that will stay there. After automation, in general, calibrations require fewer man-hours and a more uniform result is achieved. Computer generated calibrations might not be the absolute best, there might be one person who can beat the instrument consistently; however, the instrument will do better than the average person. Finally, the results tend to be fairly reliable, that is, the instrument will work day and night and can calibrate over the weekend, or can do a rush job overnight without demanding overtime pay. This is obviously applicable to a robot-type of situation, but how does it work for methods development? Methods development in spectroscopy is extremely operator-dependent. It is extremely time consuming for that operator to have to try to interpret the problem and have to understand the chemistry that is present, to have to understand the spectroscopy that is present, to have to understand what is the purpose of the measurement and trying to relate some chemical process or some chemical value back to the physical process in the plant, and then to relate that to the equipment, and then to relate that to the bottom line in terms of dollars. All of these things are very difficult to comprehend, and by automating this with a mathematical approach, a more uniform method of calibration is achieved. The way this is typically done is that the technician brings in a series of samples, runs their spectra, runs the computer program and about 8 or 9 times out of 10 they come up with a valid calibration that can be used right away, end of report. Now they can go on and develop a different application. Furthermore, if the operator leaves it only takes about a week to train somebody to do this procedure.

This area of automation is really striking in the area of methods development, but it does not get rid of the analytical spectroscopist. Rather, the spectroscopist can be relieved of routine methods work. Historically, the analytical chemist always had to be there to fine tune the instrument. Later, the analytical chemist always had to be there to do the analysis or to push the button to start the instrument. Now the analytical chemist has to be there to develop the methods for the new instruments. It would be best if the analytical chemists were to get all the way out of those things and be working on tomorrow's technologies. They could be looking towards new sensors and new approaches, and not have to be present in the lab, a slave to today's technologies or today's methods development.

In summary to this introduction to NIR, it is emphasized that this technique is a quantitative technique that is rapid, because the spectrum can be scanned very rapidly and there is no sample preparation, which also tends to make it cost effective. The method is entirely controlled by computer as far as making a prediction, and finally, it is very accurate and precise. It is accurate because a nice response in the spectrum is returned. There is very nicely designed instrumentation, because the region favors the optics, and it favors the detectors; finally, it is precise because the signal-to-noise that is achieved in the instrument is enhanced by the mathematical ability to counter non-linear effects. The calibration programs will find ways to avoid the temperature effect, or will find bands that have the same non-linearities and thus make an off-set, and these levels of corrections are beyond what the average methods developer interprets. So we see that the NIR analysis is generally more precise than an operator developed method.

Process applications

The area of near-infrared process control will be broken down into several similar types of applications. It is not possible to go through everything that has been done, because the list is extremely long. Therefore, examples in specific categories will be discussed.

NIR monitoring of chemical species can be broken down into two broad classes, agricultural and industrial. In agricultural species one might analyze for something such as protein, moisture, or oil as the chemical species. Sample matrices that might be analyzed include: wheat, barley, rye, hops, alfalfa, corn, meat or milk. These have straightforward chemical properties, not so straightforward sample matrices.

In industrial applications methane or freon, hydrazine, metal hydrides, other organic compounds, tri-butyl phosphates, moisture (moisture is the classic near-infrared application), propane, general organics and a surprising number of inorganics are the species of interest.

Where might one look for these applications? In petroleums, plastics, powders, coatings, textiles, emulsions. Once again, the sample matrices extend from very simple, clear liquids, to more dificult ones such as solids. Some difficult samples are not just solids, but are oriented solids, slurries, or are very viscous samples.

Besides chemical properties, or strictly what one might call analytical chemistry values, one can branch off a little bit with near-infrared to the area of chemical parameters. Chemical parameters are properties where the link between the parameters and chemistry is very obvious, but there is still not a species for which a concentration can be recorded. Examples of this include the unsaturation or saturation of an oil, or perhaps the cis-trans ratio in an oil. They are chemical values, but it is difficult to write a concentration term for these species. Another chemical parameter is the difference between unbound and bound moisture, that is, the amount of moisture that inherently stays in the product through the processing versus that amount of moisture which is driven off by the driving process, or is gained as it sits in a humid atmosphere. Other parameters such as iodine value, or solid fat index, are values that in themselves are monitoring a chemical effect, but their values are now of more use than the effects they originally were intended to measure. Another parameter is total solids or total inorganics. These are related to a series of chemical species but not to any one.

In the industrial area, chemical parameters include molecular weight. The relationship between a chemical or a molecule and its molecular weight is pretty obvious, yet a concentration equation for molecular weight cannot be written down in a straightforward manner. The BTU content of a material is another example of an industrial parameter where the relationship is fairly obvious. Acidity, salinity, total solids, total coating weight, are all similar chemical properties; they can be related to the chemistry fairly simply, but they are not very well related to the Beer–Lambert Law.

The final measuring possibility uses the near-infrared to model physical or process parameters. Not only can the physical values be modelled, but calibration in terms of process settings can be performed. If there is a dial on the control board and an engineer is to maintain product quality by varying the dial from 1 to 10, the near-infrared instrument can be calibrated so that it returns the number to set the wheel on, instead of some chemical value or some indirect chemical value, which then has to be interpreted.

What are some of these physical properties which can be monitored by NIR? In an agricultural situation, applications include wheat hardness, where the NIR can tell how hard it is going to be to grind a sample before it ever is ground. Digestibility is another physical property that can be measured. Observing the grass' spectrum, the nutritional value the sheep is going to get can be found before the sheep eats the grass. This is a vast improvement over the previous method, which includes weighing what went in and weighing what came out of the sheep. One of the high-tech applications, done by near-infrared is the harvest state of a forage field. The near-infrared instrument does not calculate the amount of protein in the forage but rather the calibration will predict when the harvest should be performed, given the spectrum reading, along with the current market value for protein and the current time of year. The machine does not evaluate the amount of protein in the forage alone, but it does the regression which relates the price being paid for bulk weight to the protein premium. It also can predict how much more protein is going to be gained, and how much more mass is going to be gained in the near future. These numbers are combined to determine the optimal harvest time.

Another agricultural property is bakability. The way that people used to judge bakability is that they would bake a bunch of little loaves of bread. This is, in fact, the way that many things are done. A pilot plant is set up. Based on the pilot plant experience, the real plant is designed. If the pilot plant experience can be eliminated and the near-infrared analysis used in its place, there is quite a cost savings in terms of time. Also, materials no longer need to be stockpiled in a warehouse. Total throughput of the process can be increased by using the near-infrared. Similarly, properties which can be monitored by NIR include taste or oil extractability, where the amount of oil the process will extract is predicted. With this measurement, the product is bought based on the value the company will get from it and not the value it has as a bulk commodity.

Industrial examples of these applications of physical or process parameters also include oil extractability. In this case it is a different type of oil, something like an oil sand or an oil tar or an oil shale. When an oil extraction process is used, and the process engineer is really not so much concerned with how much oil is in the oil shale, but how much oil the process is going to extract. He also needs to know how to best set the process so that it extracts the maximum amount of oil.

Another industrial physical parameter measured by NIR is density. Density can be used to give a complete history of a polymer like nylon, so that the method of processing can inferred from the NIR. Elasticity is a similar polymer parameter which can also be measured by NIR. With an elasticity measurement one can predict how far a sample can be stretched without breaking. That is a lot better than simply pulling one sample, seeing where it breaks, and expecting the next sample to break at approximately the same point. Individual samples can be analyzed for a destructive test without having to use a destructive method.

Process instrumentation

Instrumentation in the near-infrared is quite varied, and this variation of instrumentation means that there is quite a wide range of NIR applications available. The types of near-infrared instruments include the grating spectrometer, which is available in other light ranges; the Fourier Transform; filter instruments, (which are the workhorses in the near-infrared at present); and correlation type of instruments, which are in development.

In some other spectral regions there might be a big problem in having only a limited wavelength range available. Fortunately, in the near-infrared usually the information that is needed is contained in that limited range. Therefore, fixed filter instruments can also be used. These typically consist of a filter wheel which rotates normal to the light path. The one problem with this method is that one is limited by the number of filters that are in the wheel. That also may not be a fatal problem. The information may still be contained in the set of wavelengths selected. The use of filters is quite applicable to a process method, because once the grating spectrometer is used to decide which wavelengths are needed, cheap, rugged interference filters can be put into place. This gives a very cheap spectrophotometer to make measurements.

In agricultural applications, the filter instruments have been used almost exclusively. As the switch to industrial applications is made, it is necessary to get a wider wavelength range, and so one switches to a grating-type of instrument. The grating that usually is used is a holographic grating, with a very low F number. This allows a very wide spectral range to be scanned. Another type of scanning instrument is the Fourier Transform interferometer. While several people have claimed that the Fourier Transform is useless in the near-infrared, that is not quite true. It does not have quite the same advantages that it might have in the mid-infrared. It still has a very useful function in using the near-infrared spectroscopy, including obtaining complete spectra in a real-time basis. One final type of scanning spectrophotometer is the tilting filter instrument. Tilting filter spectrophotometers use an interference filter, which tilts through the light beam so that a range of wavelengths is scanned using a very limited number of interference filters. Interference filters are fairly cheap. This means that one can scan with a cheap, rugged instrument. Finally there exists a correlation-type instrument. This instrument actually uses a weighted sum of the wavelengths for the correlation. The individual responses are never assigned to specific wavelengths. This is analogous to using the interferogram for the calibration in FT-IR, rather than transforming it to the wavelength space.

Along with the number of instruments available, there are also several ways of making the measurements. It can be done on discrete samples, as would be done with any other spectrometric technique. The sample is removed from the bulk, put in some container and taken to the instrument. The spectrum can be obtained over as long a length of time as desired. The problem is that the sample must be statistically representative of future samples.

Portable spectrophotometers are one way to avoid collecting discrete samples. A worker can actually carry a spectrophotometer to the sample. The sample can be scanned *in situ*. The telephone companies have spurred the development of optical meters for their communication lines. The meter consists of a LED light source and a silicon detector. But with a little bit of modification, it can be developed into an instrument that a workman could carry on his belt. One story Tomas Hirschfeld often related was how a stockpile of material is identified. A foreman is called over who has been at the company for twenty years. He looks at the material and he says, "Oh yeah, I remember now, we got that back in '68 . . ." Well, a portable spectrometer is a quick and sure method of actually determining what a sample is made from. Using spectroscopy to classify a material is a great improvement over guesswork.

Flow-through cells would be used in a beside-line type of system. This is where the sample is diverted from the process line to the sample cell. One type of instrument which exists flows grain through the cell. A wheat sample is actually flowing through and the transmission spectrum of a solid, flowing sample is obtained. Another example is that of a powder on a conveyor belt, where the sample "flows" by an overhead detection system. A problem is the baseline shift because of the particle size differences. Also the conveyor belt sometimes shows through. These are problems that are being researched at present. Once particle size differences can be corrected, a powder could be measured as it goes by on the conveyor belt.

One nice thing about near-infrared spectroscopy is that dirt or contaminants built up on the window is a very minor portion of a spectrum. There is very little interference from contaminants building up on the window. Also, in a flow system the integrity of the window must be high. Using a sodium chloride cell is undesirable since it will probably blow out the first time the sample is run through. Quartz and sapphire are both very good optical materials in the near-infrared. Also, many times the materials will outlast the pipe. After the pipe is corroded away, the window is still there working fine.

Quality, productivity and competitive position

The near-infrared can be applied to a plant using a variety of instrumental and sampling techniques. The number of instruments that are available, coupled with the number of sampling methods, give a wide range of applications that can use near-infrared spectroscopy.

Now, to address the question of how one can apply a near-infrared to a plant situation. Tomas Hirschfeld advised that the best source of ideas were other people. Taking his advice, we will borrow the accountants' idea for the widget manufacturer in an example to help answer the above question. In accounting, a typical problem is to look at a widget factory and calculate the possible costs and profits. We will use the widget factory to show how to apply the near-infrared to the various processes.

Presently many factories take samples out of the system to be analyzed. These samples are taken into the analytical lab where a lengthy analysis is done. By the time a value is found, the received raw materials are already in the sample line. If a bad raw material has been received, it is already over in the reaction chamber, and a lot of time and money has been lost. Applying near-infrared spectroscopy to the received raw materials would prevent this. By using the calibration, all raw materials received that did not meet the specifications would be rejected. Two near-infrared instruments have been drawn in Fig. 2 to illustrate this point. In process analysis there is a need to be concerned with preventing down-time. If one of the instruments fails, lamps burn out, etc., a back-up system can take over the load. Thus, there is no lost time because of an instrument failure. The lines from the instruments in Fig. 2 are not necessarily fiber optics but can be thought of as just a connecting line to the instrument. In this case it might actually be a portable instrument, where the worker uses the instrument's response as a basis for action. Or there may be a filter instrument at the receiving dock, where the raw materials are scanned discreetly. It does not have to be a fiber-optic based system.

The next step in the widget process is mixing or blending of the raw materials. How can one tell when the mixing is complete? If a transmission measurement is made where we actually have a small cell that is dropped into the tank as the batch is mixed, continuously monitoring the mixing of the samples can be done. In this case fiber optics could be used

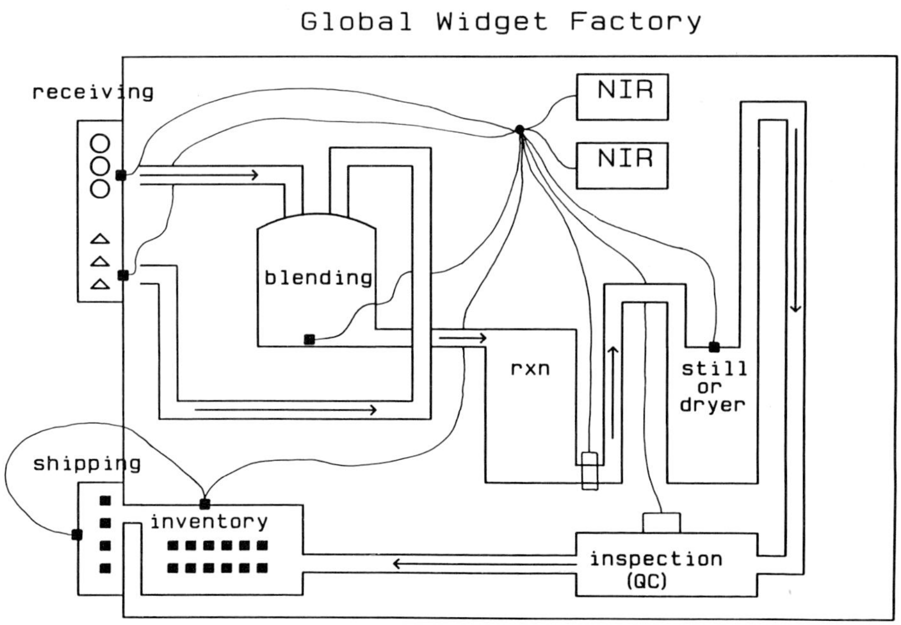

Figure 2. Floor plan of a factory showing potential application of the near-infrared to process analysis.

to do the remote measuring. This way the raw materials are mixed for the optimal amount of time. The blended material can be piped to the reaction chamber as soon as it is thoroughly mixed. The mixing process does not need to be done for a set amount of time, which usually has to be longer than minimum to make sure the raw materials are mixed. The near-infrared has a lot of utility for this. As the materials are mixed together, the spectrum should have a very clear, measurable shift that is related to the degree of mixing. Using the methods described earlier, a correlation can be made to the percentage the materials are mixed.

After the raw materials have been blended they might be sent to a reaction chamber. Measurements can be made through the pipe as the sample flows by the window. Whether the reaction has gone to completion, whether the reaction is going in the desired direction, and whether to reject reactions that for some reason are not going as planned, can be determined by NIR.

At the completion of the reaction, the sample might be piped to a distillation column. The near-infrared can be used as a feedback sensor. This has actually been done in England, where they used the near-infrared to determine how much of the distillant to feedback into the pot to get the optimum distillation process. Or NIR can be used to monitor a drying process. As a sample is dried, we can use the near-infrared to measure the amount of solvent coming off. This could be water or some organic solvent. The sample is moved along in the process only after the distillant reaches the desired purity, or the sample has dried to an acceptable level.

The near-infrared can be used to do a quality control inspection. The advantage of near-infrared in this case is that the samples are inspected as they pass by. All of the samples are monitored. There is no need to try to do a statistical sampling of the product. As the samples pass by, all the bad samples are rejected. There actually is a good example of this. When one goes to the store, there is no desire to purchase bad fruit, and most people get very upset when they pay for rotten fruit. The near-infrared is used now to reject the fruit that, while the integrity on the outside is fine, on the inside is rotten.

After a good has been manufactured there is usually a need to inventory it. We can use the near-infrared to monitor the integrity of our inventory. After the product is finished, NIR can be used to make sure that what is shipped is the same as what should be shipped, avoiding a mix-up. NIR monitoring also allows the shipping of the product before it becomes unsalable. Of course this assumes that the product is packaged so that NIR can be used. As long as our widget product is stored in vats, NIR can be used; however, complications arise when the product has been put into cans! Fiber optics may not be used in all cases. This might be the place to use a portable spectrometer, that the worker has on his belt. It seems like a very reasonable idea to give workers a portable instrument that they can use to scan things and have a spectroscopic basis for what is done. They can use spectroscopic reasons to pull a sample off the shelf and say, "this is starting to get old, we need to ship it off."

With any technique, there is a tendency to tout it as the savior of the world. That is the not the purpose of this work. The near-infrared is not perfect for everything. There are some places where the near-infrared is an excellent candidate. That is, where one wants to measure bulk chemicals. To measure minute quantities, say parts per billion, is not the near-infrared's strength. Now, if that part per billion quantity causes a massive change in some other constituent, such as polymer linking, then we can measure the parts per billion quantity. In this case the parts per billion quantity is not being measured, rather the effect of the substance on another is being measured. It is not economically feasible to develop a near-infrared technique for a few samples. A calibration should be made to replace a repetitive analysis. This releases the analyst to develop new techniques.

What are poor candidates for near-infrared? What is it not recommended that near-infrared be applied to? Trace contaminants are difficult to measure with NIR. Unless, of course, they have a very large effect, trace contaminants are not usually done. Low volume production is difficult to use NIR for, such as a special order for some company. By the time fifty samples have been synthesized and analyzed to develop the correlation, the person waiting for the project is very frustrated. It has taken so long to develop the correlation that no time was spent developing the product. So in those very low volume, special orders, near-infrared does not really have that much utility. Also, there has to be that reference technique. At the very minimum there must be people that stick up their noses and say, "we don't want that." There must be some way to quantify the effect.

What areas are the near-infrared going to be applied to in the future? There are biomedical applications, such as in-vivo monitoring of species. Also researchers are into imaging. Replacing, maybe, some x-ray techniques with near-infrared, which is not as destructive to tissue as x-ray. If some areas using x-rays can be replaced with the near-infrared, it will not be as expensive. Depth profiling the sample using near-infrared spectroscopy is a region ripe for development as very little work has been done. Another area for development is first order calibration. It has been said all along that one has to have a representative calibration set. A set that is representative of the sample must be used to develop the correlation. If things change, if the process changes in the middle, then the calibration falls apart. What needs to be done is to develop a calibration that changes with the process. If the product is within specifications after some parameter changes, then the calibration automatically changes to compensate. So that effects that do not change the process, or the end-product, are built into the calibration.

We hope that it has been shown with this homily what near-infrared can do in process analysis. Stepping through the mythical widget factory has allowed us to show how the different areas can have NIR analysis applied to them. The near-infrared is now being applied to process chemistry. With a little effort the NIR can be integrated into a plant to achieve better productivity.

Fourier transform infrared spectrometry: A tool for modern agricultural research

PETER R. GRIFFITHS and GILBERT L. PARIENTE
Department of Chemistry, University of California, Riverside, CA 92521, U.S.A.

Introduction

The concept of using a two-beam interferometer for the measurement of infrared spectra has taken almost a century to come to fruition. Michelson first devised the concept of the interferometer which now bears his name in 1887. However it was not until about sixty years later that Fellgett recognized the multiplex advantage attained for spectra measured using a Michelson interferometer, i.e., the fact that information from all wavelengths in the spectral region under investigation is sampled throughout the entire measurement. In the early 1950's, Jacquinot proved that the light-grasp of an interferometer was greater than that of a monochromator operating at the same spectral resolution. The combination of the Fellgett (or multiplex) and Jacquinot (or optical throughput) advantages of an interferometer leads, in theory at least, to a sensitivity advantage for interferometrically measured spectra of between two and three orders of magnitude for data acquired in equal times. Only recently, however, has the performance which may be predicted on theoretical grounds started to be achieved in practice, and Fourier transform infrared (FT-IR) spectrometers based on Michelson (and other two-beam) interferometers is now the most popular technique for measuring high quality infrared spectra of samples of many different types.

Several types of interferometer were developed for chemical spectroscopy in the 1960's. The group at the National Physical Laboratory in England led by Alastair Gebbie pioneered the use of simple, low-cost, slow-scanning interferometers for far infrared spectrometry. In France, the school headed by Pierre Connes developed a high-precision, stepped-scanning interferometer for ultrahigh resolution measurements in the near infrared and visible spectral regions. The impact of these two instrumental concepts on the design of modern interferometers was not as great, however, as the development of a small, low-resolution, rapid-scanning Michelson interferometer by Mertz and his colleagues at Block

Paper presented at the U.S.D.A. Beltsville Symposium in Agricultural Research. XI. Research Instrumentation for the 21st Century.

Engineering. With this instrument, interferograms could be measured at a rate of better than one scan per second.

Computation of the spectrum from the interferogram could be a time-consuming operation until the discovery of the fast Fourier transform (FFT) algorithm by Cooley and Tukey in 1964. Although the FFT dramatically reduced the number of operations required for the computation of a spectrum, computers were still large, expensive, and relatively slow at that time. With the development of small, inexpensive, fast minicomputers and, more recently, microcomputers, the full advantages of using two-beam interferometers for the measurement of infrared spectra are being realized, and FT-IR spectrometry is one of the fast-growing techniques of instrumental analysis today. It must be stated, however, that the full impact of FT-IR spectrometry has not yet been reached for agricultural research. It is the purpose of this article to introduce the concepts of FT-IR spectrometry and to summarize several applications where the impact of this technique should be felt in the field of agricultural research.

The principle of the Michelson interferometer is fairly simple. A collimated input beam of radiation is split into two paths by a semi-reflecting film, or beamsplitter. One beam is reflected back to the beamsplitter by a fixed mirror and the other by a moving mirror, see Fig. 1. The difference in path of these two beams creates a difference in phase when the two beams are recombined at the beamsplitter. For a monochromatic input beam, the amplitude of the output beam is modulated sinusoidally, with each maximum occurring when the beams from the fixed and moving mirrors are in-phase and each minimum occurring when they are out-of-phase by 180°. For a polychromatic input beam, such as the output of an incandescent blackbody source, entering an ideal Michelson interferometer, all wavelengths are in-phase when the fixed and moving mirrors are equidistant and assume various phase differences as the moving mirror is scanned. The result is a signal, or *interferogram*, which has a large central maximum at zero optical path difference but whose a.c. value rapidly decays to zero at other mirror positions. The interferogram of a Globar source measured by a rapid-scanning Michelson interferometer is shown in Fig. 2. Although ideally this interferogram should be perfectly symmetrical about the zero path difference point, in practice several sources of phase shifts lead to a small asymmetry which can be corrected for during the computation of the spectrum.

The Fourier transform of this interferogram is shown in Fig. 3(a). This is a single-beam spectrum whose amplitude depends on several factors such as the emission characteristics of the source, efficiency of the beam-

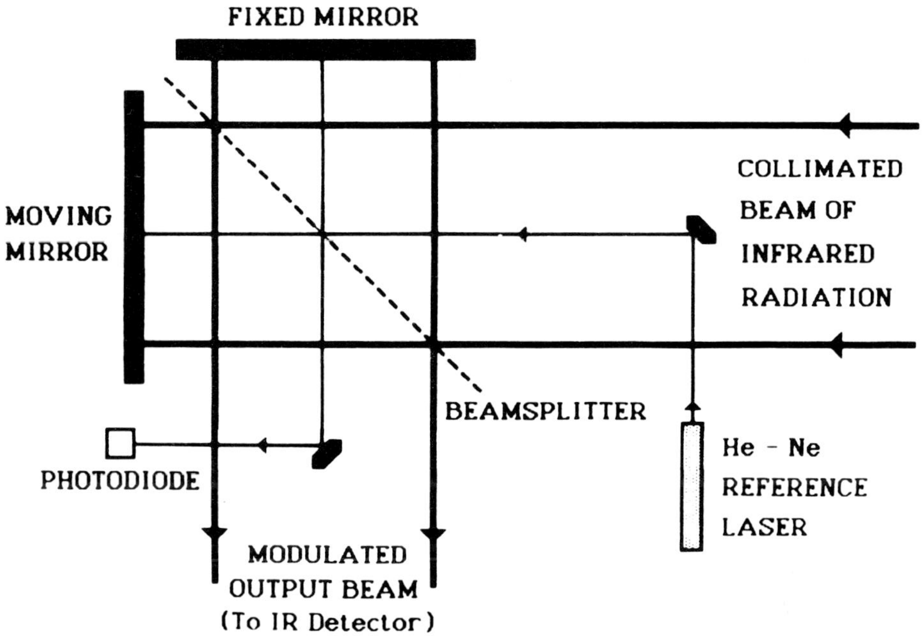

Figure 1. Schematic representation of a typical Michelson interferometer. The collimated beams from a broad-band infrared source and from a small He–Ne laser pass collinearly into the interferometer. After modulation by the interferometer, the laser beam is reflected to a photodiode. The infrared beam is passed through the sampling optics and focused on a suitable detector. The signal from the IR detector is digitized at uniformly spaced points given by the zero-crossings of the sinusoidal laser interferogram.

splitter, optical and electronic filtering, and absorption by radiation by atmospheric components such as CO_2 and water vapor. To measure an absorption spectrum, a sample is placed in the beam and a second interferogram is measured and transformed, see Fig. 3(b). The ratio of these spectra is then computed in order to obtain the transmittance spectrum, see Fig. 4. For several applications of FT-IR spectrometry such as spectral subtraction, deconvolution, and library searching (*vide infra*) it is necessary to have the spectrum in a linear absorbance format. The absorbance spectrum can be readily computed from the transmittance spectrum, see Fig. 5. It should be noted that the spectra shown in Figs. 3, 4 and 5 were all measured using a single scan of the interferometer mirror, which takes about 1 second, using a room temperature pyroelectric detector. The high signal-to-noise ratio (SNR) of these spectra attests to the power of contemporary FT-IR spectrometers.

The success of FT-IR spectrometry is reflected in the number of companies now marketing these instruments, see Table 1. Of these ten are located in North America, two in Europe and three in Asia. A few of these companies make instruments for certain specialized applica-

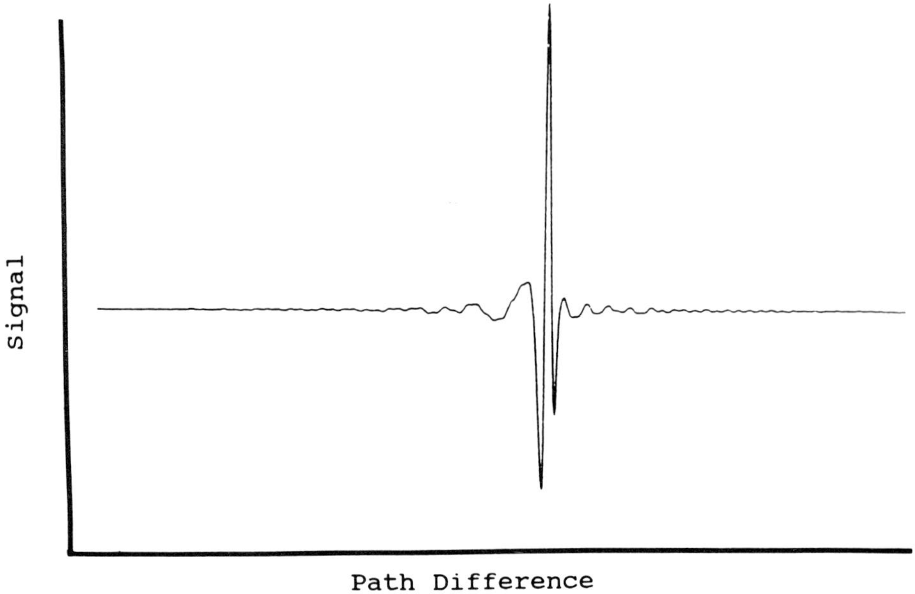

Figure 2. Typical interferogram of a broad-band infrared source measured using a rapid-scanning Michelson interferometer equipped with a pyroelectric detector, shown in the region of zero path difference.

tions, but most of them have designed instruments for laboratory spectroscopy. Choosing between these spectrometers can present a major problem for people entering the field of FT-IR spectrometry for the first time. Fortunately, however, the performance of most of these instruments is very high and the principal differences between them are in the versatility and flexibility of their optics and data systems.

The remarkable growth of FT-IR spectrometry can be assigned primarily to six factors:

a infrared spectrometry is one of the most versatile of all spectroscopic techniques, allowing characterization of samples in the gaseous, liquid and solid states and surface species;

b techniques have been developed which permit both major and minor components to be observed;

c very rapid scan speeds (up to 50 scans per second for low resolution measurements) can be achieved, permitting transient species to be studied, while signal-averaging allows spectra of stable species to be measured at an exceptionally high SNR;

d the presence of powerful microcomputer-based data systems, sometimes incorporating fast array processors, allows remarkably versatile interactive data processing;

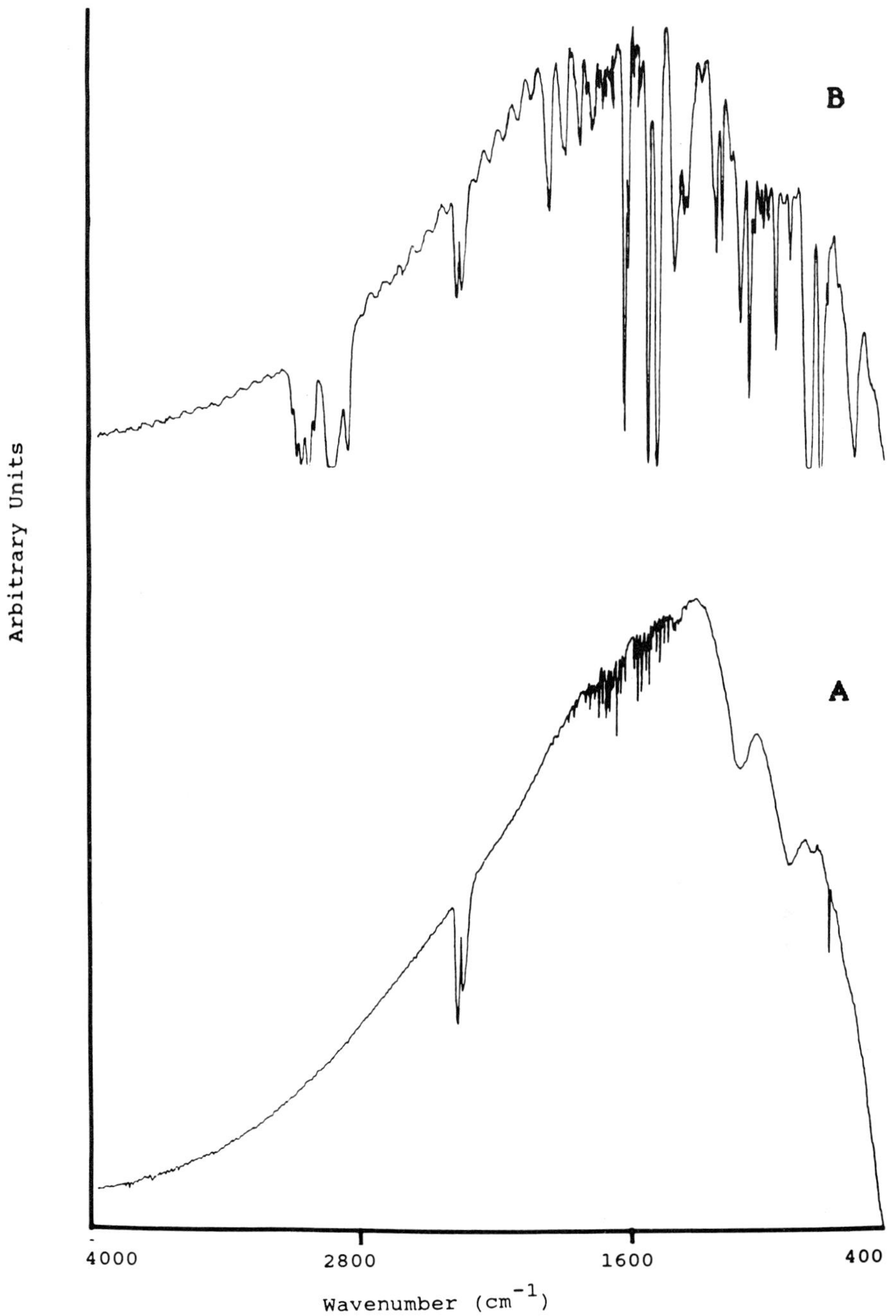

Figure 3. (a) Single-beam spectrum computed from the interferogram shown in Fig. 2. The features near 2350 and 670 cm^{-1} are due to absorption by atmospheric CO_2, and the structure centered near 1600 cm^{-1} is caused by atmospheric water vapor. This spectrum was measured at 4 cm^{-1} resolution using a single-scan of about 800 ms duration. (b) Single-beam spectrum of a polystyrene film measured under the same conditions as Fig. 3(a).

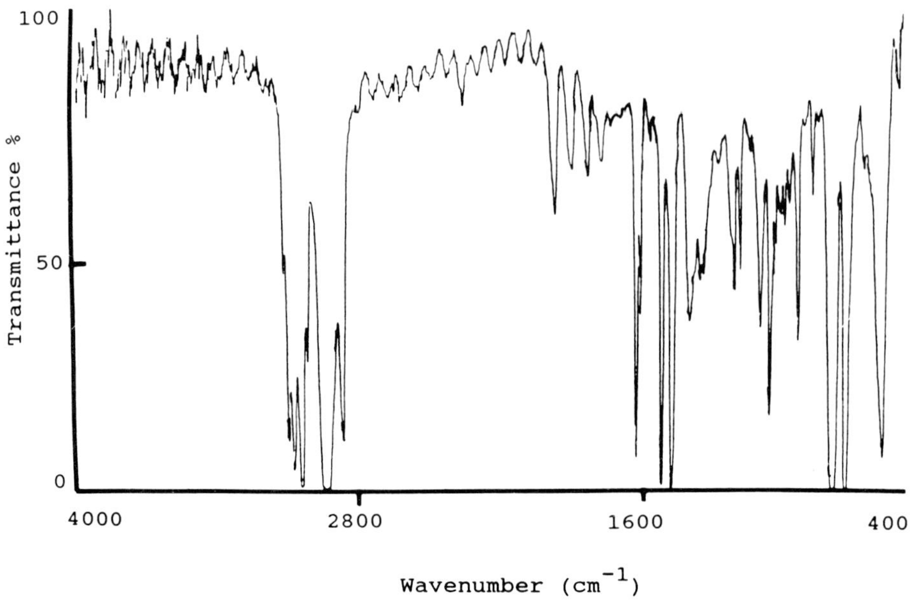

Figure 4. Transmittance spectrum of polystyrene film obtained by ratioing the spectra shown in Figs. 3(b) and 3(a).

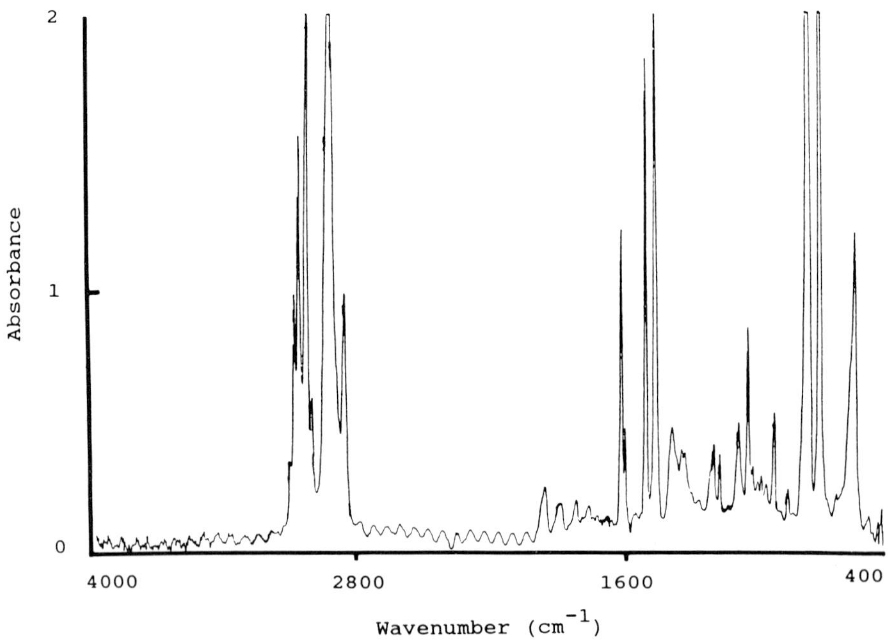

Figure 5. Absorbance spectrum of polystyrene film computed from the transmittance spectrum shown in Fig. 4.

Table 1. Vendors of FT-IR Spectrometers in 1986

Company	Country
Analect, Division of Laser Precision	U.S.A.
BOMEM	Canada
Bruker Physik	Germany
Digilab, Division of Bio-Rad Laboratories	U.S.A.
Idealab	U.S.A.
IBM Instruments (not now in FT-IR business)	U.S.A.
Hewlett-Packard	U.S.A.
Jasco	Japan
JEOL	Japan
J.J. Lloyd	United Kingdom
Mattson Instruments	U.S.A.
MIDAC	U.S.A.
Nicolet Instruments	U.S.A.
Perkin-Elmer	U.S.A.
Shimadzu	Japan

e the development of fast, sensitive detectors for mid infrared spectrometry, especially the liquid nitrogen cooled mercury cadmium telluride photoconductive detector, has led to a further improvement in SNR of about a factor of ten compared with measurements made using room temperature pyroelectric detectors;

f the use of a small helium-neon laser allows the position of the moving mirror of the interferometer to be known accurately at all times, leading to an unsurpassed wavenumber precision.

A good account of the theory and instrumentation of FT-IR spectrometry is found in the recent book by Griffiths and de Haseth [1].

Sampling techniques for fluids

There is a quite limited number of sampling techniques for materials in the gaseous or liquid phase. For gases, cells have been developed with pathlengths of anywhere from a few millimeters to over a kilometer, depending whether the analyte is at or above ambient pressure or whether it is present at a concentration of a few parts per billion. Similarly, liquids can be held in cells of path lengths from about 15 μm (for neat liquids) to several millimeters (for solutes at concentrations of parts per million or less in weakly absorbing solvents).

An alternative sampling technique for liquid samples is attenuated total reflectance (ATR) [2]. For ATR measurements, radiation is passed into an internal reflection element (IRE) constructed from a material with a high refractive index, typically between 2.5 and 4.0. A beam of radiation hits one face of the IRE at normal incidence. A fraction of the incident radiation is reflected from this surface but most of the beam is transmitted into the crystal. The IRE is cut so that the beam hits the internal surface at an angle greater than the critical angle. In this case the beam is totally reflected. The beam then passes to another parallel internal surface where it is again totally reflected. This process continues until the beam hits a face at an angle which is less than the critical angle, at which point the beam can reemerge from the IRE. A typical shape of the IRE might be trapezoidal, see Fig. 6(a). At each internal reflection, the beam actually penetrates outside the IRE to an effective distance of about one-tenth of the wavelength. If a sample is in contact with the outside surface of the crystal, part of this *evanescent* wave may be absorbed. Although the effective pathlength per reflection is quite small, the fact that there are several internal reflections (often as many as 25) means that ATR spectrometry will often yield a spectrum of approximately the same intensity as a conventional transmission cell of short pathlength. Since short pathlength cells can often be quite difficult to fill with viscous liquids, ATR spectroscopy can be a convenient method of obtaining the spectra of solutes in strongly absorbing solvents. It is difficult, however, to obtain the spectrum of solutes at concentrations much below 0.1% (w:v) in almost any solvent by ATR spectrometry.

ATR spectrometry can be used to monitor processes occuring in aqueous solution, such as fermentation broths. This can be done in a simple manner by silvering one end of the IRE so that the beam reemerges from the face it entered, as shown in Fig. 6(b). Several types of ATR flow-cells have been developed. One of the simplest and most popular is the cylindrical internal reflection cell (or CIRCLE cell). The IRE of a CIRCLE cell is circular in cross-section with conical ends, see Fig. 6(c). Its circular cross-section makes this IRE compatible with the round beam of FT-IR spectrometers and makes it very easy to seal with O-rings.

An interesting technique that can be used for trapping species present in the vapor phase is to add a considerable excess of argon and to freeze the mixture at a temperature below 20 K. This technique is applicable to gases and vapors, and has even been used for the identification of components of solid samples after sublimation using a high temperature oven. Because argon atoms are typically at about a thousandfold excess, each analyte molecule is effectively isolated in the argon matrix. Band broadening due to intermolecular interactions and rotational motion is minimized, so that the width of absorption bands in matrix isolation (MI) spectra is often less than $1\,\mathrm{cm}^{-1}$. Not only is the specificity of infrared spectrometry of stable species increased through the application

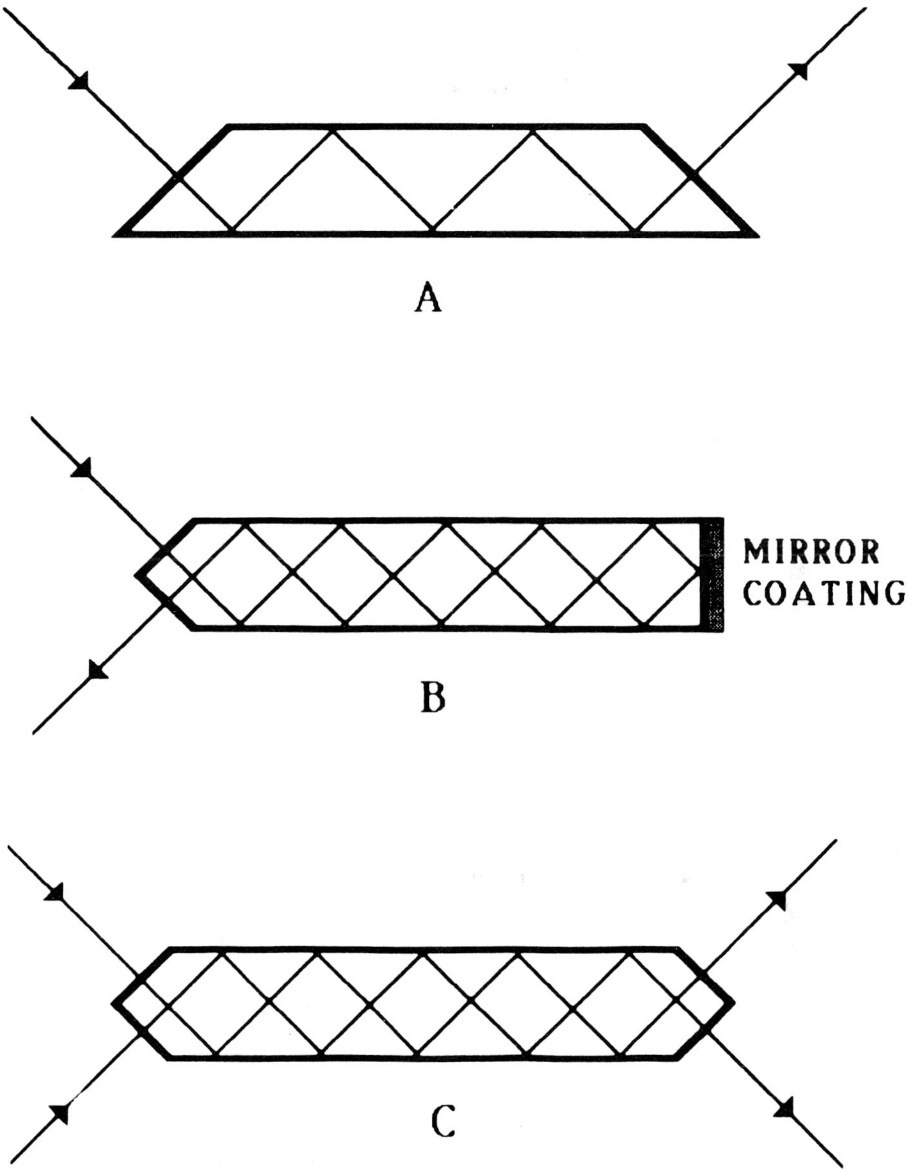

Figure 6. Three types of internal reflection element used for ATR spectrometry; in each case the beam is multiply internally reflected before emerging (a) Trapezoidal IRE; (b) IRE with one end silvered; (c) Cylindrical IRE with conical ends.

of MI techniques, but it is also possible to trap reactive intermediates in gas-phase reactions in the argon matrix. A particularly important application of MI spectrometry is in the identification of eluates from gas chromatographic columns, *vide infra*.

Sampling techniques for solids

Unlike the case for fluid samples, a plethora of techniques exists for solid samples, each of which has certain strengths and weaknesses. Of course, solids may be sampled in the time-honored ways by preparing them as KBr disks, as mineral oil mulls, in solution, or casting them from solution onto the surface of an internal reflection element for ATR spectrometry. In the past decade, several other sampling techniques have become popular for FT-IR spectrometry. These include microscopy, diffuse reflectance, and photoacoustic spectrometry, each of which will be described in this section.

(a) FT-IR microscopy. Until recently if the infrared spectrum of a sample less than 500 μm in diameter was to be obtained, it was necessary to place a mask of the appropriate diameter over the sample to prevent radiation which had not passed through the analyte from reaching the detector. This very time-consuming step can now be eliminated through the use of one of several microscopes which have been specifically designed for installation on an FT-IR spectrometer. In FT-IR microscopes, radiation is focused at the plane of the sample and then reimaged at an adjustable aperture, the area of the sample which is observed by the detector can be controlled. The shortest dimension which can be observed by FT-IR microscopy is determined by diffraction. The average wavelength in the mid infrared region of the spectrum is approximately 10 μm, so that observation of smaller samples than this is difficult.

When microscopes of the design described above are used for the measurement of fibers less than 20 μm in diameter, a large stray light level is often observed. This may be larger than 50% of the measured signal, even in regions of intense absorption by the sample. To minimize this effect, a new microscope was recently introduced by Spectra-Tech (Stamford, CT) which incorporates the principle of redundant aperturing. In this microscope, apertures are located both above and below the sample. Although the energy transmitted to the detector by redundantly apertured microscopes is slightly less than the energy of single-aperture microscopes, the quantitative accuracy of the former device is considerably better for narrow fibers or very small specks. Typical spectra of natural and synthetic fibers measured using this microscope are shown in Fig. 7.

(b) Diffuse reflectance spectrometry. Mid infrared reflectance (DR) spectra of finely powdered solid samples can be readily measured at high SNR using FT-IR spectrometers. A large number of different designs of optics for DR spectrometry have been described, most of which operate

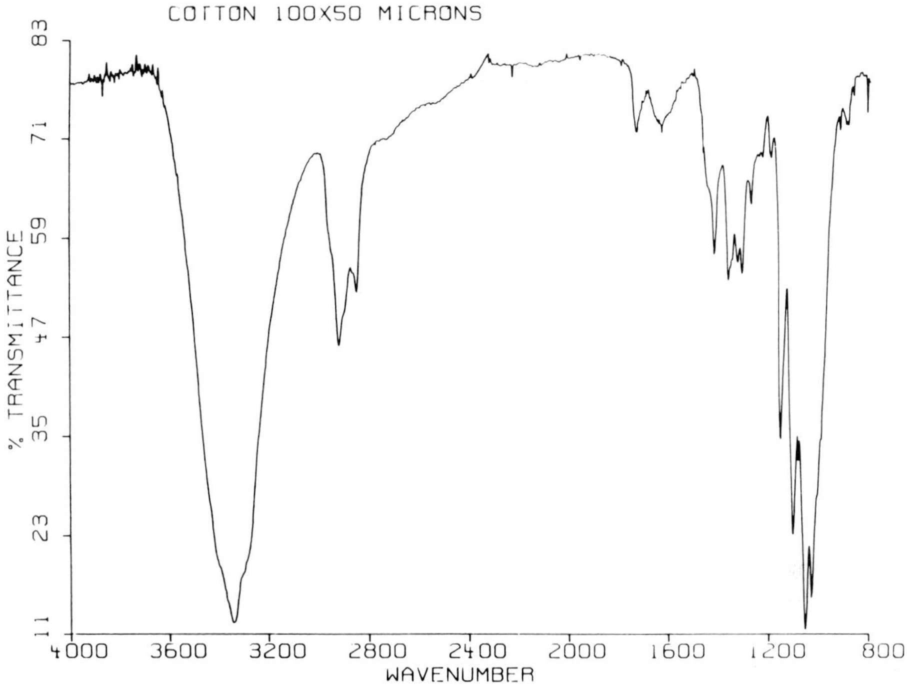
COTTON 100X50 MICRONS
% TRANSMITTANCE
83
71
59
47
35
23
11
4000
3600
3200
2800
2400
2000
1600
1200
800
WAVENUMBER

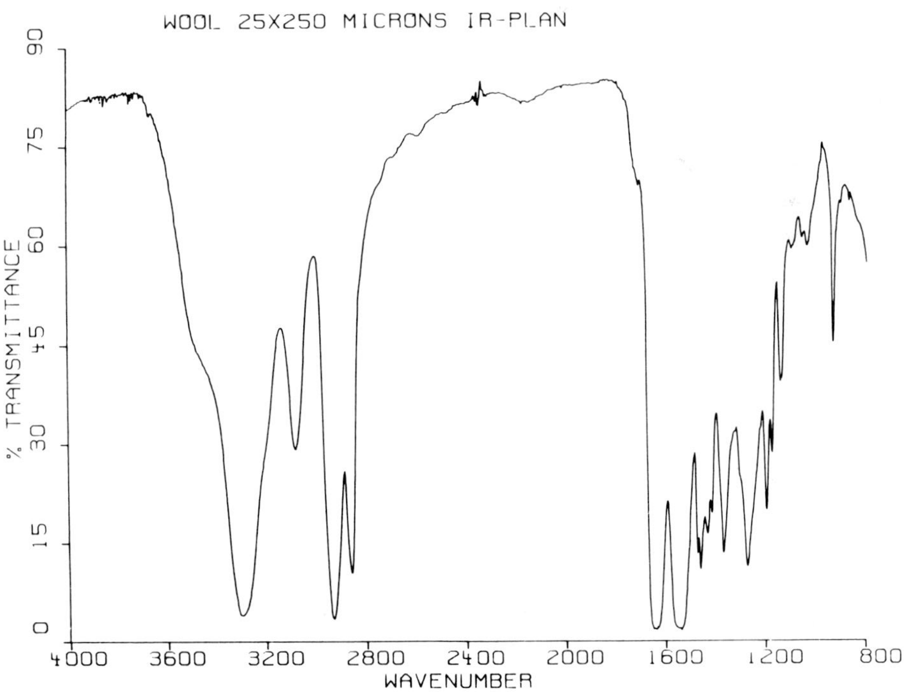
WOOL 25X250 MICRONS IR-PLAN
% TRANSMITTANCE
90
75
60
45
30
15
0
4000
3600
3200
2800
2400
2000
1600
1200
800
WAVENUMBER

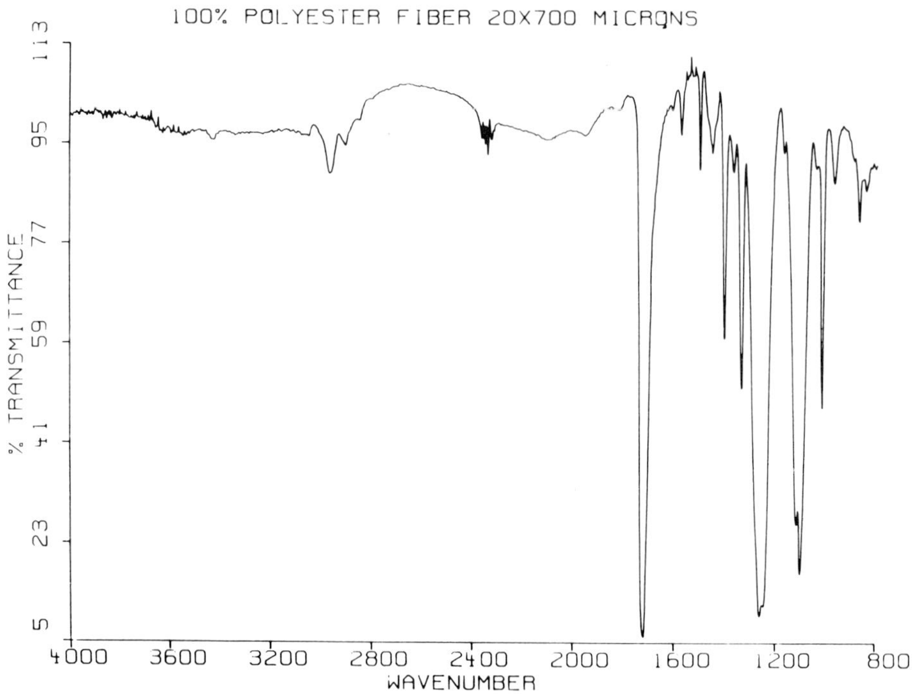

Figure 7. Spectra of fibers measured on an FT-IR spectrometer equipped with a redundantly apertured microscope (a) Cotton fiber, 50 μm in diameter; (b) Wool fiber, 25 μm in diameter; (c) Polyester fiber, 20 μm in diameter.

with approximately the same optical efficiency [1, 3]. Although high quality DR spectra of certain samples (such as powdered coals) may be obtained directly, it is usually necessary to dilute the analyte in a non-absorbing matrix, such as powdered KCl. The effect of dilution on the spectrum of all-purpose flour is shown in Fig. 8.

Diffuse reflectance spectra in the mid infrared are usually converted to the Kubelka–Munk function, $f(R_\infty)$,

$$f(R_\infty) = \frac{(1 - R_\infty)^2}{2R_\infty} = \frac{2.303\,ac}{s} \tag{1}$$

to give a spectrum whose band intensities vary linearly with sample concentration, c. In this expression, R_∞ is the ratio of the reflectance of the sample to the reflectance of a nonabsorbing powdered reference, with both measured at "infinite depth", i.e., each sample is sufficiently thick that any further increase in thickness does not change its spectrum. Even for very weakly absorbing samples, a thickness of 3 mm generally satisfies the infinite depth criterion. a is the absorptivity of the analyte and s is a scattering coefficient.

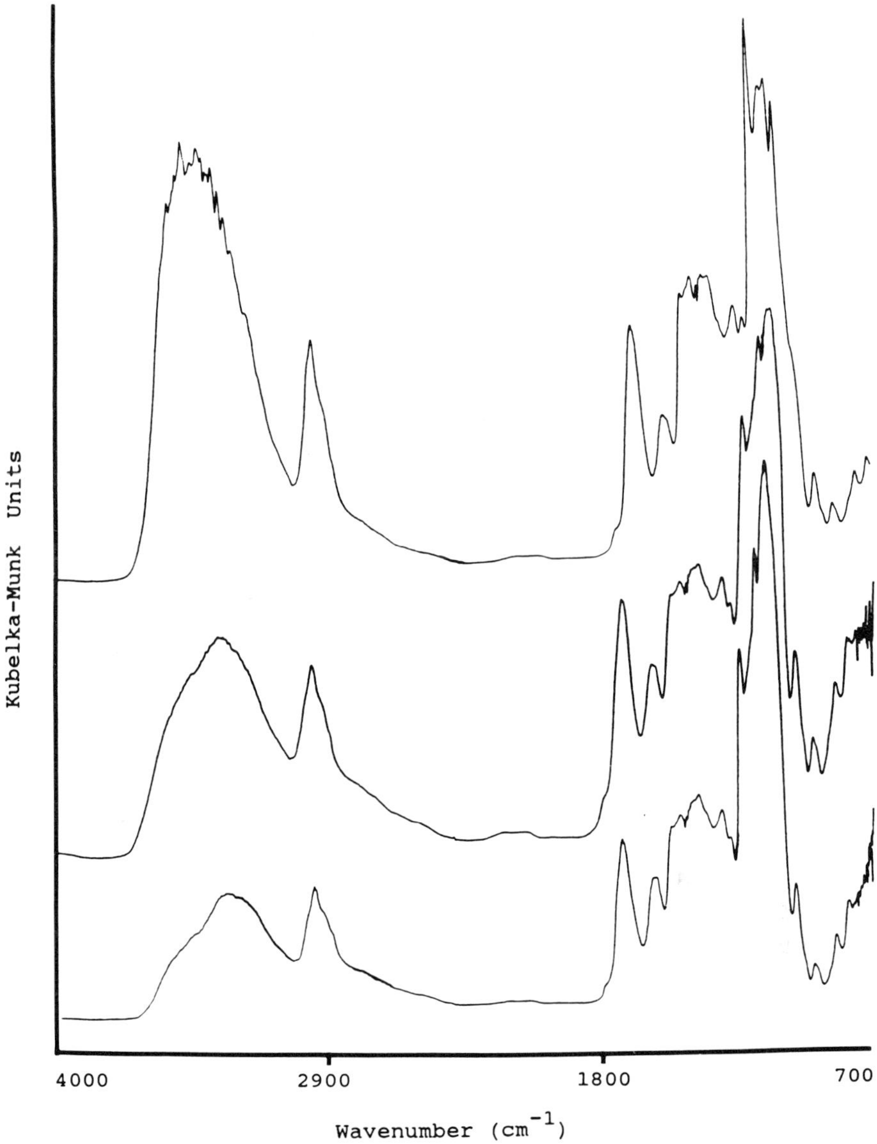

Figure 8. Diffuse reflectance spectra of an all-purpose flour; (upper trace) without dilution; (middle trace) as a 10% dispersion in KCl powder; (lower trace) as a 1% dispersion in KCl powder. The relative band intensities in the spectrum of the 1% dispersion follow the true absorbance spectrum of the sample more accurately than bands in the other two spectra.

The derivation of the Kubelka–Munk equation (4) tacitly assumes that the front surface reflection from the upper surface of the sample is minimal. While this is a fairly good approximation for weakly absorbing or nonabsorbing samples, it is known that the reflectance can increase dramatically in the region of strong absorption bonds. The effect of front surface reflection can cause considerable deviations from linearity in plots of $f(R_\infty)$ vs. sample concentration, especially for strong bands in the spectra of samples at high concentration. Nevertheless by selecting weak bands in the DR spectra of major components and strong bands in the spectra of minor components in weakly absorbing matrices, it is possible to cover up to six orders of magnitude in concentration by mid infrared diffuse reflectance spectrometry. Indeed it is noteworthy that DR spectrometry is often the sampling technique of choice for submicrogram quantities of solid analytes [5].

This property has proved beneficial in the design of several interfaces between high performance liquid chromatographs and FT-IR spectrometers. As a general rule, the very low concentration of solute and the strong absorption by the solvent makes it extremely difficult to obtain real-time HPLC/FT-IR measurements simply by passing the column effluent through a flow cell or past an internal reflection element. However by continuously evaporating part of the solvent at elevated temperature and dripping the solution remaining onto cups containing a powdered infrared transparent matrix, the solvent may be completely evaporated while each eluate is deposited. By measuring the DR spectrum of the contents of each cup, identifiable spectra of submicrogram quantities of each eluate can be obtained in only a few seconds [1, 6, 7].

(c) Photoacoustic spectrometry. For a rapid-scanning Michelson interferometer whose moving mirror travels at a velocity V cm s^{-1}, radiation of wavenumber $\bar{\nu}$ cm^{-1} is modulated at a frequency of $2V\bar{\nu}$ Hz. The interferometer can therefore be thought of as a wavenumber-selective chopper. If a sample has an absorption band at wavenumber $\bar{\nu}_0$, radiation of this wavenumber is absorbed and the sample returns to the ground vibrational state through the emission of heat. If radiation modulated by a Michelson interferometer is passed onto the sample, the temperature of the sample near its surface will be modulated at a range of frequencies between $2V\bar{\nu}_{max}$ and $2V\bar{\nu}_{min}$ Hz, where $\bar{\nu}_{max}$ and $\bar{\nu}_{min}$ are maximum and minimum wavenumbers in the spectral range. When the sample is enclosed in a chamber of constant volume, the temperature, and hence the pressure, of the surrounding gas will also be modulated in this frequency range.

Since the range of modulation frequencies generated by a typical rapid-scanning interferometer falls in the audio-frequency region, the effect of absorption by the sample can be detected by a microphone. Indeed the signal from a totally absorbing sample, such as carbon black, has the appearance of the interferogram shown in Fig. 2. Transformation of the interferogram of a typical sample gives the absorption spectrum directly, and the spectrum is usually corrected for the instrumental response by ratioing it against the spectrum of a carbon black reference.

The photoacoustic (PA) spectrum of the neat flour whose DR spectrum was shown in Fig. 10(a) is illustrated in Fig. 9(a). Like the DR spectrum of the neat flour, the PA spectrum also shows distortions, which in this case are caused in part by photoacoustic saturation and in part by front surface reflection. To minimize the effect of photoacoustic saturation, a very thin sample of flour must be prepared. Although photoacoustic spectra cannot usually be measured with as good SNR as DR spectra for powdered samples at low concentration, PA spectra will sometimes give information in cases where all other sampling techniques draw a blank. Samples falling into this category can include bulk polymers with a variety of morphologies [8], minerals, and species on the surface of strongly absorbing matrices. An example of measurement falling into the latter category is an investigation of the adsorption of a carbamate insecticide on a clay carrier [9].

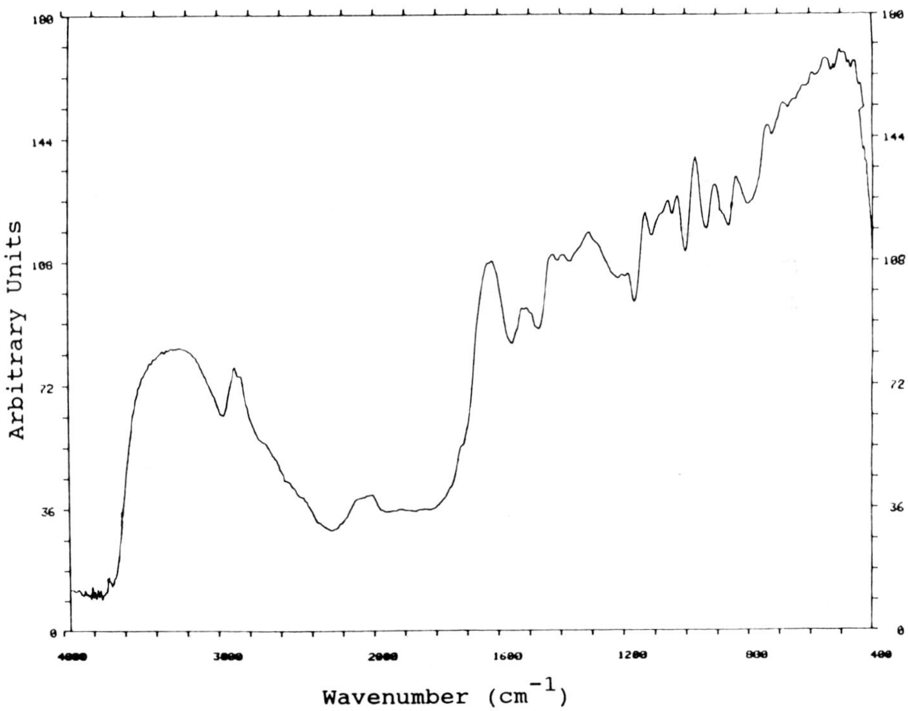

Figure 9. Photoacoustic spectrum of a thick sample of the same flour used for the upper trace of Fig. 8.

Data processing techniques

The availability of inexpensive yet powerful computers for FT-IR spectrometers has further enhanced the usefulness of these instruments. A complete list of all the data processing techniques which have been developed for FT-IR spectrometry is beyond the scope of this paper. Nevertheless several algorithms of direct applicability to agricultural research will be summarized here.

For quantitative analysis, several methods based on curve-fitting and/or factor analysis have been developed. Two techniques for multicomponent analysis based on matrix expressions of Beer's Law have started to become popular. In the so-called *K*-matrix method [10], Beer's Law is applied in its classical form, namely:

$$A = KC + E$$

where A is the absorbance (or data) matrix, K is a matrix based on the product of absorptivity and pathlength of each component C is the concentration matrix, and E is an error matrix. The alternative method is the so-called *P*-matrix technique which is based on the inverse expression of Beer's Law [11]:

$$C = PA + E$$

where P is the proportionality matrix. Of the two methods it is probable that the *K*-matrix method will become the more popular, since it has been believed to be less susceptible to errors due to the presence of unsuspected interferents. Surprisingly perhaps, the methods based on multiple linear regression which have become so vital for near infrared analysis have not been widely applied to mid infrared spectrometry.

The high wavenumber precision which can be achieved on contemporary FT-IR spectrometers is not readily taken advantage of when absorption bands are very broad. However Cameron et al. [12] have shown that it is possible to obtain exceptionally accurate values for the center-of-gravity of an absorption band from several points located near the band maximum. They have shown that it is possible to obtain an estimate of the center-of-gravity within a few hundredth of a wavenumber even when the band is several wavenumbers wide. This algorithm has been used to monitor small changes in band position occurring during thermotropic changes in lipids [13] and even live bacteria [14].

Several methods involving data processing in the Fourier domain have become popular. These include derivative spectrometry, smoothing, and spectral interpolation by zero-filling. The most important of these techniques, in the opinion of these authors is band narrowing using the technique known as *Fourier self-deconvolution* (FSD) [15]. In complex spectra many neighboring bands are so broad that they are completely unresolved. Through the use of FSD the width of each band in the spectrum may be narrowed to the point that the components of many multiplets become either completely or partially resolved.

For FSD, the spectrum must be in a format in which the shape of the absorption bands is known. For condensed phase samples measured by transmission spectrometry, the spectrum must therefore be in a linear absorbance format, in which case each band is, to a very good approximation, Lorentzian. Similarly for diffuse reflectance measurements, the spectrum must be converted to the Kubelka–Munk function. A Lorentzian band can be considered to be the convolution of a delta function (a monochromatic line) and a Lorentzian line shape function. The Fourier transform (FT) of the convolution of two functions in the product of the FT of each function. The FT of the delta function is a cosine wave while the FT of a Lorentzian function is an exponential decay. Thus the FT of a Lorentzian band is a decaying cosine wave.

Analytically, a Lorentzian band has the form:

$$A(\bar{\nu}) = A_0 \frac{\gamma^2}{\gamma^2 + (\bar{\nu} - \bar{\nu}_0)^2} \tag{2}$$

where $A(\bar{\nu})$ is the absorbance at wavenumber $\bar{\nu}$, A_0 is the maximum absorbance (at $\bar{\nu}_0$) and γ is the half-width of half-height (HWHH) of the band. The FT of this function is:

$$y(x) = \frac{A_0 \gamma}{2} \cos(2\pi\bar{\nu}_0 x) \exp(-2\pi\gamma x) \tag{3}$$

It should be noted that the unit of the abscissa of the Fourier domain array, x, is the reciprocal of the unit of the abscissa of the spectral array, $\bar{\nu}$. Thus x has the units of length (cm), and has been called the *spatial frequency*. $y(x)$ can be seen to be composed of three terms. The coefficient, $0.5\,A_0\gamma$, is directly proportional to the area of the band. The cosine wave, $\cos(2\pi\bar{\nu}_0 x)$, has a frequency which is directly proportional to the wavenumber of maximum absorption, ν_0. The exponent of the decay, $\exp(-2\pi\gamma x)$, is dependent on the width of the band.

If $y(x)$ is multiplied by an exponentially increasing function, $\exp(2\pi\gamma' x)$, the resultant function has the form:

$$y'(x) = \frac{A_0 \gamma}{2} \cos(2\pi\bar{\nu}_0 x) \exp(-2\pi(\gamma - \gamma')x)$$

The first two terms of $y'(x)$ are the same as those of the original function, $y(x)$, but the rate of decay has been reduced. On computing the reverse FT, the area and center wavenumber of the band remain unchanged, while its width is decreased to a new value, $\gamma - \gamma'$.

For spectra of complex samples, such as many samples of agricultural origin, the application of Fourier self-deconvolution can dramatically enhance the apparent information content of the spectrum. To illustrate the effect of FSD, the diffuse reflectance spectrum of the 1% dispersion of all-purpose flour shown in Fig. 8(c) is shown in Fig. 10 after mild deconvolution. The increased spectral structure observable after deconvolution is readily apparent, but the neophyte must be warned against excessive deconvolution (i.e. using a value of γ' which is too large) because of the deleterious effects of noise and side-lobes.

It is a general rule in spectroscopy that it is rare to get something for nothing, and this is certainly applicable to FSD. What is lost after resolution enhancement by FSD is signal-to-noise ratio. For broad band spectra measured at 4 cm^{-1} resolution, the values of the Fourier domain array, $y(x)$, at high spatial frequency have decayed to a level which may be quite close to the noise. On multiplication by $\exp(2\pi\gamma' x)$, these values are greatly amplified relative to the values at low spatial frequency. The result can be an enormous degradation in SNR. Thus the original spectrum should always be measured at as high sensitivity as possible.

For spectra containing relatively narrow bands, it may be found that $y(x)$ has not decayed below the noise level at the final point of the Fourier domain array. In this case there is a strong possibility that the spectrum after FSD will exhibit side-lobes, which can make it difficult to distinguish weak bands near strong ones. The possibility of obtaining side-lobes is minimized by multiplying $y'(x)$ by yet another function, known as an apodization function, which monotonically decreases with spatial frequency, usually having a value of zero at the maximum spatial frequency.

The use of mid infrared spectrometry in conjunction with FSD is just starting to have an impact on agricultural research. Susi and coworkers have shown that the amide I and II bands in the spectra of proteins can be resolved using this technique [16]. By curve fitting the deconvolved spectra of several proteins, they found that the content of various conformational forms of the proteins correlated well with the corresponding data obtained by X-ray diffraction of the crystalline proteins. We believe that the combination of FT-IR spectrometry with innovative sampling techniques and Fourier self-deconvolution will become increasingly important in agricultural research.

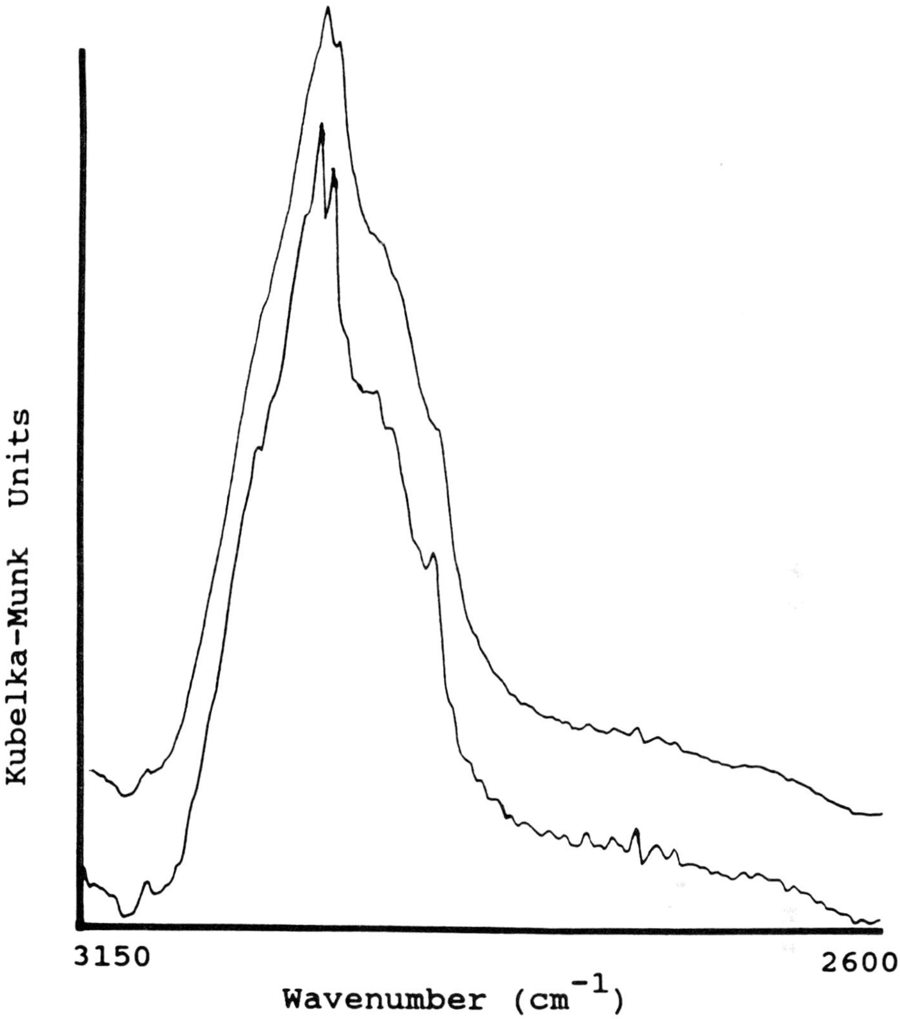

Figure 10. (Upper trace) Spectrum of the 1% dispersion of flour shown in Fig. 8, plotted in the region of the C–H stretching bands. (Lower trace) The same spectrum after deconvolution with $\gamma' = 2.5\,cm^{-1}$. Only 85% of the Fourier domain array was multiplied by the exponential weighting function and apodization function, with values at the remaining spatial frequencies being replaced by zeros, in order to minimize noise at high spatial frequencies where the spectral information content is low.

Coupled techniques

For about two decades, Fourier transform infrared spectrometers have been coupled with gas chromatographs [1, 17–19]. The usual technique for GC/FT-IR involves continuously passing the effluent from a GC column through a long (10–20 cm) narrow (1–2 mm internal diameter) heated (250–300°C) *light-pipe*, which is an internally gold-coated borosilicate glass capillary tube. Ten years ago, the GC/FT-IR interface was a technique of undeniable academic interest, but with barely enough sensitivity to be of use for the identification of peaks eluting from capillary gas chromatographs. Recently, however, GC/FT-IR interfaces based on the use of light-pipes have been improved to the point that they can be used to obtain identifiable spectra of low nanogram quantities of species eluting from wall-coated open tubular (WCOT) GC columns.

GC/FT-IR detection limits have been further reduced by trapping each eluate in a matrix of solid argon at a temperature of less than 20 K. In this matrix isolation technique, argon is usually added to the column effluent, although the chromatography can be performed using argon as the mobile phase. The argon is then frozen out on the surface of a slowly rotating disk, trapping each eluate in the argon matrix. Subsequently (usually after the GC run is completed) the "memorized chromatogram" is rotated through the infrared beam at the focus of a reflection accessory and the spectrum of each peak is measured sequentially [1, 19–21].

Using this device, GC/FT-IR detection limits have been decreased by about two orders of magnitude in comparison to the use of light-pipes. This increase in sensitivity is largely derived from three factors: (a) the widths of bands in the spectra of matrix isolated species are less than the widths of the corresponding bands in the spectra of non-isolated samples at ambient temperature, leading to an increase in the peak absorbance; (b) the diameter of the spot in which a given peak is trapped is considerably smaller than the internal diameter of a light pipe, which means that the effective pathlength of the sample is increased; (c) since this technique no longer requires real-time measurement, the signal-to-noise ratio of the spectrum can be improved by signal-averaging.

Of these advantages, (b) and (c) are probably the most important, and should be applied in high sensitivity interfaces between all types of chromatographs and FT-IR spectrometers. The primary disadvantage of the GC/MI-FT-IR interface is its mechanical complexity and hence high cost (well over $100 000 at the time of this writing). We have investigated the feasibility of trapping eluates from capillary GC columns onto an infrared window which is held at a temperature which can be easily reached using thermoelectric (Peltier effect) cooling or liquid cryogens. We found that eluates could be trapped in spots of 100–200 μm diameter. By passing this window through the focus of an FT-IR microscope, detection limits in the low nanogram to subnanogram range can be achieved for molecules whose volatility is less than or equal to that of nitrobenzene [19, 22]. To be applicable to molecules of greater volatility than nitrobenzene it appears to be necessary to further reduce the temperature of the substrate, possibly to 77 K.

The same principle can be applied to molecules which are either too non-volatile or too thermally labile to be separable by gas chromatography. Of particular relevance in this respect are separations effected by capillary supercritical fluid chromatography (SFC). Capillary SFC has one great advantage over high performance liquid chromatography (HPLC) for the separation of nonvolatile or thermolabile molecules, and this relates to the high resolution of this technique. The advantages of SFC for high resolution chromatography result from a combination of the properties of supercritical fluids. Above their critical temperature and pressure, the density of many fluids is high enough that large molecules can be solvated. Conversely, the viscosity of supercritical fluids is sufficiently low that there is a minimal pressure drop even when the fluid is contained in a 20 meter long capillary whose internal diameter (50–100 μm) is slightly less than the columns used for capillary GC. Finally, diffusion coefficients of solutes in supercritical fluids are typically ten times higher than in liquids. These three properties combine to make it possible to use supercritical fluids as the mobile phase for capillary chromatography using narrow-bore WCOT colums for the high resolution separation of nonvolatile compounds.

The most common fluids used as the mobile phase for capillary SFC are gaseous at ambient temperature and pressure. (CO_2 is a very commonly used material in this respect). Supercritical fluid chromatographs are equipped with a restrictor of about 2 μm in diameter at the end of the column to maintain the pressure in the column. On emerging from the restrictor, the mobile phase immediately evaporates. If a substrate is placed close to the restrictor, nonvolatile eluates can be trapped in a very small area. By constructing the substrate from an infrared-transparent material, the spectrum of each eluate can be measured under an infrared microscope in an exactly analogous manner described above for GC/FT-IR. Detection limits in the low nanogram range have been found in preliminary studies using this interface, and these detection limits can be expected to decrease once the deposition process has been optimized [23, 24]. This interface has been applied to the characterization of the components of cooking oil and shortening, where it was shown that components emerging at essentially identical retention times could exhibit quite different degrees of unsaturation [24].

Supercritical fluids have also been used to extract certain molecules from plant matter. The nature of these extractions can be readily followed by FT-IR spectrometry, often without the need for a chromatographic separation. A simple example of this type of analysis is found in the extraction of caffeine from coffee by supercritical CO_2. For this investigation, a 1 inch length of 1/8″ o.d. stainless steel tube fitted with a steel frit at one end was packed with ground coffee. The tube was fitted in a heating block and high pressure CO_2 was passed into the tube by means of a microcomputer-controlled pneumatic amplifier pump [25]. The extract was collected on a moving window held immediately below the restrictor. For this work the window was transferred to an FT-IR microscope so that spectra could be measured in an analogous manner to the SFC/FT-IR interface described above.

The results of caffeine extractions performed at various pressures were quite unexpected. The solvent strength of supercritical fluids is generally believed to be dependent on the density of fluid, which in turn depends on its pressure. The spectrum between 1800 and 1600 cm^{-1} of the caffeine extract produced by collecting the material extracted by CO_2 at a density of 0.7 g/ml for 10 minutes is shown in Fig. 11(a). The two absorption bands observed are due to the two carbonyl groups of caffeine. The density was then increased to 0.8 g/ml and the extract collected for the same period of time. The spectrum, which is shown in Fig. 11(b), indicates another component is extracted along with the caffeine. Corresponding data taken at a density of 0.9 g/ml, shown in Fig. 11(c) suggest that the principal component of the material being extracted at this pressure is a lipid.

At first glance these results appear to show that the caffeine starts to be extracted at 0.7 g/ml and by the time higher densities are reached, all the caffeine is extracted. Subsequent experiments where a fresh sample was extracted isobarically for a time equal to the total time for the densities in Figs. 11(a), (b) and (c), showed that the efficiency of the extraction of caffeine is maximized at a density of 0.7 g/ml and is decreased as the density is increased [25]. This result suggests that the specificity at which materials can be extracted from natural matrices with supercritical fluids may be greater than has previously been recognized. It can certainly be stated that the simplicity and specificity of the FT-IR detection and identification greatly facilitated this measurement.

Figure 11. Spectra of the extract from a ground roasted coffee obtained using supercritical CO_2 at a temperature of 40°C and densities of (a) 0.700 g/ml (1653 psi); (b) 0.800 g/ml (2360 psi); (c) 0.900 g/ml (4068 psi). The extract was condensed on a ZnSe window and spectra were measured using an FT-IR spectrometer equipped with a microscope accessory.

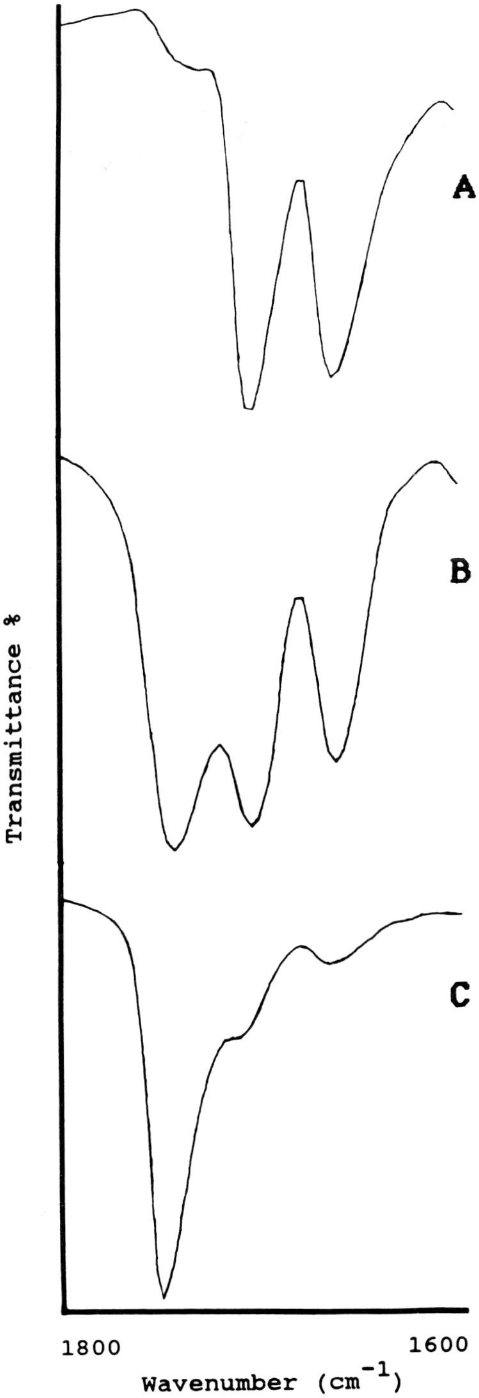
A
B
C
Transmittance %
1800
1600
Wavenumber (cm^{-1})

Conclusion

The results described in this paper have merely served to illustrate a few applications of FT-IR spectrometry to areas of potential agricultural interest. A surprisingly small number of applications in this area has been reported in the past decade despite the enormous rise in popularity of FT-IR spectrometry in other fields. We hope that this brief review will stimulate the interest of agricultural research scientists in this useful, sensitive, and versatile instrumental technique.

References

1. P.R. Griffiths and J.A. de Haseth. 1986. *Fourier Transform Infrared Spectrometry*. Wiley-Interscience, New York.
2. N.J. Harrick. 1967. *Internal Reflection Spectroscopy*. Wiley-Interscience, New York; now available from Harrick Scientific, Ossining, NY.
3. P.R. Griffiths and M.P. Fuller. 1982. *Mid Infrared Spectrometry of Powdered Samples*. In *Advances in Infrared and Raman Spectroscopy*, Vol. 9, (R.J.H. Clark and R.E. Hester, eds.), Heyden Publishing Co., London, pp. 63–129.
4. P. Kubelka and F. Munk. 1931. Z. Tech. Phys. 12: 593.
5. M.P. Fuller and P.R. Griffiths. 1980. Appl. Spectrosc. 34: 533.
6. P.R. Griffiths and C.M. Conroy. 1986. Adv. Chromatogr. 25: 105.
7. C.M. Conroy, P.J. Duff, P.R. Griffiths and L.V. Azarraga. 1984. Anal. Chem. 56: 2636.
8. D.W. Vidrine. 1980. Appl. Spectrosc. 34: 314.
9. S.R. Lowry, D.G. Mead and D.W. Vidrine. 1982. Anal. Chem. 54: 546.
10. D.M. Haaland and R.G. Easterling. 1982. Appl. Spectrosc., 36: 665.
11. C.W. Brown, P.F. Lynch, R.J. Obremski and D.S. Lavery. 1982. Anal. Chem. 54: 1472.
12. D.G. Cameron, J.K. Kauppinen, D.J. Moffatt and H.H. Mantsch. 1982. Appl. Spectrosc. 36: 245.
13. D.G. Cameron, H.L. Casal, H.H. Mantsch, Y. Boulanger and I.C.P. Smith. 1981. Biophys. J. 35: 1.
14. D.G. Cameron, A. Marin and H.H. Mantsch. 1983. Science 219: 180.
15. J.K. Kauppinen, D.J. Moffatt, H.H. Mantsch and D.G. Cameron. 1981. Appl. Spectrosc. 35: 271.
16. D.M. Byler, J.N. Brouillette and H. Susi. 1986. Spectroscopy 1: 29.
17. M.D. Erickson. 1979. Appl. Spectrosc. Rev. 15: 261.
18. P.R. Griffiths, J.A. de Haseth and L.V. Azarraga. 1983. Anal. Chem. 55: 1361A.
19. P.R. Griffiths and D.E. Henry. 1986. *Prog. Anal. Spectrosc.*, 9: 455.
20. J.F. Schneider, G.T. Reedy and D.G. Ettinger. 1985. J. Chromatogr. Sci. 23: 49.
21. G.T. Reedy, D.G. Ettinger, J.F. Schneider and S. Bourne. 1985. Anal. Chem. 57: 1602.
22. R. Fuoco, K.H. Shafer and P.R. Griffiths. 1986. Anal. Chem., 58: 3249.
23. S.L. Pentoney, K.H. Shafer, P.R. Griffiths and R. Fuoco. 1986. J. High Res. Chromatogr. Chromatogr. Commun. 9: 168.
24. S.L. Pentoney, K.H. Shafer and P.R. Griffiths. 1986. J. Chromatogr. Sci. 24: 230.
25. G.L. Pariente, S.L. Pentoney and P.R. Griffiths. 1987. Anal. Chem., 59: 808.

Application of Raman spectroscopy to molecular agricultural research

J.R. SCHERER and B.A. BOLTON
Western Regional Research Center, ARS, USDA, Albany, CA 94710, U.S.A.

Introduction

American agriculture has played a major role in contributing to the well-being of the world, both from subsistence and economical points of view. Chemistry has shared in this success by developing new agricultural products, increasing agricultural production and improving utilization of farm products. Today, there is every reason to believe that chemistry will play an increasingly sophisticated role in discovering new biochemical mechanisms that will lead to the development of new products and increase agricultural productivity (Pimentel, 1985). Geisbühler (1984) has pointed out the irony that one of the barriers to future progress in agricultural research is the successfulness of the application of older, "more conventional procedures which, until recently, have been acceptable enough to discourage systematic attempts at alternative approaches". He further notes that there is "a considerable and continuing lack of basic scientific knowledge in spite of the present rapid acquisition of biochemical and biomolecular information".

It seems clear that those methods that seek to answer basic questions about agricultural processes from a molecular viewpoint will be increasingly successful in solving today's agricultural problems as well as those anticipated in the 21st century. Modern physio-chemical techniques, such as those being considered at this symposium, are uniquely suited to probing the molecular basis of agricultural biochemical processes. The present challenge to agriculture is to make the commitment to utilize the newest technological advances to explore the frontiers of molecular agricultural research.

There has been enormous growth in the number of applications of Raman spectroscopy to biological problems in the last ten years and it would be impractical to review here all the various forms of Raman spectroscopy that are currently being used. Today, the number of biological applications most likely outnumbers (as measured by publications) nonbiological application (Gerrard, 1984). In this paper, we will present selected examples that demonstrate uses of Spontaneous Raman (SR), Resonance Raman (RR), Surface Enhanced Raman (SER),

Raman Microprobe (RM), and Coherent Anti-Stokes Raman (CAR) spectroscopies in biological applications. For reviews on these subjects, the reader is referred to the following: SRS- general [Carey (1982), Tu (1982), Gerard, (1984)], lipids [Levin (1984)]; RRS- photosynthesis [Lutz (1984)], heme proteins, porphyrins, enzymes [Tu (1982) Spiro (1985), Friedman, Rosseau and Ondrias (1982), Rousseau and Ondrias (1983)]; SERS- [Campion (1985)]; RM- (Rosasco (1980)]; CARS- [Harvey (1978), Morris and Wallan (1979)]. Areas of research not covered in this review are: Raman Optical Activity (ROA), Magnetic Raman Optical Activity (MROA), Hyper Raman spectroscopy (HRS), or Raman Induced Kerr Effect spectroscopy (RIKES).

Raman scattering processes

In Raman spectroscopy we consider scattering processes that involve changes in the vibrational (and rotational) energy levels of the molecule being studied. The electric moment induced by an electric field **E**, oscillating with a frequency ω_0, can be expressed as a series expansion in **E** [Demtröder (1981)]

$$\mathbf{P} = P_1\mathbf{E} + P_2\mathbf{E}^2 + P_3\mathbf{E}^3 + \ldots \tag{1}$$

where $\mathbf{E} = \mathbf{E}_0 \cos\omega_0 t$. For small electric fields, only the first term is important, but for larger field strengths such as are encountered with pulsed lasers, the higher order terms give rise to dominant non-linear processes. The first term P_1 is a three-dimensional tensor (sometimes labeled α) whose elements may be expanded in terms of the vibrational coordinates q_n of the molecule.

$$P_1 = [\alpha_{ij}] = [\alpha_{ij}(0)] + \sum_{n=1}^{3m-6} [\partial\alpha_{ij}/\partial q_n]q_n \tag{2}$$

where $\alpha_{ij}(0)$ is one of 9 elements of the polarizability of the molecule at the equilibrium (nonvibrating) position. The time dependent vibrational coordinate (one of $3m - 6$ vibrational coordinates where m is the number of atoms in the molecule) may be expressed in terms of a stationary state normal coordinate q_{no} and t.

$$q_n(t) = q_{no}\cos\omega_n t \tag{3}$$

substitution of (3) and (2) into (1) gives rise to an expression having terms in $\cos\omega_0 t$ and $\cos\omega_0 t \cos\omega_n t$, which may be rearranged into the form

$$p = [\alpha_{ij}(0)]\mathbf{E}_0\cos\omega_0 t + \tfrac{1}{2}\sum_{n=1}^{3m-6} [\partial\alpha_{ij}/\partial q_n]\mathbf{E}_0 q_{no}\cos(\omega_0 \pm \omega_n)t \tag{4}$$

The first term gives rise to Rayleigh scattering (inelastic), the remaining terms to anti-Stokes scattering at $\omega_0 + \omega_n$ and Stokes scattering at $\omega_0 - \omega_n$. These processes are illustrated by the energy level diagram in Fig. 1. ω_L is the frequency of the incident laser beam. In Rayleigh scattering, the molecule moves to a virtual state (dashed line) well below any low-lying electronic transition, and immediately falls back to the $v'' = 0$ ground state emitting light with frequency ω_L. In normal Raman scattering (Stokes), the molecule makes a vibrational transition to $v'' = 1$ by absorbing ω_n of the energy from the ω_L beam and consequently the scattered light has frequency $\omega_L - \omega_n$. This process is many orders of magnitude weaker than Rayleigh scattering. In anti-Stokes scattering, a vibrationally excited molecule $v'' = 1$ moves to a virtual state. The molecule makes a transition to its ground vibrational state $v'' = 0$ and gives up ω_n energy to ω_L. The intensity of this scattering is very much weaker than Stokes scattering because of the low population (Boltzmann distribution) of molecules in the $v'' = 1$ state relative to the $v'' = 0$ state. For vibrational frequencies greater than $400\,\text{cm}^{-1}$ the anti-Stokes lines are very weak and not useful for molecular characterization.

As the frequency of the laser increases and approaches the first low-lying electronic excited state, Stokes scattering of vibrations that have

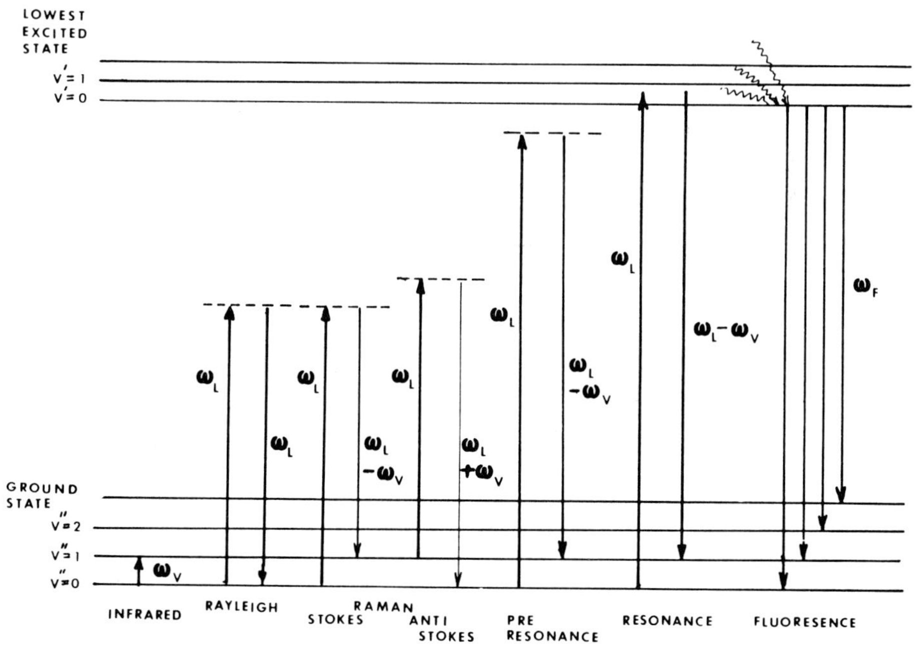

Figure 1. Schematic energy level picture of (linear) Raman scattering processes.

the same symmetry as the excited state become more intense. This phenomenon is labeled pre-resonance. Finally, as the incident laser frequency moves into the absorption band of the excited state, the resonance becomes strong and the corresponding Raman bands increase their intensity by as much as 10^4 times their non-resonant intensities. Thus, it is possible to observe scattering from very dilute solutions of chromophores or from specific isolated chromophores that may be embedded in large macromolecules (i.e., heme, chlorophylls, proteins, etc.). If the exciting line is near or within an electronic absorption band, fluorescence may be observed from transitions from the $v' = 0$ vibrational level of the excited state to the various vibrational states in the ground state. These transitions are at fixed frequencies and do not move with the exciting line as the laser frequency is scanned through the electronic absorption band. Thus they may be easily distinguished from Raman scattering. However, this process can be more efficient than Raman scattering, and frequently is intense enough to completely mask Raman scattering. The radiationless decay processes that move the molecule to the $v' = 0$ state can take a longer time than the Raman scattering process ($ps = 10^{-12}$s), which makes it possible, in some cases, to time discriminate against the fluorescence by using pulsed lasers and gated photo-detection schemes.

As the laser power increases, a substantial fraction of the molecules can be moved from the initial state to the final state and the intensity of the Stokes wave increases dramatically and becomes strongly directional (see Fig. 2). The intensity of this stimulated stokes wave ω_s can become comparable in magnitude to the laser wave. This process populates the higher v'' energy state which then gives rise to strong stimulated anti-Stokes scattering at $\omega_a = 2\omega_1 - \omega_s$. The generation of a strong ω_s wave depends on the scattering cross section of a vibration $(\partial\alpha_{ij}/\partial q_n)$ and Stokes and anti-Stokes scattering is only observed for those few vibrations of a molecule that have large cross sections.

On the other hand, if an initial intense laser beam ω_1 is scanned with a second laser beam ω_2, such that $\omega_1 - \omega_2$ corresponds to the frequency of a Raman active vibration ω_v, the ω_1 and ω_2 waves can mix and produce strong directional scattering at $2\omega_1 - \omega_2 = \omega_a$ in the anti-Stokes region (higher in frequency than either ω_1 or ω_2) which is spatially separated from the pump ω_1 and scanned ω_2 beams. This technique is called Coherent anti-Stokes Raman spectroscopy (CARS). Phase matching requirements in condensed media require that the incident beams be crossed at a specific angle that depends on the refractive index of the medium. The reader is referred to the literature for experimental details. The intensity of the ω_a beam, its spatial separation and the freedom from

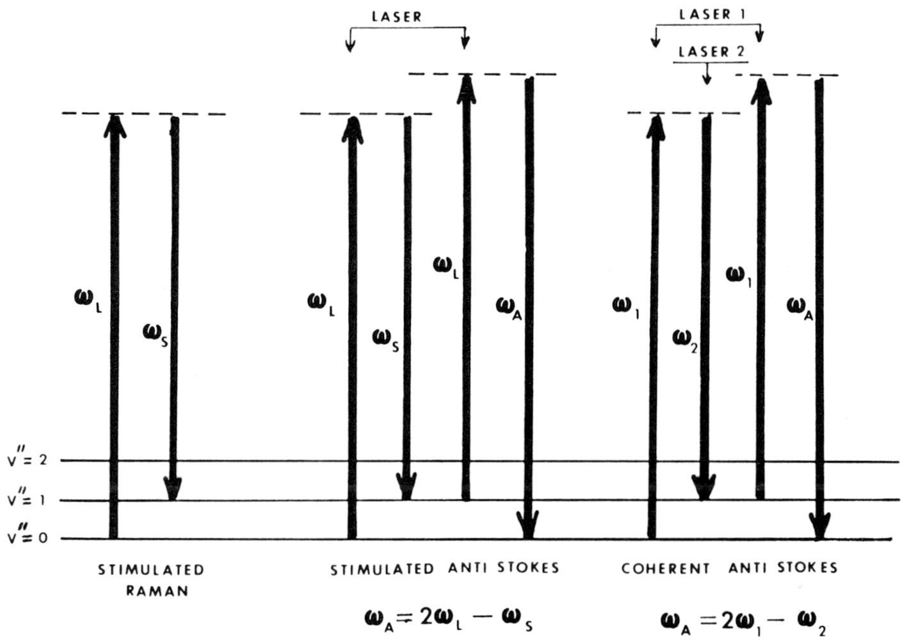

Figure 2. Schematic energy level picture of nonlinear Raman scattering processes.

interferring fluorescence makes CARS a powerful tool in biological investigations.

Instrumentation

The Raman experiment has three distinct parts: excitation of the sample, dispersion of the scattered light and detection of the spectral intensities of the dispersed scattering.

Many different types of lasers are used in Raman spectroscopy and the choice depends on the application [Demtröder (1981)]. In Fig. 3 we show wavelengths accessible with several laser systems. For conventional spontaneous Raman studies on colorless materials, continuous wave (CW) Argon or Krypton ion lasers produce a few discrete wavelengths in the visible spectral region, each with up to a few watts of power. Ion lasers are used to optically pump dye lasers whose output can be tuned from 400 nm to 800 nm with up to 1 W of CW power. Argon and Krypton ion lasers also have lines near 360 nm that can be used to pump dye lasers.

Pulsed lasers are used to study time dependent photochemical processes on time scales of ms to ps [Hochstrasser and Johnson (1985)]. Ion lasers can be mode locked and cavity dumped to produce ps pulses with KW peak powers. Pulsed nitrogen lasers with output at 337 nm can be

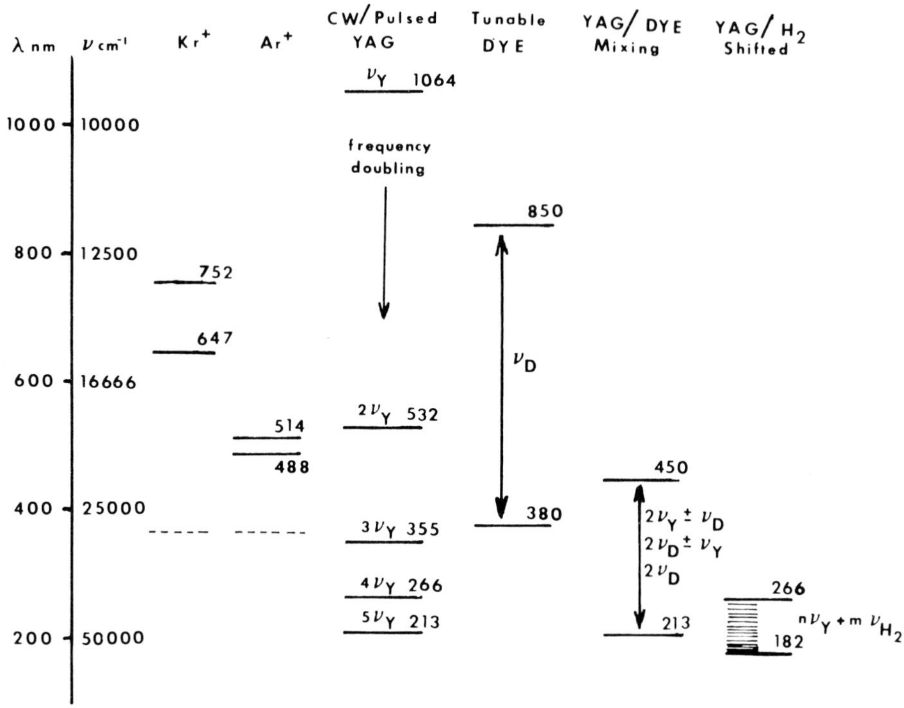

Figure 3. Excitation wavelengths for a few selected laser systems.

used to pump dye lasers to produce pulse widths of 1–10 ns and peak powers of 10^4 to 10^5 W. Excimer pumped dye lasers can achieve an order of magnitude greater peak powers with similar pulse lengths. YAG lasers can pump dye lasers to comparable power levels with pulse widths of 5–30 ns. The high peak powers of excimer and YAG lasers enable them to interact with non-linear crystals to produce stimulated waves (non-linear optical mixing) at $n\omega_y$. Thus, the YAG laser with output at 1064 nm $\times$ ω_y (9398 cm^{-1}) can be frequency doubled to 532 nm (18797 cm^{-1}) which can be mixed again with the 1064 nm wave to produce $3\omega_y$ at 355 nm (28195 cm^{-1}) (see Fig. 3). In turn, the second harmonic of a YAG laser can be mixed with the output of a second harmonic YAG pumped dye laser to produce tuneable laser radiation in the 450 nm to 217 nm range. Alternatively, the higher order harmonics of the YAG laser can be passed into a cell containing H_2 gas at several atmospheres pressure to produce stimulated waves shifted by the hydrogen vibrational stretching frequency of 4155 cm^{-1} [Hudson (1986)]. These wave mixing processes can produce a series of excitation lines from 266 nm to 184 nm. The high excitation frequencies of these beams allows us to excite into UV absorption bands of higher energy transitions.

Spontaneous Raman scattering is emitted from the sample in all directions, however, this light does not necessarily maintain the same polarization characteristics as the incident laser beam because of the complexity of the three-dimensional tensor α. Typical Raman measurements involve collecting the scattered light with lenses or mirrors in solid angles whose centers are 90° to the incident laser beam (right angle scattering), 180° to the incident laser beam (back scattering), or even at intermediate angles. The collected light from the sample area is imaged on the entrance slit of a monochromator. The scattered light is dispersed by gratings and projected onto the exit plane of the monochromator which either contains a slit or an aperture. If it contains a slit, a small wavelength interval is passed through the slit and movement of the grating moves the spectrum across the exit slit, allowing the detector to measure the light intensity as a function of wavelength. If it contains an aperture, a large part of the spectrum can be collected at once on the surface of a linear array detector or multichannel analyzer.

Because of the four to six orders of magnitude weaker strength of spontaneous Raman scattering compared to Rayleigh scattering, it is necessary to use photomultipliers or devices with high quantum efficiency to detect the photons at the monochromator exit slit. Photodiode array detectors (multichannel detectors) placed at the exit plane measure the intensity distribution over a large spectral range in a short time period (single pulses of laser excitation) or build up the intensity by integration over extended time periods. This spectrograph mode of operation is very important for probing the states of photo-biological processes as functions of time.

Spontaneous Raman

An enormous amount of work has been done to understand the structure/frequency correlations in spontaneous Raman spectra of proteins, nucleic acids, lipids and carbohydrates. The reader is directed to books by Carey [1982] and Tu [1982] and the references therein for review material on research using conventional Raman spectroscopy.

Our own current interests involve looking at the changes in the fluidity of membranes caused by reduction of the degree of hydration in the headgroup region of lipid bilayers. The CH stretching bands in the Raman spectra of acyl chains of 1,2 diacyl phosphatidylcholines exhibit thermotropic behavior when passing from a gel phase to a liquid crystal phase [Levin (1984)] and are sensitive probes of membrane phase transitions. Membranes must maintain fluid liquid-crystal properties at ambient temperatures in order to maintain viability of membrane regulated phenomenon. Measurements of Raman spectra of a model membrane of 1,2 dimyristyol phosphatidylcholine, held in controlled temperature and humidity environments, have shown that the gel liquid crystal phase transition is raised by $\sim 0.44°C/RH$ unit on going from 100% RH to 40% RH. Thus, physiological changes that can reduce the water activity to 0.8 may well be expected to shift the phase transition by as much as 9°C. Raman studies of the effect of lowered humidity on other non-bilayer transitions are currently being planned.

We have also studied the hydration of thin protein films using integrated optics and Raman spectroscopy. Figure 4 illustrates a method of coupling a laser beam polarized along E_y or (E_x, E_z) into a thin film of refractive index n_2 supported on a substrate with refractive index n_3 and exposed to air ($n_1 = 1.0$). The laser beam is focused at the corner of a high index prism at an angle of ϕ from the horizontal plane. At all but a discrete few angles, depending on the thickness of the film and the refractive index, the beam will undergo total internal reflection at the corner. However, at those particular angles, the laser beam will enter and propagate along the film [Scherer and Bailey (1983)]. When the film is exposed to a controlled humidity environment, the water uptake by the film can be determined from changes in the refractive index and thickness, and the degree of hydrogen bonding of the water and conformational changes in the protein can be monitored from changes in the Raman spectrum obtained from the propagating laser beam.

Conventional Raman spectroscopy has been used to study the mechanism of retrogradation of starches following gelatinization [Bulkin and Kwak (1986)]. Raman data were used to distinguish 4 stages in the retrogradation process: conformational ordering (possibly formation of double helices in amylopectin branches); an induction time for the onset

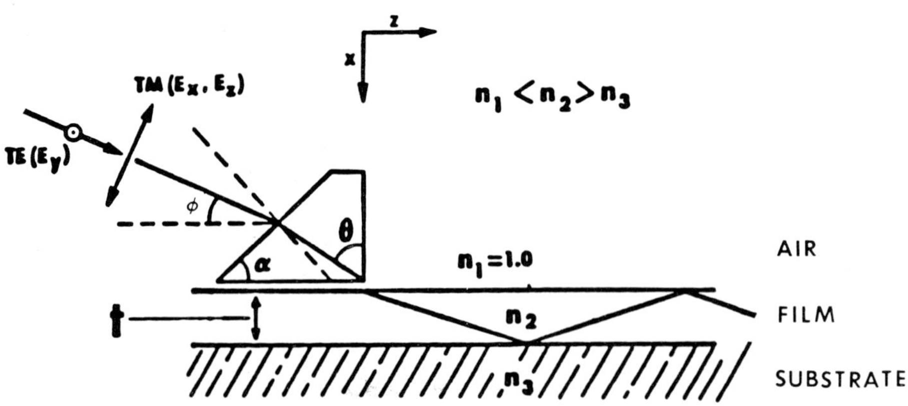

Figure 4. Illustration of wavelength excitation of a thin film using the prism coupler.

of crystal growth; a primary crystalization step; and slow introduction of long-range order (crystalline phase propagation).

Resonance Raman

Virtually all biorelevant molecules are rapidly becoming accessible to study by resonance Raman methods. Much early work on protein bound chromophores, such as those found in hemoglobins, cytochromes, chlorophylls, retinals, or protein-complexed metal ions was facilitated by low-lying electronic transitons in the 400 to 800 nm region and the utilization of dye lasers [Carey (1982), Tu (1982)]. Lutz and coworkers (1984) have carried out extensive research to characterize resonance spectra of chlorophylls, pheophytins, cytochromes and carotenoids from plants, photosynthetic bacteria and algae. Careful choice of excitation wavelength in the Soret region ($\sim$400 nm) leads to selective enhancement of resonance Raman spectra of various chlorophylls, both in the antenna light harvesting system and reaction centers. Resonance Raman spectroscopy will continue to be an important tool in characterizing and clarifying the role of various important pigments on the photosynthetic apparatus.

Recently, resonance Raman spectroscopy has been extended into the ultraviolet spectral region. This has made it possible to obtain important structural information either about proteins, nucleic acids, and other complex molecules that do not have low-lying electronic transitions in the visible spectral region, or about other structural regions of molecules containing visible chromophores. Johnson et al. (1984) has obtained Raman spectra of myoglobin without contributions from the heme, which dominates the spectrum obtained with visible excitation.

Aromatic amino acids (tryptophan, tyrosine and phenylalanine) exhibit strong fluorescence when the excitation wavelength is near 266 nm and this interferes with observation of Raman scattering. However, lowering the excitation wavelength to 200–218 nm places the whole Raman spectrum between 200 to 232 nm which is well below the region of interference from fluorescence [Rava and Spiro (1984)]. Figure 5 from Rava and Spiro [1985] shows marked differences on the UV absorption spectra of tyrosine and tryptophan in the 200 nm region. In Fig. 6 are resonance Raman spectra of tyrosine, tryptophyltyrosine and tryptophan which show the selective enhancement of tryosine or tryptophan by using either 200 or 218 nm excitation [Rava and Spiro (1984)]. Surface exposures of aromatic residues in aqueous solutions of insulin and α-lactalbumin as a function of pH, have also been determined [Rava and Spiro (1984)] with excitation in this region. Amide I, II and III vibrations of the peptide backbone also become enhanced with excitation below

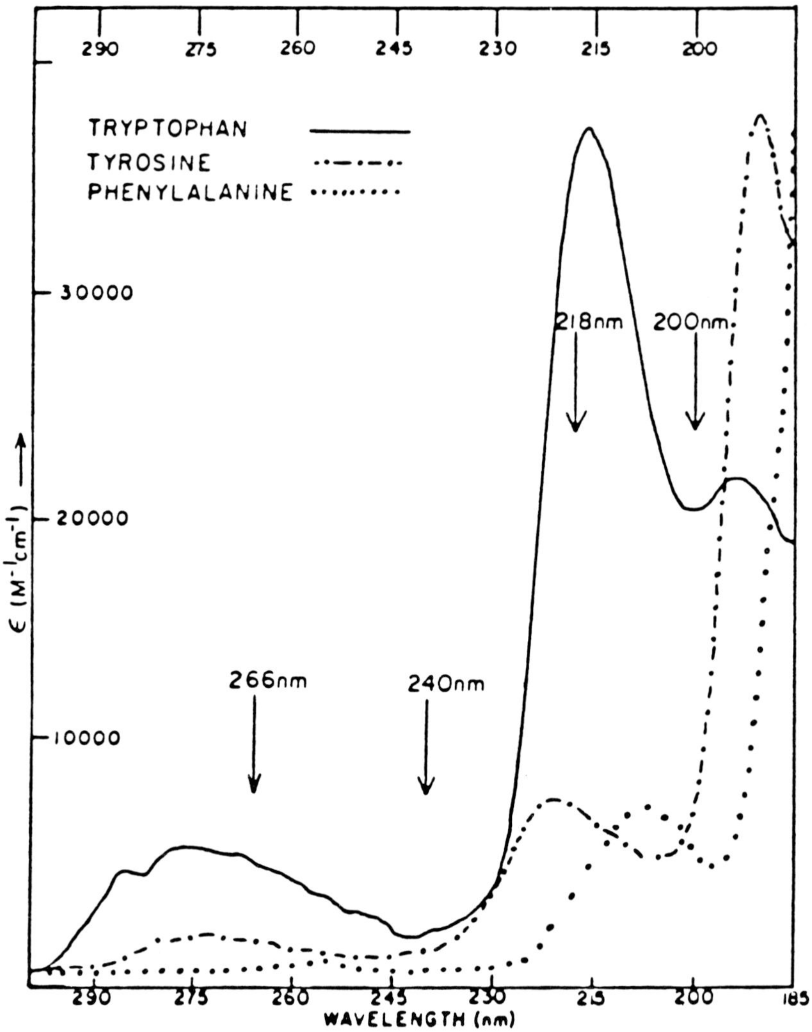

Figure 5. UV adsorption spectra of tryptophan, tyrosine and phenylalanine obtained with 30 μM samples in a 1-cm cell. Reproduced from Rava and Spiro (1985) with permission of the copyright holder.

218 nm, making it possible to obtain information about changes in the secondary structures of proteins. Differences in ferricytochrome *c* from tuna and horse heart as a function of pH have been determined using excitation at 200 and 218 nm [Copeland and Spiro (1985)].

Figure 6. Raman spectra for aqueous solutions of (a) tyrosine (10^{-3}M), (b) and (c) tryptophyltyrosine (10^{-3}M) and (d) trypthophan (10^{-3}M). Reproduced from Rava and Spiro (1984) with permission of the copyright holder.

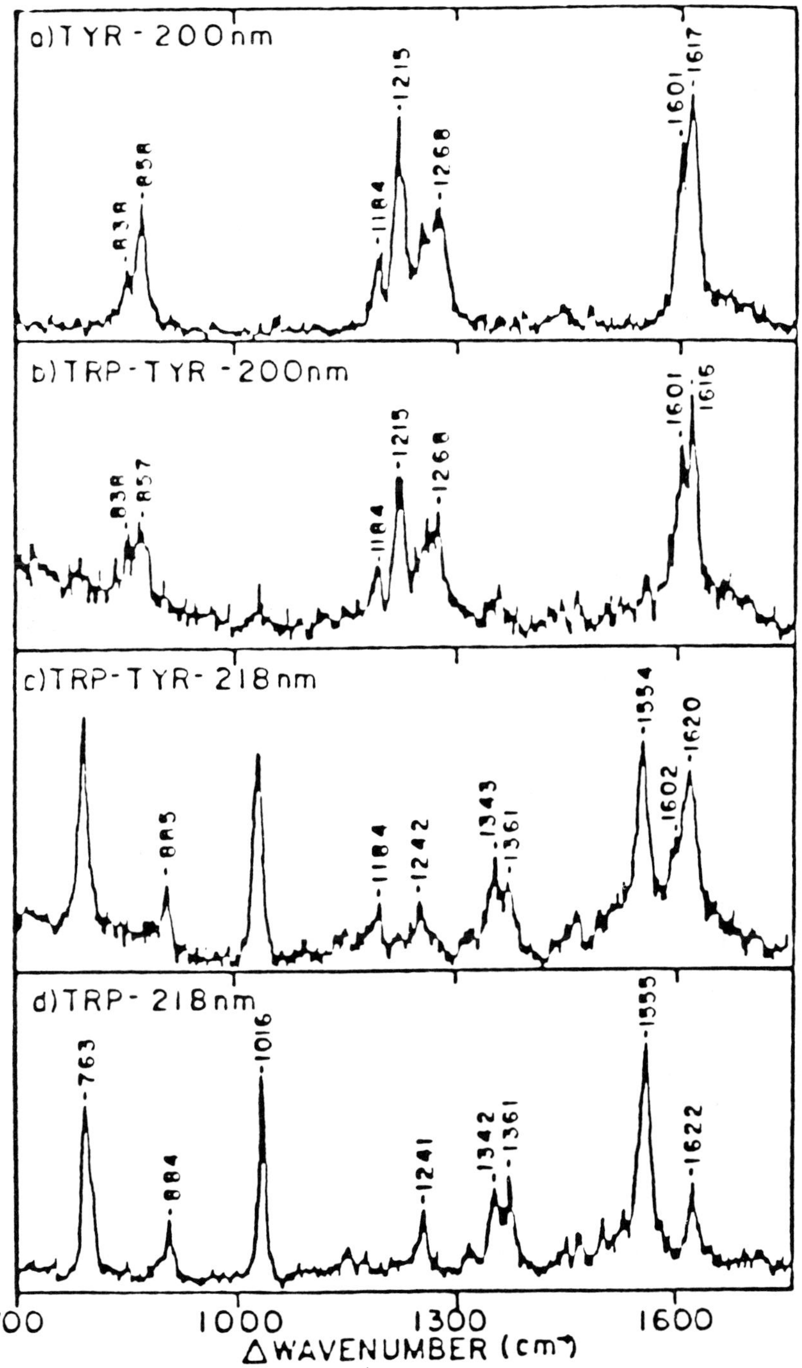
a) TYR - 200nm
1184
1215
1268
1601
1617
b) TRP - TYR - 200nm
1184
1215
1268
1601
1616
c) TRP - TYR - 218nm
1184
1242
1343
1361
1554
1602
1620
d) TRP - 218nm
763
884
1016
1241
1342
1361
1555
1622
700
1000
1300
1600
ΔWAVENUMBER (cm⁻¹)

There have been many resonance Raman studies of nucleic acids excited by radiation in the 250 to 300 nm region [Carey (1982), Tu (1982)]. Compared to proteins, nucleic acids are only weakly fluorescent and it is relatively easy to obtain resonance Raman spectra without serious background interference. However, the observation of resonance spectra of these structures in the presence of proteins with excitation above 250 nm has serious limitations. Spectra of the nucleotides show considerable differences in their spectra excited at 266, 253, 240, 229, 218, 209 and 200 nm [Kubasek et al. (1985), Fodor et al. (1985)].

This new technique of selective tuning of laser excitation into the absorption bands of the molecular groups of interests has great promise in sorting out structural perturbations in biopolymers.

Raman microprobe

The Raman microprobe combines an optical microscope (with bright and dark field illumination) with a Raman spectrometer. This important technique enables us to obtain (structural and/or analytical) information that is not directly available from other micro-analytical techniques. Several arrangements of both illumination and detection (multichannel or monochannel) systems have been reported in the literature [Delhaye and Dhamelincourt (1975), Abraham and Etz (1979), Rosasco (1980)]. Figure 7 shows the schematic diagram of the commercially offered microprobe MOLE (Molecular Optical Laser Examiner). Two illumination schemes (point or global) can be employed. In point illumination, the microscope objective focuses the laser beam on the sample (resolution $\sim 1\,\mu m$) and also collects the scattered light for transfer to a monochromator equipped with either monochannel or multichannel detectors. Global illumination exposes an area to the laser beam and the microscope transmits the image to a monochromator which then translates the intensified image at any selected wavelength to a low level TV camera. To illustrate the effect of global illumination, Fig. 8 shows a sample illuminated with monochromatic radiation. In a standard experiment, the Raman spectrum would contain vibrational bands from substances A and B as well as the matrix C. Using the MOLE, one adjusts the spectrograph to pass one band in the Raman spectrum for one of the components (A or B or C). A micrographic image of the distribution of that component in the area of illumination will be observed on the TV screen. Mapping the distribution of other components of the sample can be done by tuning the spectrograph to a characteristic Raman band for each component.

This technique has been used for spatial analysis of the distribution of carotenoids in carrot root cells. Carotenoids consist of several different

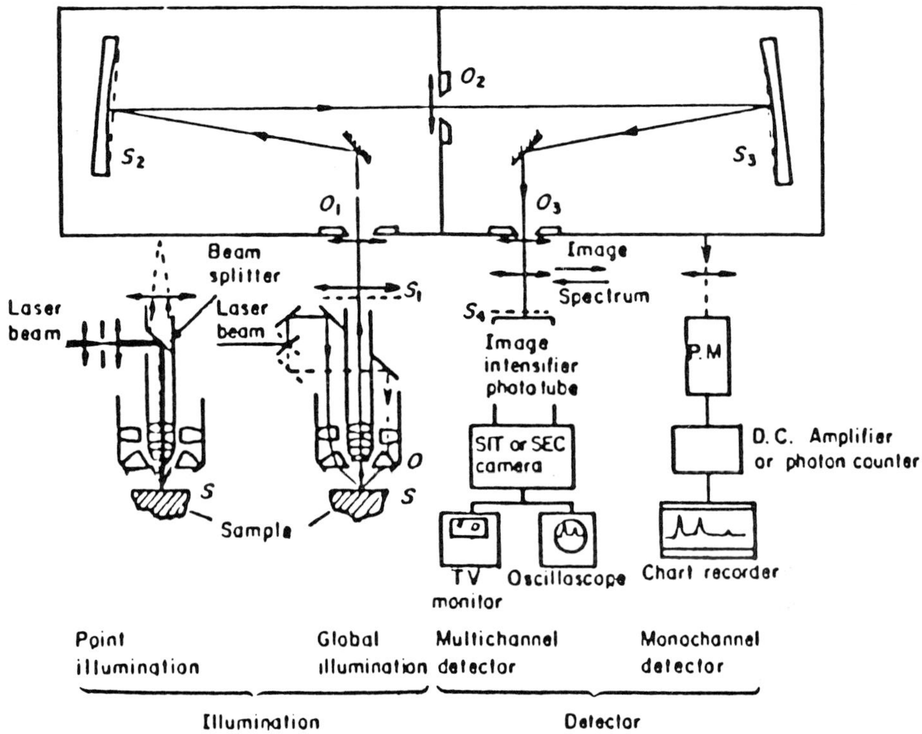

Figure 7. Schematic diagram of the laser Raman microprobe showing components of the illumination, dispersion and detection systems.

polyene compounds, all of which have bands that contribute to a Raman spectrum. However, by selecting one of the Raman bands for a particular carotenoid, one can obtain an image of the localization of that component [Cavagnat, Cruege and Pham (1981)].

Some biological and biochemical problems that have been investigated with the Raman microprobe are: orientation of monodomains of phospholipid multilayers [Peticolas, Harrand and Dupeyrat (1982)], mineralization of tooth enamel and embryonic chick Tibia [Etz and Blaha (1980)], identification of purines in standard histological fixed thin sections from organs of various animals [Ballan-Dufrancus et al. (1979)], identification of particulates of pesticides in ciliated protoza [Delhaye et al. (1979)], fluorescence spectra from the cytoplasm of a single microscopic bioluminescent algae, Pyrocystis lunula [Dupaix et al. (1982)], and chromophore orientation in single cell photoreceptor rods from goldfish and toads using Resonance Raman spectroscopy [Barry and Matheis (1982)].

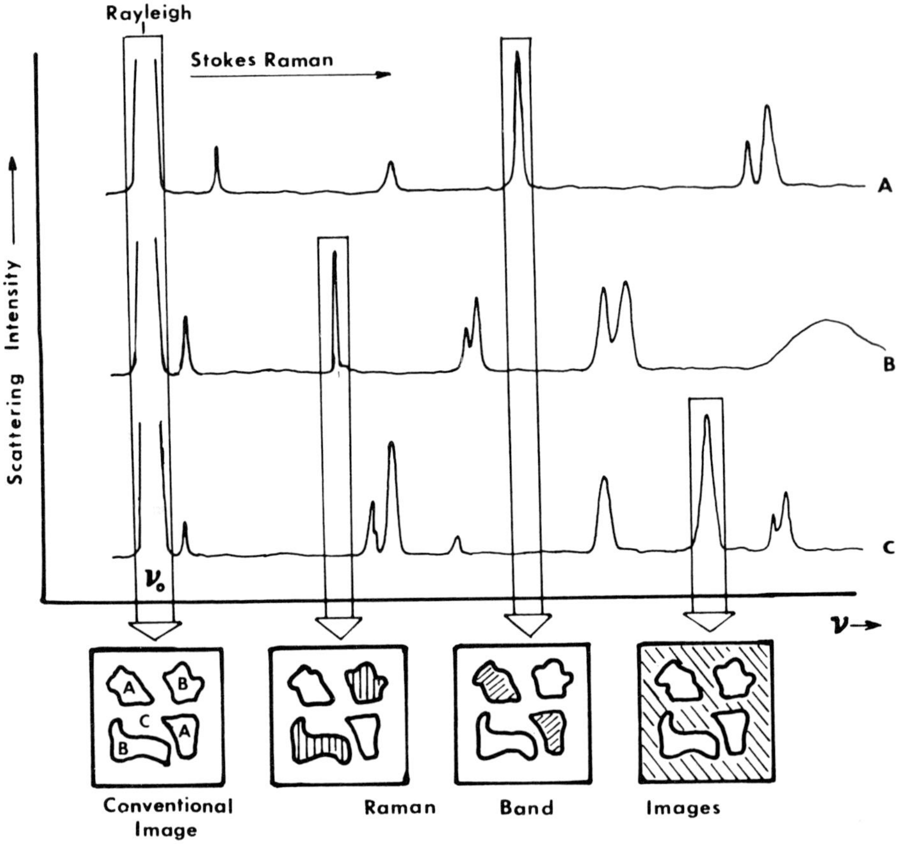

Figure 8. Illustration of selective imaging by using wavelengths corresponding to isolated Raman bands from different objects (A, B, C) in the field of view. (adapted from Dhamelincourt (1979)).

Surface-enhanced Raman spectroscopy (SERS)

The interaction of metal surfaces, in the form of electrodes, colloids or island films on glass substrates, with certain chemical groups on the surface of biological macromolecules can give rise to large intensity increases in Raman bands from these groups [Otto (1983)]. The potential of the SERS technique for probing molecular structure of biomolecules *in situ* was demonstrated with flavoproteins adsorbed on silver colloids [Copeland, Fodor and Spiro (1984)]. Figure 9 compares the SERS and RR spectra of riboflavin binding proteins (RBP). The RR spectrum is resonance enhanced because the excitation laser frequency of 488 mm falls within the first $\pi - \pi*$ flavin absorption band. However, the SERS spectrum has a better signal to noise ratio than the RR spectrum. There is a one-to-one correspondence of all the frequencies of the flavin bands,

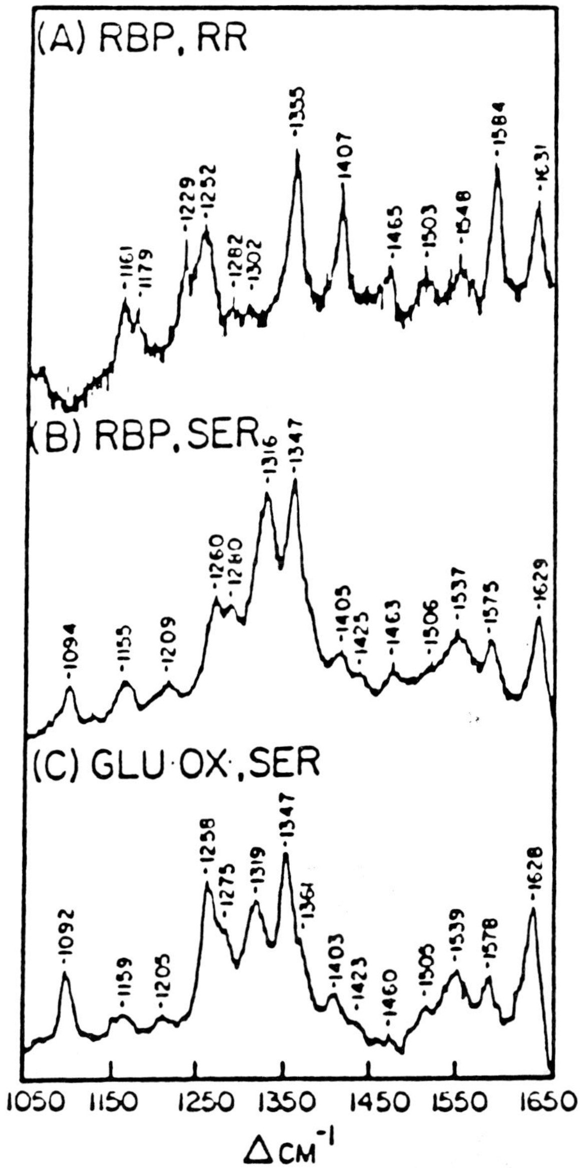

Figure 9. Flavin Raman spectra, with 488 nm a$^+$ laser excitation for (a) RBP (0.50 mM-flavin), (b) RBP (0.58 μM flavin) on Ag sol and (c) GO (0.58 μM flavin) on Ag sol. Reproduced from Copeland et al. (1984) with permission of copyright holder.

but the relative intensities are different. The frequency and intensity differences between spectra are attributed to the interaction of the flavin chromophores with the silver particles. It was suggested that the uracil end of the flavin chromophore (N-3, 0-2 & 0-4 hetero-atoms) is involved

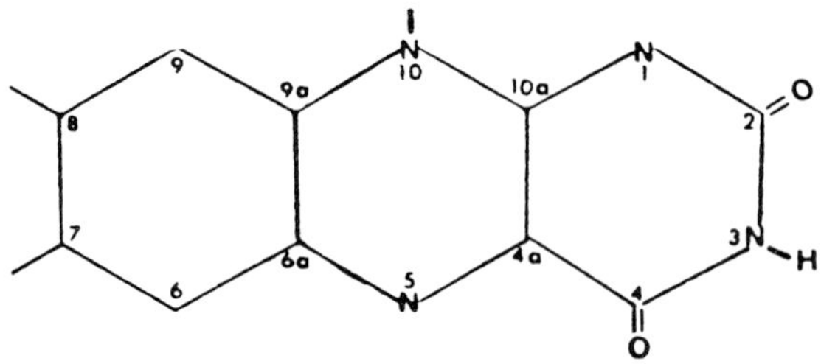

Isoalloxazine

in the adsorption of silver. In the case of glucose oxidase, this interaction did not significantly diminish the enzymatic activity (86%), implying that the active site remains intact in the associated enzyme.

The concentration of the chromophore can also affect the quality (signal to noise ratio) of the SERS spectra. For glucose oxidase concentrations greater than 0.43 μM, the Raman spectra was nearly obscured by fluorescence [Copeland, Fodor and Spiro (1984)]. This behavior was interpreted in terms of insufficient silver interaction sites for the association of the excess glucose oxidase.

Antitumoral properties of certain platinum coordination compounds are believed to result from the stereochemistry of platinum binding to DNA within the cell. J.M. Sequaris et al. [1984] have used SERS to study DNA modified by antitumoroid active cis-$Pt(NH_3)_2Cl_2$ and the antitumorid inactive [Pt-(dien)Cl]Cl platinum coordination compounds. The Pt-N stretching vibration (500–660 cm^{-1}) was only observed in the SERS spectra for [Pt(dien)Cl]Cl complex. Since this technique depends on proximity of the complex to the silver surface, the results suggest two different stereochemical configurations for the Pt-DNA complexes: the antitumoral inactive [Pt-(dien)Cl]Cl forms an outer complex, whereas the antitumoral active $Pt(NH_3)_2Cl_2$ is localized inside the DNA helical structure.

SERS has been used to study surface characteristics of bile pigments [Lippitsch (1981)] and nucleic acids [Koglin et al. (1984)]. While these studies are encouraging for some biological applications, additional studies are needed to assure that the interaction with metal surfaces does not alter the natural structure. Such structural perturbations have been observed in SERS spectra of heme proteins hemoglobin and cytochrome b_5. Spectral changes indicate the formation of surface-bound hemin μ-oxo dimers, which suggests that, in some cases, the heme groups can be extracted from their binding sites [Smulevich and Spiro (1985)].

Coherent anti-Stokes Raman spectroscopy (CARS)

As mentioned earlier, anti-Stokes lines in spontaneous Raman scattering are weaker than Stokes lines because the upper vibrational levels are less populated than lower levels at normal temperatures. However, the CARS nonlinear optical technique produces a coherent laser-like beam of high intensity (in the anti-Stokes region) which can easily be separated spatially from the isotropic fluorescence. The technique requires two pulsed laser beams. The pump laser remains at the constant frequency ω_1, whereas the second laser is scanned with a variable frequency $\omega_2(\omega_1 > \omega_2)$. These two focused laser beams are crossed in the sample,

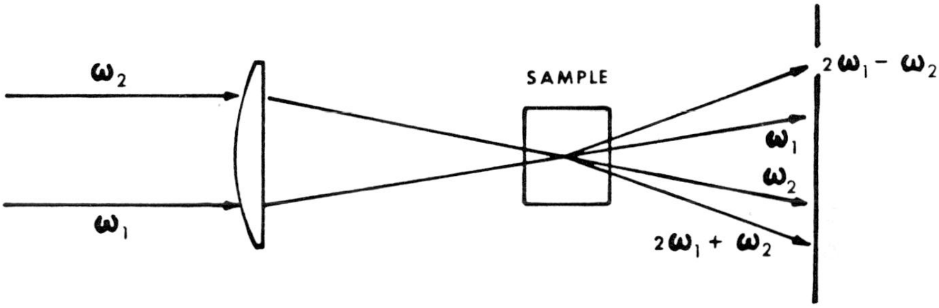

creating new frequencies at $2\omega_1 + \omega_2$ and $2\omega_1 - \omega_2$. The intensity of anti-Stokes emission $(2\omega_1 - \omega_2)$ is many orders of magnitude greater than that obtained with normal RS. If the laser frequencies coincide or are in the proximity of an electronic transition, the CARS signal will be resonance enhanced.

Resonance CARS has been used to distinguish between semiquinone and oxidized forms of flavin nucleotides [Dutta and Spiro (1980)]. The

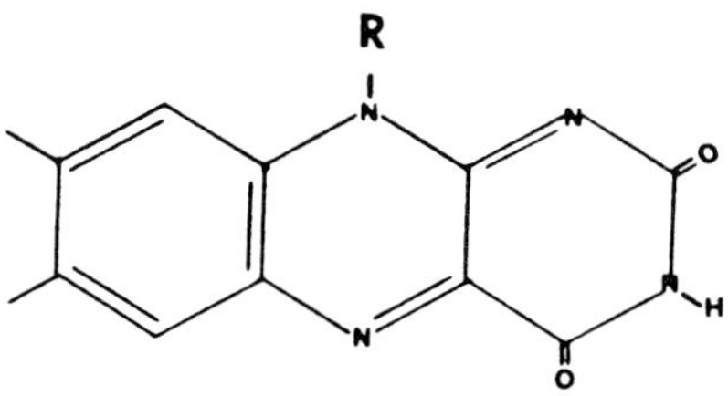

Oxidized

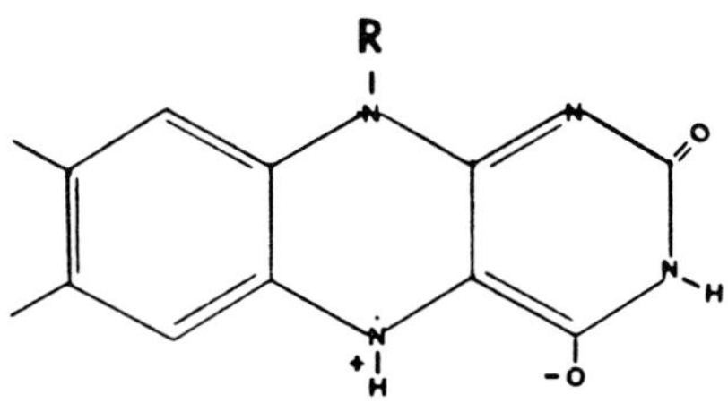

Semiquinone

semiquinone form has an absorption in the 600 nm region and the oxidized form has its primary absorption in the 450 nm region. Dutta and Spiro chose pump laser excitation at 480 nm to obtain resonance CARS spectrum of the oxidized form and excitation at 620 nm to obtain the CARS spectrum of the semiquinone form. The different CARS spectra permit a clear identification of the state of the isoalloxazine ring. Similarly, highly fluorescent biomolecules such as flavin adenine dinucleotide (FAD) and glucose oxidase [Dutta, Nestor and Spiro (1977)] have been analyzed by resonance enhanced CARS spectra. Presently, the most serious drawback for CARS is the problem of interferences caused by the nonresonant background which limits the sensitivity of the technique to 1% in solution.

F.T. Raman

Historically Raman spectroscopy has had one constant drawback: minor amounts of fluorescing material (often the sample itself) can totally obliterate the weak Raman scattering. The CARS process avoids

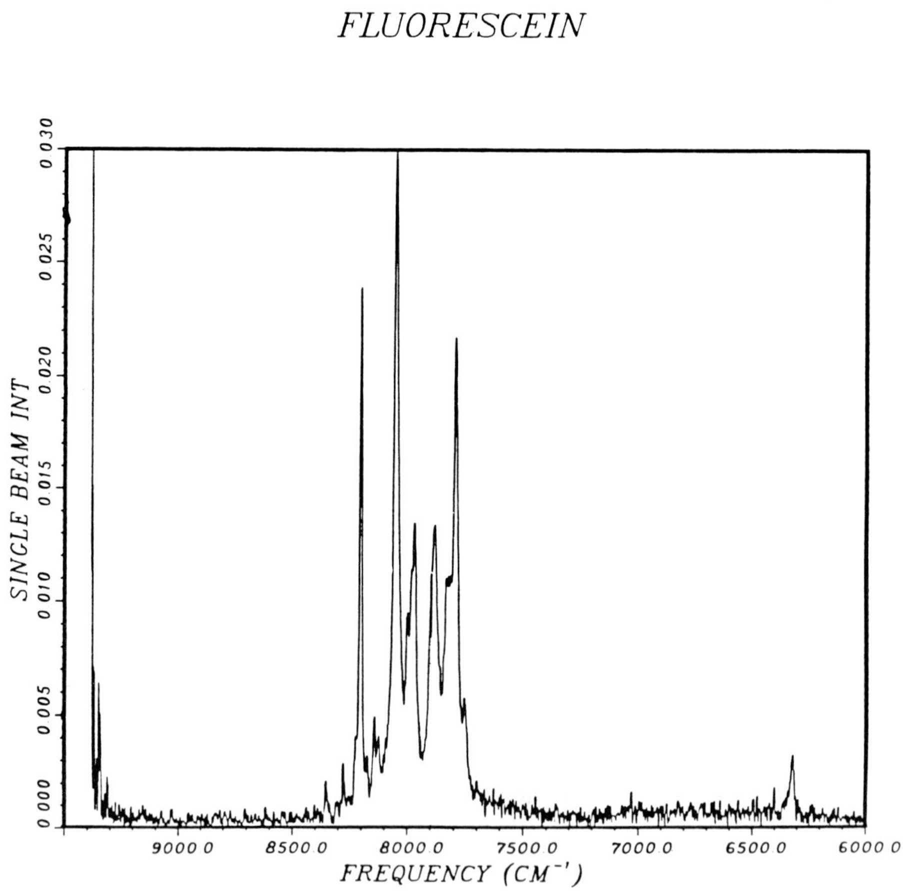

Figure 10. FT-Raman spectra of fluorescein obtained with 1064 nm YAG excitation kindly supplied by B. Chase.

this problem, as does far UV resonance Raman. However, in the latter case, many non-resonant modes will be undetected. It is now possible to obtain ordinary Raman spectra of fluorescing samples by moving the excitation wavelength well into the near infrared, out of the region of low-lying electronic absorption.

The development of highly attentuating optical notch filters and sensitive near infrared detectors has made it possible to obtain Raman scattering using continuous YAG laser excitation at 1.064 μm and the Fourier transform near infrared spectrometer. The efficiency of Raman scattering has a basic $1/\lambda^4$ dependence which means that the scattering efficiency is 20 times less than that at 500 nm or 7800 times less that at 200 nm. These factors can be recovered with the increased power available in the output of the YAG laser at 1064 nm, the accumulation of spectral scans and efficient near infrared detectors. The net result has been that reasonably high quality "ordinary" Raman scattering has been obtained with samples that would normally be considered impossible [Hirschfeld and Chase (1986)]. In Fig. 10 we show the "ordinary" Raman spectrum of fluorescein obtained by B. Chase [(personal communication)] with 1064 nm excitation. Besides the obvious lack of interference from fluorescence, this technique has the added advantage of high wavelength accuracy provided by the FT spectrometer which makes it possible to obtain accurate Raman difference spectra.

Conclusions

Raman spectroscopy has evolved considerably since Raman's discovery of the effect in 1928. The development of the optical laser gave new life to what was considered a spectroscopically useful but not widely practical technique. The development of tunable dye lasers made it possible to easily examine perturbations in the molecular structure around visible chromophores with low-lying electronic transitions. Pulsed laser techniques have permitted time-resolved studies of transient molecular states involved in complex biochemical cycles. The development of high energy pulsed lasers has accelerated research into the UV and far UV region. These developments have pushed spectroscopic research through the "fluorescence barrier". Interestingly, the most recent development of FT-Raman has also broken this barrier by moving the excitation into the near infrared spectral region. Thus, growth in the field of Raman scattering has expanded spectrally from its "visible" origins. The implications of these developments for biological applications are enormous. With proper instrumentation the spectroscopist/biochemist can choose the appropriate wavelength to obtain desired molecular information instead of choosing the problem on the basis of instrumental limitations.

It is our view that the various Raman scattering techniques we have been describing could be important tools in molecular agricultural research from now to the end of the 20th century. Need we wait for the 21st century?

References

1. Abraham, J.L. and E.S. Etz. 1979. *Molecular microanalysis of pathological specimens in situ with a laser Raman microprobe*. Science 206: 716–718.
2. Ballan-Dufrancais, C., Truchet, M. and P. Dhamelincourt. 1979. *Interest of Raman laser microprobe (MOLE) for the identification of purinic concretions in histological sections*. Biol. of the Cell 36: 51.
3. Barry, B. and R. Mathies, 1982. *Resonance Raman microscopy of rod and cone receptors*. J. Cell Biol. 94: 479.
4. Bulkin, B.V. and Y. Kwak. 1987. *Retrogradation Kinetics of waxy corn and positive starches – A rapid Raman spectroscopic study*. Carbohydrate Research 160: 95–112.
5. Carey, P.R. 1982. *Biochemical applications of Raman and resonance Raman spectroscopies*. Academic Press, N.Y.
6. Carey, P.R. and A.C. Storer. 1984. *Characterizations of transient enzyme-substrate bonds by resonance Raman spectroscopy*. Ann. Rev. Biophys. Bioeng. 13: 25–19.
7. Campion, A. 1985. *Raman spectroscopy of molecules adsorbed on solid surfaces*. Ann. Rev. Phys. Chem. 36: 549–572.
8. Cavagnat, R., Cruege, F. and V.H. Pham. 1981. *Biological applications of resonance Raman spectroscopy and resonance micro-Raman spectroscopy*. Biochemie 63: 927.
9. Copeland, R.A., Fodor, S.P.A. and T.G. Spiro. 1984. *Surface-enhanced Raman spectra of an active flavo enzyme: Glucose oxidase and riboflaven binding protein on silver particles*. J. Am. Chem. Soc. 106: 3872.
10. Copeland, R.A. and T.G. Spiro. 1985. *Ultraviolet resonance Raman spectra of cytochrome C conformational states*. Biochem. 24: 4960–4968.
11. Delhaye, M. and P. Dhamelincourt. 1975. *Raman microprobe and microscope with laser excitation*. J. Raman Spect. 3: 33–43.
12. Delhaye, M., Dhamelincourt, P. and F. Wallart. 1979. *Analysis of particulates by Raman microprobe*. Toxicol. Environm. Chem. Rev. 3: 73–87.
13. Demtröder, W. 1981. *Laser Spectroscopy*, Springer Series in Chemical Physics. Springer Verlag, Vol. 5.
14. Dhamelincourt, P. 1979. Thesis. Universite de Lille, France (unpublished).
15. Dupaix, A., Arrio, B., Lecuyer, B., Fresneau, C., Volfin, P., Merlin, J.C., Dhamelincourt, P. and B. de Bettignies. 1982. *Intracellular spectroscopic studies of a bioluminescent cell: Pyrocystis lunula*. Biol. of the Cell 43: 157.
16. Dutta P.K., Nestor, J.R. and T.G. Spiro. 1977. *Resonance coherent anti-Stokes Raman scattering spectra of fluorescent biological chromophores: vibrational evidence for hydrogen bonding of flaven to glucose oxidase and for rapid solvent exchange*. Proc. Natl. Acad. Sci. 74: 4146–4149.
17. Dutta, P.K. and T.G. Spiro. 1980. *Resonance coherent anti-Stokes Raman scattering spectra of oxidized and semiquinone forms of clostridium MP flavodoxin*. Biochemistry 19: 1590.
18. Etz, E.S. and J.J. Blaha. 1980. *Scope and limitations of single particle analysis by Raman microprobe spectroscopy*. NBS Publ. (USA) 533: 153–197.
19. Fodor, S.P.A., Rava, R.P., Hays, T.R. and T.G. Spiro. 1985. *Ultraviolet resonance Raman spectroscopy of the nucleotides with 266-, 240-, 218-, and 200-nm pulsed laser excitation*. J. Am. Chem. Soc. 107: 1520–1529.

20. Friedman, J.M., Rousseau, D.L. and M.R. Ondrias. 1982. *Time-resolved resonance Raman studies of hemoglobin.* Ann. Rev. Phys. Chem. 33: 471–491.
21. Gerrard, D.L. 1984. *Raman Spectroscopy.* Anal. Chem. 56: 219–225.
22. Geissbühler, H. 1984. *Biorational reflections in agricultural chemical research.* Chimia 38: 307–316.
23. Harvey, A.B. 1978. *Coherent Antistokes Raman Spectroscopy.* Anal. Chem. 50: 905A–912A.
24. Hirschfeld, T. and B. Chase. 1986. *FT-Raman spectroscopy: Development and Justification.* App. Spectrosc. 40: 133–137.
25. Hochstrasser, R.M. and C.K. Johnson. 1985. *Lasers in biology.* Laser Focus, Vol. 21, 100–118.
26. Hudson, B. 1986. *Resonance Raman spectroscopy in the far-ultraviolet region.* Spectroscopy 1: 22–31.
27. Johnston, C.R., Ludwig, M., O'Donnell, S. and S.A. Asher. 1984. *UV resonance Raman spectroscopy of the aromatic acids and myoglobin.* J. Am. Chem. Soc. 106: 5008–5010.
28. Koglin, E., Séquaris, J.-M., Fritz, J.-C. and P. Valenta. 1984. *Surface enhanced Raman scattering (SERS) of Nucleic acid Bases Adsorbed on silver colloids.* J. Mol. Struct. 114: 219–223.
29. Kubasek, W.L., B. Husdon, and W.L. Peticolas. 1985. *Ultra violet resonance Raman excitation profiles of nucleic acid bases with excitation from 200 to 300 nm.* Proc. Nat. Acad. Sci. (USA), 82: 2369–2373.
30. Levin, I.W. 1984. *Vibrational spectroscopy of membrane assemblies*, in *Advances in infrared and Raman spectroscopy* (Clark, R.J.H. and Hester, R.E., eds.), Wiley Heyden, Vol. 11, pp. 1–48.
31. Lippitsch, M.E. 1981. *Surface enhanced Raman spectra of biliverdine and pyrromethenone adsorbed to silver colloids.* Chem. Phys. Lett. 79: 224.
32. Lutz, M. 1984. Resonance Raman studies in photosynthesis, in *Advances in infrared and Raman spectroscopy* (Clark, R.J.H. and Hester, R.E., eds.), Wiley Heyden, Vol. 11, pp. 211–300.
33. Morris, M.D. and D.J. Wallan. 1979. *Resonance Raman Spectroscopy*, Anal. Chem. 51: 182A.
34. Otto, A. 1983. *Light scattering in solids* (M. Cordona and G. Guntherodt eds.), Vol. 4.
35. Peticolas, W.L., Harrand M. and R. Dupeyrat. 1982. *Polarized Raman spectra of oriented monodomains of phospholipid multilayers.* J. Raman Spectrosc. 12: 130.
36. Pimentel, G.C. 1985. Committee chairman to survey opportunities in the chemical sciences, *Opportunities in Chemistry*, National Academy Press.
37. Rosasco, G.T. 1980. *Raman microprobe spectroscopy in Advances in Infrared and Raman Spectroscopy.* R.J. Clark and R.H. Hester, eds., Heyden, Vol. 7, pp. 223–282.
38. Rava, R.P. and T.G. Spiro. 1984. *Selective enhancement of tyrosine and tryptophan resonance Raman spectra via ultraviolet laser excitation.* J. Am. Chem. Soc., 106: 4062–4064.
39. Rava, R.P. and T.G. Spiro. 1985. *Ultraviolet resonance Raman spectra of insulin and a-lactalbumin with 218 and 200 nm laser excitation.* Biochem. 24: 1861–1865.
40. Rava, R.P. and T.G. Spiro. 1985. *Resonance enhancement in the ultraviolet Raman spectra of aromatic amino acids.* J. Phys. Chem. 89: 1856–1861.
41. Rousseau, D.L. and M.R. Ondrias. 1983. Ann. Rev. Biophys. Bioeng. 12: 357–380.
42. Séquaris, J.-M., Koglin, E. and B. Malfoy. 1984. *Inner and outer complexes of Pt-coordination compounds with DNA probed by SERS spectroscopy*, FEBS 173, 95.
43. Scherer, J.R. and G.F. Bailey. 1983. *Water in polymer membranes I; Water sorption and refractive index of cellulose acetate.* J. Membrane Sci. 13: 29–41.
44. Smulevich, C. and T.G. Spiro. 1985. *Surface Advanced Raman Spectroscopic Evidence that Adsorption on Silver Particles can Denature Heme Proteins*, J. Phys. Chem. 89: 5168–5173.
45. Spiro, T.G. 1985. *Resonance Raman spectroscopy as a probe of heme protein structure and dynamics*, in *Advances in protein chemistry.* Academic Press Inc., Vol. 37, pp. 111–159.
46. Tu, A.T. 1982. *Raman spectroscopy in biology*, Wiley.

NMR: New techiques for chemical analysis and biological investigation

EDWIN D. BECKER and CHERIE L. FISK
National Institutes of Health, Bethesda, MD 20892, U.S.A.

Abstract Nuclear magnetic resonance (NMR) has become an increasingly valuable technique over the last three decades for the study of molecular structure of organic compounds. Within recent years a number of major advances in experimental methods have led to an explosion in the information elicited from NMR spectra and to a spectacular increase in the kinds of samples that can be studied by NMR. With the use of higher magnetic fields and greatly improved electronic techniques, NMR has undergone orders of magnitude increase in sensitivity, so that low concentrations of materials can be studied in chemical samples and in living systems, from cells to whole animals and humans. The use of two-dimensional NMR methods permits the unravelling of complex spectra, and innovative pulse techniques provide high resolution NMR spectra in solids. NMR imaging furnishes information on spatial distribution of materials and permits discrimination among various kinds of tissues. This chapter reviews some fundamentals of NMR and presents an elementary account of the theory and methods used in modern NMR applications.

Basic principles of NMR

This chapter is intended to provide a summary of the fundamentals of nuclear magnetic resonance (NMR) and to describe some of the techniques currently in use. It has been planned in conjunction with the following chapter by Dr. Pfeffer, which describes applications to problems of interest in agricultural chemistry.

Although some aspects of NMR require a quantum mechanical formulation, many features of the technique can be adequately and conveniently discussed by means of a classical picture, such as that given in Fig. 1. We see that the macroscopic magnetization, **M**, is made up of an ensemble of individual nuclear magnetic moments ($\boldsymbol{\mu}_i$), each precessing about the magnetic field $\mathbf{B}_0$, which is oriented along the z axis. At equilibrium the phase of each precessing nucleus is random and thus **M** is oriented along the z axis. The whole purpose of an NMR experiment is to tip the magnetization away from the z axis and to generate a component in the xy plane in which all the nuclei are in phase. This occurs when a radio frequency (RF) field $\mathbf{B}_1$ is introduced at the proper (resonance) frequency. When the RF field is removed, the component M_{xy} is left to precess in the xy plane, and it generates a signal in the radio frequency pickup coil placed in this plane. The component M_{xy} will not precess indefinitely, but decays with time. The magnetization returns to its equilibrium position in which **M** lies along the z axis with no component in the xy plane. This return to equilibrium is a result of relaxation

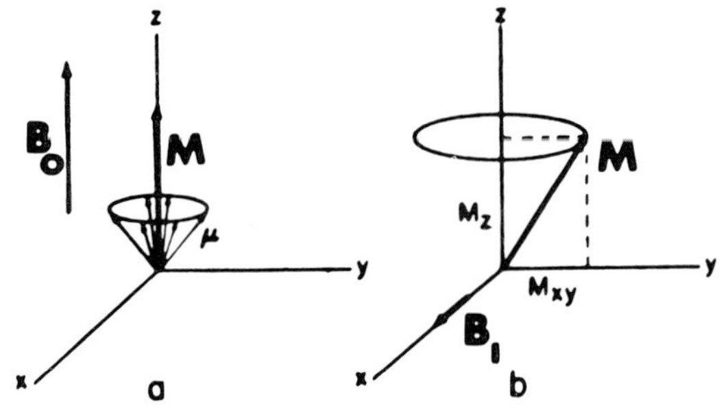

Figure 1. (a) Ensemble of identical nuclei (with magnetic moments, μ) precessing about the magnetic field $\mathbf{B}_0$, and the resultant macroscopic magnetization $\mathbf{M}$. Only the excess of spins in the lower energy state (parallel to $\mathbf{B}_0$) are depicted. (b) $\mathbf{M}$ tipped in the yz plane by interaction with radio-frequency field $\mathbf{B}_1$.

processes characterized by two times: T_1, the spin-lattice or longitudinal relaxation time, for return along the z axis; and T_2, the transverse relaxation time, which results from the dephasing of all the nuclei in the xy plane. T_1 relaxation is a critical parameter in an NMR experiment. Only when **M** has been returned to its equilibrium position can the experiment be repeated with maximum sensitivity. This becomes crucial in experiments on insensitive nuclei such as ^{13}C, which are often repeated many times to improve the signal to noise ratio. T_2 relaxation is important in determining the lineshapes of NMR signals. Both types of relaxation are of importance in the study of molecular structure and dynamics.

The fundamental relationship of NMR is the well known Larmor equation

$$\nu_0 = \frac{\gamma}{2\pi} B_0$$

showing that the precession frequency (ν_0) is directly proportional to the applied magnetic field, with the constant of proportionality γ, the magnetogyric ratio, characteristic of the particular nucleus we are studying. If this were all there is to NMR, it would be of little value in chemistry. The factors that make NMR of chemical interest are primarily the following four:

1. Chemical shift
2. Scalar (spin-spin) coupling
3. Magnetic dipole-dipole interactions
4. Electric quadrupole interactions

All four of these interactions are anisotropic – i.e., the values depend upon orientation of molecules relative to the applied magnetic field. We shall discuss each of these interactions in turn, beginning with molecules in the solid state, since the major effects are seen most readily in this state. We shall describe the methods by which one can study NMR in the solid; then see what happens as molecules tumble around more rapidly in the liquid phase; and finally inquire about the different techniques that are used in the study of liquid phase NMR, including both one- and two-dimensional Fourier transform methods.

Nuclear spin interactions in solids

Chemical shifts arise from the effects of different and distinctive electronic configurations (shieldings) on the nuclei in a molecule. Shielding (σ) affects the Larmor equation causing a frequency shift:

$$\nu_0 = \frac{\gamma}{2\pi} B_0(1 - \sigma)$$

Figure 2 depicts a simple molecule, formaldehyde, in two different orientations relative to the applied magnetic field $\mathbf{B}_0$. The ^{13}C chemical shift clearly depends upon the orientation of that molecule relative to the magnetic field because the electron motions induced by the magnetic field, which cause the chemical shielding, vary with orientation. Thus the ^{13}C chemical shift varies over a significant range depending upon the orientation of the molecule (chemical shift anisotropy), and the spectrum of the molecule in a polycrystalline or amorphous solid provides a superposition of the spectra from all the orientations. For example, Fig. 3 shows the ^{13}C spectrum, not of formaldehyde, but of polycrystalline polyethylene, where the CH_2 groups can be in various orientations. Overall, the spectrum covers a range of 2 to 3 kilohertz in width,

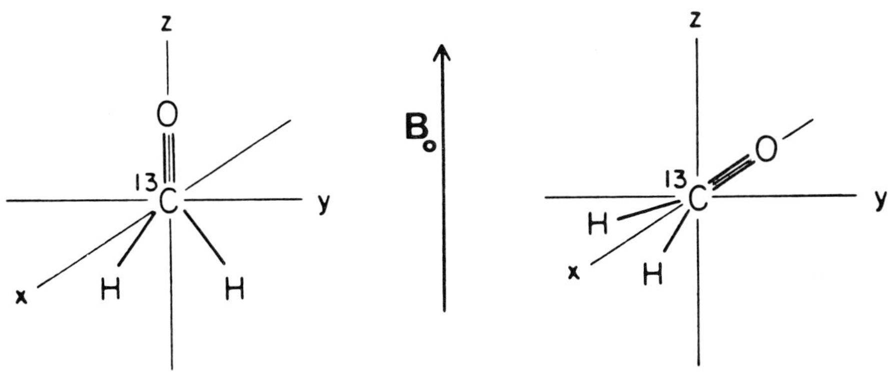

Figure 2. Illustration of chemical shift anisotropy. The chemical shift of the ^{13}C nucleus of formaldehyde varies depending on the relative orientation of the molecule with respect to the magnetic field $\mathbf{B}_0$.

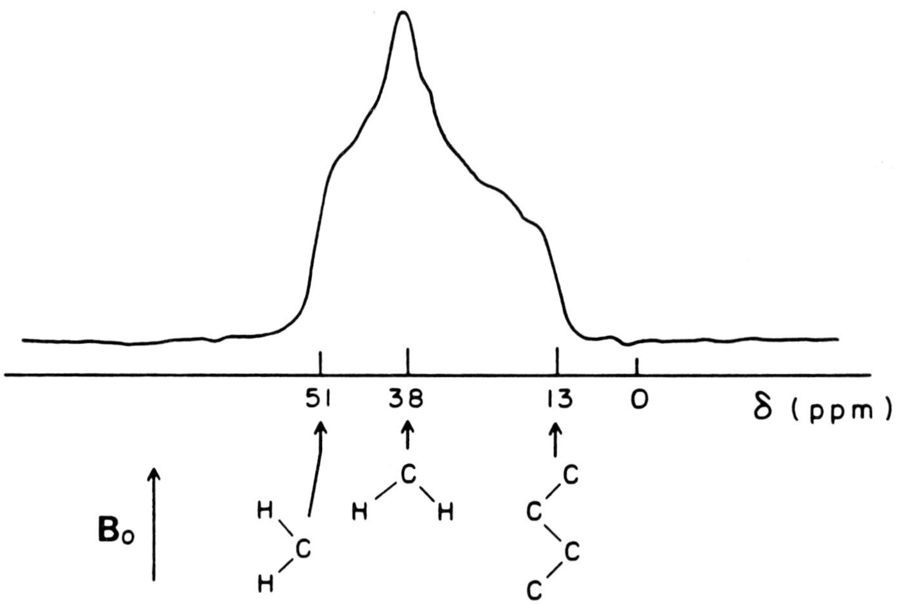

Figure 3. ^{13}C spectrum of polycrystalline polyethylene. The molecular orientations leading to certain chemical shifts are identified below the spectrum.

depending upon the magnetic field that is used. We shall return to this subject later.

The second factor mentioned above, the scalar coupling, is familiar to everyone who has observed multiplet splitting of NMR lines in liquid-phase spectra. The magnetic moments of all nuclei orient in at least two directions with respect to the magnetic field $\mathbf{B}_0$, and it is the sensitivity of one nucleus to the spin state of a neighbor that gives rise to scalar coupling. Scalar interactions are mediated by the electrons in chemical bonds, and thus are often called "indirect" spin-spin coupling. Like chemical shifts, scalar couplings are anisotropic and might be expected to vary with the orientation of molecules in the solid relative to the magnetic field. However, the effects of scalar coupling are rarely observed in solids since they are usually masked by the very much larger dipolar and quadrupolar interactions described below. We shall not consider further any small effects of the anisotropy of scalar coupling.

There is another kind of interaction between nuclear magnetic moments which occurs directly, not by means of the bonding electrons. It is usually termed the direct dipole-dipole interaction and is simply a classical interaction through space. For example, if two nuclei of spin 1/2 are lined up as shown in Fig. 4(a), there is an attractive force between their north and south poles, whereas if they are lined up beside each

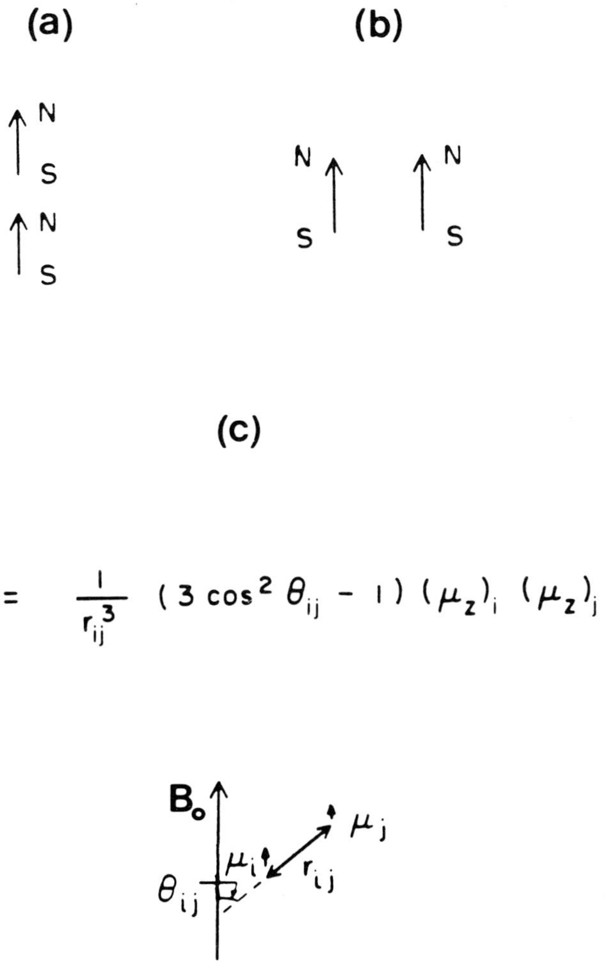

Figure 4. Illustration of the interaction between two magnetic moments. (a) An orientation of two spin-1/2 nuclei resulting in an attractive force between north and south poles, and (b) an orientation resulting in a repulsive force. (c) The dipolar Hamiltonian operator for the two nuclear magnetic moments oriented in the magnetic field $\mathbf{B}_0$.

other, as in Fig. 4(b), there is a repulsive force. Obviously the energy of interaction (dipolar coupling) between those two depends upon the strength of their magnetic (dipole) moments and upon the orientation they make relative to each other. That can be described in Fig. 4(c) in terms of the angle θ between the magnetic field and the vector joining the two nuclear dipoles. The magnitude of the interaction varies with the inverse cube of the distance, so it is nuclei that are close together that interact most strongly. The angular factor, as given in Fig. 4(c), turns out to be $(3\cos^2\theta - 1)$, a quantity that enters equations of this type repeatedly. When molecules tumble rapidly, as they do in liquids, then an

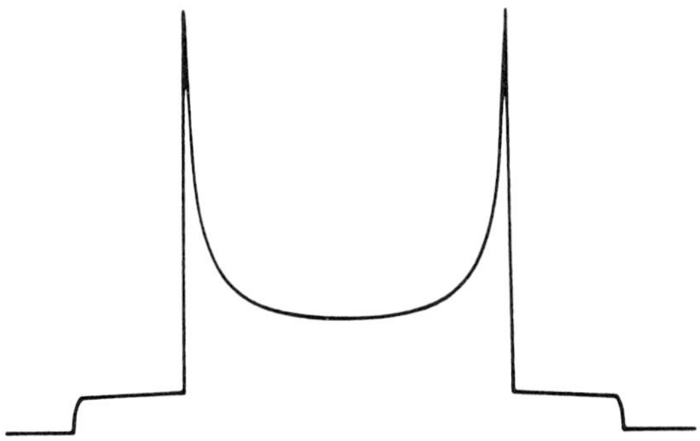

Figure 5. Schematic representation of the theoretical NMR spectrum arising from the dipolar interaction of two nuclei in a polycrystalline solid. Electric quadrupolar interactions lead to a similar lineshape.

individual molecule quickly sees all values of θ, and the NMR spectrum results from the value of $\cos^2\theta$ averaged over space, which turns out to be 1/3. Thus the term $(3\cos^2\theta - 1)$ neatly goes to zero, the dipole interactions vanish, and lines in liquids are sharp. In a polycrystalline solid, however, we see a superposition of all values of θ, much as in the case of the chemical shift. The result is a broadening, and a line shape of the sort shown in Fig. 5, in which the width is many kilohertz. Where many nuclei, not just two, interact with each other, this line shape loses its sharp features but retains the same overall width.

We next consider nuclei which do not necessarily have large magnetic dipole moments but have large electric quadrupole moments. Interactions of a quadrupole moment with an electric field gradient around it formed by the electrons in the molecule leads to very much the same sort of lineshape as shown in Fig. 5, dependent on the now familiar factor of $(3\cos^2\theta - 1)$. Such lines may be extremely broad, even for deuterium, which has a small quadrupole moment. For example, Fig. 6 shows the ^{2}H NMR spectrum of collagen, which covers a range of about 100 kHz. As the temperature is raised and the molecules begin to move, there is partial averaging of $\cos^2\theta$, and the lineshape changes but still covers a very wide range.

High resolution spectra in solids

Broad line spectra of the sort shown in Figs. 3 and 6 have been obtained since the beginning of NMR studies. But in the last decade or so innovative ways have been found to obtain sharp lines in solids and thus extract information on chemical details in which we are interested. To see how these methods work, we look first at the effect of dipolar

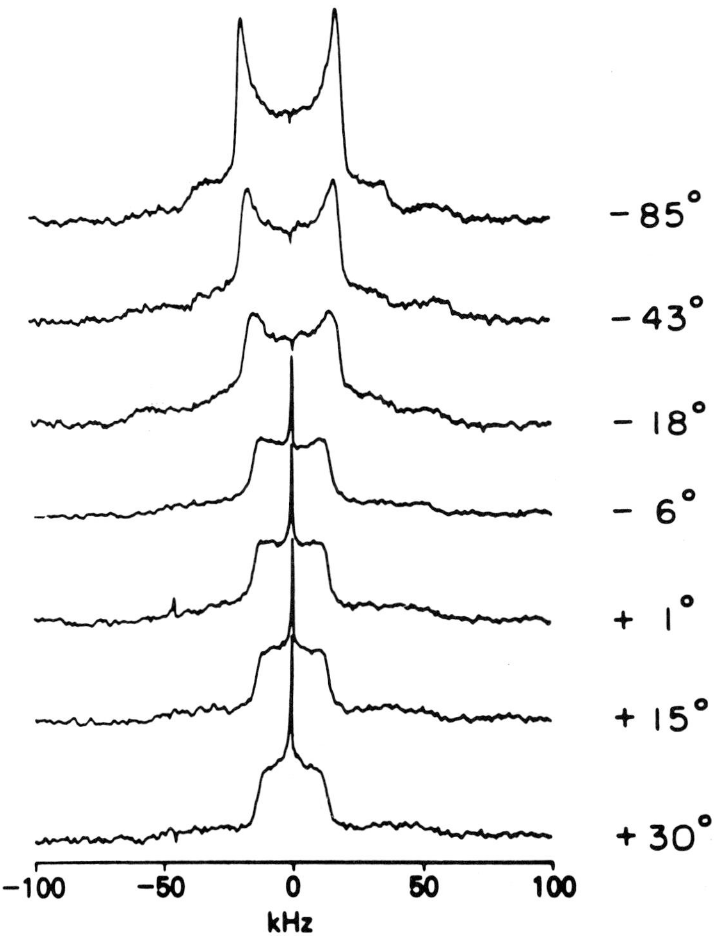

Figure 6. ^{2}H NMR spectra of collagen containing ^{2}H-enriched leucine.

interactions, as summarized in Fig. 4(c). For a given molecule in which r is fixed, there are two factors that can be manipulated. One is the angular factor, and the other is the relative orientations (spin "up" or spin "down") of the z components of the nuclear magnetic moments. These can be dealt with separately. Let us consider first the magnetic moments. Suppose these are from two different species of nuclei, such as ^{1}H and ^{13}C, and we wish to observe ^{13}C. One simple and effective way to remove the effect of the ^{1}H moment is to introduce an additional strong RF field at the proton frequency. This "decoupling" field causes the proton nuclei to flip up and down so rapidly that the carbons in effect don't know if they're coupled to protons that have spin up or spin down. Consequently, each doublet arising from different spin orientations col-

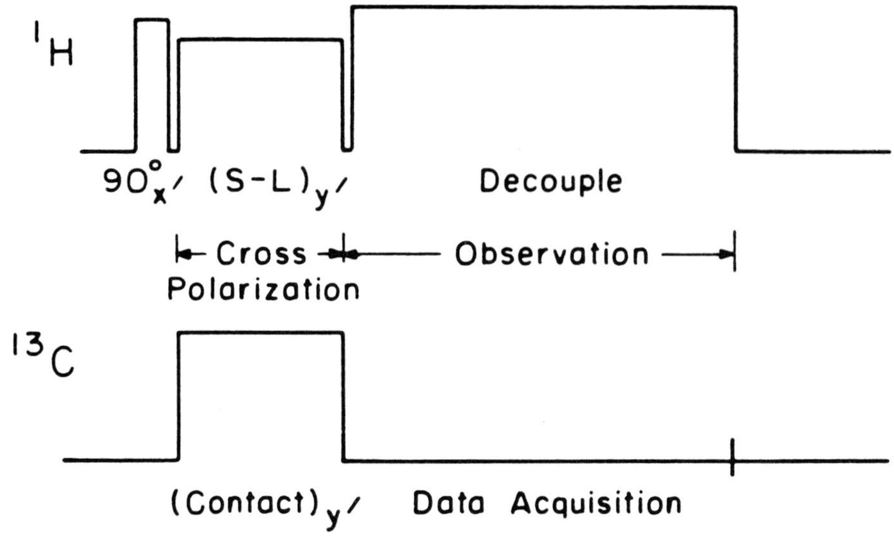

Hartmann-Hahn Condition: $\gamma_H B_1(H) = \gamma_C B_1(C)$

Figure 7. Pulse sequence for cross polarization of ^{13}C and ^1H. During the "Cross Polarization" period, both ^1H and ^{13}C spins are irradiated at their Larmor frequencies using $\mathbf{B}_1$ field strengths to satisfy the Hartman-Hahn condition.

lapses to a singlet, and overall what would have been a broad line is narrowed substantially by this *dipolar decoupling*. This technique is analogous to decoupling techniques used to remove the effect of scalar couplings and resultant splittings from high resolution NMR spectra in liquids. The major difference between the two techniques is the substantially higher RF power needed for dipolar decoupling in solids.

A second technique that can be applied concurrently, *cross polarization*, causes transfer of some of the polarization from the large magnetic moment of the proton to the smaller magnetic moment of ^{13}C (or another nucleus with low magnetogyric ratio and low abundance such as ^{15}N). Figure 7 illustrates a common method by which both cross polarization and dipolar decoupling are carried out together. The ^1H and ^{13}C are irradiated simultaneously during a "contact" period, with the magnitudes of the RF fields for ^1H and ^{13}C adjusted so that the Hartman–Hahn condition specified in Fig. 7 is met. Under those circumstances the carbon and proton nuclei are in effect indistinguishable, and there is equilibration of the magnetization among all of the spins. As a result, the ^{13}C's act as though they have magnetic moments four times as large as they do in reality and hence have a much improved sensitivity. More-

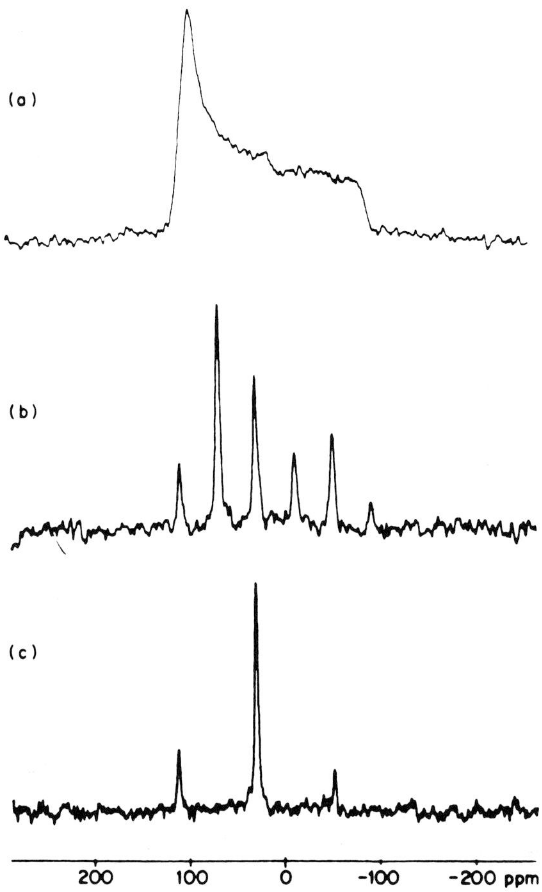

Figure 8. Solid-state NMR spectra of triphenylphosphine oxide. (a) Stationary sample; (b) Slow magic angle spinning (ν_0 = 1.5 kHz); (c) Magic angle spinning (ν_0 = 3.0 kHz). [From Ref. 7.]

over, relaxation of the system becomes dependent primarily on ^1H relaxation times, which are usually shorter than those for ^{13}C, so that the experiment may be repeated more quickly. By leaving the ^1H RF field on during the acquisition period, dipolar decoupling also occurs with resultant narrowing of the otherwise broad lines.

Application of cross polarization and dipolar decoupling still does not lead to very sharp lines in solids because the chemical shift anisotropy effects that we dicussed previously are not eliminated by decoupling. Once again the angular factor $(3\cos^2\theta - 1)$ becomes important. Spinning the entire sample at an angle θ of 54.7° relative to $\mathbf{B}_0$ reduces this term to zero. This unique angle is referred to as the *magic angle*. The technique of magic angle spinning (MAS) is illustrated in Fig. 8, where the broad line changes first into a number of spinning side bands, and as

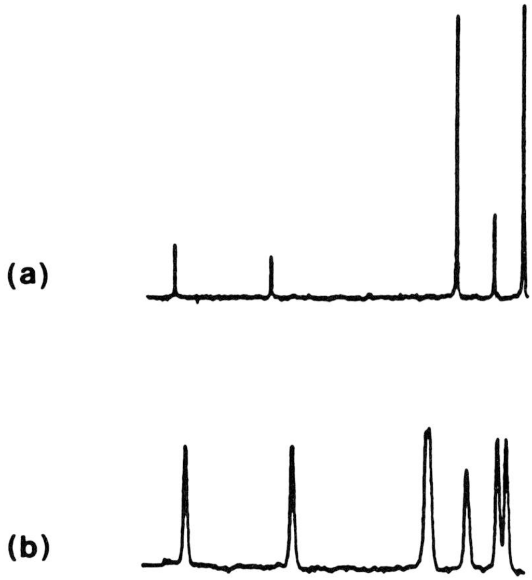

Figure 9. The use of cross polarization (CP) and magic angle spinning (MAS) in ^{13}C NMR spectra. Aromatic and carbonyl region of (a) normal solution spectrum of tyrosine (pH 12.6). (b) CP/MAS spectrum of tyrosine powder. [From Ref. 8.]

the spinning speed increases to a rate comparable to the chemical shift anisotropy in frequency units, the high resolution spectrum appears. In practice, the two techniques of cross polarization (including high power, dipolar decoupling) and magic angle spinning are usually carried out simultaneously. In many cases, as illustrated in Fig. 9, the spectra thus obtained from solids have lines that are almost as sharp as those in liquids.

For nuclei that have quadrupole moments, cross polarization and dipolar decoupling are inapplicable, and a different approach is needed. Figure 10 illustrates the energy level pattern for a quadrupolar nucleus with a spin of 1, such as deuterium. The spacings of the three energy levels vary with the angle between the principal axis of the electric field gradient and $\mathbf{B}_0$. Three specific angles are illustrated in Fig. 10. The spectrum in the solid, which results from the superposition of spectra from all angles, gives a very broad line of the sort indicated previously. Note, however, that while the normally allowed individual transitions described by the solid lines cover a wide range of frequencies, the dashed lines connecting the lowest and highest energy levels are all the same length regardless of orientation. That transition, the *double quantum* transition going from the lowest to the highest energy level, is forbidden by the usual selection rules. However, by means of appropriate experiments that we shall not describe here, it is possible to detect double

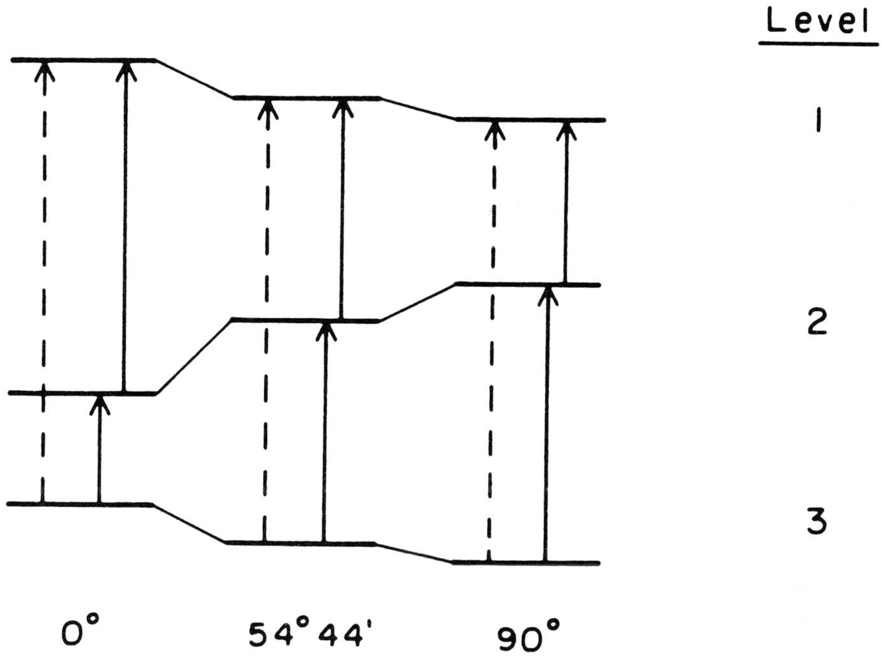

Figure 10. Energy levels and transitions for a nucleus with spin 1 as a function of molecular orientation. Solid lines indicate single quantum transitions, and dotted lines indicate double quantum transitions.

quantum (and in some cases multiple quantum) transitions. For ^2H NMR the result is solid spectra with relatively narrow lines. (Multiple quantum transitions are often also used in ^1H NMR spectra of liquids to discriminate against various species. For example, an experiment to detect three-quantum transitions (with nuclei of spin 1/2) will not permit observation of two-spin systems, since a two-spin system cannot give rise to a three-quantum transition. This technique can be employed to filter out very effectively strong solvent H_2O lines).

Relaxation in liquids

We now turn from solids to a discussion of the realm of the rapid, random motion of liquids. By "rapid" we mean that the *correlation time* τ_c, i.e., the average time which a molecule takes to tumble through one radian or to translate roughly a molecular diameter, is short compared with the reciprocal of the linewidth in the solid. The effect of this rapid motion is to average to zero each of the anisotropic interactions we discussed earlier – chemical shift anisotropy, dipole-dipole interactions, quadrupole interactions – and sharp lines are observed. It is important to remember, however, that even though these interactions have been averaged over a sufficiently long time period, they have not gone away. Although the spectrum displays only the average effect, the interactions are in fact present; and as the molecule tumbles, all can cause local fields which fluctuate (with components at the Larmor frequency) and cause the nuclei to relax. These interactions – dipolar, quadrupolar and chemical shift anisotropy – are the most common mechanisms of relaxation.

We wish to focus on one of these mechanisms, the dipolar interaction. It can be shown that the effectiveness of this interaction for relaxation depends on the square of the static dipolar interaction described in Fig. 4(c). Thus

$$1/T_1^d \propto \gamma_I^2 \gamma_S^2 / r_{IS}^6$$

Because of the quadratic dependence on the magnetogyric ratios, those nuclei with large magnetic moments are most effective in relaxation. Likewise, the inverse sixth power of the distance between the nuclei (r_{IS}), implies that nuclei which are relatively close to each other contribute

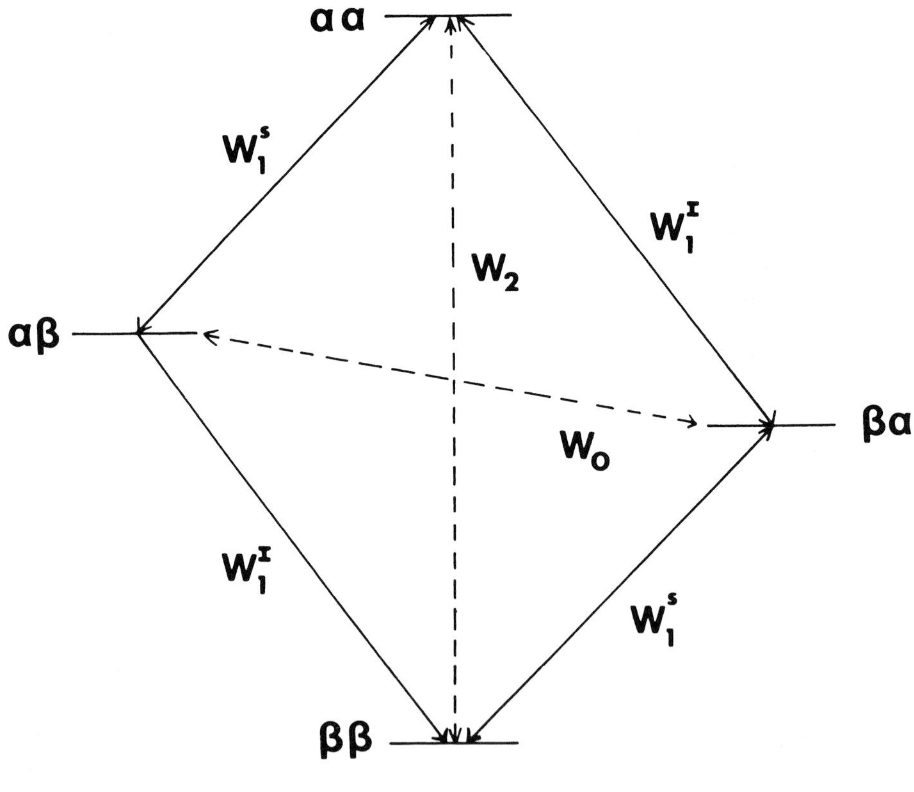

Figure 11. Energy levels for two spin-1/2 nuclei (S and I) differing substantially in chemical shift. Spin states are denoted by α and β. Zero (W_0), single (W_1) and double (W_2) quantum transitions are identified by arrows.

most significantly to relaxation. Usually nuclei that relax each other are within the same molecule, but they also can be in separate molecules that come close together. The dependence of T_1 on internuclear distance can be exploited by means of the *nuclear Overhauser effect* (NOE). The NOE is easily understood and is illustrated in Fig. 11 by a diagram of two nuclei (with spin states α and β) interacting with each other by means of dipolar relaxation. The four solid lines correspond to the allowed one-quantum transitions that may be induced by radio-frequency fields, whereas the dashed lines correspond to zero-quantum and double-quantum transitions that do not appear under normal circumstances in the NMR spectrum. However, all six pathways, including zero- and double-quantum (*non*radiative transitions), are possible paths for relaxation. Interplay of all of the six coupled rate equations that describe these paths is the mechanism that leads to the equilibrium distribution of nuclei among the energy levels described by the usual Boltzmann distribution. By perturbing that population either through double resonance (i.e., saturating the transitions of one of the nuclei), or by applying a so-called 180° pulse (described below) to interchange the populations of two energy levels, the Boltzmann distribution is upset, and as a result the intensity of a signal is changed. From measurements of NOEs, information on interatomic distances in complex molecules can be obtained, as we describe later.

Improvements in NMR sensitivity

Thus far we have concentrated on describing the effects of several important interactions on NMR spectra in solids and their influence on NMR spectra in liquids. We now turn to ways in which these effects have been exploited in studying biological systems. The tremendous explosion during the last few years in the use of NMR to study biological systems clearly has resulted from technological developments that fall into three categories: (1) higher magnetic fields; (2) improvement in sensitivity; and (3) development of pulse methods. Over the last 20 or 25 years the commonly available magnetic fields and resonance frequencies have increased by about a factor of 10, with a corresponding spreading out of spectra to permit the extraction of information that was masked previously. However, during the same period, we have experienced an improvement of a factor of about 600 in sensitivity, as shown in Table 1. The gain in sensitivity is partly due to the factor of ten increase in Boltzmann distribution from the higher magnetic field, but it is also due to better probe design and higher quality electronic components. In addition to the sensitivity gain of almost three orders of magnitude

Table 1. Signal/noise (S/N) for various spectrometers

Year	Spectrometer frequency (MHz)	S/N*
1961	60	6
1965	100	30
1969	220	80
1978	200	300
1978	360	800
1985	500	3600

* Strongest peak in the methylene proton signal of ethylbenzene, 1% by volume. Single scan or single pulse.

illustrated in Table 1, NMR has benefited from the replacement of the older continuous wave (cw) techniques with pulse Fourier transform (FT) methods. FT methods provide at least another order of magnitude gain in sensitivity through far more efficient gathering of data. Since the mid-1970s, pulse FT methods have become routine and provide great versatility in NMR experiments.

We have already alluded to the use of pulse methods in our discussion of NMR spectra of solids. All pulse FT methods progress in much the same way shown in Fig. 12. The magnetization is first rotated from its equilibrium position into the xy plane. The angle of rotation, shown here as 90°, can be selected by varying the pulse length. The different nuclear moments then dephase due to T_2 processes and magnetic field inhomogeneities, and the result is a decaying signal, a *free induction decay* (FID).

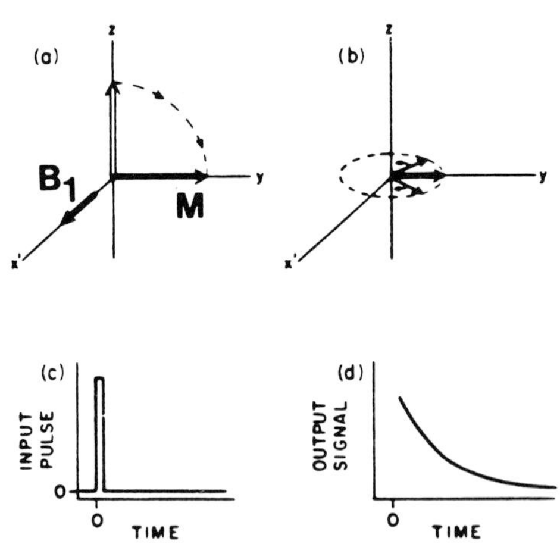

Figure 12. Typical pulse sequence for pulse FT NMR. (a) a 90° pulse along x rotates **M** from the equilibrium position to the y axis. (b) **M** decreases as magnetic moments dephase. (c) The input signal, a 90° pulse, corresponding to (a). (d) Exponential free induction decay, corresponding to (b).

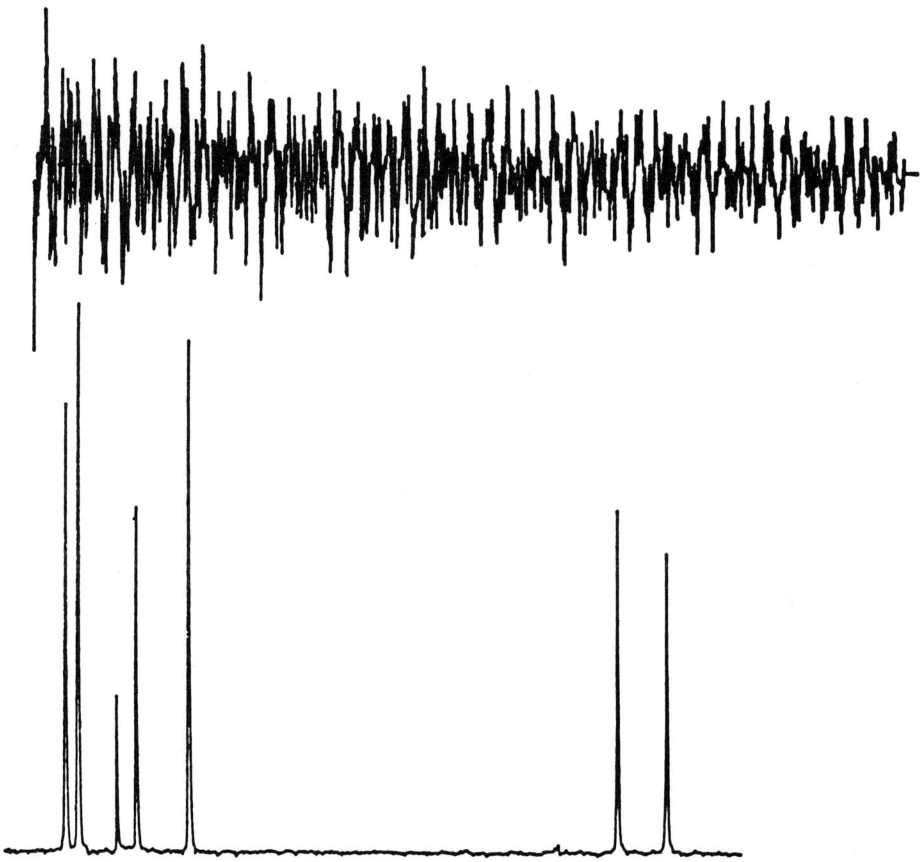

Figure 13. The proton-decoupled ^{13}C free induction decay (top) and frequency spectrum (bottom) of 3-ethylpyridine.

Figure 13 shows a typical FID for ^{13}C in an organic molecule. It is complex – essentially an interference pattern resulting from the different frequencies of all the chemically shifted nuclei in the molecule. By the process of Fourier transformation, the FID is converted into the usual frequency spectrum. The FT process extracts from the time response of the FID all the characteristic frequencies (and lineshapes) that are present. FT methods permit us to obtain a spectrum in a much shorter time than by old-fashioned scanning continuous wave methods (e.g., in about a second, rather than in many minutes). Moreover, the use of short pulses to move the magnetization quickly (in a matter of microseconds) through a chosen angle (e.g., 90° or 180°) provides the versatility needed for the methods that we have already discussed, such as cross polarization in solids.

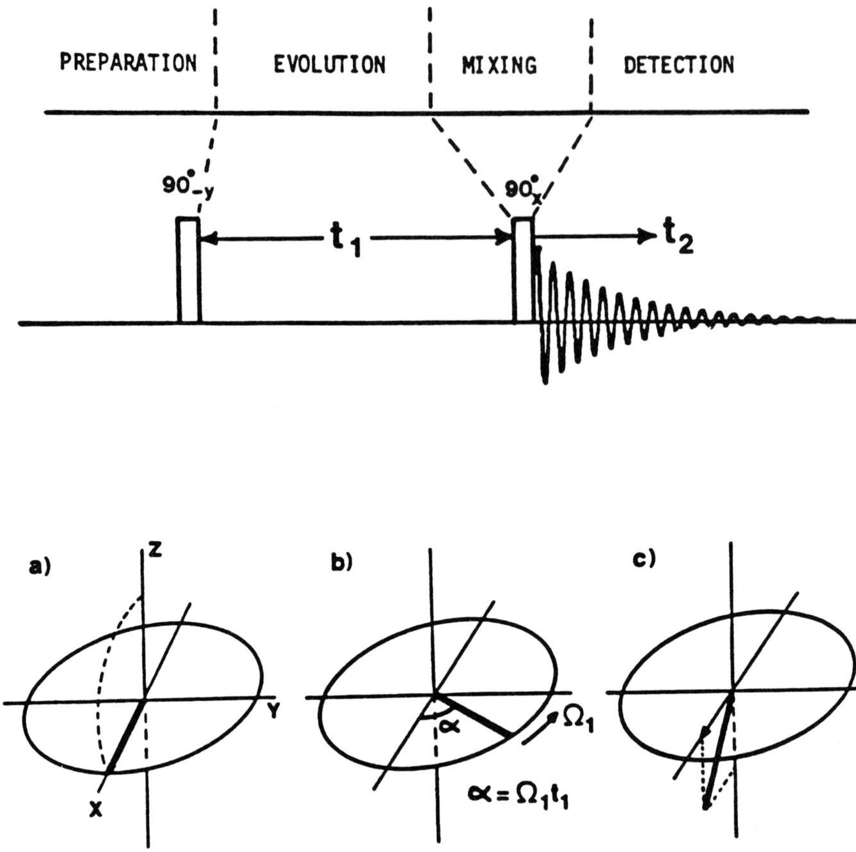

Figure 14. The evolution of magnetization during a generalized pulse sequence used for two-dimensional NMR. (a) The first 90° pulse rotates the magnetization from the z axis to the x axis. (b) During the evolution period, t_1, the magnetization rotates in the xy plane through an angle that depends on its Larmor frequency, Ω_1. (c) The second 90° pulse rotates the magnetization from the xy into the xz plane, leaving a component proportional to sin(α) parallel to the x axis.

Two-dimensional NMR methods

The success of pulse FT methods in many different applications has also given rise to another whole series of experiments based on *two* successive Fourier transformations. There are now several hundred different *two-dimensional* (2-D) FT techniques, designed for a wide range of purposes. Although the individual pulse sequences may be rather complex, the basic principle behind all such 2-D methods is simple and can be illustrated with the aid of Fig. 14. The nuclear spin system to be studied is first "prepared" in some way; usually this simply involves allowing the system to attain equilibrium. The experiment then begins by application of some pulse, in this example a 90° pulse. During the evolution time t_1 (taken to be short compared with the relaxation time T_2) the magnetization then precesses in the xy plane. Another pulse or series of pulses (in Fig. 14 shown as just a single 90° pulse) allows magnetizations from different nuclei to "mix" for a short period, and then we observe a free induction decay, just as we normally would.

Figure 14 also shows what happens to the magnetization as a result of the pulse sequence. Suppose in this simple case that the spin system consists of only one chemically distinct type of nucleus (e.g., ^1H in H_2O). The first 90°pulse is applied with its phase chosen such that the $\mathbf{B}_1$ field lies along the negative y axis. The effectof $\mathbf{B}_1$ is to rotate the magnetization $\mathbf{M}$ from the z axis down along the x axis. $\mathbf{M}$ then precesses in the xy plane at its resonance frequency, and after a time t_1 it has moved through an angle α. If at time t_1 a second 90° degree pulse is applied, this time along the x axis, then $\mathbf{M}$ is rotated down into the xz plane. In this particular experiment we are interested in the projection of $\mathbf{M}$ along the x axis, since it is that projection that determines the magnitude of the signal acquired during the time t_2. If this pulse sequence is repeated several times with different values of t_1, the initial amplitudes of the FIDs will vary, depending on the value of α, that is, depending on the *phase* of the magnetization at time t_1. Fourier transformation of each of these FIDs with respect to t_2 gives a series of spectra, modulated in amplitude as a function of t_1 as illustrated in the left portion of Fig. 15. Viewed from the side, this set of spectra appears very much like a FID; and in fact, to obtain the frequency of the modulation we have only to carry out a Fourier transformation with the variable t_1. The result of this process,

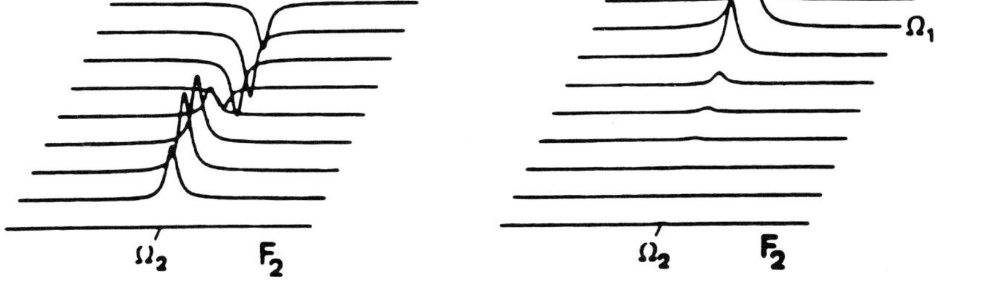

Figure 15. (a) A set of spectra obtained with the sequence of Fig. 14 with increasing durations of the evolution period, t_1. (b) The two-dimensional result of Fourier transformation of data in (a).

which has involved overall two FTs, is shown on the right of Fig. 15 as a two-dimensional frequency plot. In this particularly simple example, we chose a sample (H_2O) with a single type of nucleus and thus only one magnetization, which is precessing at one single frequency. Thus the position of the peak on the two frequency axes F_1 and F_2 is identical, and 2-D plot gives just one peak at center of the plot and is of little practical interest.

Suppose, however, we have two nuclei that are precessing at different frequencies and are spin-coupled to each other. Again, the resonance frequencies of the two nuclei determine the phases of their magnetizations in the xy plane for a given value of t_1, and this information is extracted during the Fourier transformations. For reasons that we shall

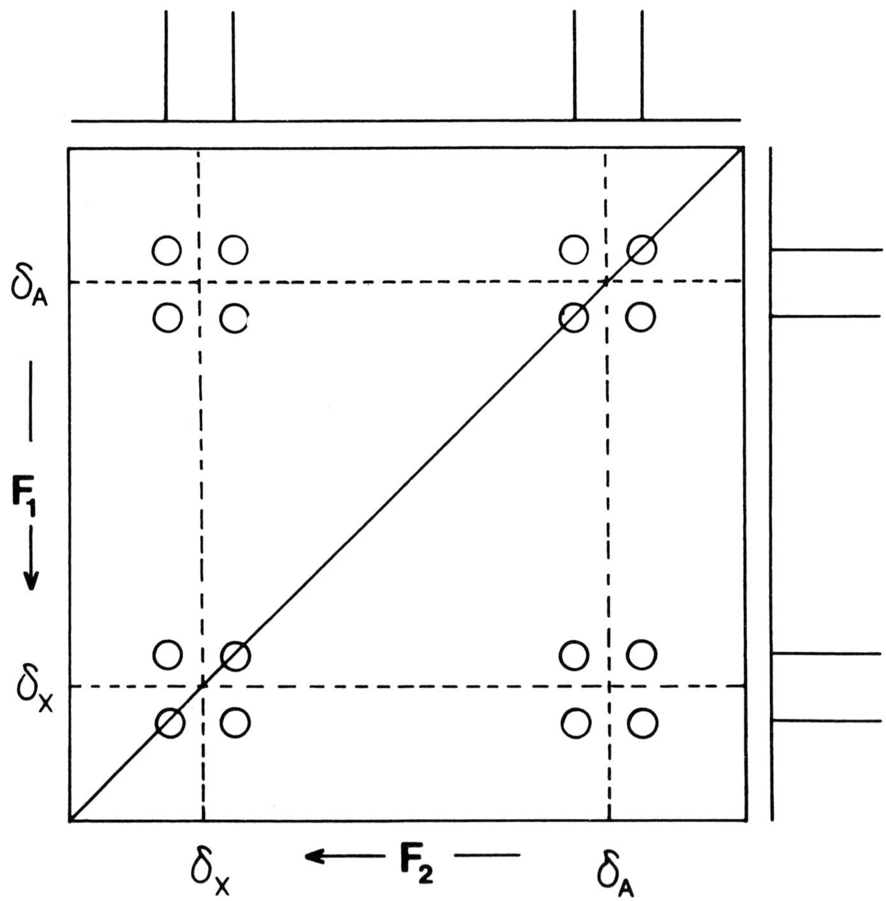

Figure 16. Schematic diagram of a two-dimensional COSY spectrum of two coupled nuclei. The peaks are represented as contours on the plot. Peaks on the diagonal line corespond to the normal one-dimensional spectrum which is also shown along the edges of the plot. Cross peaks appear off-diagonal at the frequencies characteristic of the coupled nuclei.

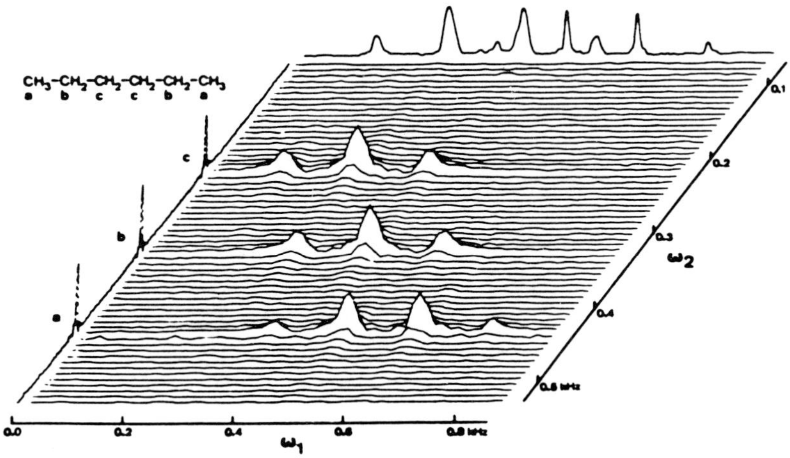

Figure 17. Two-dimensional FT ^{13}C Spectrum of *n*-hexane. A representation of the one-dimensional undecoupled spectrum appears along the top, while a representation of the one-dimensional decoupled spectrum appears along the left side of the plot. The coupled multiplets corresponding to each carbon type are clearly discernible in the two-dimensional plot. [From Ref. 9.]

not discuss in detail, the second 90° pulse in Fig. 14 causes a transfer of magnetization (mixing) between two nuclei, but only if they are spin-coupled. This magnetization transfer results in a 2-D spectrum showing off-diagonal or "cross" peaks at the intersections of the frequencies characteristic of the two coupled nuclei. For example, Fig. 16 shows a schematic diagram of the 2-D plot that would result from such an experiment. Peaks are shown as contours in the $F_1 F_2$ plane. Just as in our previous simple example, there are peaks along the diagonal of this plot which are of little interest. However, in addition we now also have off-diagonal peaks giving the frequencies of the two nuclei which are spin-coupled to each other. This technique is known as correlated spectropcopy, or COSY for short, and is one that has become quite popular in providing information about coupling patterns in complex molecules. It has the great advantage of providing information about all the coupled nuclei of a molecule in a single experiment.

A second example of correlated spectroscopy, here involving nuclei of different types (e.g., ^{1}H and ^{13}C) is illustrated in Fig. 17, which shows the ^{13}C spectrum of *n*-hexane. The spectrum depicted along the axis to the left is the normal decoupled ^{13}C spectrum, in this case consisting of just three lines, while the projection along the rear is the spin-coupled spectrum in which all ^{13}C's are coupled to protons. This spectrum shows several overlapping spin multiplets. The two-dimensional experiment

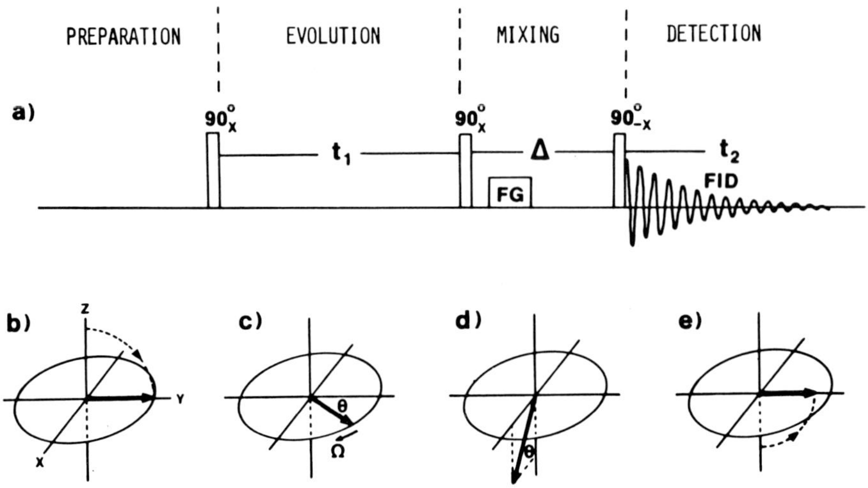

Figure 18. (a) Example of a 2-D NOESY pulse sequence. FIDs (free induction decays) are collected for a large number of incremental t_1 durations, then Fourier transformed with respect to both t_1 and t_2 to yield a 2-D plot. Magnetization vector picture: (b) just after the first 90° pulse, (c) just before the second 90° pulse, (d) just after the second 90° pulse, and (e) just after the final 90° pulse.

shown in Fig. 17 uses a different pulse sequence from that of COSY, which we will not describe here. Clearly it is effective in spreading out the individual multiplets for easy identification and assignment.

Another simple and popular 2-D NMR method makes use of three consecutive 90° pulses, as shown in Fig. 18. The first two pulses cause the magnetization to move as we saw in Fig. 14. However, instead of focusing on the magnetization component in the *xy* plane, as we did for COSY, we deliberately destroy the *xy* component during the mixing period Δ by applying a magnetic field gradient so as to cause a rapid dephasing of all magnetization in the *xy* plane. In this experiment we pay attention to the component of magnetization along the *z* axis and note its decay by spin-lattice relaxation. If this process is dominated by dipolar interactions, as is usually the case for nuclei of spin 1/2, then, as we saw earlier, it is primarily nearby nuclei that bring about relaxation. The final 90° pulse in Fig. 18 leads to a free induction decay that contains information about the frequencies of those nuclei that relax each other, hence are usually close to each other. This experiment is a 2-D analog of the nuclear Overhauser effect (NOE), and by analogy to COSY it is usually called NOESY. Figure 19 shows the kind of data that can be obtained. All these cross peaks are not noise; they are indications of nuclei that are close to each other. The frequencies of these cross peaks can be analyzed to identify the neighboring nuclei. In addition, the peak intensities provide some semi-quantitative information as to the distances between the nuclei.

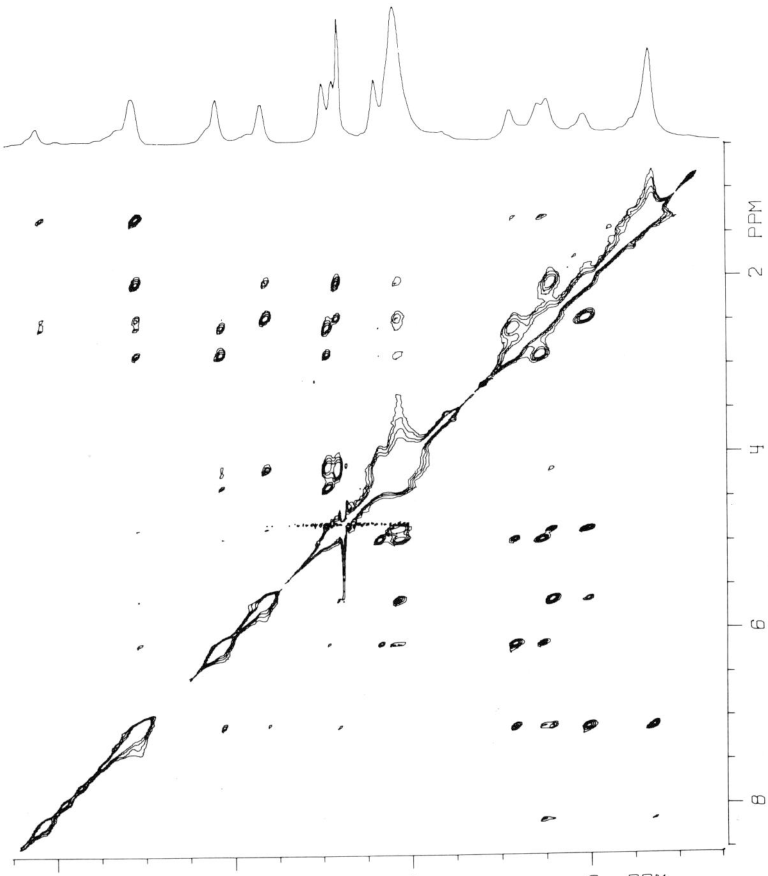

Figure 19. 500 MHz 2-D proton NOESY spectrum of a synthetic DNA duplex, $d(AT)_{20}$. Each off-diagonal cross peak indicates protons that are close in space. The coordinates of the cross peak give the chemical shifts of the two neighboring protons, and the cross peak's intensity is a function of the inter-nuclear distance.

NMR imaging

Another application of 2-D NMR is NMR imaging. Some applications of imaging were given in an earlier chapter by Lauterbur. The principle of NMR imaging is to apply a magnetic field gradient across the sample and thus make the NMR resonance frequency dependent upon the specific location in the sample where the nuclei reside. Lauterbur's discussion focussed on one method of image reconstruction – back-projection. However, the most common method by which NMR images are obtained uses 2-D FT techniques. A simple way of obtaining data is illustrated in Fig. 20. Initially, a gradient is applied along the *x*

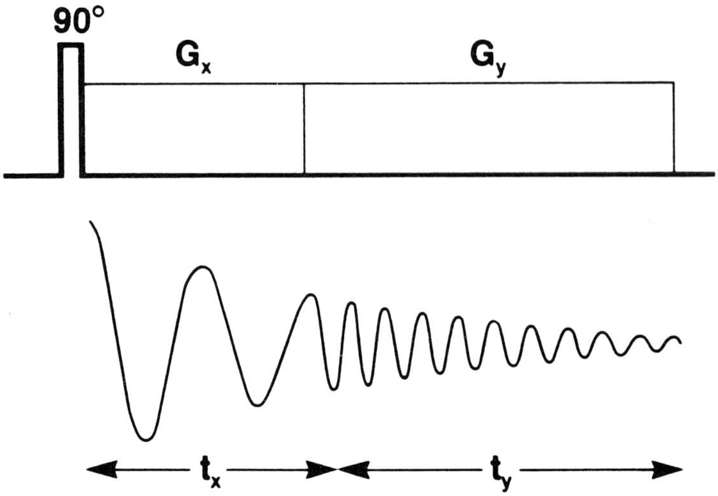

Figure 20. General pulse sequence for NMR imaging using 2-D NMR.

direction, so that the nuclei precess with a frequency that depends on their location along the *x* axis. That information is preserved via the phase of the signal, as we have seen, during the time that the gradient is applied along the *y* direction. Thus, a two dimensional Fourier transformation provides information on overall location in the *xy* plane. The method can easily be extended to three dimensions. In practice, there are a number of improvements that have been incorporated into this simple approach, but the use of 2-D NMR remains as the basic method.

Conclusion

NMR was discovered 40 years ago. Today it remains as a vibrant technique with applications in chemistry, physics, biology and medicine. Applications to problems of significance in agriculture are discussed in the following chapter.

Acknowledgments

The authors wish to thank a number of individuals for providing figures for this chapter, including Dennis Torchia (Figs. 3 and 6), Ad Bax (Figs. 14, 15, 16 and 18), and Ad Bax, Vladimir Sklenar and Jerry Zon (Fig. 19).

References

1. A. Bax. 1982. *Two-Dimensional NMR in Liquids*, (Reidel, Boston, 1982); A. Bax and L. Lerner. 1986. *Two-Dimensional Nuclear Magnetic Resonance Spectroscopy*. *Science* 232: 960.

2. E.D. Becker. 1980. *High Resolution N.M.R., Second Edition*. Academic Press, New York.
3. T.C. Farrar and E.D. Becker. 1971. *Pulse and Fourier Transform NMR*. Academic Press, New York.
4. R.K. Harris. 1983. *Nuclear Magnetic Resonance Spectroscopy. A Physicochemical View*. Pitman, London.
5. C.P. Slichter. 1978. *Principles of Magnetic Resonance, Second Edition*. Springer-Verlag, Berlin.
6. J. Valk, C. MacLean and P.R. Algra. 1985. *Basic Principles of Nuclear Magnetic Resonance Imaging*. Elsevier, Amsterdam.
7. R.E. Wasylishen and C.A. Fyfe. 1982. *Annual Review of NMR Spectroscopy* 12: 1.
8. M.H. Frey and S.J. Opella. 1980. *Journal of the Chemical Society, Chemical Communications*, p. 474.
9. L. Muller, A. Kumar and R.R. Ernst. 1975. *Journal of Chemical Physics* 63: 5490.

Agricultural/biological applications of NMR

PHILIP E. PFEFFER
Eastern Regional Research Center,[†] Philadelphia, PA 19118, U.S.A.

Abstract This chapter is designed to introduce the neophyte to a range of nuclear magnetic resonance (NMR) techniques and experiments that have or will have an impact on the solution of agriculture/biological problems. Throughout the various sections I have attempted to use examples from the recent literature as well as work from our own laboratory to illustrate the diversity of applications. The four main topic areas discussed are: (1) multinuclear high field/high resolution methods including 2-D methods for structure determination; (2) *in vivo*, multinuclear high resolution studies of plant material; (3) solid state cross polarization magic angle spinning (CPMAS) NMR which will cover the study of intact polymers, soil matrices, seeds and plant tissue metabolism, and (4) magnetic resonance imaging (MRI) as it is applied to plant and animal tissues and some examples of how it is utilized in conjunction with spatially resolved high resolution surface coil *in vivo* techniques.

Introduction

The advent of the computer and superconducting magnet technology has immeasurably advanced the state of the art of the modern Nuclear Magnetic Resonance (NMR) spectrometer in terms of its (1) inherent sensitivity, (2) multinuclear capabilities, (3) capability of routinely identifying resonances to solve complex structural problems, and (4) ability to control nuclear spin physics, to evaluate molecular dynamics and elucidate biochemical phenomenon in solution and solid states.

Todays commercial, pulsed superconducting, Fourier transform, (FT) spectrometer operating at magnetic fields as high as 11.8T can obtain spectra of submicrogram quantities of material within reasonable periods of time. In addition the extended frequency range and rapid digitization make it possible to explore a seemingly inexhaustible array of magnetically active nuclei, dipolar and quadrupolar, *in vitro* as well as *in vivo*. With the innovative development of RF controlled multipulse programming to produce nuclear spin "acrobatics" comes the realization that the previously unseen communication between related nuclei could be revealed in a second time dimension, hence the birth of two dimensional NMR (2-D NMR). Presently, the complete assignment of a ^{13}C resonance spectrum representing a complex molecule as well as detailed stereochemical, conformational and through space nuclear distances are available with a days instrument time. What's more, the

[†] Agricultural Research Service, U.S. Department of Agriculture.

information is most relevant since unlike the data derived from crystallographic methods, it is descriptive of the chemistry in the native solution state.

Similarly, spin "acrobatics" or nuclear spin manipulation has been exploited to enhance the inherent sensitivity of the NMR experiment in solid state applications. Further refinement of this technique has led to methods for identifying specific molecular entities as well as defining their respective relationships with one another within intact multi component assemblies.

The innovative development of magnetic field gradient technology has given birth to magnetic resonance imaging (MRI) and a new era in medical applications. In addition the combination of MRI and spatially resolved high resolution NMR are just beginning to establish new frontiers in the exploration of a variety of living tissues.

In the previous chapter Dr. Becker has described the many innovations of the modern NMR spectrometer and given us an overview of the new techniques and methods that are reputed to be some of the most potent scientific tools we have today. In this contribution I will attempt to touch on examples from the recent literature and our own work that illustrate the application of the NMR experiment to the solution of Agricultural/Biological problems. The four main topics that will be discussed will include, (a) multinuclear high field/high resolution methods including 2-D methods for structure determination, (b) *in vivo*, multinuclear high resolution studies of plant material, (c) solid state cross polarization magic angle spinning (CPMAS) NMR which will cover the study of intact polymers, soil matrices, seeds and plant tissue metabolism and (d) MRI as it is applied to plant and animal tissues and some examples of how it is utilized in conjunction with spatially resolved high resolution surface coil *in vivo* techniques.

This chapter is designed to introduce the neophyte to a sampling of illustrative examples that demonstrate the Agricultural/Biological applications of NMR spectroscopy. It is by no means to be considered an exhaustive review of the NMR literature. In most instances the material will be confined to the study of agriculturally relevant topics. In some areas where general application of a method is recognized, pertinent review articles will be cited. The reader is referred to the following comprehensive texts which provide excellent in depth coverage of the field of biological NMR (Knowles, Marsh and Rattle, 1976; Shulman 1979; Jardetzky and Roberts, 1981; Gorenstein, 1984).

High resolution NMR

(a) Structure determination (analytical applications)

To establish the structures of organic biomolecules, ^{1}H and ^{13}C, are, because of their ubiquitous nature, the most widely used nuclei for study by NMR. Although many other nuclei, such as ^{31}P, ^{2}H, ^{15}N and ^{17}O to name a few, can be examined routinely their usefulness in structural studies is limited and beyond the scope of this chapter.

Early NMR instrumentation was limited in its application to relatively large samples coupled with poor dispersion due to the low fields that were available (1.4–2.5T). With the development of superconducting magnets and higher field sensitivity, resonance shift dispersion for protons was significantly improved. The innovation of pulsed Fourier transform methods and computer averaging made dilute spin nuclei such as ^{13}C, (1.1% natural abundance) important for structure elucidation. However, the most significant advance that has brought NMR spectroscopy to the fore today is the development of pulse controlled nuclear spin physics and its innovative application to resonance identification, structure determination and molecular dynamics. A comprehensive review of these methods is given by Benn and Günther (1983) and full treatment of exclusively 2-D implementation is found in a book by Bax (1982).

Previously difficult to interpret off resonance or gated decoupling experiments had to be performed in order to establish the identity of a carbon type (1°, 2°, 3° or 4°) in complex ^{13}C spectra. In special cases such as carbohydrate molecules, carbon resonance assignments could be made with the aid of deuterium isotope shifts (Pfeffer, Valentine and Parrish, 1979). Today the modern pulse programmed spectrometer can do ^{13}C resonance multiplicity assignments automatically and more accurately and in a fraction of the time required in the past. Figure 1 illustrates an example of the use of one of these new pulse sequences called Insensitive Nuclei Enhanced by Polarization Transfer (INEPT) for ^{13}C multiplicity sorting of a ^{13}C spectrum of ATX-I (I), a mycotoxin produced by Alternaria mold growing on wheat or rice. (Stinson, Osman and Pfeffer, 1982). To implement this experiment, one employs different delay times or evolutionary periods in the INEPT pulse sequence based on an average ^{13}C–^{1}H coupling constant for the molecule under study. Varying the evolutionary periods corresponding to different fractions of the average 1/J value produces different dephasing responses in the (1°, 2°, 3°) proton populations and ultimately the generated ^{13}C spectra of their bound carbons induced by cross polarization. Quaternary carbon resonances are assigned by difference with this technique since they have

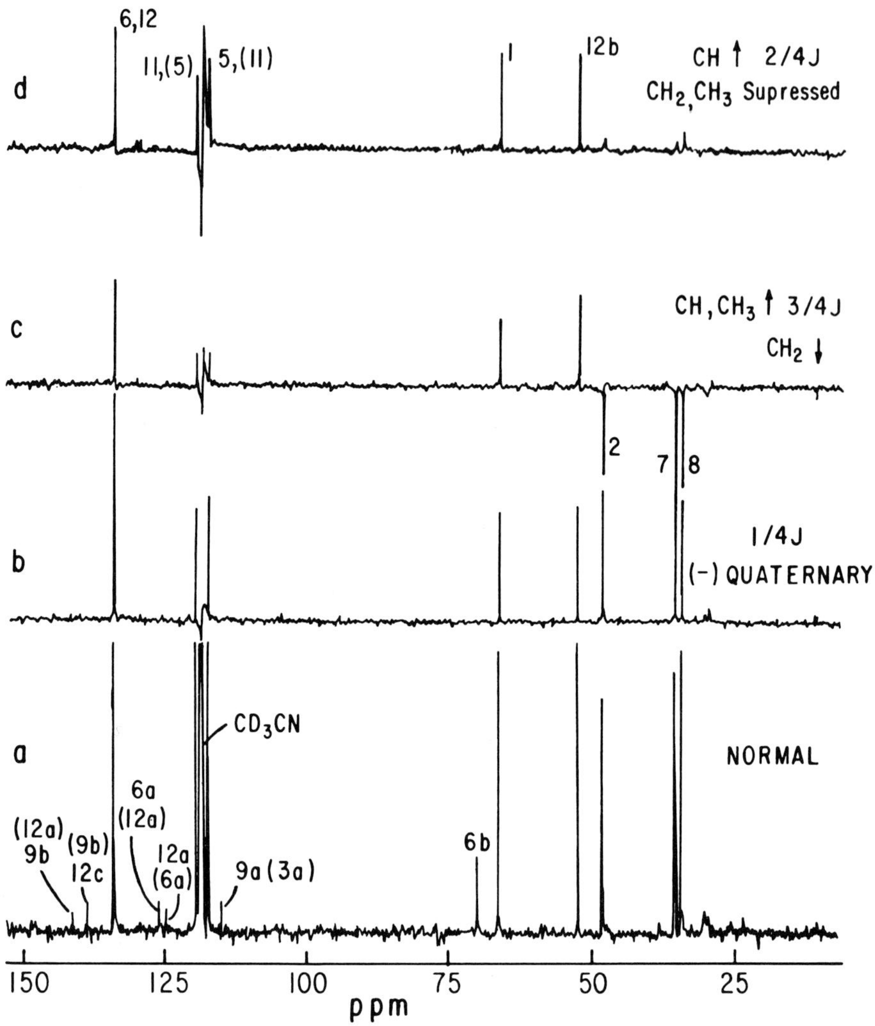

Figure 1. (a) Normal upfield portion of the 67.2 MHz ^{13}C NMR spectrum of I in CD_3CN, (b) polarization transfer spectrum with delay = 1/4J, (c) delay = 3/4J, (d) delay = 2/4J. Stinson et al., (1982).

no directly bound protons with which to cross polarize and are not seen in the spectra. Other variations on this sequence such as DEPT or APT (Doddrell, Pegg and Bendall, 1982; Patt and Shoolery, 1981) have also been reported and offer certain advantages.

To establish the assignments of each of these carbon types, identification of the corresponding proton spectrum and a molecular weight and empirical formula are essential. In spite of present day larger magnetic fields, proton resonances often overlap and are rarely observed in first

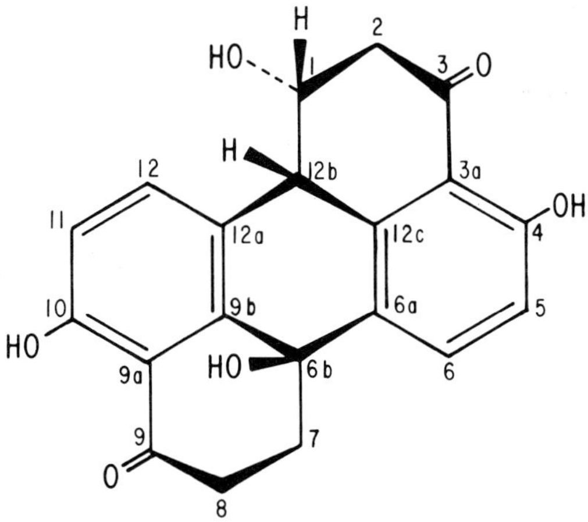

1

order spectra in complex molecules. (It is often said "you can never get a high enough magnetic field for proton NMR"). At the bottom of Fig. 2 is the one dimensional proton spectrum of ATX taken at 270 MHz (Stinson et al., 1982). Note that the severe overlapping resonances make it impossible to analyze this spectrum by inspection. Yet if we apply different time evolution periods in the two-dimensional J resolved pulse sequence (Aue, Karhan and Ernst, 1976), we can separate the coupling and chemical shifts and project them, respectively, on two separate axes. The combination of this detailed proton coupling information, dihedral angle calculations, carbon sorting and computer structure generation has made it possible to establish both the solution structure and conformation of ATX (1).

In other types of structures such as biopolymers, the basic composition may be known and the information that is required relates to a sequence of residues and their secondary structure in solution. An excellent example of the application of two-dimensional proton NMR methodology is given in a sequence and conformational study of a sphingolipid by Prestegard et al., (1982). Here, both 2-D SECSY (spin echo correlation spectroscopy) and 2-D NOESY (nuclear Overhauser enhancement spectroscopy) are combined to establish proton connectivities (through bond interaction) in the molecules as well as to determine the dipolar through space interresidue interactions, respectively. Figure 3 shows both the projection of the one dimensional, 2-D-SECSY, and 2-D-NOESY spectrum of the glycolipid derived from the glycolipid gangliotriaosylceramide.

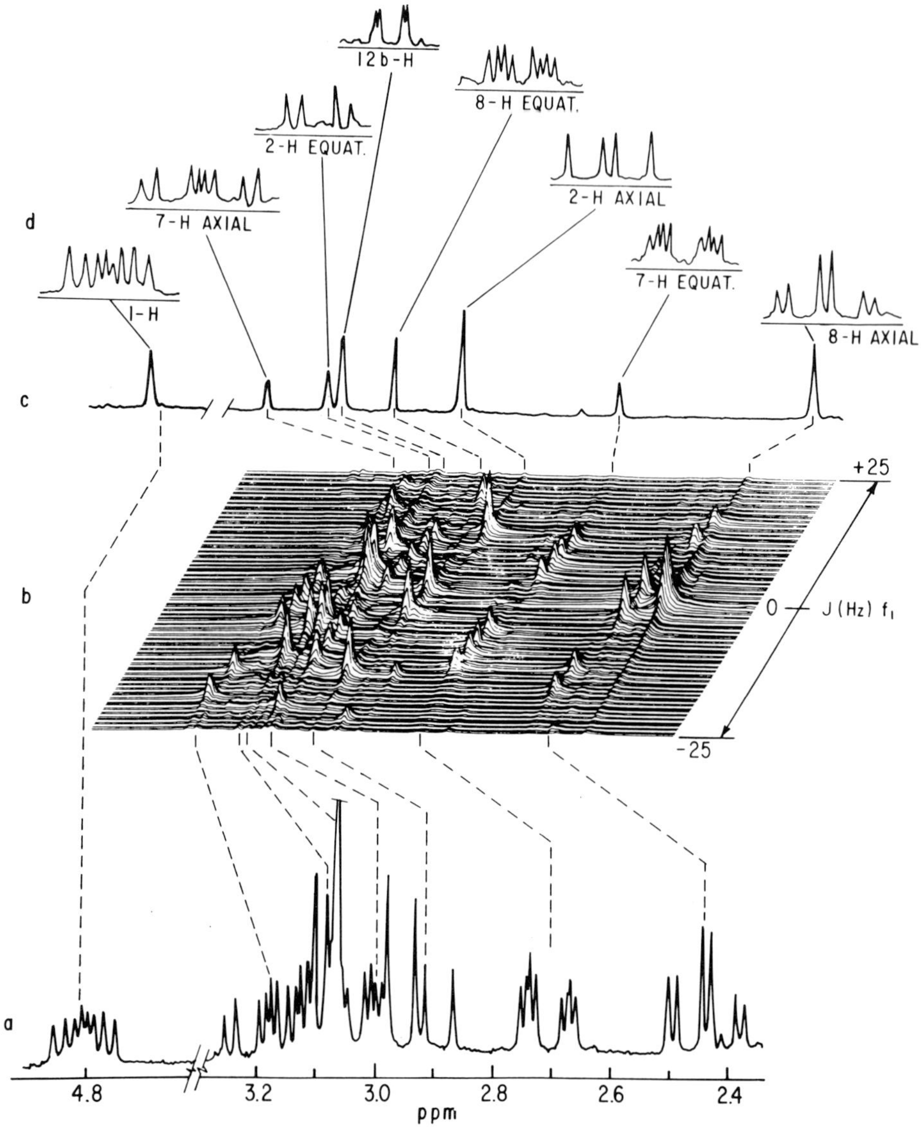

Figure 2. Upfield portion of the ^{1}H spectrum of I in $CDCl_3$ measured at 270 MHz. (a) Conventional NMR spectrum, (b) two-dimensional J spectrum (130 traces are shown; the methine H-1 at δ 4.77 has been omitted from this plot), (c), projected spectrum of *b* onto the horizontal (f_2) axis, (d) J spectra (cross sections of the stacked plot). Stinson et al., (1982).

The solid lines equidistant from the one-dimensional spectrum displayed on the central axis define the through bond scaler J coupled connectivities between vicinal and germinal protons, (SECSY correlation), and the structure of each carbohydrate residue, i.e. glucose, galac-

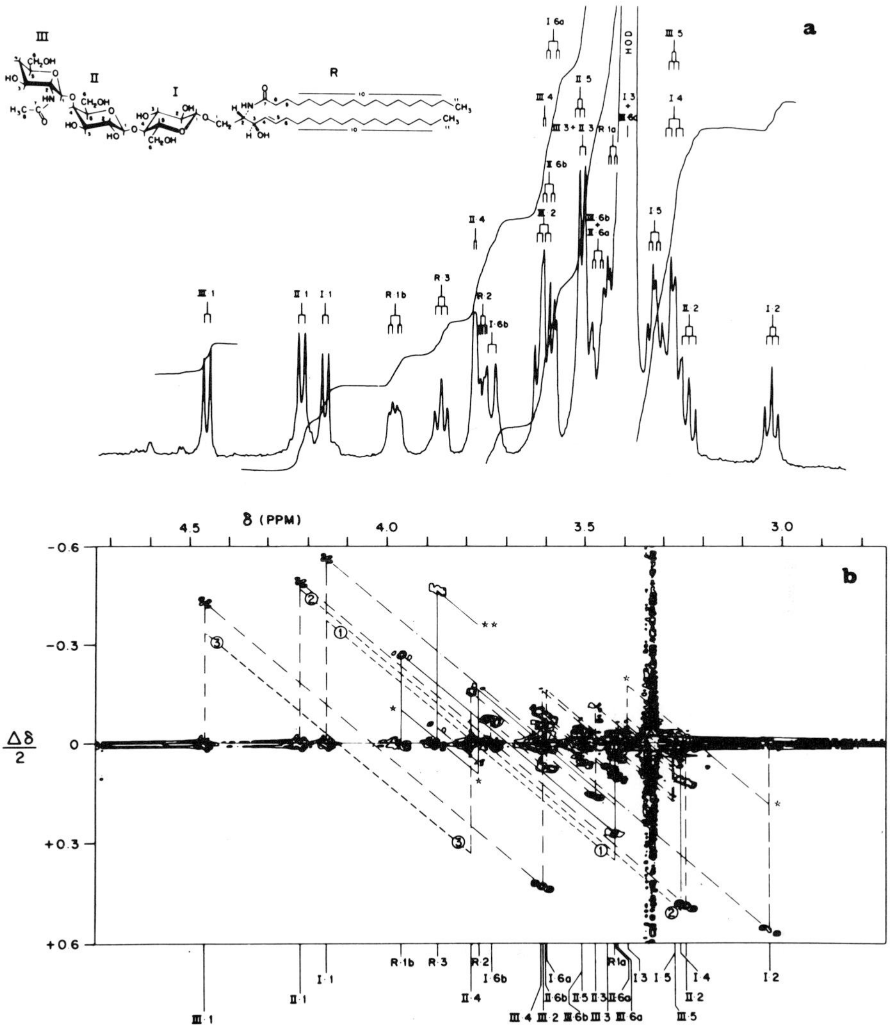

Figure 3. Structure and proton NMR spectra of gangliotriaosylceramide (1): (a) Integrated, one-dimensional spectrum of oligosaccharide ring proton region obtained at 30°C, after 420 pulses in an 8 K data set with a repetition rate of 10 s. Me_4Si was used as internal reference. (b) Two-dimensional SECSY spectrum obtained at 40°C. A total of 16 pulses in a 256 × 1024 data set were required, taking approximately 2 h. J connectivities are labeled as follows: I, β-galactopyranosyl (– – –); III, 2-acetamido-2 deoxy-β-galactopyranosyl (– – –); R. ceramide (– –). Also shown are the three (numbered) interresidue NOE couplings (– – –), obtained from the 2-D-NOE spectrum (which used 256 × 88 acquisitions and required 20.5 h). Other notations are for J connectivities observed at slightly lower contour thresholds (*) and to protons outside the observed window (**). A slight temperature dependence in chemical shift is to be noted, especially for the III-3 and II-3 resonances. Prestegard et al., (1982).

tose, mannose, etc. In addition the dotted line connectivities define the common NOE responses resulting from dipolar through space interactions of protons on adjacent residue carbons attached at glycosidic linkage pairs, e.g. (OCH_2-I1); (I4-II1); (II4-III1). Also the intensity of the Overhauser effect can be used to determine the proximity of the adjacent residue protons with respect to each other and thereby the conformational relationship of one residue with another. Other workers have exploited the two-dimensional carbon-proton heteronuclear COSY experiment, after establishing the assignments of the proton spectrum, to identify carbon connectivities for the metabolite vomitoxin produced by a fungus growing on cereal crops (Blackwell, Greenhalgh and Bain, 1984). These methods of analysis have also been used for structural studies of oligosaccharides (Morris and Hall, 1981). A complete review of these methods and applications to polysaccharide containing structures is given in a special issue of the journal of carbohydrate chemistry (Pfeffer and Binkley 1984).

NMR methods are clearly challenged to the limit when confronted with the task of evaluating complex protein structure. Before such a study is undertaken one must be well acquainted with the primary structure and often an X-ray crystallographic map. Why then do we need to do an NMR experiment? In essence, the homonuclear proton 2-D COSY experiment can assign residue proton resonances which then can be examined with the 2-D NOESY methodology to establish secondary through space interresidue interactions. This data can then be used to generate three-dimensional computer model structure. Wüthrich (1984), has made major contributions in the 2-D studies of protein. Some of these recent findings suggest that protein conformational preferences as defined by X-ray crystallographic methods in the solid differ from those in solution and membrane bound (MB) state. Figure 4 illustrates the sheer complexity of the proton 2-D COSY spectrum of a relatively simple 29 amino acid MB-polypeptide hormone glucagon (Wüthrich, 1984). While the analysis of this spectrum might take up to a year, new computer methods may shorten this time significantly in the near future (Billeter, Engeli and Wüthrich, 1985).

Analysis of the 2-D NOESY experiment yielded many of the through space interactions leading to the conformational calculations of this structure. Final calculation of the three dimensional structure of the membrane bound molecule based on a detailed analysis of the NMR data revealed that overall the distance from the C-terminal helical region to the N-terminus is appreciably longer than in the glucagon single crystals.

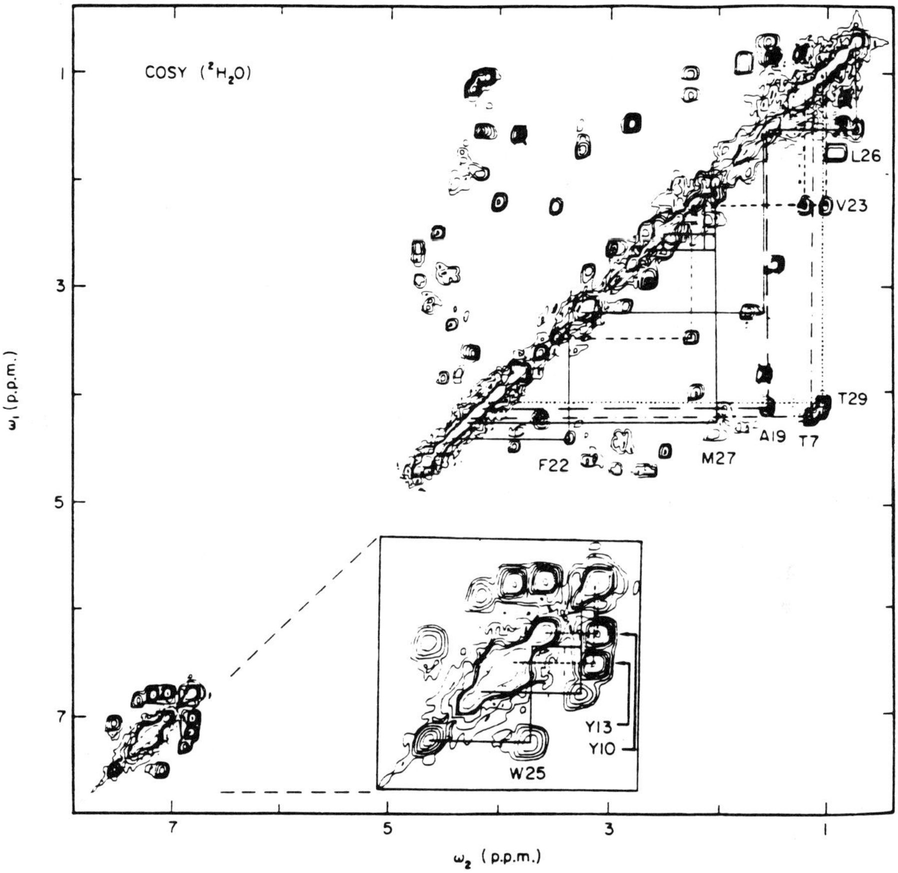

Figure 4. Contour plot of a 360 MHz ^{1}H COSY spectrum of glucagon bound to perdeuterated dodecylphosphocholine micelles in ^2H_2O. The sample contained 0.015M glucagon. 0.7M[$^2H_{38}$]-dodecylphosphocholine, 0.05M phosphate buffer, p^{2}H 6.0, $T = 37$°C. Under these conditions the predominant species in the solution are mixed micelles of 1 glucagon molecule and ~40 detergent molecules with a molecular weight of about 17 000. The spectrum was recorded in 24 h, the digital resolution is 5.88 Hz/point. The symmetrized absolute value spectrum is shown. The aromatic region is also presented on an expanded scale. Proton-proton J-connectivities are indicated for the following residues: Thr 7 (–·–·–), Ala 19 (——), Phe 22 (——), Val 23 (– – – –), Leu 26 (——), Met 26 (——), Thr 29 (· · · ·) and the aromatic rings of Tyr 10 (· · · ·), Tyr 13 (– – –) and Trp 25 (——). In order not to overcrowd the figure, only the C$^{\alpha}$H connectivity with the lower field C$^{\beta}$H line is shown. Crosspeaks originating from residual protons in the perdeuterated dodecylphosphocholine are marked (×). Wüthrich, (1984).

(b) Biosynthetic pathways and molecular interactions

Multinuclear NMR methods have revolutionized the study of biosynthetic pathways and molecular interaction as they relate to enzyme active sites and biochemical processes. Because of the voluminous literature available in this area, I will confine this discussion to some key review articles, references and a single example, in hopes that the interested reader will be stimulated to pursue these topics in further detail. In general the great majority of the studies involving macromolecular structure and interactions require the enrichment of a dilute spin nucleus such as ^{13}C, ^{15}N, ^{2}H, or the use of a particular metal ion as a probe. A recent study in our laboratory has utilized the incorporation of ^{13}C acetate label to elucidate the biosynthetic pathway for the production of the metabolite alternariol from Alternaria, a fungus that grows on wheat or rice (Stinson et al., 1986). After we established the complete and unambiguous ^{13}C resonance assignments for this molecule, the fungus was fed $^{13}CH_3{-}^{13}\overset{\overset{O}{\|}}{C}{-}ONa$ to produce alternariol containing intact incorporated acetate units. To elucidate the pathway for its production the enriched compound was subjected to a 2-dimensional multiple quantum coherence analysis-INADEQUATE (Incredible Natural Abundance Double Quantum Transfer Experiment) (Bax et al., 1981). The pulse sequence in this experiment generates a ^{13}C spectrum (Fig. 5) of those ^{13}C nuclei coupled directly to adjacent ^{13}C nuclei. In the natural abundance compound the probability of having a ^{13}C nucleus next to another ^{13}C is 1/10 000. Essentially the double quantum filtering experiment suppresses the signals from any ^{13}C nuclei that would be positioned next to a ^{12}C and so the results are diagnostic for detecting a synthetic pathway requiring the splitting of dilabeled acetate. The $^{13}C{-}^{13}C$ connectivities given in the contour plot shown in Fig. 5 defines the sequential labelling pattern and biosynthetic polyketide pathway for alterneriol given in the structure in Fig. 5. A number of examples that use ^{13}C and ^{2}H enrichment techniques for elucidating biosynthetic pathways is given by (Steyn, 1980).

^{15}N enrichment has been used to study enzymic binding sites in carboxypeptidase (Bachovehin et al., 1982) the pk_a values of specific amino acid residues in serine proteases (Kanamori and Roberts, 1983) and base pairing and motion in DNA (James, James, and Lapidot, 1981) and tRNA (Gonnella et al., 1982). Often because of the difficulties of incorporating an ^{15}N isotope into a large biomolecule, the label is put into a small substrate and its dynamic properties can be examined in the macromolecule's active binding site (Morishima and Inubushi, 1978). High resolution ^{15}N and ^{13}C NMR have also been effectively used to characterize the structure of wheat proteins as a function of concentration and heating (Baianu, Johnson and Waddell, 1982). Because of the low sensitivity and negative NOE, ^{15}N NMR has not seen the wide

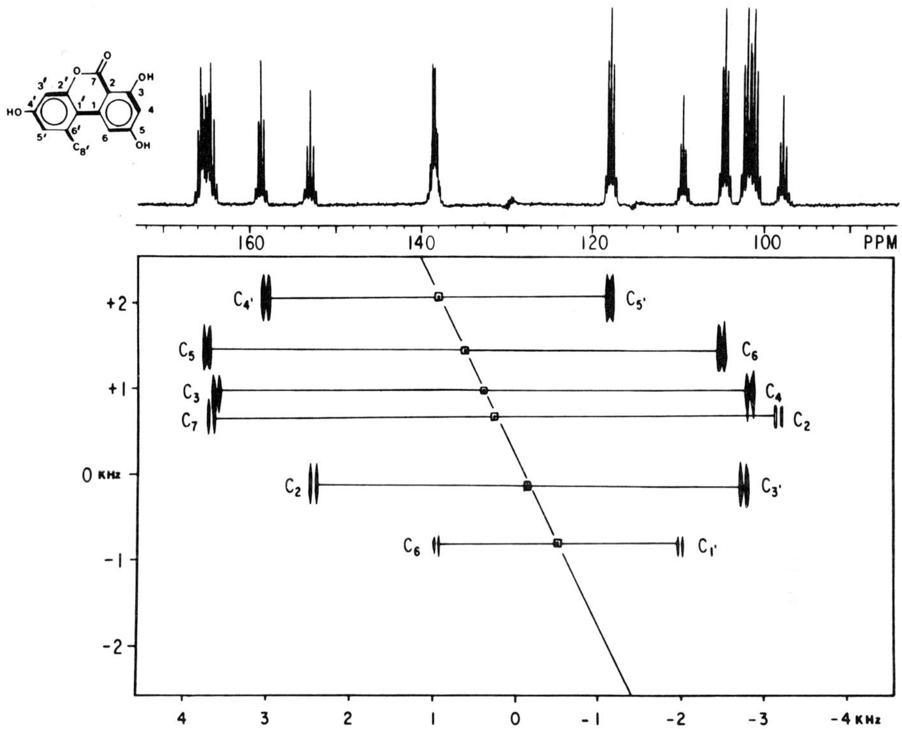

Figure 5. 100 MHz ^{13}C 2-D INADEQUATE contour plot for a 0.47 *M* solution of alternariol (biosynthetically prepared from [1,$2^{13}C$] acetate). Only the 5 KHz region in the vicinity of $F_1 = 0$ is shown. Total acquisition time for this spectrum was 38 h. Stinson et al., (1986).

application in *in vitro* biological systems. However, its use in *in vivo* studies because of its unusual correlation time (Tc) dependent NOE has made it a good probe for monitoring cellular viscosity, (this will be described in the next section).

Although ^2H has low inherent NMR sensitivity its ease of incorporation and usefulness of its quadrupolar relaxation for probing motion are responsible for its most popular application in studies of protein mobility (Wooten and Cohen 1979; Brown, Kumosinski and Pfeffer, 1983) and membrane lipid and lipid protein interactions (Mantsch, Saito and Smith 1976; Jacobs and Oldfield, 1981). In order to obtain motional information for highly oriented, relatively rigid structures as found in membranes, large quadrupolar splittings must be measured with specialized instrumentation capable of rapid digitization of spectral widths in the MHz range. In addition high resolution ^2H NMR has been used to examine the molecular interactions of dietary fiber components with digestive bile acids as a means of evaluating a proposed mechanism suggested for the lowering of serum cholesterol (Pfeffer et al., 1981).

Metal ion NMR spectroscopy has become an important tool of the biologically inclined spectroscopist since the late 70's. In general both high natural abundance metals such as ^{23}Na, ^{113}Cd and the less sensitive quadrupolar nuclei ^{25}Mg, ^{39}K and ^{43}Ca have found important applications to studies of macromolecular equilibria, exchange rates and the structure and dynamics of metal ion-ligand interactions. The breath of these studies runs the gamut from proteins and enzymes to membranes and charged polysaccharides. A most comprehensive review of the up-to-date literature in this area is given by Braunlin, Drakenberg and Forsèn (1985).

High resolution *in vivo* NMR of plant materials

(a) ^{31}P studies

High resolution NMR has become the most widely utilized method for examining functioning isolated cells, organelles from plant and animal tissues and microorganisms, primarily because it is non-invasive i.e., it does not require ionizing radiation, dehydration of cell contents or isolation and or fractionation of cellular components. NMR applications to the study of mammalian tissues and tissue disorders has been thoroughly reviewed (Roberts and Jardetzky, 1981; Gadian, 1982; Barany and Glonek, 1984).

The two most often used nuclei for whole tissue and cell studies are ^{31}P and ^{13}C, however more interests in ^{14}N, ^{15}N and ^{23}Na has recently been sparked (Martin, 1985). Although the ^{1}H nucleus is the most sensitive, it has seen only limited use in *in vivo* applications because of the dominance of the large tissue H_2O resonance in spectra, the relatively narrow chemical shift range (~ 10 ppm) and broad overlapping lines of a multitude of components. Brown and Campbell (1980) have dealt with these problems by employing an elegant spin echo technique for examining the red blood cell.

This section will deal with the use of ^{31}P NMR, drawing on some examples from our ongoing work on metal ion transport in maize root tissue. Work dealing with other nuclei such as ^{13}C, ^{15}N, ^{14}N and ^{23}Na will also be mentioned. For additional references concerning plant metabolism and general NMR application to plant materials the reader is referred to reviews by Roberts (1984), and Loughman and Ratcliffe (1984), respectively.

^{31}P combines the advantages of 100% natural abundance, relatively high sensitivity, and a chemical shift range for metabolic compounds of 30 + ppm which make it most attractive for the examination of plant tissue and cells. In addition because of the occurrence of the relatively few key phosphorus containing metabolites such as sugar phosphates, various nucleotides, (NTP, NDP, NAD and UDPG) and intracellular orthophosphate (P_i), the important energetic status of the cell can be monitored from multistacked experiments in relatively short periods of time. Also, since the position of the P_i resonance is strongly pH dependent in the range of 5.0–8.0, one can easily measure ongoing changes in intracellular pH directly from the spectra (Roberts et al., 1980). While spectral analysis is often relatively straightforward one must keep in mind that we can only examine mobile, low molecular weight compounds with standard high resolution instrumentation. This means that a large fraction of the phosphorus signal is NMR invisible, e.g. signals from membrane phospholipids and relatively immobile nucleic acids and phosphoproteins cannot be observed. This can be a serious problem e.g., if one wants to follow the pathway of the production of polymeric materials in which fluxes are rapid through low steady state concentrations of metabolites (Jacob et al., 1985). Also, if metabolites are immobilized by precipitation or by tight binding to cell walls or membranes, they will be undetectable. In terms of establishing quantitative relationships between resonance responses, careful attention must be paid to the relaxation properties of each resonance in the spectrum in order to account for ongoing changes and molecular interconversions within the cells (Schleich, Willis and Matson, 1984; Pfeffer et al., 1986).

To keep plant tissue viable it must be continually perfused with a solution containing sufficient oxygen to maintain adequate metabolic activity. Recent designs (Lee and Ratcliffe, 1983) with additional modifications (Pfeffer et al., 1986) allow experiments to proceed up to 72 hours with little change in viability of the tissue. Figure 6 shows spectra of excised maize root tips including the chemical shift assignments. Spectrum *A* demonstrates the loss of viability due to inadequate replenishing of O_2 while *B* shows full viability after 23 hrs provided by a recycling perfusate (45–50 ml/min) saturated with oxygen. Under optimum aerobic conditions two P_i resonances are detectable corresponding to the vacuolar compartment pH $\sim$ 5.5 (high field P_i resonance) and the cytoplasmic compartment pH $\sim$ 7.6 (Roberts et al., 1980). pH values were obtained from a calibration curve of P_i pH vs. chemical shift (Pfeffer et al., 1986). Under anaerobic conditions cytoplasmic acidification occurs leading to a shift in the P_i (cyt) peak corresponding to a pH of 6.2 and a loss of viability due to local CO_2 buildup around the tissue. A similar

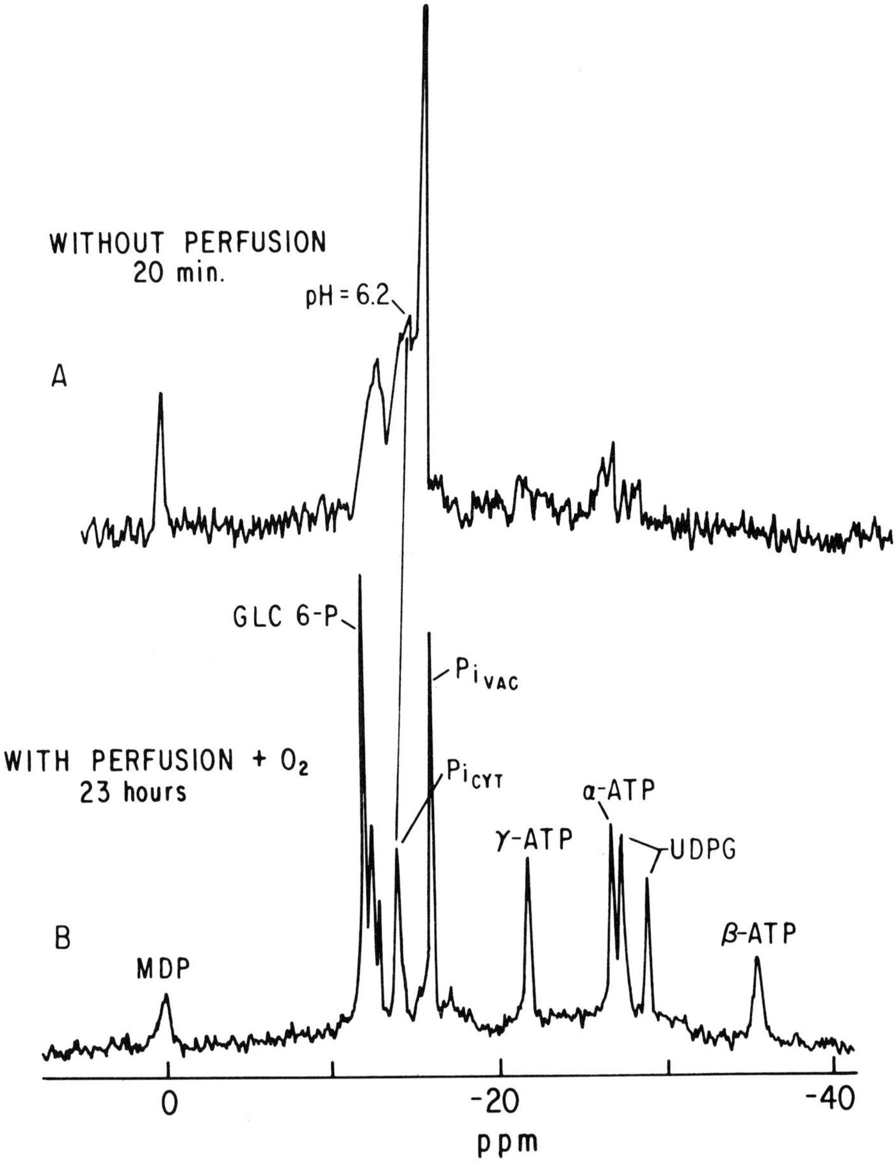

Figure 6. 161.7 MHz ^{31}P spectrum of approximately 900 excised root tips (5–7 mm) taken under fast acquisition conditions: (a) 5000 scans, after remaining in the NMR tube with no perfusion for 20 minutes; (b) 10 000 scansafter 23 hours of perfusion with oxygen saturated perfusate.

cytoplasmic acidification will take place, however, without loss of viability if hypoxic conditions are maintained, i.e., if rapid circulation is continued with nitrogen saturated perfusate (Roberts et al., 1984; Roberts, Andrade and Anderson, 1985). This condition is rapidly reversed to the normal state with reoxygenation of the perfusate even after periods of 20 hrs, except for certain genetically deficient varieties that cannot control their internal pH (Roberts, et al., 1984).

The pH gradient between the cytoplasm and vacuolar (across the tonoplast) is maintained through proton pumping driven by ATPase. When this tissue is exposed to a protonophore such as 2,4-dinitrophenol (DNP) or the protonophore, FCCP the loss of intracellular pH gradients occurs (Kime, Loughman and Ratcliffe, 1982; Pfeffer unpublished results) signifying a lack of mitochondrial control and membrane integrity, respectively. Kime *et al.* (1982) and we (1986) have observed vacuole trapping of paramagnetic Mn^{2+} in the vacuole of maize root tips. The phenomenon is shown in Fig. 7 in which the vacuolar P_i resonance fails to regenerate after extended washing of the tissue. In contrast, only trace amounts of Mn^{2+} are transported into the tissue when it is exposed in a hypoxic state (Pfeffer et al., 1986) and full restoration of the P_i vacuolar resonance is seen on reoxygenation. Similar observations are made when this tissue is energetically stressed with 2-deoxyglucose or denied an exogenous carbohydrate source (Pfeffer et al., 1986a). Clearly, movement of such divalent metals as Mn^{2+} is a facilitated process dependent on the level of ATP available in the cell.

The P_i (nutrition) uptake in various plant tissues has been extensively studied (Lee Ratcliffe, 1983a). Over the range of phosphorus nutrition investigated, the quantity of vacuolar P_i per unit fresh weight of root increased considerably, whereas cytoplasmic P_i remained constant. Roberts (1984a) has demonstrated that salt stress can cause stimulation of P_i uptake and the loss of regulation of cytoplasmic P_i levels. We have observed that in the absence of exogenous P_i maize root tips generate up to 125% additional mobile P_i, (primarily found in the vacuole) over a period of 34 hours (Pfeffer et al., 1986a). The source of this phosphate is not clear, however suppression of this additional generated signal area with aluminum ion at low pH suggests that P_i may be migrating from locations in the cell wall. Aluminum has been shown to have an inhibitory effect on the acid phosphatase generation of extracellular P_i in plant cell cultures of *Agrostis capillaris* (McCain and Davies, 1984). Figures 8 and 9 illustrate the effect of aluminum ion on P_i generation in maize tips over a period of 20 hrs.

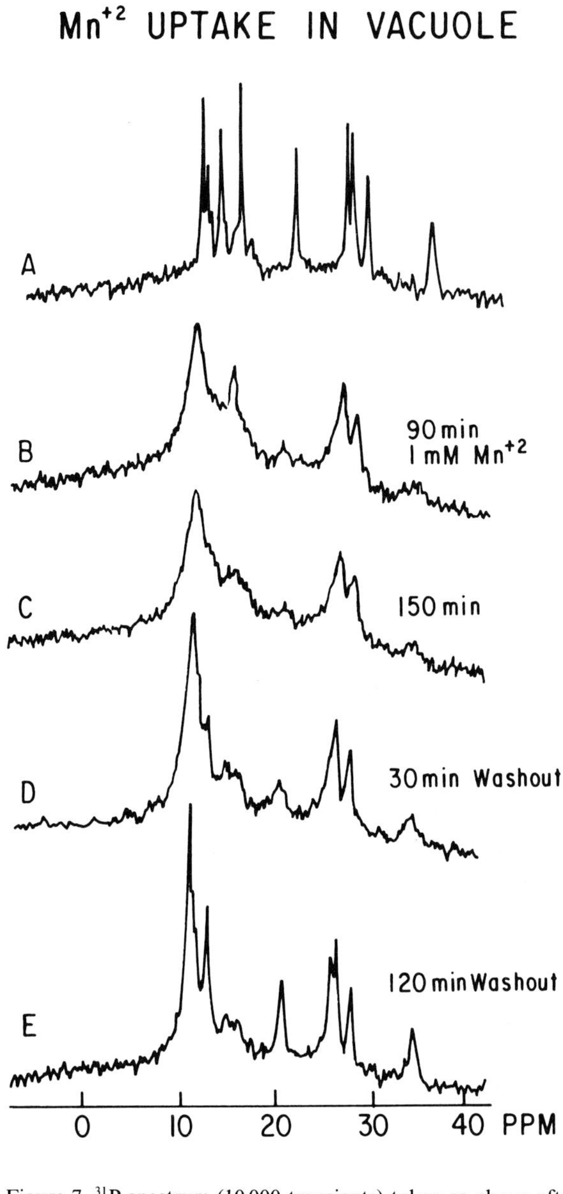

Figure 7. ^{31}P spectrum (10 000 transients) taken as above after 3 h of perfusion with a solution of 0.1 mM $CaSO_4$, 50 mM glucose, 10 mM Mes buffer (pH 6.0), O_2. B, same as above except for the addition of 1.0 mM $MnCl_2$ and exposure for 54 to 81 min. C, Same as (B), exposure 108 of 135 min. D, Sample perfused with buffer as in (A) for 0 to 27 min. E, Same as (D) for 81 to 108 min. Pfeffer et al., (1986).

The presence of 0.1 mM Cd^{2+} in the perfusion medium decreased the cytoplasmic levels of ATP in corn roots by 50% in $1\frac{1}{2}$ hours and the cytoplasmic pH dropped from 7.6 to 6.8 (Pfeffer et al., 1986a). Addition of 1.0 mM Ca^{2+} prevents this acute toxicity by competing favorably with

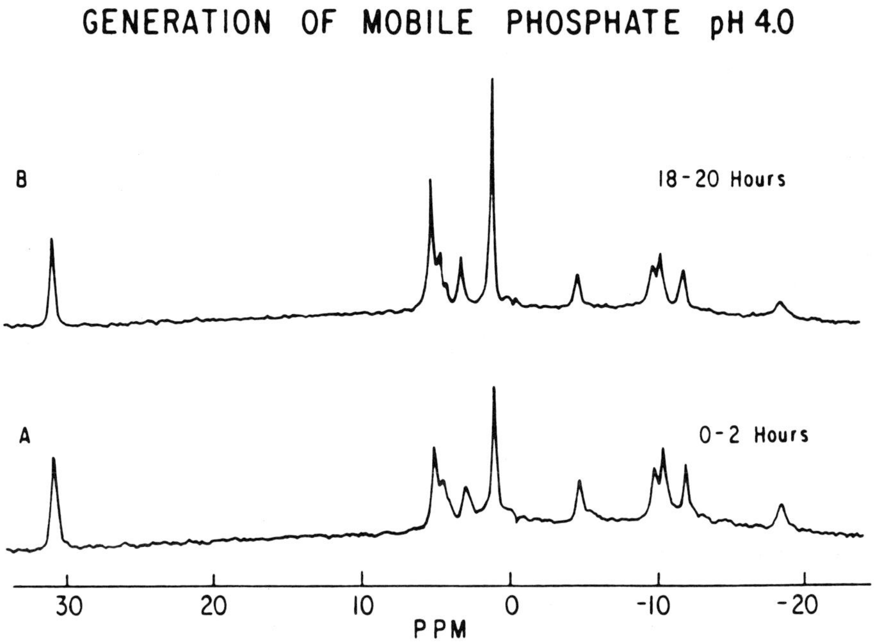

Figure 8. Below, 161.7MH$_2$ ^{31}P spectrum of approximately 900 excised root tips (5–7 mm) taken under the fast acquisition conditions, 30° pulse. (12 μsec) 2000 data points zero filled to 16 000, recycling time 0.162s. 20 000 transients, frequency of 16 KHz, 15 Hz linebroadening. Perfusate containing 0.1 mM $CaSO_4$ and 50 mM glucose pH 4.0. (a) 0–2 hours; (b) 18–20 hours. Internal standard of 120 mM HMPA in a capillary referenced as 30.74 ppm relative to 85% H_3PO_4 taken as 0.0 ppm.

Cd^{2+} for entry through the cell membrane. A similar inhibition of Mn^{2+} entry is also observed in the presence of excess Ca^{2+} (Pfeffer et al., 1986a).

To obtain quantitative information on the state of metabolism as it pertains to ATP synthesis rates and oxygen consumption we must be able to measure *in vivo* exchange processes. Saturation transfer is an NMR technique whereby low-power rf field saturation of a spin on a nucleus, will transmit that saturation to any other nucleus with which it is exchanging (Shulman, 1979). In order to measure unidirectional reaction rates the systems must be in a steady state; the exchanging partners, e.g. γ-ATP and P_i must have separate and detectable NMR signals and the rate constants for both forward and reverse reactions must be of the magnitude of the spin lattice relaxation rate of each of the exchanging nuclei (Roberts, 1984). Using oxygen tension measurements Roberts & Wemmer and Jardetzsky, (1984) have demonstrated that the P/O ratio which is 3 in normal metabolizing root tips drops to 2 when succinate is substituted for glucose. This observation indicates that a secondary

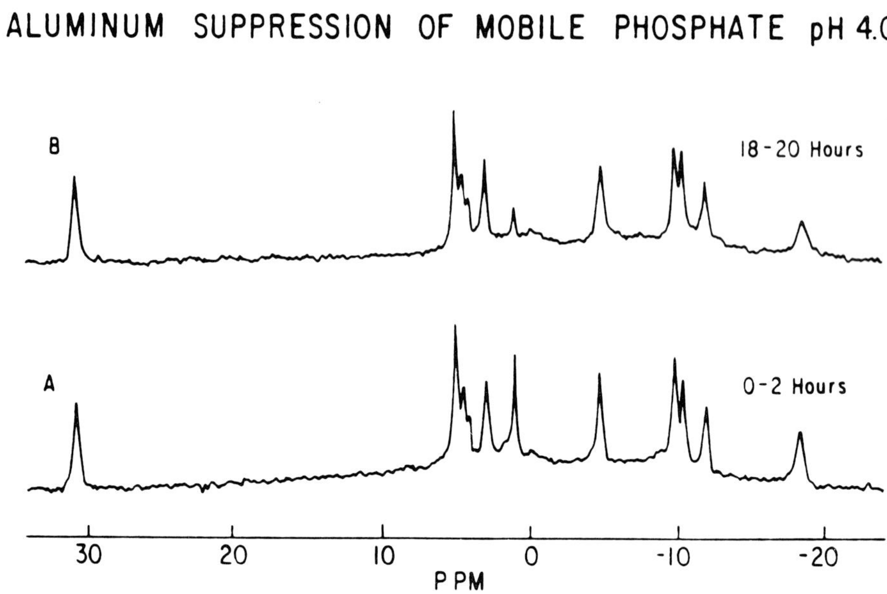

Figure 9. Above. Same as in Fig. 8 except perfusate contained, in addition to 0.1 mM $CaSO_4$ and 50 mM glucose, 2.5 mM $Al_2(SO_4)_3$ at pH 4.0.

pathway in the electron transport system has been activated in which $FADH_2$ is the principal electron donor. Addition of cyanide with succinate, suppresses the rate of ATP synthesis completely while cyanide and glucose still give a P/O ratio of 1. It is hypothesized from these results that this plant tissue contains a cyanide resistant respiration path in which electrons travel to a terminal oxidase from a point between the first and second coupling sites of the mitochondrial electron transfer chain and, therefore only the first coupling site of the electron transfer chain is used. Figure 10 illustrates the saturation transfer experiment in which the γ-ATP resonance is saturated and the transfer is seen at the cytoplasmic P_i resonance.

A study by Kallas and Dahlquist, (1981) demonstrated that ^{131}P NMR could be used to follow the alkalinization of the cytoplasm of the photosynthetic cyanobacterium *Synechococcus* under light irradiation. This was followed by an extension of this method to the investigation of spinach protoplasts and chloroplasts (Foyer et al., 1982). Unlike the previously examined intact bacteria (Kallus and Dahlquest, 1981), these workers observed that the intracellular pH is not tightly controlled but determined by the pH of the suspending medium. Most recently, Waterton, Bridges and Irving (1983) have been able to lessen the effects of bulk magnetic susceptibility and obtain ^{31}P spectra of intact photosynthetic

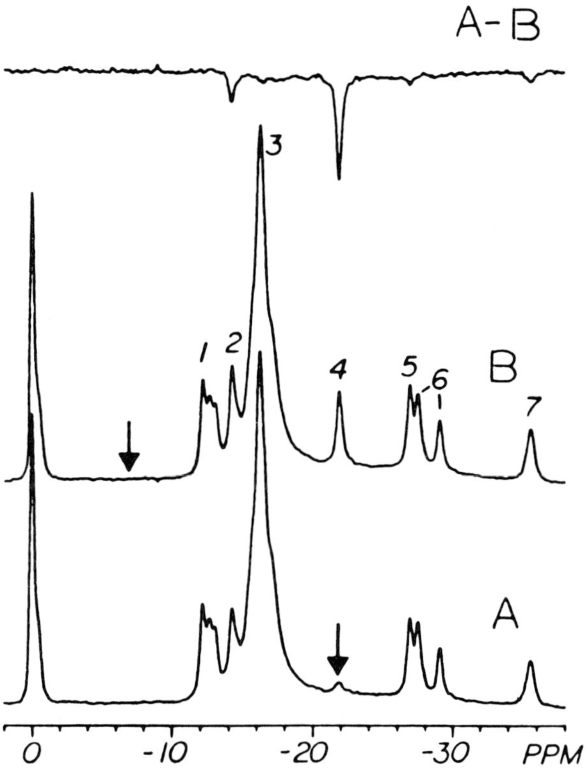

Figure 10. Saturation transfer from γ-ATP to cytoplasmic P_i, in aerobic maize root tips. Saturation transfer spectra obtained with root tips perfused with O_2-saturated 50 mM Glc + 0.1 mM $CaSO_4$, 40 ml/min. Spectrum A obtained with selective presaturation of the γ-ATP resonance; spectrum B obtained with selective irradiation of a point equidistant from the cytoplasmic P_i line (peak 2) on the low field side of the spectrum (as indicated by arrows). Spectrum A–B is the difference between spectrum A and spectrum B, showing the transfer of saturation from the γ-ATP line to the cytoplasmic P_i line. Roberts et al. (1984).

wheat-leaf tissue using (1) magic-angle spinning, (2) saturation recovery and (3) infiltrating of water into the tissue air spaces. These spectra also demonstrate the extrusion of protons from the chloroplast stroma and alkalinization as reported by the extravascular P_i signal.

Although the investigation of less complex plant cell cultures offer certain advantages over the relatively heterogeneous tissues, maintaining viability, good S/N and good spectral resolution has been problematic (Lee and Ratcliffe 1983). Rébeillé et al. (1985) have obtained excellent ^{31}P spectra of suspended sycamore cells with simple air bubbling. Their experiments demonstrated when cells are deprived of sucrose, starch hydrolysis cannot maintain the high phosphate esters levels and respiration rates synonymous with high cellular metabolism.

(b) Plant material studies with other nuclei

^{13}C in natural abundance (1.1%) has been used almost exclusively for examining the oil composition of soybeans (Schaefer and Stejskal, 1975) and seeds (Rutar et al., 1977), however recently, Colnago and Seidle (1983) have observed raffinose and stachyose in germinating soybeans. Use of ^{13}C enrichment has dominated the voluminous numbers of reports describing carbon fluxes in microorganisms (Den Hollander and Shulman 1983; Dickinson et al., 1983; Baxter, 1985). This methodology has not yet attained widespread applications in the area of *in vivo* plant studies although a number of reports on algae and fungi have recently appeared and have been reviewed (Martin, 1985). Stidham et al., 1983 have successfully used the ^{13}C signal from malate in intact *Kalanchoe tubiflora* leaves (introduced by exposure to $^{13}CO_2$) to follow the decarboxylation of malate of this CAM plant in the light. The position of the ^{13}C resonance was also useful for determining the pH of the vacuole during light and dark periods. Also, a substantial level of carbonic anhydrase activity has been detected in the CAM plant using double labeled ^{13}C $^{18}O_2$. Since isolated malate from the CAM plant leaves contained only a small amount of ^{18}O this experiment demonstrated that the intervention of carbonic anhydrase facilitated the exchange of ^{18}O with cellular water (Holtum et al., 1984). Ashworth and Mettler (1984) have done metabolic studies in both cultured tobacco and corn cells to establish intracellular glycine and serine pool sizes. These workers utilized 2-^{13}C glycine and observed signals from 2-^{13}C glycine, 2-^{13}C serine and 3-^{13}C serine and 2,3-^{13}C serine *in vivo*. Assignments were verified by ^{13}C homonuclear COSY experiments. The results of these studies suggest that serine formation from glycine is the result of a tight coupling between glycine decarboxylase and serine hydroxymethyl transferase. The metabolism of 1-^{13}C glucose has also been followed in the elaboration of cell wall polysaccharide biosynthesis in protoplasts from cell suspensions of millet (Heyser and Sillerud, 1984).

Natural abundance ^{14}N NMR has gotten very little attention in *in vivo* studies because of an extremely rapid relaxation and broad lines due to its quadruple moment. However, if certain symmetrical probe molecules such as NH_4^+ or NO_3^- are used, relatively narrow line spectra are attainable (Richards & Thomas, 1974). With this in mind Belton et al., (1985) examined the intracellular uptake of nitrate and ammonium in maize, barley and pea root tissue. Nitrate was found to accumulate in the vacuole of all these tissues while ammonium-grown barley roots showed a biphase buildup of the ammonium resonance, consistent with possible intracellular compartmentation.

Despite the need for isotopic enrichment ^{15}N is generally the nucleus of choice for most NMR applications. Its spin of $\frac{1}{2}$ gives it narrow lines and its dominant dipolar relaxation make it useful for examining micro-viscosities. Kanamori et al., (1982) capitalized on the dipolar relaxation of intracellular ^{15}N labelled glutamine, alanine and arginine to establish that the viscosity of the cytoplasm of *N. crassa* is substantially less than its vacuole due to polyanion interactions in the latter. ^{15}N has also been used as a intracellular pH probe for *N. crassa* mycelia. Both linewidths as well as chemical shifts are diagnostic for the pH environment of ^{15}N labelled amino acids, histidine, arginine, alanine and proline used in the study (Legerton et al., 1983). The chemical shift pH dependence of N_τ of histidine was found to be most sensitive in the pH range of 5–7, (characteristic of the vacuole) whereas, P_i whose pK is closer to 6.8 is relatively insensitive for reporting the relatively acidic vacuole pH.

Movement of metal ions in and out of plant root tissue is an important phenomenon to monitor. Gupta and Gupta (1982) demonstrated that dysprosium tripolyphosphate shift reagents could be used to shift the resonance of external sodium ions relative to the internal sodium ions of intact cells and tissues. Sillerud and Heyser (1984) took advantage of this method to measure the ^{23}Na efflux kinetics of adapted (adapted for rapid growth at 130 mM NaCl) and non-adapted Proso millet cell suspensions. The non-adapted cells showed little measurable Na efflux after preloading, while the adapter cells demonstrated biphasic efflux kinetics characteristic of vacuole and cytoplasmic emptying. Similarly, excised corn root tips have been investigated by ^{23}Na NMR for their uptake of sodium ions. At the same time the energy status and intracellular pH associated with the sodium ion influx was concurrently monitored with ^{31}P NMR (Gerasinowicz, Tu and Pfeffer, 1986). Figures 11(a) and (b) show the concurrently obtained ^{23}Na and ^{31}P influx spectra of corn root tips taken under normal aerobic conditions. The rate of sodium ion influx decreased from the normal aerobic state under the following conditions cyanide > hypoxia > FCCP. Although treatment of maize roots tips with cyanide and succinate has been demonstrated by saturation transfer to completely inhibit oxidative phosphorylation (Roberts, Wemmer and Jardetzsky, 1984), sodium influx was only slightly inhibited (Gerasimowicz et al., 1986). These findings suggest that an alternate, non-ATPase dependent pump i.e., perhaps an NADH linked proton pump is responsible for the sustained facilitated sodium influx.

Solid state NMR

With the innovations of cross polarization, magic angle spinning and high power decoupling, solid state NMR spectra of dilute spin nuclei are

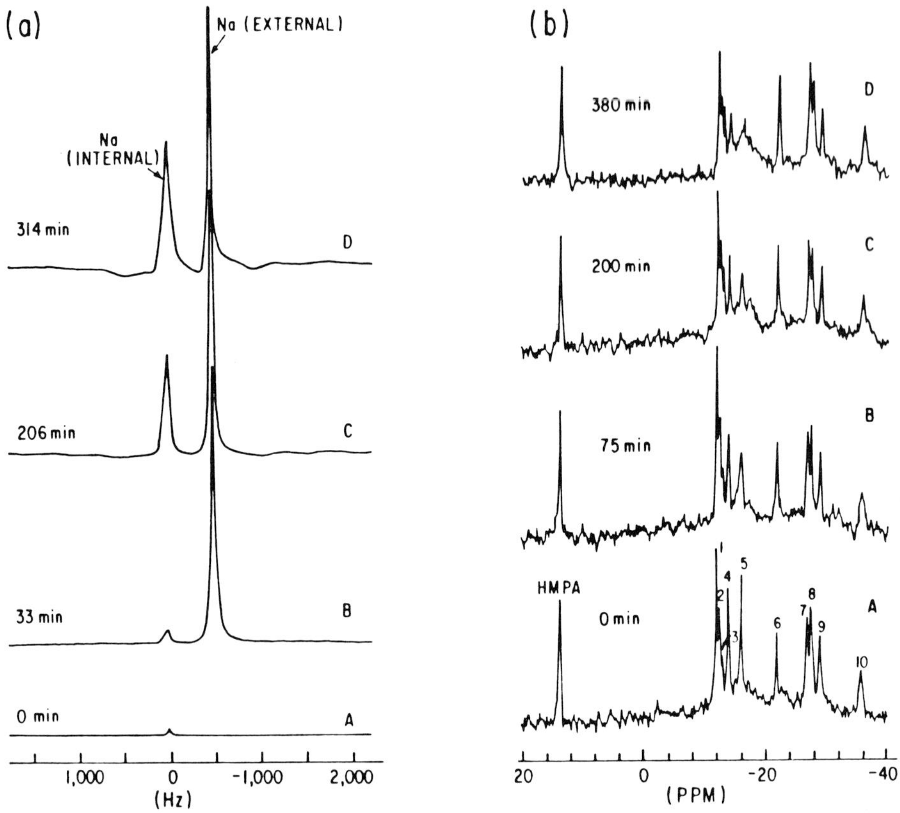

Figure 11a. (left) Sodium Influx into Aerobic Root Tissue. 3.0 mM of $Dy(PPP_i)_2^{-7}$ was added along with the appropriate amount of NaCl to keep the concentration of Na^+ at 160 mM. Spectrum (A) represents the endogenous Na^+ concentration. Spectra (B), (C), and (D) represent the distribution of Na^+ between cellular and external spaces at the indicated time after the addition of shift reagent and NaCl. Each spectrum was time averaged for 10.83 minutes. Other experimental conditions used were: frequency range 8 KHz; 16 K data points zero filled to 16 K; 548 transients per spectrum; repetition time 1.124 sec; 28.5 μsec (90° pulse); zero broadening factor. 11b. (right) Effects of Na^+ Stress on Aerobic Roots. The roots were perfused for 2 hours (A). Then, 160 mM NaCl was added to the perfusion medium. (B), (C), and (D) represent the status of roots at the indicated time after the addition of NaCl. The ^{31}P spectra were obtained sequentially on disk, and each spectrum was normalized in terms of S/N so that the relative concentrations of the components could be compared to the initial spectrum of each series. Each spectrum was signal averaged for 27 minutes. Other parameters used were; frequency range 16 KHz; 2 K data points zero filled to 16 K; 10 K transients per spectrum; repetition time 162 msec; 12 μsec (30° pulse); broadening factor 15 Hz. Gerasimowitz et al. (1986).

now attainable with a resolution that can approach that observed in solution studies. A detailed description of this methodology is beyond the scope of this chapter, however those interested in the basic concepts and aspects of quantitation are directed to works by Yannoni, (1982) and Fyfe, (1983). Biological applications are covered in a review by

Ganish (1984), the carbohydrate and polysaccharide literature has been covered by Pfeffer (1984), Fyfe et al., (1983) and Pfeffer (1985), and proteins by Hatfield and Maciel (in press).

(a) Composition and structural studies

As an analytical tool solid state NMR made some early contributions to studies of the composition of intact seeds. In 1974 Schaefer and Stejskal utilized dipolar decoupling to generate an oil seed ^{13}C spectrum showing the oil, starch and protein resonances. Later with the application of the cross polarization technique, the signals from the oil were suppressed to afford a clear ^{13}C spectrum of the immobile protein and carbohydrate (Schaefer and Stejskal, 1975; Rutar and Blinc, 1980).

A greater refinement in resolution was subsequently attained by the addition of magic angle spinning. This combination of techniques has greatly refined ^{13}C solid state methodology, now called cross polarization magic angle spinning (CPMAS), to the point where the spectra of seeds can be directly used for the evaluation of protein relative to starch content (O'Donnell, Ackerman and Maciel, 1981; Rutar, 1982). Careful attention to cross polarization parameters can also allow one to examine the liquid-like triglyceride fraction as well as the immobile components in intact soy beans (Haw and Maciel 1983).

The relative signal intensity from mobile liquid-like and immobile components is very sensitive to the cross polarization contact time used in the CPMAS experiment. At short contact times the rigid components are favored for cross polarization while longer contact times cause an attenuation of the signals from rigid components through spin lattice relaxation in the rotating frame and favor the signals from the mobile components. Figures 12(a) and (b) illustrate the phenomenon in which the rigid components in the seed (carbohydrates and proteins) are observed (12a) with a short 100 microsecond contact time while only the lipid components are preferentially observed with a 20 millisecond contact time (Fig. 12(b)).

With careful attention to the parameters used to generate spectra, ^{13}C CPMAS has become an important analytical technique for studying the structures, composition and degradation of plant tissues *in vitro*. The use of dipolar dephasing, a method by which preferential loss of ^{13}C signals attributed to protonated carbons, (except for methyl carbons) simplifies the spectra of multi component matrices is illustrated. Figure 13 shows the use of dipolar dephasing which reveals the spectrum of the non-protonated carbons and OCH_3 groups associated with the minor lignin component in intact wood cellulose (Gerasimowicz, Hicks and Pfeffer, 1985). Note the striking similarity of the dipolar dephased spectrum with

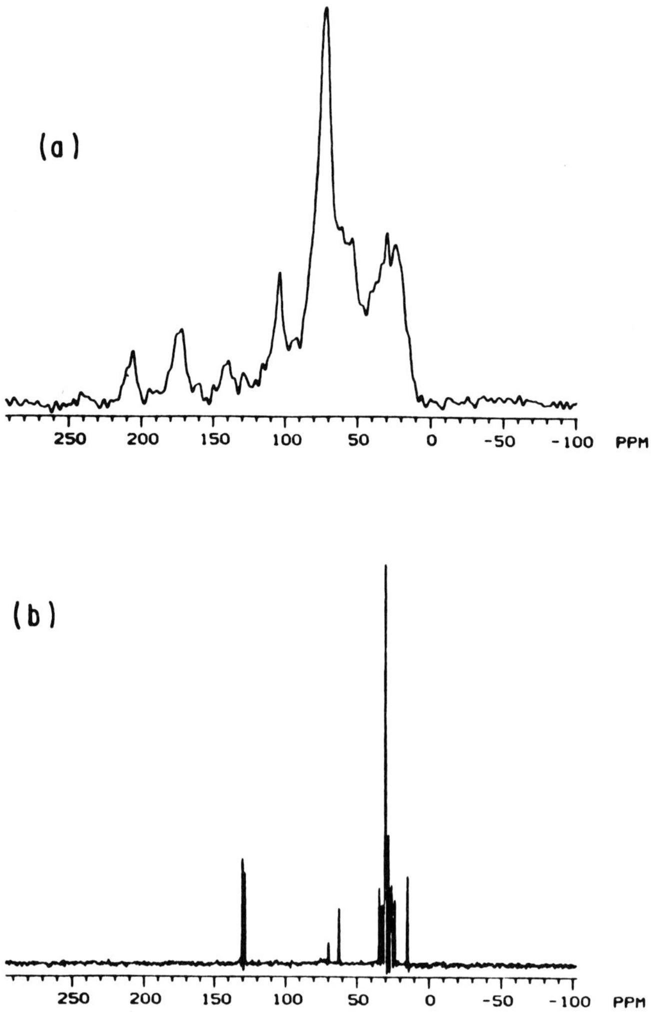

Figure 12(a). A 50.3-MHz ^{13}C CPMAS spectrum of two intact soybeans: 100-μs contact time, 8000 2-s repetitions. The Hartmann–Hahn match was carefully adjusted. 12(b). A 50.3-MHz ^{13}C CPMAS spectrum of two intact soybeans; 20-ms contact time, 2080 3-s repetitions. The Hartmann–Hahn match was carefully adjusted. Haw and Maciel, (1983).

that of the wood cellulose sample, exhaustively treated with cellulase. Relaxation methods were also utilized in this study to evaluate the level of phase homogeneity present in this lignin-carbohydrate matrix. Haw et al., (1984) have thoroughly addressed the question of quantitation for evaluating the levels of lignin in wool pulps which have undergone different types of processing. Compositional changes brought on by microbial modification of lignin or degradation of wood has been qualitatively explored by Schaefer et al., (1981) and Hedges et al., (1985)

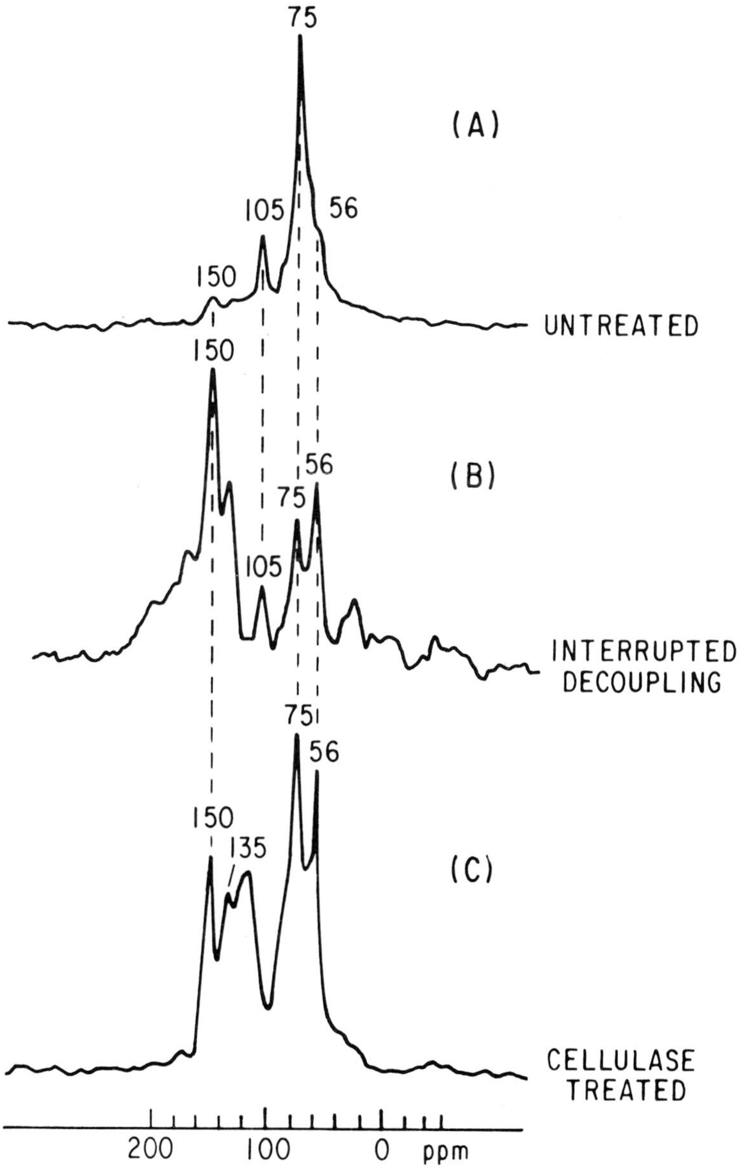

Figure 13. *Piceae glauca* pulp ^{13}C CPMAS spectra. Spectra A, B and C were obtained with 0.5 msec contact times and decoupling was interrupted in spectrum B for 40 μsec. Gerasimowicz et al., (1984).

respectively. Eflonson et al., (1984) have examined numerous forages by ^{13}C CPMAS and dipolar dephasing methods to distinguish between signals attributed to protein, lignin and possibly fructose resonances emanating from easily digestible fructoside polymers. A similar study aimed at determining the ratio of carbohydrate, lignin and protein in grass species has also been reported (Himmelsback, Barton and Windham, 1983).

Schaefer (1985) has recently used ^{13}P solid state NMR to determine in part the composition of the phosphorous-containing compounds of intact lyophilized alfalfa tissue. In addition, when the tissue was grown in culture under both nonregenerating and regenerating conditions, the ^{31}P spectra showed that a low average intracellular pH accompanied the latter.

Polysaccharide breakdown or disordering in ripening apple cell wall tissue was monitored as a function of change in molecular dynamics of the polymer matrix (Irwin et al., 1984). Spin lattice relaxation times were used to estimate changes in polymer mobilities as a function of ripening. The polyuranide carbonyl resonances were of particular interest since they showed a 63% drop in their proton T_1 over a period of 21 days. These data correlate well with a decrease in fruit tissue firmness over this period indicative of the pectin matrix breakdown and overall disordering.

Two-dimensional methods are beginning to be exploited in solid state NMR to obtain more detailed information on protein structure by monitoring through space dipolar interactions between dilute spins (Cross, Frey and Opella, 1983). Two-dimensional spin-exchange experiments conducted on uniformly labeled and selectively ^{15}N labelled coat protein (containing 50 amino acids) showed several cross peaks or sites of interaction consistent with the proposed α-helical structure.

Perhaps the most complex and intractable agricultural material to be examined by solid state NMR is soil, amended soils and soil components. Of particular importance in these studies is the composition and disposition of the organic matter as the soil undergoes different degradative processes. Wilson et al., (1983) have characterized the degradation of the organic matter in whole soils with ^{13}C CPMAS methods. With the aid of dipolar dephasing techniques, estimates of non-protonated carbons and the degree of condensation of the aromatic rings or alkyl ring substitution and demethylation were made for different periods of humification. Figure 14 shows the emphasis of different parts of the spectrum of aged pine leaves with and without dipolar dephasing. As with studies of wood (Gerasimowicz et al., 1985), the non-protonated resonances associated with the carbonyl and condensed and substituted aromatic carbons dominate the dipolar dephased spectrum 14b. As time proceeds preferential microbial degradation of the carbohydrate components enrich this material in lignin. A comprehensive review of the CPMAS ^{13}C spectra, composition and diagnosis of humic substances, fulvic acids and humins derived from different soil samples is given by

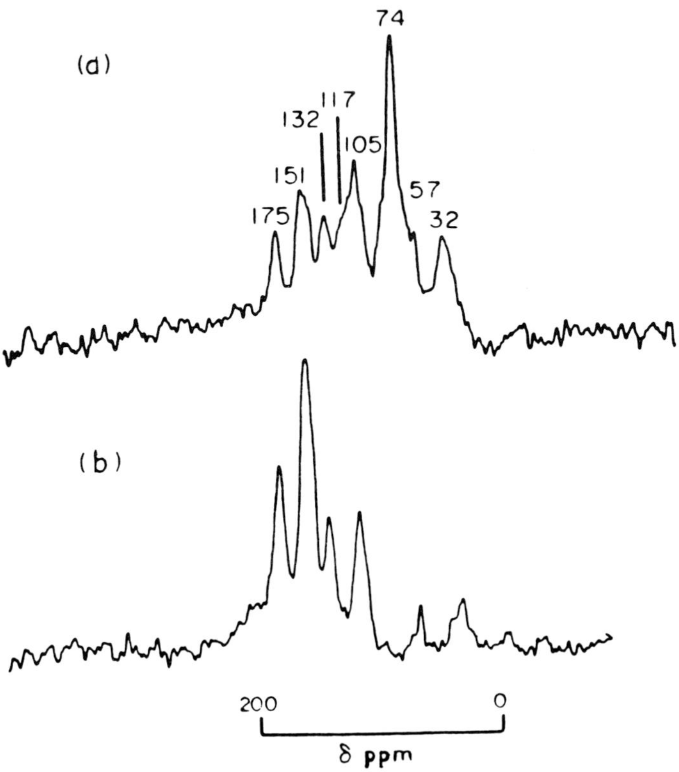

Figure 14. Dipolar dephased spectra of aged pine leaves. (a) $T_{dd} = 1\,\mu$sec; (b) summation of 15 spectra with T_{dd} incremented by $5\,\mu$sec from 40–$190\,\mu$sec. Wilson et al., (1983).

Hatcher et al., (1983). The accompanying table summarizes the ^{13}C resonance absorption positions for the different functional carbons found in spectra of soils (Hatcher et al., 1983). Mixtures of representative components present in amended soils and composts were used to simulate the spectra of these complex matrices (Piotrowski, Valentine and Pfeffer, 1984). Additionally, a careful examination of the relaxation parameters associated with the relative magnetization responses of these various components uncovered inherent spectral distortions in the presence of paramagnetic Fe^{3+} (Pfeffer, Gerasimowicz and Piotrowski, 1984). Figure 15 illustrates the loss of signal intensity of the carbon resonances representing the hydrophilic cellulosic components. This phenomenon results from a significant shortening of T_ρ (spin lattice relaxative time in the rotating frame) due to preferential Fe^{+3} interactions. Clearly, to obtain reliable spectra one must be fully aware of the presence of paramagnetic metals in such complex materials. Also the complete relaxation profile of the components should be evaluated before a quantitative assessment of these spectra is undertaken. Later

Table 1. ^{13}C-NMR chemical-shift regions for various types of carbons typically present in humic substances and soils[a]

Region I (0 to 50 ppm)	Region II (50 to 110 ppm)	Region III (110 to 150 ppm)	Region IV (160 to 190 ppm)	Region V (190 to 220 ppm)
paraffinic C = C–C*–C	alcohols C–OH	olefinic-C	carboxyl C(=O)–OH	aldehyde C(=O)–H
(C)(C)C*–C	amines (C–NH$_2$, C–NHR, C–NR$_2$)	aromatic-C	ester C(=O)–OR	ketone C=O
	carbohydrates		amide C(=O)–N	
C–C*(C)(C)–C	ethers (C–O–C)			
C–CH$_3$	methoxyl (OCH$_3$)			
	aetals O–C–O			

[a] Hatcher et al., (1983).

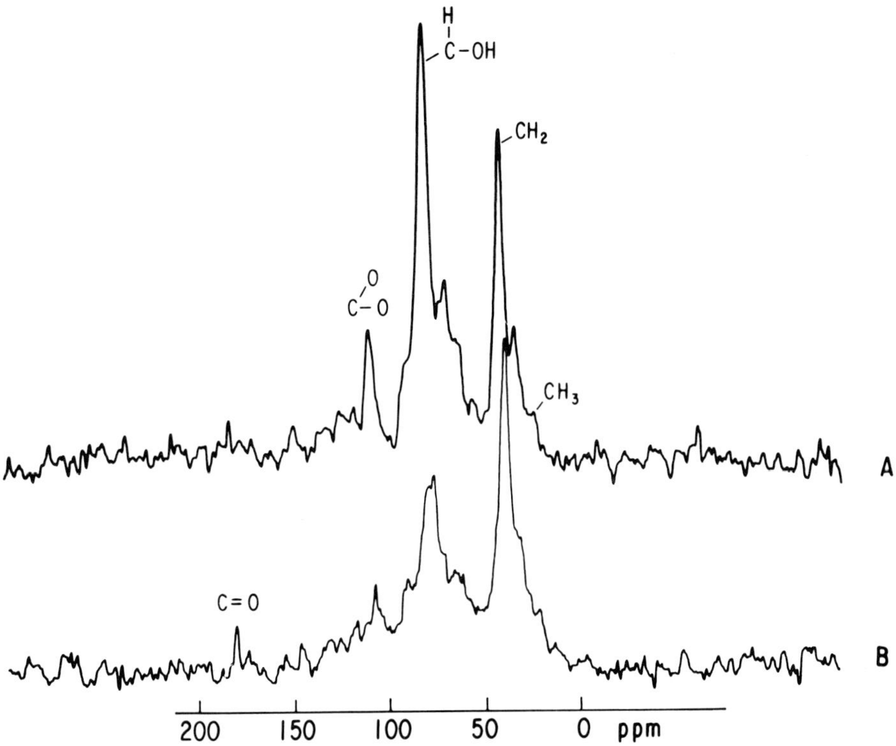

Figure 15. Comparison of ^{13}C CPMAS NMR spectra of model sludge mixtures (a) with no Fe^{3+} and (b) containing 1.9% Fe^{3+}. Each spectrum was obtained under the optimized conditions, 14 000 scans, 16-s pulse delays, a 0.5-ms contact time, and 20-Hz line broadening. Pfeffer et al., (1984).

reports of the spectra of organic soil samples which contain large natural amounts of copper (Copper mucks) verified the preferential paramagnetic effects on quantification resulting from the loss in intensity of the cellulosic resonances (Preston et al., 1984). Complementary ^{13}C solid state NMR and FTIR spectroscopic studies of humic acids derived from different stages of sludge treatment and composting gave more refined information on the nature of the carbonyl components (amides and carbonyl groups) than could CPMAS alone (Gerasimowicz and Byler, 1985).

(b) Metabolic studies

Over the past five years Schaefer and co-workers have effectively used ^{13}C and ^{15}N labeling techniques in solid state CPMAS NMR to complement high resolution ^{13}C and ^{15}N NMR which measures the flux of stable isotopes through soluble pools of cellular systems *in vivo*.

Jacob et al., (1980) demonstrated the efficacy of ^{15}N CPMAS methodology for examining ^{15}N incorporation into the amino acids resulting via nitrogen reductase activity on ^{15}N nitrate in lyophilized mycelia from *Neurospora Crassa.* Other applications of this method led to the investigation of the amount of amide and α-amino nitrogens produced in the pod and seeds of a soybean plant at various stages of development (Schaefer et al., 1979). Soybean cotyledons grown on asparagine showed preferential incorporation of amino nitrogen of asparagine into storage proteins over amide nitrogen (Skokut et al., 1982).

Subsequent development of an elegant technique called double cross polarization ^{13}C and ^{15}N NMR has expanded the scope of metabolic studies in whole plant tissue. In these experiments, the magnetization is transferred first from ^{1}H to ^{13}C or ^{15}N and subsequently from ^{13}C to ^{15}N or ^{15}N to ^{13}C. The double cross polarization mechanism, in fact, labels all ^{13}C signals or all ^{15}N signals coming from a ^{13}C–^{15}N bond. By subtracting the signals from ^{13}C or ^{15}N derived through non-mutual cross polarization from the mutually cross polarized signal a resonance area is obtained which gives a direct quantitative measure of the uptake and metabolism of intact double labels (Schaefer et al., 1984). An example of this method is given in Fig. 16. Here we see the single $^{1}H \rightarrow {}^{15}N$ cross polarization ^{15}N spectra given for plants grown on ^{15}N enriched fertilizer with and without $^{13}CO_2$. The difference spectra resulting from the double cross polarization spectra minus the single cross polarization experiment show almost complete cancelling of the resonance line in the single ^{15}N labelled plant due to the almost complete absence of ^{13}C–^{15}N species in the protein, whereas a significant amount of double labelled species is observed in the others grown with ^{15}N and $^{13}CO_2$. Exploitation of this methodology was demonstrated in the evaluation of protein turnover in soybean leaves (Schaefer et al., 1981). These experiments involved exposing soybean plants massively labeled with ^{15}N (from fertilizer) to a pulse of $^{13}CO_2$. The isolated and lyophilized leaf tissue was then subjected to the double cross ^{13}C–^{15}N NMR experiment as a function of time. The findings indicate that the proteins in the mature leaf, as measured by incorporation of double label are very stable before senescence, and have a turnover constant of about 30 days. The double label method was also utilized to assess the fate of (4-^{13}C, amide-^{15}N) asparagine in developing cotyledons (Schaefer et al., 1981a). Here 50% of all the asparagine residues were incorporated intact within the soybean protein. In contrast no intact double label was observed from the metabolism of ^{13}C–^{15}N labeled allantoin in soybean cotyledons (Coker and Schaefer, 1985). These data are consistent with allantoin's degradation to two molecules of urea and one two-carbon fragment.

In general all of the solid state methods, while more technically de-

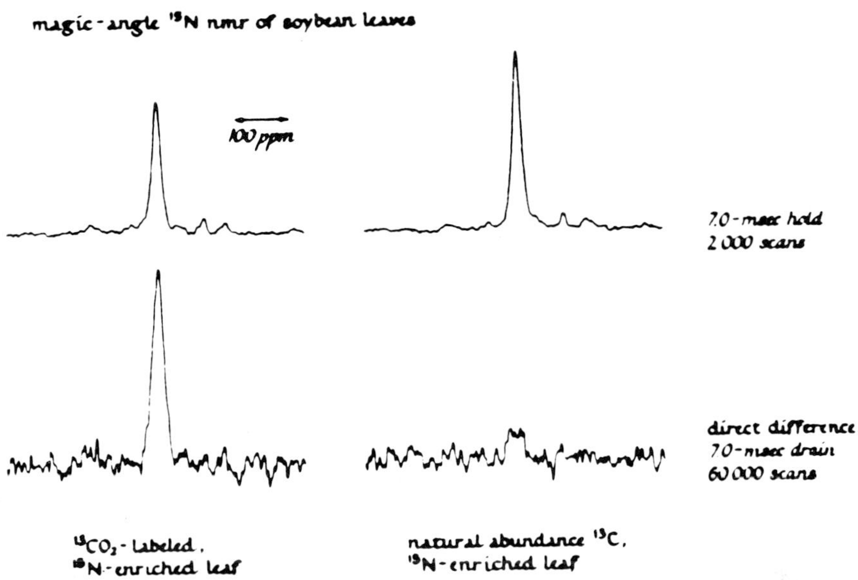

Figure 16. Magic angle cross-polarization ^{15}N NMR spectra of ^{15}N-enriched lyophilized soybean leaves exposed to $^{13}CO_2$ for 7 days (left) and exposed only to normal CO_2 (right) during active photosynthesis. The spectra at the top of the figure were obtained using the pulse sequences with a nitrogen spin lock of 7 ms and the carbon rf field off-resonance by 60 kHz. The spectra at the bottom of the figure are the result of double cross-polarization direct difference experiments in which the carbon rf field is first off-resonance and then on-resonance. This procedure was repeated 30 000 times. The positive signal which accumulates is a measure of the concentration of ^{15}N–^{13}C pairs in the labeled leaf protein. Schaefer et al., (1981).

manding represent an excellent complement to the high resolution *in vivo* techniques for tissue studies as well as the fractionation-dependent *in vitro* experiments necessary for examining solubilized polymeric matrices.

Magnetic resonance imaging

In contrast to conventional NMR spectroscopy, magnetic resonance imaging (MRI) is concerned with applications to heterogeneous specimens, for example parts of the human body, whole plants, etc., which are not small, and which are placed deliberately in a nonuniform magnetic field. The purpose of the nonuniform field, as described by Dr. Lauterber in his chapter is to label different parts of the specimen with different field strengths so that they respond with recognizably different NMR frequencies, enabling the structure and internal processes of the specimen to be derived and displayed. Moreover, besides giving morphological information MRI gives additional diagnostic insights through relaxation parameters, which are not available from other imaging methods. Excellent reviews of the imaging techniques and applications have been given by Bottomly (1982) and Andrew (1983).

Since hydrogen is the most abundant element in all living organisms, proton NMR is used as the principle nuclide for all images. Applications in agricultural area such as studies of plants and animal embryology are just beginning. The first images of an intact plant, (bean *Vicia faba*), from root to top in soil, with a spectral resolution of 0.56 mm × 0.56 mm was recently described by Bottomley, Rogers and Foster (1986). Their images were able to detect water transport in roots with light stressed foliage using water doped with a paramagnetic NMR contrast agent. In addition the basic root structure and pathology as evidenced by partial decay of hypogeal cotyledons were also indicated. Figure 17 shows the ^{1}H images of the *faba* root system in which the effects of light stress are evident. Cure (in press) has examined the same plant system but with much greater spatial resolution (0.15 mm × 0.15 mm). These improved images allow for discrimination between tissues of the main root and visualization of the vascular connection of branch roots to the central stele. Application of proton imaging with simultaneous spatially resolved high resolution NMR has been carried out during the embryonic development of the desert locust (Gassner and Lohman, in press). These images taken over a period of 8 days show the dynamics of pattern formation, compartmentation, and biochemical changes within and between compartments within the embryo. Submicroliter cube resolution allowed for the assessment of distribution changes in water and lipid during development. Figure 18 shows the changes in lipid water concentrations in two distinct sections of the embryo in addition to the overall image. Other reports of spatially resolved high resolution NMR for examining the *in vivo* chemistry of living organisms by ^{31}P NMR have been reported by Belton et al. (1982) for embryonic development of hen's eggs and by Storey et al. (1984) for gall fly larvae. A full review of topical magnetic resonance is given by Gordon, Hanley and Shaw (1982).

Perspectives

While high speed computers have made 2-D NMR a more or less routine technique for establishing the shift assignments and structure of small to medium sized organic molecules, an investment of as much as 1–2 years time is often required to analyze the complex data derived from the spectra of relatively small protein molecules. Future developments in the area of computer assisted data analysis and molecular modeling promise to shorten this time appreciably.

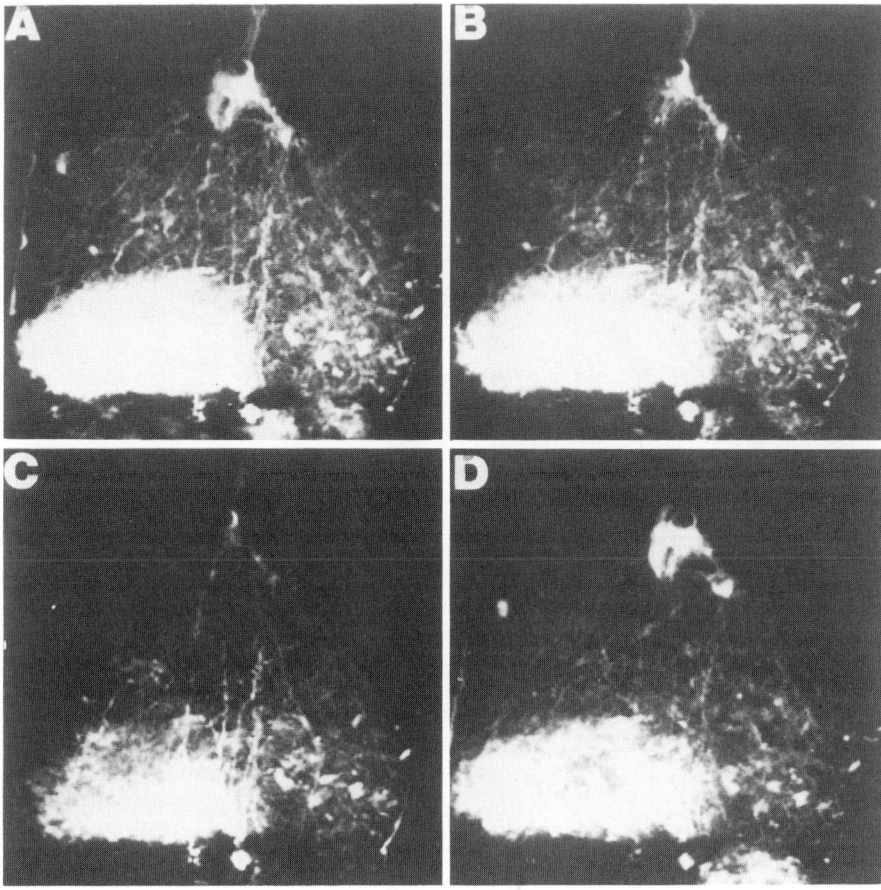

Figure 17. (a) Image of the V. faba root/soil system, with a 20-ml bolus of water injected into the soil at the base. (B) The root system image after 87 min of exposure to light stress shows water leaving the cotyledonary area. (c) After 200 min of light exposure, water has almost cleared from this area. (d) The image after 275 min, light having been fully removed at 200 min, shows recovery (return of water) in the cotyledonary area. All images were obtained in 3.4 min, with a 0.2-s pulse repetition period. (Scale = 1:2.7). Bottomly et al., (1986).

In vivo work on plant materials is advancing at a rapid pace and intact tissue studies (^{31}P, ^{1}H, ^{13}C, ^{15}N) of whole unexcised plant sections will be expanding as large bore magnets become more accessible. This approach to spatially resolved or topical spectroscopy will be conveniently coupled with imaging methodology in a single commercial instrument. Presently, a few of these prototypes have been utilized with relatively good success. Images based on ^{23}Na, ^{39}K and ^{31}P nuclei offer promise as an alternate to the established ^{1}H images because their differences in relaxation behavior and diverse distribution yield a perspective on cellular structure.

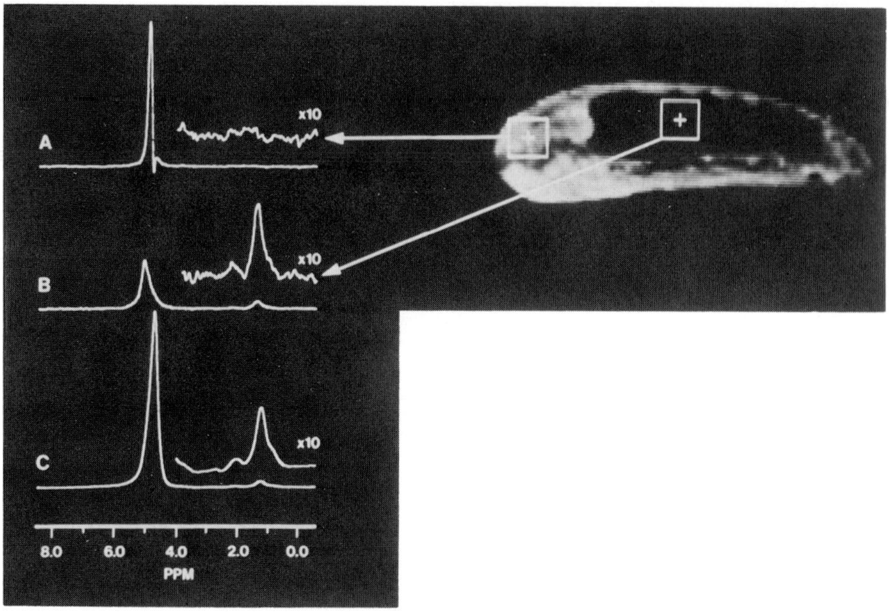

Figure 18. Image selected *in vivo* spectroscopy (ISIS) spectra from the image of locust shown, indicate that the submicroliter volume dorsal to the stomatodeum (a) at 83% development is essentially void of lipid while the haeomocoel (b) contains lipid. Note the overall spectrum (c). Gassner and Lohman, (1986).

Two-dimensional methods are expanding into *in vivo* and solid state spectroscopy. Its principal application will be in the area of determining metabolic exchange processes, (unidirectional flux between metabolites), in living tissue as well as molecular diffusion rates in solid materials. Greatly enhanced resolution in solid state spectra of heterogeneous materials is also envisioned through the development of a new technique called zero field NMR.

All of the above innovations promise to expand the capabilities of NMR and its impact on all biologically oriented disciplines including agriculture.

References

1. Andrew, R.E. 1983. *NMR Imaging*. Acc. Chem. Res. 16: 114–122.
2. Ashworth, D.J. and I.J. Mettler. 1984. *Direct observation of glycine metabolism in tobacco suspension cells by Carbon-13 NMR spectroscopy*. Biochem. 23: 2252–2257.
3. Aue, W.P., J. Karhan and R.R. Ernest. 1976. *Homonuclear broad band decoupling and two-dimensional J resolved NMR spectroscopy*. J. Chem. Phys. 64: 4226–4227.
4. Bachovchin, W.W., K. Kanamori, B.L. Vallee and J.D. Roberts. 1982. *Catalytic mechanism of serine proteases: Reexamination of the pH dependence of the histidyl $^{1}J^{13}C2$-H coupling constant in the catalytic triad of δ-lytic protease*. Biochem. 21: 2885–2893.

5. Baianu, I., L.F. Johnson and D.K. Waddell. 1982. *High resolution Proton, Carbon-13 and nitrogen-15 nuclear magnetic resonance studies of when proteins at high magnetic fields: Spectral assignments, changes with concentration and heating treatments of flinor gliadins in solution-comparison with gluten spectra.* J. Sci. Food Agric. 33: 373–383.
6. Bárány M. and Glonek. 1984. *Identification of diseased states by phosphorus-31 NMR.* in Phosphorus-31 NMR principle and applications – Academic Press.
7. Bax, A. 1982. *Two-Dimensional Nuclear Magnetic Resonance in Liquids.* Delft University Press. D. Reidel Publishing Company, Dordrecht, Holland.
8. Bax, A., R. Freeman, T.A. Frenkiel and M.H. Levitt. 1981. *Assignment of carbon-13 NMR via double-quantum coherence.* J. Magn. Res. 43: 478–483.
9. Baxter, R.L. 1985. *Microbiological applications of NMR spectroscopy.* Microb. Sciences 2: 340–345.
10. Belton, P.S., R.E. Gordon, J.M. Jones and D. Shaw. 1983. *A ^{31}P topical magnetic resonance study of embryonic development in hen's eggs*, Brit. Poultry Sci. 24: 429–433.
11. Belton, P.S., R.B. Lee and R.G. Ratcliffe. 1985. *A ^{14}N nuclear magnetic resonance study of inorganic nitrogen metabolism in barley, maize and pea roots.* J Exp. Bot. 36: 190–210.
12. Benn, R. and H. Günther. 1983. *Modern Pulse Methods in High-Resolution NMR Spectroscopy.* Angew. Chem. Int. Ed. Engl. 22: 350–380.
13. Billeter, M., M. Engelo and K. Wüthrich. 1985. *Interactive program for investigation of protein structures based on proton NMR experiments.* J. Mol. Graphics *β*: 79–83, 97–8.
14. Blackwell, B.A., R. Greenhalgh and A.D. Bain. 1984. *Carbon-13 and proton nuclear magnetic resonance spectral assignments of deoxynivalenol and other mycotoxins from ferscerium graminearum.* J. Agric. Food Chem. 32: 1078–1083.
15. Bottomley, P.A. 1982. *NMR imaging techniques and applications. A Review*, Rev. Sci. Instr. 53: 1319–1340.
16. Bottomley, P.A., H.H. Rogers and T.H. Foster. 1986. *NMR imaging shows water distribution and transport in plant root systems in situ.* Proc. Natl. Acad. Sci. USA 83: 87–89.
17. Braunlin, W.H., T. Drakenberg and S. Forsén. 1985. *Metal ion NMR: Application to biological systems.* Current topics in bioenergetics 14: 97–147. Academic Press, New York.
18. Brown, F.F. and I.D. Campbell. 1980. *NMR studies of red blood cells.* Phil. Trans. Roy. Soc. Lond. B. 289–406.
19. Brown, E.M., T.F. Kumosinski and P.E. Pfeffer. 1983. *Relative exposure and dynamics of lysine residues in globular proteins.* Biophys. J. 41: 110a–112.
20. Coker, G.T. and J. Schaefer. 1985. *^{15}N and ^{13}C NMR determination of allantoin metabolism in developing soybean cotyledons.* 1985. Plant. Physiol. 77: 129–135.
21. Colnago, L.A. and P.R. Seidle. 1983. *Application of carbon-13 nuclear magnetic resonance to the germination of soybean seeds in vivo.* J. Agric. Food Chem. 31: 459–461.
22. Cross, T.A., M.H. Frey and S.J. Opella. 1983. *^{15}N spin exchange in a protein.* J. Amer. Chem. Soc. 105: 7471–7473.
23. Cure, W.J. In press. *Nuclear magnetic resonance imaging of plant roots in site.* Plant Physiol.
24. Doddrell, D.M., D.T. Pegg and M.R. Bendall. 1982. *Distortionless enhancement of NMR signals by polarization transfer.* J. Magn. Res. 48: 323–327.
25. Efloson, R.M., J.A. Ripmeester, N. Cyr, L.P. Milligan and G. Mathison. 1984. *Nutritional evaluation of forages by high-resolution solid state ^{13}C NMR.* Can. J. Amim. Sci. 64: 93–102.
26. Fyfe, C.A. 1983. *Solid State NMR for Chemists.* C.F.C. Press, Guelph, Ontario, Canada.
27. Fyfe, C.A., R.L. Dudley, P.J. Stephenson, Y. Deslandes, G.K. Hamer and R.H. Marchesscault. 1983. *Application of high-resolution solid state NMR with cross-polarization/magic angle spinning (CP/MAS) techniques to cellulose chemistry.* Rev. Macromol. Chem. Phys. 23: 187–216.
28. Gadian, D.G. 1982. *NMR and Living Systems.* Oxford: Clarendon Press.
29. Ganesh, K.N. 1984. *Biological applications of high resolution solid state NMR spectroscopy.* Appl. Spect. Rev. 20: 107–157.
30. Gannella, N.C., T. Birdseye, M. Nee and J.D. Roberts. 1982. *^{15}N NMR study of a mixture of uniformly labeled tRNAs.* Proc. Natl. Acad. Sci. USA 79: 4834–4837.

31. Gassner, G. and J.A.B. Lohman, In press. *Proton NMR imaging and spectral analysis of locust embryogenesis*. Science.
32. Gerasimowicz, W.V. and D.M. Byler. 1985. *Carbon-13 CPMAS NMR and FTIR spectroscopic studies of humic acids*. Soil Science 139: 270–277.
33. Gerasimowicz, W.V., K.B. Hicks and P.E. Pfeffer. 1984. *Evidence for the existence of associated lignin – carbohydrate polymers as revealed by carbon-13 CPMAS solid state NMR spectroscopy*. Macromolecules 17: 2597–2603.
34. Gerasimowicz, W.V., S.-I. Tu and P.E. Pfeffer. 1986. *Energy facilitated Na^+ uptake in excised corn roots via ^{31}P and ^{23}Na NMR*. Plant Physiol. (in press).
35. Gordon, R.E., P.E. Hanley and D. Shaw. 1982. *Topical magnetic resonance*. Progress in nuclear magnetic resonance spectroscopy 15: 1–47.
36. Gorenstein, D.G., ed. 1984. *Phosphorus-31 NMR*. Academic Press Inc., New York.
37. Gupta, R.K. and P. Gupta. 1982. *Direct observation of resolved resonances from intra- and extracellular sodium-23 ions in NMR studies of intact cells and tissues using dysposium (III) tripolyphosphate as a paramagnetic shift reagent*. J. Magn. Res. 47: 344–350.
38. Hatcher, P.A., I.A. Breger, L.W. Dennis and G.E. Maciel. 1983. *Solid State ^{13}C-NMR of sedimentary humic substances: New revelations on their chemical composition*. Aquatic and terrestrial humic materials. Ann Arbor Science Publishers Ann Arbor, Michigan.
39. Hatfield, G.R., G.E. Maciel, in press. *High resolution NMR experiments for solid protein samples*. Methods for Protein analysis published by Am. Oil Chem Soc.
40. Haw, J.F. and G.E. Maciel. 1983. *Carbon-13 nuclear magnetic resonance spectroscopy of soil seeds with cross polarization and magic-angle spinning*. Anal. Chem 55: 1262–1267.
41. Haw, J.F. and G.E. Maciel. 1984. *Carbon-13 nuclear magnetic resonance spectrometric study of wood and wood pulping with cross polarization and magic angle spinning*. Anal. Chem. 56: 1323–1329.
42. Hedges, J.I., G.L. Cowie, J.R. Estel, J. Barbour and P. Hatcher. 1985. *Degradation of carbohydrates and lignins in buried woods*. Geochem. Cosmochem. Acta. 49: 701–711.
43. Heyser, J.W. and L.O. Sillerud. 1984. *^{13}C NMR monitoring of in vivo polysaccharide biosynthesis by cell suspensions and protoplasts of Proso millet (Panicum miliaceaum)*. Plant Physiol. 75 suppl., S595.
44. Himmelsbach, D.S., F.E. Barton II and W.R. Windham. 1983. *Comparison of carbohydrate, lignin, and protein ratios between grass species by cross polarization – magic angle spinning carbon-13 nuclear magnetic resonance*. J. Agric. Food Chem. 31: 401–404.
45. Den Hollander, J.A. and R.G. Shulman. 1983. *^{13}C NMR studies of in vivo kinetic rates of metabolic processes*. Tetrahedron, 39: 3529–3538.
46. Holtum, J.A., R. Summons, C.A. Roeske, H.N. Cinubs and M. O'Leary. 1984. *Oxygen-18 incorporation into malic acid during nocturnal carbon dioxide fixation in crassulacean acid metabolism plants. A new approach to estimating in vivo carbonic anhydrase activity*. J. Biol Chem. 259: 6870–6881.
47. Irwin, P.L., P.E. Pfeffer, W.V. Gerasinowicz, R. Pressey and C.E. Sams. 1984. *Ripening-related perturbations in apple cell wall nuclear spin dynamics*. Phytochem. 23: 2239–2242.
48. Jacob, G.S., E.O. Schaefer and R.A. McKay. 1980. *Magic-angle N-15 NMR study of nitrate metabolism of Neurospora Crassa*. Biochem. Biophy. Res. Comm. 97: 1176–1182.
49. Jacob, G.S., J. Schaefer, E.O. Stejskal and R.A. McKay. 1985. *Solid-State NMR determination of glyphosate metabolism in a pseudomonas sp.* J. Biol. Chem. 260: 5899–5905.
50. Jacobs, R.E. and E. Oldfield. 1981. *NMR of membranes*. Prog. in NMR spectroscopy 14: 113–136. Pergamon Press.
51. James, T.L., J.L. James and A. Lapidot. 1981. *Structure and dynamic information about double-stranded DNA from nitrogen-15 NMR*. J. Am. Chem. Soc. 103: 6748–6752.
52. Jardetzky, O. and G.C.K. Roberts. 1981. *NMR in Molecular Biology*. Academic Press Inc., New York.
53. Kallas, T. and F.W. Dahlquist, 1981. *Phosphorus-31 nuclear magnetic resonance analysis of internal pH during photosynthesis in the cyanobacterium Synechoccus*. Biochem. 20: 5900–5907.

54. Kanamori, K., T.L. Legerton, R.L. Weiss and J.D. Roberts. 1982. *Nitrogen-15 spin lattice relaxation times of amino acids in Neuropora crassa as a probe at intracellular environment.* Biochem. 21: 4916–4920.
55. Kanamori, K. and J.D. Roberts. 1983. *^{15}N NMR studies of biological systems.* Acc. Chem. Res. 16: 35–41.
56. Kime, J.J., B.C. Loughman and R.G. Ratcliffe. 1982. *The application of ^{31}P nuclear magnetic resonance to higher plant tissue. 2. Detection of intracellular changes.* J. Exp. Bot. 33: 670–681.
57. Knowles, P.F., D. Marsh and H.W.E. Rattle. 1976. *Magnetic Resonance of Biomolecules.* John Wiley and Sons Inc., New York.
58. Lee, R.B. and R.B. Ratcliffe. 1983a. *Phosphorus nutrition and the introcellular distribution of inorganic phosphate in pea root tips. A quantitative study using NMR.* J. Exp. Bot. 34: 1224–1244.
59. Lee, R.B. and R.G. Ratcliffe. 1983. *Development of an aeration system for use in plant tissue NMR experiments.* J. Exp. Bot. 34: 1213–1221.
60. Legerton, T.L., K. Kanamori, R.L. Weiss and J.D. Roberts. 1983. *Measurement of cytoplasmic and vacuolar pH in Neurospora using nitrogen-15 nuclear magnetic resonance spectroscopy.* Biochem. 22: 899–903.
61. Loughman, B.C. and R.G. Ratcliffe. 1984. *Nuclear magnetic resonance and the study of plants.* Advances in plant nutrition Volume I. Praeger. Publishers New York.
62. Morishima, I. and T. Inubuski. 1978. *^{15}N nuclear magnetic resonance studies of iron-bound $C^{15}N$ in ferric low-spin cyanide complexes of various porphyrin derivatives and various hemoproteins.* J. Am. Chem. Soc. 100: 3568–3572.
63. Martin, F. 1985. *Monitoring plant metabolism by ^{13}C, ^{15}N and ^{14}N nuclear magnetic resonance spectroscopy. A review of the applications to algae, fungi and higher plants.* Physiologie Végétal 4: 463–490.
64. McCain, S. and M.S. Davies. 1984. *Effects of pretreatment with phosphate in natural populations of Agrostis Capillaris L. II Interactions with aluminum on the acid phosphatase activity and potassium leakage of intact roots.* New Phytol. 96: 589–599.
65. Mantsch, H.H., H. Saitö and I.C.P. Smith. 1976. *Deuterium magnetic resonance applications in chemistry, physics and biology.* Progress in NMR spectroscopy 10: 3–276.
66. Morris, G.A. and L.D. Hall. 1981. *Experimental chemical shift correlation maps from heteronuclear two dimensional NMR spectroscopy. 1. Carbon-13 and proton chemical shifts of raffinose and its submits.* J. Am. Chem. Soc. 103: 4703–4711.
67. O'Donnell, D.J., J.J.H. Ackerman and G.E. Maciel. 1981. *Comparative study of whole seed protein and starch content via cross polarization magic angle spinning carbon-13 nuclear magnetic resonance.* J. Agric. Food Chem. 29: 514–518.
68. Patt, S.L. and J.N. Shoolery. 1982. *Attached proton test for C-13 NMR.* J. Magn. Res. 46: 535–539.
69. Pfeffer, P.E. 1984. *High resolution solid state ^{13}C NMR and its application in carbohydrate chemistry.* J. Carb. Chem. 3: 613–639.
70. Pfeffer, P.E. 1985. *Investigations of polysaccharides in the solid state by ^{13}C CP-MAS NMR spectroscopy.* In new developments in industrial polysaccharides. Gordon and Breach Science publishers New York, NY.
71. Pfeffer, P.E. and R. Binkely. *Special topics issue on 2-D and CP-MAS NMR spectroscopy.* 1984. J. Carb. Chem. 3: 493–593.
72. Pfeffer, P.E., L.W. Doner, P.D. Hoaland and G.G. McDonald. 1981. *Molecular interactions with dietary fiber components. Investigation of the possible association of pectin and bile acids.* J. Agric. Food Chem. 29: 455–461.
73. Pfeffer, P.E., S.-I. Tu, W.V. Gerasimowicz and R.T. Boswell. 1986a. *In vivo ^{31}P NMR studies of simulated environmental conditions affecting metal ion migration and toxicity in corn root tissue.* 27th Expth. NMR Conference. Abstract. Balt. MD April 12–17.
74. Pfeffer, P.E., S.-I., Tu, W.V. Gerasimowicz and J.R. Cavannaugh. 1986. *In vivo ^{31}P NMR studies of corn root tissue and its uptake of toxic metals.* Plant Physiol., 80: 77–84.
75. Pfeffer, P.E., W.V. Gerasimowiez and E.G. Piotrowski. 1984. *Effect of paramagnetic iron on*

quantitation in carbon-13 cross polarization magic angle spinning nuclear magnetic resonance spectroscopy of heterogeneous environmental matrices. Anal. Chem. 56: 734–741.

76. Pfeffer, P.E., K.M. Valentine and F.W. Parrish. 1979. *Deuterium induced differential isotope shift ^{13}C NMR. I. Resonance reassignments of mono and disaccharides.* J. Amer. Chem. Soc. 101: 1265–1274.
77. Piotrowski, E.G., K.M. Valentine and P.E. Pfeffer. 1984. *Solid state ^{13}C, cross-polarization, "magic angle" spinning NMR spectroscopy studies of sewage sludge.* Soil Science 137: 194–203.
78. Prestegard, J.H., T.A.W. Koerner Jr., P.C. Demon and R.K. Yu. 1982. *Complete Analysis of Oligosaccharide Primary structure using two-Dimensional high-field proton NMR.* J.A. Chem. Soc. 104: 4993–4995.
79. Preston, C.M., R.L. Dudley, C.A. Fyfe and S.P. Mathur. 1984. *Effects of variation in contact time and copper content in a ^{13}C CPMAS NMR study of four organic soils.* Geoderma 33: 245–53.
80. Rébeillé, F., R. Bligny, J.B. Martin and R. Douce. 1985. *Effect of sucrose starvation on sycamore (acer pseudoplatanus) cell carbohydrate and Pi status.* Biochem. J. 226: 679–684.
81. Richards, R.E. and N.A. Thomas. 1974. *A ^{14}N NMR study of amino acids, peptides and other biologically interesting molecules.* T. Chem. Soc. Perkin Trans. 2: 368–374.
82. Roberts, J.K.M. 1984. *Study of plant metabolism in vivo using NMR spectroscopy.* Anm. Rev. Plant Physiol. 35: 375–386.
83. Roberts, J.K.M., F.H. Andrade and I.C. Anderson. 1985. *Further evidence that cytoplasmic acidosis is a determinant of flooding intolerance in plants.* Plant Physiol. 77: 492–494.
84. Roberts, J.K.M., J. Callas, D. Wemmer, V. Walbot and O. Jardetzky. 1984. *The mechanism of cytoplasmic pH regulation in hypoxic maize root tips and its role in survival and hypoxia.* Proc. Natl. Acad. Sci., 81: 3379–3383.
85. Roberts, J.K.M. and O. Jardetzky. 1981. *Monitoring of Cellular metabolism by NMR.* Biochem. Biophys. Acta 639: 53–76.
86. Roberts, J.K.M., C.S. Linker, A.G. Benoit, O. Jardetzky and R.H. Nieman. 1984a. *Salt stimulation of phosphate uptake in maize root tips studied by ^{31}P nuclear magnetic resonance.* Plant Physiol. 75: 947–950.
87. Roberts, J.K.M., D. Wemmer and O. Jardetzky. 1984b. *Measurement of mitochondrial ATP ase activity in maize root tips by saturation transfer ^{31}P nuclear magnetic resonance.* Plant Physiol. 74: 632–639.
88. Rutar, V. 1982. *A new possibility for nondestructive protein content determinations in viable seeds.* Appl. Spectroscopy 36: 259–260.
89. Rutar, V., M. Burger, R. Blinc and L. Ehrenberg. 1977. *^{13}C NMR determination of the oil composition in individual plant seeds.* J. Mag. Reson. 27: 83–90.
90. Rutar, V. and R. Blinc. 1980. *Nondestructive determination of protein content of viable seeds by proton enhanced ^{13}C NMR.* 2. Naturforsch 35c: 12–15.
91. Schaefer, J. 1985. *Regeneration in alfalfa tissue culture.* Plant Physiol. 79: 584–589.
92. Schaefer, J., M.D. Sefaik, E.O. Stejekal, R.A. McKay and P.L. Hall. 1981. *Characterization of the catabolic transformation of lignin in culture using magic-angle carbon-13 nuclear magnetic resonance.* Macromolecules 14: 557–559.
93. Schaefer, J., T.A. Skokut, E.O. Stejskal, R.A. McKay and J.E. Varner. 1981. *Estimation of protein turnover in soybean leaves using magic angle double cross-polarization nitrogen 15 magnetic resonance.* J. Biol. Chem 256: 11574–11579.
94. Schaefer, J., T.A. Skokut, E.O. Stejskal, R.A. McKay and J.E. Varner. 1981a. *Asparagine amide metabolism in developing cotyledons of soy beans.* Proc. Natl. Acad. Sci. USA 78: 5978–5982.
95. Schaefer, J. and E.O. Stejskal. 1974. *Carbon-13 nuclear magnetic resonance measurement of oil composition in single viable soy beans.* J. Amer. Oil Chem. Soc. 51: 210–213.
96. Schaefer J. and E.O. Stejskal. 1974. *Determination of oil, starch and protein content of viable intact seeds by carbon-13 nuclear magnetic resonance.* J. Amer. Oil Chem. Soc. 51: 562–563.
97. Schaefer, J. and E.O. Stejskal. 1975. *Carbon-13 nuclear magnetic resonance analysis of intact oilseeds.* J. Am. Oil Chem. Soc. 98: 1031–1032.

98. Schaefer, J., E.O. Stejskal, J.R. Garbow and R.A. McKay. 1984. *Quantitative determination of the concentrations of ^{13}C–^{15}N chemical bonds by double cross-polarization NMR*. J. Mag. Res. 59: 150–156.
99. Schaefer, J., E.O. Stejskal and R.A. McKay. 1979. *Cross polarization NMR of N-15 labeled soy beans*. Biochem. Biophys. Res. Comm. 88: 274–280.
100. Schleich, T.J., A. Willis and G.B. Matson. 1984. *Longitudinal (T_1) relaxation times of phosphorus metabolites in the bovine and rabbit lens*. Exp. Eye Res. 39: 455–468.
101. Shulman, R.G. ed. 1979. *Biological Applications of Magnetic Resonance*. Academic Press Inc., New York.
102. Sillerud, L.O. and J. Heyser. 1984. *Use of ^{23}Na-Nuclear magnetic resonance to follow sodium uptake and efflux in NaCl-adapted and nonadapted millet (panicum miliaceum) suspensions*. Plant Physiol. 75: 269–272.
103. Steyn, P.S. ed. *The biosynthesis of mycotoxins. A study of secondary metabolism*. 1980. Academic Press, New York.
104. Stidham, M.A., D.E. Moreland and J.N. Siedow. 1983. *^{13}C nuclear magnetic resonance studies of Crassulacean Acid Metabolism in intact leaves of Kalanchse tubiflora*. Plant Physiol. 73: 517–520.
105. Stinson, E., S. Osman and P.E. Pfeffer. 1982. *Structure of altertoxin-I, a mycotoxin from Alternaria*. J. Org. Chem. 47: 4110–4113.
106. Stinson, E.E., W.B. Wise, A.J. Jurewicz, R. Moreau and P.E. Pfeffer. 1986. *Aternariol: Biosynthesis from Norlichexamn thone investigated by 2-D^{13}C NMR*. Can. Jour. of Chem. (In press).
107. Storey, K.B., M. Miceli, K.W. Butler, I.C.P. Smith and R. Deslauriers. 1984. *^{31}P-NMR studies of the freeze-tolerant larvae of the gall fly, Eurosta solidaginis*.
108. Waterton, J.C., I.G. Bridges and M.P. Irving. 1983. *Intracellular Compartmentation detected by ^{31}P NMR in intact photosynthetic wheat leaf tissue*. Biochem. Biophys. Acta. 763: 315–320.
109. Wilson, M.A., R.J. Pugmire and D.M. Grant. 1983. *Nuclear Magnetic resonance spectroscopy of soils and related materials. Relaxation of ^{13}C nuclei in cross polarization nuclear magnetic resonance experiments*. Org. Geochem. 5: 121–129.
110. Wooten J.B. and J.S. Cohen. 1979. *Protein mobility and self-association by deuterium nuclear magnetic resonance*. Biochem. 18: 4188–4191.
111. Wüthrich, K. 1984. *Three dimensional structure of non-crystalline polypeptides by nuclear magnetic resonance*. Biomed. Res. 5: 151–160.
112. Yannoni, C. 1982. *High resolution NMR in solids: The CPMAS experiment*. Acc. Chem. Res. 15: 201–208.

Applications of electron paramagnetic resonance spectroscopy to the study of bio-membranes

JOHN J. WINDLE
U.S. Department of Agriculture, Agricultural Research Service, Berkeley, CA 94710, U.S.A.

Abstract Conventional and saturation transfer electron paramagnetic resonance spectroscopy (EPR and ST-EPR) in combination with EPR spin labeling are the EPR techniques most widely used in biological research. They are especially well suited for the study of molecular dynamics. In this review some applications of EPR techniques to the study of molecular motions in bio-membranes wil be discussed as examples. Basic EPR concepts and instrumentation will be reviewed and EPR and ST-EPR methodologies compared. Spin labels designed as membrane probes will be described and their use in investigations of membrane lipid-lipid and lipid-protein interactions discussed and illustrated with some examples from the current literature. A particular advantage of EPR is that it can be used with living systems. Spin-lable spectra of the plasma membrane of living, whole-rye cells and of rye plasma membrane-lipid vesicles were obtained as a function of temperature. In each case, the spectra were found to be composed of two components: a fluid lipid fraction and an immobilized lipid fraction. Temperature profiles for each specimen are presented and discussed in terms of lipid-lipid and lipid-protein interactions. Problems of spin label distribution and decay in living cells are also discussed.

Introduction

Electron paramagnetic resonance (EPR) was first demonstrated in the USSR by E. Zavoisky in 1945 (Zavoisky, 1945). The first EPR spectrometer was developed by Beringer and Castle in 1950 (Beringer and Castle, 1950). The first EPR application to biology was made by Commoner, Townsend and Pake in 1954 (Commoner, Townsend and Pake, 1954), and the first EPR spectrometer in the Agricultural Research Service, USDA was designed and constructed at the Western Regional Research Laboratory and became operational in 1957 (Windle and Thurston, 1957). The particular features of EPR which have made it an especially useful tool for biological studies are: it is a non-destructive technique, it does not require purification of the samples in order to make a measurement, it exerts negligible influence on the processes being studied and it can be applied to integrated systems such as living cells. Several thousand biological EPR papers have been written since the first biological application. With the advent of spin labeling (Ohnishi and McConnell, 1965), biological applications of EPR have greatly increased and are expected to continue at a high level well into the 21st century.

EPR requires the presence of a paramagnetic center in the spectrum before a spectrum can be obtained. A paramagnetic center is an atom or molecule containing one or more unpaired electrons. Before the development of spin labeling, only specimens containing naturally occurring paramagnetic centers were available for study; the one exception being specimens in which paramagnetic centers were formed by irradiation. Paramagnetic centers naturally encountered in biological samples are: (1) paramagnetic ions such as Fe^{3+}, Cu^{2+}, Mn^{2+}, Mo^{3+}, Ni^{2+} and Co^{2+} and (2) molecules with one or two unpaired electrons, viz., free radicals or biradicals and triplet states, respectively. Paramagnetic ions occur biologically as chelates or complexes with biomolecules, proteins and enzymes. A large and diverse literature has developed on the EPR of metalloproteins and enzymes (Beinert, H. 1985). Even more extensive is the EPR literature on free radicals which includes free radicals as intermediates in enzyme mediated electron transport and in oxidation and reduction reactions (Edmondson, 1985), in photosynthesis (also triplet states) (Evans, Ford and Atkinson, 1985), and in radiation damage to bio-molecules (Copeland, 1972). No attempt will be made in this paper to discuss these applications. A good discussion of the biological applications of EPR, in addition to the above references, can be found in the text edited by Swartz, Bolton and Borg (1972).

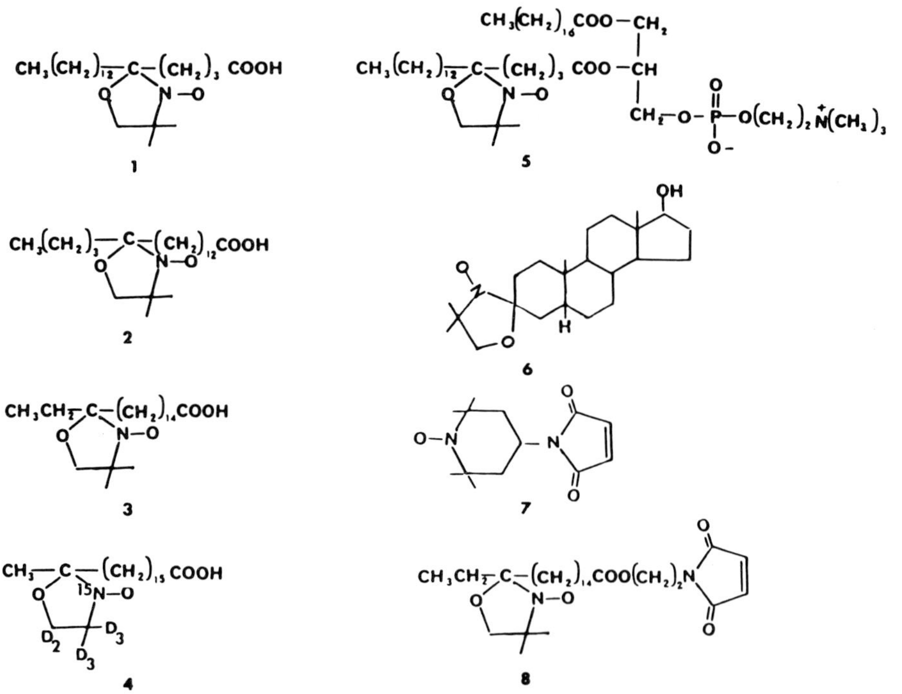

Figure 1. Spin labels referred to in this paper. (1) 5NS, (2) 12NS, (3) 16NS, (4) PD^{15}N-17NS, (5) 5PCSL, (6) 3DA, (7) M-TEMPO, and (8) M-16NS.

Spin labels are stable free radicals which can be incorporated into a biological specimen to provide a paramagnetic center and serve as a probe of the center's immediate environment. Spin labels have greatly increased the application of EPR to biology. The investigator is no longer limited only to specimens containing naturally occurring paramagnetic centers. He now can study any system in which a spin label can be incorporated, and he can select from a large number of different spin labels tailored for specific applications. See examples in Fig. 1. In addition new spin labels are being developed every year. These factors have resulted in spin labeling becoming the most widely applied EPR method in biological research. In this review the basic EPR concepts and instrumentation will be discussed, conventional EPR and saturation transfer EPR compared and applications of spin labeling to bio-membranes described. Illustrations of the techniques will be drawn from the spin-label literature on bio-membranes and from our own work on plant cell plasma membranes.

Basic principles of EPR

Resonance condition

The nitroxide free radical is probably the most commonly used paramagnetic center for spin labels. An example of a nitroxide free radical is di-t-butyl nitroxide, DTBN*, illustrated in Fig. 2. The free radical is stable because it has no α-hydrogens and is protected by the *t*-butyl groups. It is referred to as a protected radical. The paramagnetism is derived from the unpaired electron which is localized in the nitrogen $2p\pi$ orbital. When placed in an external magnetic field, H, the magnetic moment or spin of the electron can assume only two energies: $+1/2g\beta H$ and $-1/2g\beta H$, where β is the Bohr magneton and g is the spectroscopic splitting factor. This is shown in Fig. 3 where the energy difference between the two levels increases linearly with increasing magnetic field. Transitions between the two levels can be induced by irradiation at an energy $h\nu = g\beta H$. This is the basic equation for magnetic resonance. To detect absorption of energy by the spin system there must be a difference in the spin populations of the two energy levels. The lower energy level

*Abbreviations: 5NS, 5-DOXYL-stearic acid; 14NS, 14-DOXYL-stearic acid; 16NS, 16-DOXYL-stearic acid; PD^{15}N-17NS, 17-(perdeutero, ^{15}N)-DOXYL-stearic acid; 5PCSL, 1-steroyl,2-(5-DOXYL-steroyl)-phosphatidylcholine; 3DA, 3-DOXYL-17β-hydroxy-5-α-androstane; M-TEMPO, maleimide-TEMPO; M-16NS, maleimide-16NS; TEMPO,2,2,6,6-tetramethylpiperidinyl-1-oxyl; TEMPOL, 4-hydroxy-2,2,6,6-tetramethyl piperidinyl-1-oxyl; TEMPAMINE, 4-amino-2,2,6,6-tetramethylpiperidinyl-1-oxyl; TEMPONE, 4-oxo-2,2,6,6-tetramethyl piperidinyl-1-oxyl; TEMPYO, 2,2,5,5-tetramethyl, 3-carbamido pyrroline-1-oxyl; DTBN, di-t-butyl nitroxide; PDDTBN, perdeutero-di-t-butyl nitroxide; DPPC, dipalmitoylphosphatidylcholine; DMPE, dimyristoylphosphatidylethanolamine.

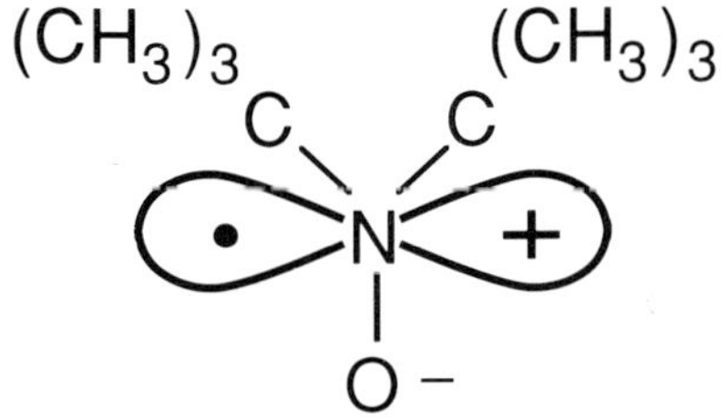

Figure 2. Di-t-butyl nitroxide. The molecular *x*-axis lies along the N–O bond, the *z*-axis along the *p*-orbital containing the unpaired electron. The absence of α-hydrogens and the presence of the *t*-butyl groups contribute to the stability of the radical.

will have a slightly larger spin population, determined by the Boltzmann factor. At room temperature the excess population of the ground state is about 0.07%. If the intensity of the irradiation energy, $h\nu$, is increased until the populations are equalized the spin system is said to be saturated. The excited spins return to the ground state mainly by two mechanisms: (1) spin-lattice, T_1, relaxation, in which the energy from the excited state is transferred to vibrational modes of the same frequency in the lattice. This mechanism is only important at temperatures below 4°K. At high temperatures a Raman process dominates; a photon is absorbed and reemitted at a different energy as the spin transition occurs; (2) spin-spin, T_2, relaxation, which occurs when the unpaired electron interacts with other magnetic dipoles nearby thus spreading the local energy over the entire system and enhancing the spin-lattice transition (Alger, 1968).

Hyperfine interaction and g-value

An important feature of the nitroxide radical is the hyperfine interaction between the unpaired electron and the nitrogen nucleus. The un-

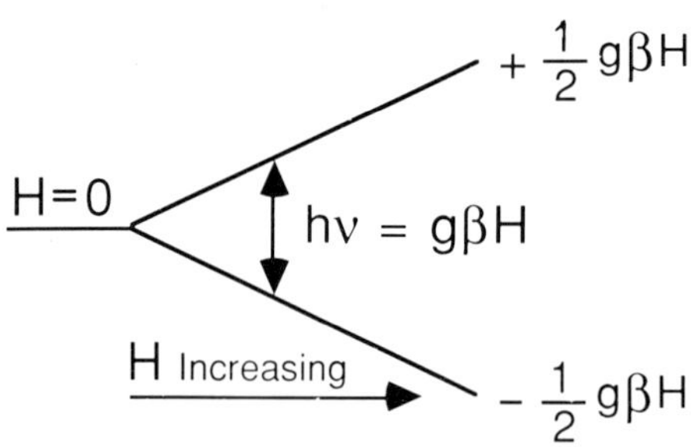

Figure 3. Energy level diagram for an electron (spin 1/2 particle) in an applied magnetic field, *H*. In the absence of the field, the two energy levels are degenerate. The degeneracy is lifted when the field is applied. Transitions between the levels are induced by irradiation at a frequency, ν, such that $h\nu = g\beta H$.

paired electron in the nitrogen p-orbital experiences a dipole-dipole interaction with the nitrogen nuclear magnetic moment, $I = 1$. The hyperfine interaction splits the electron resonance absorption into three equally intense, equally spaced components. The spacing between the components (hyperfine splitting) depends upon the magnetic field direction and is therefore said to be anisotropic. This is illustrated in Fig. 4 for the magnetic field H oriented along the three axes of the molecule. The z-axis lies along the p-orbital containing the unpaired electron and shows the greatest splitting. The x-axis lies along the N–O bond. The splittings along the x and y axes are smaller and almost equal. To a first approximation then, the nitroxide can be treated as a molecule exhibiting axial symmetry. The hyperfine splittings become $A_{xx} \cong A_{yy} = A_{\perp}$ and $A_{zz} = A_{\parallel}$.

The g-values for almost all free radicals fall within a few tenths of one percent of each other, thus all free radical spectra will occur essentially at the same magnetic field. The g-value for the nitroxide radical also is

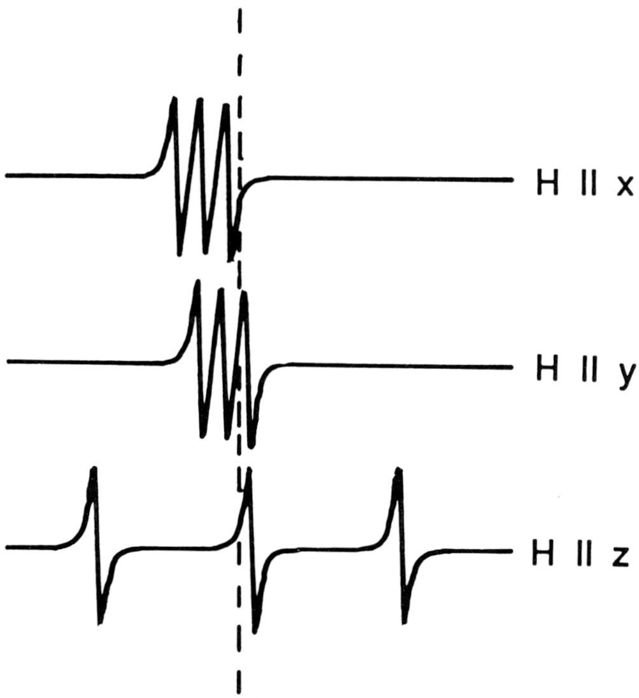

Figure 4. Spectra for di-t-butyl nitroxide at the principle orientations with respect to the applied magnetic field, H. The molecular axes, x, y and z are defined in Fig. 2. (Modified from Griffith, O.H. and P.C. Jost, 1976).

anisotropic but with a maximum difference of about 0.3%. This g-value anisotropy is the reason why the three spectra shown in Fig. 4 are not all centered along the same line. The orientation of the molecule can be determined from the g-value and hyperfine splitting anisotropies.

Molecular motion

The spin label molecules undergo rotational diffusion. If the rate of molecular motion of the nitroxide group is greater than the anisotropies expressed in frequency units (30–70 KHz) then a degree of motional averaging results. At a very fast rate, all anisotropic effects are averaged to zero and only an isotropic spectrum results. For very slow motion a "powder" spectrum results from the summation of spectra from all possible fixed orientations of the nitroxide in the magnetic field. For intermediate rates, a series of spectra characteristic of the amount of motion are observed. Such a series is shown in Fig. 5. The spectra in Fig. 5 are simulated lineshapes for randomly oriented samples of lipid spin labels with different values of the order parameter (Griffith and Jost, 1976). In the lower right hand corner a completely isotropic spectrum is shown indicating fast motion ($> 10^{10}\,s^{-1}$). As the motional rate decreases anisotropic effects come into play and characteristic changes in the lineshape are observed. At the upper left hand corner a completely immobilized "powder" spectrum is shown. Immobilization of the conventional EPR spectrum will occur at rates less than $10^7\,s^{-1}$. Thus the conventional EPR nitroxide radical spectrum is sensitive to rates of molecular motion between 10^{11} to $10^7\,s^{-1}$. Rates between 10^7 to $10^3\,s^{-1}$ are accessible with saturation transfer EPR.

Rotational correlation time

Another spectral parameter reflecting rotational diffusion is the spin-spin relaxation time, T_2 or the spectra linewidth, T_2^{-1}. From changes of T_2^{-1}, rotational correlation times τ_c, in the fast motion range, $10^{-11} \leq \tau_c \lesssim 3 \times 10^{-9}$s, can be determined. T_2^{-1} refers to the intrinsic homogenous linewidth in the absence of any contributions from inhomogenous broadening. For most nitroxide radicals used as spin labels, the linewidth of the three nitrogen hyperfine lines are inhomogenously broadened by unresolved, nuclear hyperfine interactions resulting in a Gaussian rather than Lorentzian line shape and a linewidth greater than the intrinsic linewidth. The true linewidth can be determined by computer simulation of the envelope linewidth (Bales, 1980) when the hyperfine coupling constants of the contributing nuclei are known. In a recent paper the hyperfine coupling constants for eight small nitroxide spin probes were determined in a polar (water) and non-polar (carbon

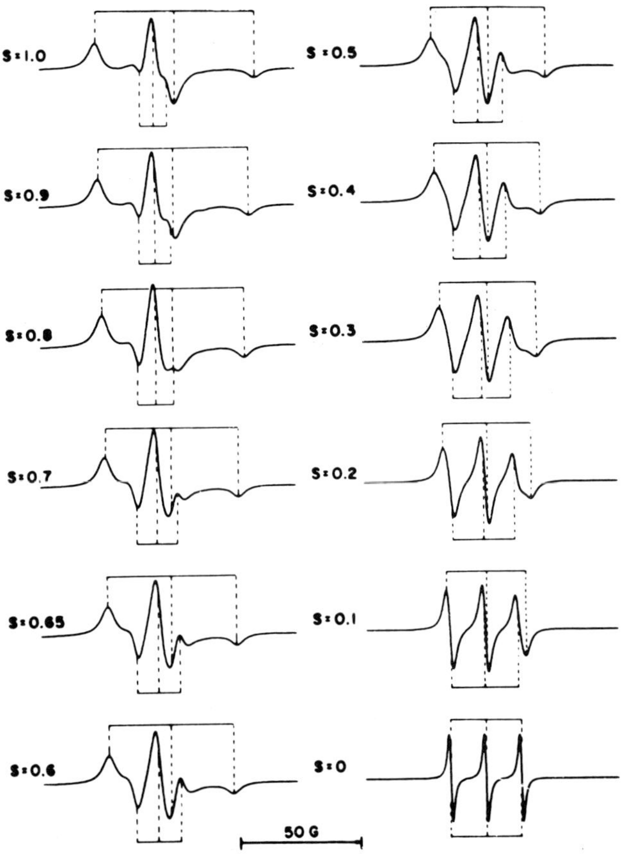

Figure 5. Calculated ESR line shapes for randomly oriented samples as a function of the order parameter S. The dashed lines indicate the precise positions of the three ESR lines defined by $A_{\parallel}$ and $g_{\parallel}$ (large splittings) and $A_{\perp}$ and $g_{\perp}$ (small splittings). (Modified from Griffith, O.H. and P.C. Jost 1976).

tetrachloride) solvent (Windle, 1981). The hyperfine splittings are known to be dependent upon the polarity of the medium. Hyperfine splittings at intermediate polarities could be obtained by interpolation. The results are summarized in Table 1.

When the intrinsic linewidths are unavailable, the effects of inhomogeneous broadening can be lessened by perdeuteration of the nitroxide radical, thereby narrowing the linewidth and reducing the difference between the inhomogeneously broadened line and the intrinsic linewidth. When the inhomogenously broadened line is used, an effective correlation time can be estimated which is satisfactory for comparative purposes and for qualitative analysis. For isotropic motion the correlation

Table 1. Hyperfine coupling constants for nitroxide spin probes in water and carbon tetrachloride at 22°C

Probe	Hyperfine coupling constants (G) ± 1.0 mG							
	a_z	a_N	a_{CH_3} (eq)	a_{CH_3} (ax)	a_{CH_2} (eq)	a_{CH_2} (ax)	a_H (4)	a_N (4)
				H_2O				
PDDTBN[a]	36.9	17.20	0.010	–	–	–	–	–
DTBN[a]	36.9	17.21	0.060	–	–	–	–	–
TEMPO[b]	38.0	17.32	0.175	0.175	0.400	0.400	0.160	–
TEMPONE[c]	36.6	16.13	0.060	0.060	0	0	–	–
PDTEMPONE[c]	35.8	16.07	0.016	0.016	0	0	–	–
TEMPOL	36.7	17.13	0.440	0.060	0.450	0.360	0.150	–
TEMPYO[d]	35.7	16.07	0.200	0.200	–	–	0.350	–
TEMPAMINE	36.0	16.89	0.436	0.038	0.450	0.315	0.090	0.120
				CCl_4				
PDDTBN[a]	34.1	15.38	0.020	–	–	–	–	–
DTBN[a]	34.1	15.36	0.110	–	–	–	–	–
TEMPO[b]	33.3	15.63	0.230	0.230	0.390	0.390	0.180	–
TEMPONE[c]	32.5	14.53	0.115	0.115	0	0	–	–
PDTEMPONE[c]	33.4	14.51	0.023	0.023	0	0	–	–
TEMPOL	33.7	15.44	0.460	0.030	0.430	0.290	0.070	–
TEMPYO[d]	33.0	14.21	0.240	0.240	–	–	0.440	–
TEMPAMINE	34.1	15.83	0.510	0.040	0.420	0.290	0.040	0.050

[a]All six methyl groups are equivalent.
[b]Chair form
[c]Twist form
[d]Planar

time can be calculated by the equation

$$\tau_c = 6.5 \times 10^{-10} \Delta H_0 \left(\sqrt{\frac{h_0}{h_1}} + \sqrt{\frac{h_0}{h_{-1}}} - 2 \right)$$

where ΔH_0 is the linewidth of the central, $m = 0$, line. $h_{-1} h_0$ and h_{+1} are the peak to peak amplitudes of the low field, central and high field lines, respectively.

The rotational correlation times of small nitroxide molecules can be used to determine the effective viscosities of the aqueous cytoplasm of cells (Marsh, 1985a).

Instrumentation for EPR

EPR and saturation transfer, ST-EPR, are the the most widely used techniques applied to biological problems. Advanced EPR techniques such as electron-electron double resonance (ELDOR), electron-nuclear double resonance (ENDOR), time domain EPR (TD-EPR), saturation recovery EPR (SR-EPR), electron spin echo EPR (ESE-EPR), magnitude saturation transfer EPR (M-ST-EPR) and magnetization hysteresis EPR (MH-EPR) are not used extensively in biological applications. The reasons are: a general lack of understanding of how to extract useful information from the spectra, an attainable signal to noise ratio which is less than that for conventional EPR and ST-EPR and the requirement for instrumentation which is expensive and generally not commercially available. Thus in this paper only conventional EPR and ST-EPR will be discussed.

Basic EPR spectrometer

Almost all EPR applications to biology are carried out with commercially available EPR spectrometers. The major supplies have been Varian Associates, Palo Alto, JEOL, Japan and Bruker, Germany*. Today only Bruker remains in the field. Bruker is sold in the United States under the IBM label. (IBM Instruments, Danbury, CT.). Most of the biological spectra are obtained at X-band, 9.3 GHz. This requires a DC magnetic field of about 3.3 kg. Microwave bridges operating at other than X-band frequences such as 35 GHz also have been used. Higher operating frequencies required proportionally higher magnetic fields. A block diagram of the basic EPR spectrometer is shown in Fig. 6. At 9.3 GHz, the microwaves are conducted in hollow, rectangular, electrically conducting pipes called waveguides. A variable attenuator is placed between the Klystron tube (the microwave source) and the circulator to control the microwave power level to the sample. The circulator is a microwave device which directs the flow of microwaves to the sample and out to the detector. Microwaves flowing into port # 1 are circulated counter clockwise to port # 2 and out to the microwave cavity containing the sample. The cavity is placed between the poles of the DC electromagnet which provides the magnetic field, H. The cavity is a high Q, resonant, reflection device. The Q or quality of the cavity is proportional to the ratio of the energy stored to the energy dissipated per cycle. Microwaves enter the cavity through a hole called the iris, are internally reflected until they are reflected out to the circulator, *via* port # 2 and out port # 3 to the crystal diode detector where they are detected and rectified.

In EPR spectroscopy, the microwave frequency is held constant because the resonant cavity is a fixed-frequency device. The resonance condition is achieved by sweeping the DC magnetic field through the resonance. Superimposed upon the DC field is a 100KHz modulation field *via* the modulation coils. For spin-label spectra, a field sweep of 100 Gauss about the resonance field value and a modulation amplitude of about 1 Gauss are typically employed. The 100 KHz modulation also is

*Mention of a specific product name by the USDA does not constitute an endorsement and does not imply a recommendation over other suitable products.

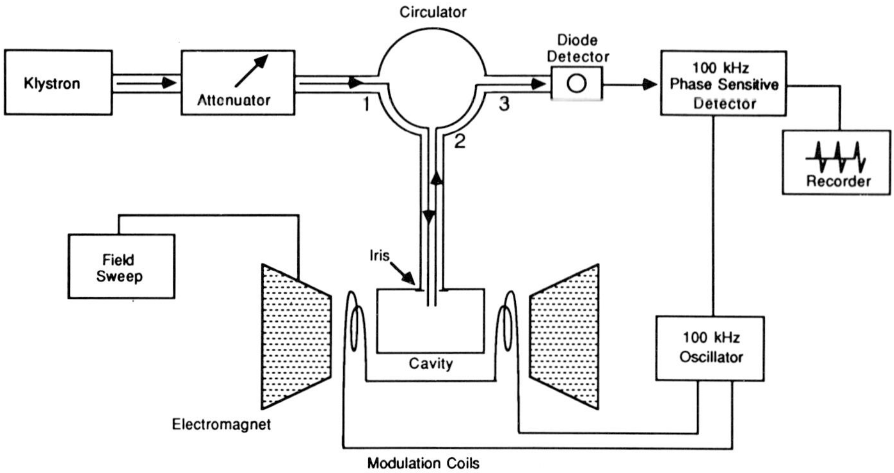

Figure 6. Block diagram of a basic conventional EPR spectrometer.

fed to a 100 KHz phase sensitive detector, PSD, and combined with the rectified microwave signal which is modulated at a 100 KC rate. Only 100 KC signals from the crystal detector which are in phase with the 100 KHz signal fed to the PSD by the oscillator will be detected. As the DC magnetic field is swept through the resonance condition the spectrum will be recorded on the strip chart recorder. If the sample contains paramagnetic centers, microwaves will be absorbed as the unpaired electron spins go from the $-1/2\,g\beta H$ to $+1/2\,g\beta H$ energy level (Fig. 3). The absorption is detected as a change in the magnetic susceptibility, x. x is a complex quantity, $x = x' + ix''$, where x' is the dispersion mode and corresponds to a change in frequency and x'' is the absorption mode and corresponds to a change in Q. By using high frequency modulation and phase sensitive detection, a first derivative presentation of the spectrum is obtained which accounts for the characteristic shape of the EPR signal. Today's EPR spectrometers are all under some degrees of computer control. In the most advanced machines the computer controls almost all the spectrometer operations. The spectra are digitized and stored for future analysis. Frequently, spectrum averaging is used to enhance the signal-to-noise ratio. Other computer manipulations include spectral simulation, addition and subtraction of spectra, spectral integration (double, for first derivativespectra) to obtain areas and spectral smoothing and other signal enhancing modalities. Computer assisted spectroscopy is probably the most important recent advance in all types of spectroscopy.

Saturation transfer EPR

Conventional EPR spectroscopy is sensitive to molecular motions over a range of correlation times of $10^{-11} < \tau_c < 10^{-7}$ s. The need for sensitivity to slow rotational motion ($10^{-7} \leq \tau_c \leq 10^{-3}$ s), such as encountered with proteins in membranes, was a primary motive for the development of the saturation transfer EPR, ST-EPR (Hyde and Thomas, 1973). Saturation transfer EPR experiments employ a conventional EPR spectrometer, the only requirement being the availabilty of more than one modulation frequency. Modern EPR spectrometers such as the Bruker 300 series offer up to 7 modulation frequencies. The two main ST-EPR techniques are: U_1', dispersion, first harmonic detection, 90° out of phase, and V_2', absorption, second harmonic detection, 90° out of phase.

U_1' ST-EPR

Spectra obtained by U_1' technique are largely independent of the modulation amplitude, H_m, of the amplitude of the microwave radiation, H_1 and of T_1, the spin-lattice relaxation time. They show some dependence on T_2 but are determined mainly by the product of the modulation frequency, ω_m and the rotational correlation time τ_c. Motions can be probed by varying the modulation frequency. Maximum sensitivity to motion is in the range of 10^{-5} to 10^{-6} s; 10^{-7} s can be reached but the sensitivity is low. At 10^{-4} s the observed spectra do not agree with the simulated spectra, making interpretation uncertain. The advantages of this technique compared to V_2', are: U_1' signals are 3 to 10 times stronger, U_1' spectra do not saturate as easily so higher microwave power can be used and the U_1' lineshape is less complex so that spectral simulations require less time. Another advantage of the dispersion mode is the suppression of weakly immobilized signals in the presence of strongly immobilized labels. This is predicted theoretically because the signal from the immobilized probes increases with saturation while the saturation decreases with probe mobility, i.e., with increased saturation transfer. The suppression is greatest in the dispersion because immobilized signals increase more rapidly with incident power in this mode. A modulation frequency of 25 KHz was found to further suppress the weakly immobilized component. This is an example of probing motion by varying the modulation frequency (Hyde and Dalton, 1979). The main drawback with the U_1' method is that dispersion spectra obtained with a conventional, reflection-type microwave bridge are quite noisy so special cavities are needed at the high powers required to saturate nitroxide spin labels.

One such cavity which shows great promise is the newly developed Loop-Gap resonator (Froncisz and Hyde, 1982). The advantages of the Loop-Gap resonator compared to the standard resonant cavity are: (1) 80 times less power needed to achieve the same H_1, (2) a greater filling i.e., the volume of sample to cavity is higher and (3) the Loop-Gap resonator utilizes much less material for the same optimum signal (Thomas et al., 1983). The U_1' spectra for the Loop-Gap resonator displays a relative sensitivity (signal/noise/sample volume) of 10^4 times higher than the standard cavity (Hyde and Dalton, 1979). The Loop-Gap resonator is not yet commercially available.

V_2' ST-EPR

V_2' ST-EPR is sensitive to a wider range of rotational correlation times compared to U_1' method. The best sensitivity is in the 10^{-4} to 10^{-5} s range making the two methods complementary. V_2' has a better signal to noise ratio for standard EPR cavities than U_1'. The second harmonic is detected using 50 KHz modulation and 100 KHz phase sensitive detection. The spectra are observed in the normal way with high frequency radiation, but with detection 90° out of phase with respect to the field modulation. In the absence of saturation, the 90° signal should be zero. This fact is used to set the phase angle. Using low power, usually 1 mW, to insure no saturation, the 90° out of phase condition is obtained by adjusting for a null in the conventional signal. All instrument settings are otherwise set to the values to be used to record the V_2' signal. The microwave power is then increased to between 45 to 63 mW to give an $H_1 = 0.25$ G. The modulation amplitude H_m, is set to 5 G. These conditions have been found to be moderately saturating and to give the nitroxide signal its maximum height. Theoretical simulations and experimental standard calibrations have been performed with these spectrometer settings. Using a maleimide derivative of the TEMPO spin-label rigidly bound to hemoglobin in aqueous glycerol solution, Thomas, Dalton and Hyde (1976) established the calibration curves and instrument conditions that have become the standards used in virtually all the V_2' ST-EPR applications. One such calibration curve for the ST-EPR peak height ratio, L''/L, and for the conventional EPR motion parameter $2A_{\|}$ vs. rotational correlation time is shown in Fig. 7.

The rotational correlation time was calculated from the Debye equation for isotropic Brownian motion, $\tau_c = 4\pi\eta R^3/3kT$, where R is the molecular radius and η is the viscosity. To use this calibration the same spectrometer settings must be used and carefully controlled. Particularly crucial are H_1, the amplitude of the microwave field and the setting of the reference phase of the phase sensitive detection. A new approach to

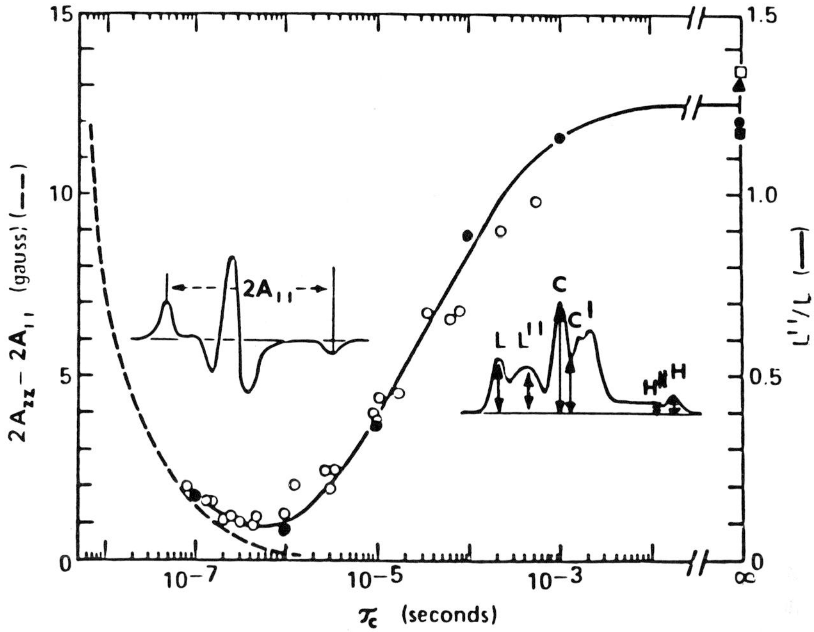

Figure 7. Calibration curve obtained from M-TEMPO-hemoglobin in aqueous glycerol. Diagnostic peak-heights, L, L'', C, C' and H'', H are indicated on the V_2', ST-EPR spectrum. Also shown is a conventional EPR spectrum. The difference between $2A_{zz}$ (hyperfine splitting for totally immobilized spectrum) and $2A_{\parallel}$ and peak-height ratio L''/L are plotted vs. correlation time, τ_c. (Modified from Thomas, D.D. 1985).

the latter problem is the use of a digital phase sensitive detector instead of a conventional analog one (Vistnes et al., 1984). Both the in phase and the phase quadrature signals of the first or second harmonic are recorded simultaneouly eliminating all errors due to drift. The in phase and phase quadrature spectra are separated by exactly a 90° phase difference eliminating the calibration necessary in the analog system. This device is being implemented on the Bruker ESR spectrometers in Europe, but as of this writing is not available yet in the U.S. At the present time practically all ST-EPR applications apply the V_2' mode.

V_2' ST-EPR: applications to membranes

Two examples of ST-EPR applied to bio-membranes will be described. In the first example, the determination of the rotational correlation time for membrane proteins will be discussed. In the second example, the motion of membrane lipids in the gel states is investigated.

Slow molecular motions in membranes are associated with rotational diffusion of membrane proteins and with membrane lipids in the gel state. The correlation times for such motions are outside the sensitivity range of conventional EPR but should be accessible to ST-EPR. To study protein diffusion, a spin label which is attached covalently and which rotates with the protein is required. The covalent binding of the spin label, maleimide-TEMPO to protein SH groups below pH 7 provides such a label. One example is the maleimide-hemoglobin system described above and used to provide calibration curves for diagnostic peak-height ratios: L''/L, C'/C and H''/H (Thomas et l., 1976), see Fig. 7. A biological example is the covalent labeling of Ca^{2+}-ATPase from sarcoplasmic reticulum. This integral membrane protein transports Ca^{2+} across the membrane. The question was whether the rotational motion of the enzyme correlated with enzymatic activity. Hildago and Thomas (1977) developed procedures for labeling the enzyme in the membrane with only a slight effect upon enzyme activity. Thomas and Hildago (1978) then determined that the probe was bound rigidly enough to the protein to characterize its overall rotational motion. An ST-EPR peak-height ratio, L''/L of 0.75 was obtained at 4°C in suspensions of membrane vesicles and in membrane pellets. An effective correlation time of 60 μsec was obtained from the calibration curve (Fig. 7). When conditions were varied, the motion of the probe increased with temperature exhibiting a dependence qualitatively similar to the enzymatic activity of Ca^{2+}-ATPase.

To investigate the phospholipid motion on the gel phase of lipid bilayers, Marsh (1980) incorporated the phospholipid spin label, 5PCSL, into bilayers formed from dipalmitoylphosphatidylcholine (DPPC) and dimyristoylphosphatidyl ethanolamine (DMPE). ST-EPR spectra were obtained above and below the liquid crystal to gel phase transition temperature. DPPC in the gel phase exhibited a pre-transition at 25°C in addition to the main transition at 41°C, whereas no unusual behavior was found for DMPE, which is known not to exhibit a pre-transition. The usual approach of equating the diagnostic peak-height ratios to those from isotropic spectra, i.e., the calibration curves, cannot be automatically applied to lipids because the motion of the spin label is anisotropic and non-Brownian (the spin label and solvent molecules are similar in size). Marsh (1980) found that the peak-height ratio, C'/C, in the central region of the spectrum of spin labeled DPPC was the only ratio to vary significantly. The transition in C'/C at 25°C was interpreted as the onset of rapid rotation of the long axis of the phospholipid molecule. In this case the motion parameters obtained from the C'/C calibration curve are *effective* correlation times because the lipid motion is anisotropic while the calibration spectra correspond to isotropic motion. Johnson, Lee and Fung (1982) investigated slow anisotropic rotational diffusion at both 9 and 35 GHz. At 35 GHz better spectral resolution can be expected. In summary, they found that anisotropic diffusion within membranes can be characterized by peak-height ratios from different parts of the ST-EPR spectrum.

Conventional EPR spectroscopy of membranes

Conventional EPR spectroscopy of spin labeled membranes has played a major role in the study of lipid-lipid and lipid-protein interactions. Membrane fluidity, phase transition and separations and flexibility gradients are examples of lipid-lipid interactions. Many papers have been written on lipid-lipid interactions and a good review can be found in Marsh (1981) and references therein. The topic will be discussed in more detail in a later section describing our own research on plant membranes.

Lipid-protein interactions

Lipid-protein interactions involve the effects of integral membrane proteins on the mobility and order of membrane lipids. The study of lipid-protein interactions by EPR has been one of the most important and active recent applications (Jost and Griffith, 1982), (Marsh, 1985b), (Devaux, 1985) and (Devaux and Seigneuret, 1985). The interactions between integral membrane proteins and membrane lipids appear to be of two kinds: (1) A motionally restricted lipid component associated with the hydrophobic protein surface (boundary layer) and (2) a strongly immobilized lipid component associated with lipids trapped between protein aggregates or protein oligomers. In early work, partially delipidated membranes enriched in cytochrome oxidase and labeled with the fatty acid spin labels 16NS, 5NS and 3DA, gave a two component spectrum consisting of a strongly immobilized spectrum and a bilayer type spectrum (Jost et al., 1973). Originally it was thought that the immobilized component corresponded to boundary layer lipids. However, it is now thought that the immobilized component in such protein-rich systems corresponds to trapped lipids because the low lipid to protein ratio used resulted in protein aggregation (Devaux, 1985). Protein aggregation is known to occur spontaneously in a few natural membranes. The purple patches of *Halobacterium halobium* are an example where bacteriorhodopsin forms aggregates with low lipid to protein ratios (Devaux, 1985), (Chignell and Chignell, 1975). The trapping of lipids in protein oligomers was demonstrated in spin labeling studies of the Ca^{2+}-ATPase by using a maleimide derivative of 16-DOXYL stearic acid (Anderson et al., 1981). Although this label was covalently attached *via* the maleimide to the protein it was not strongly immobilized because of the flexible stearic acid acyl chain. Solubilized Ca^{2+}-ATPase was a

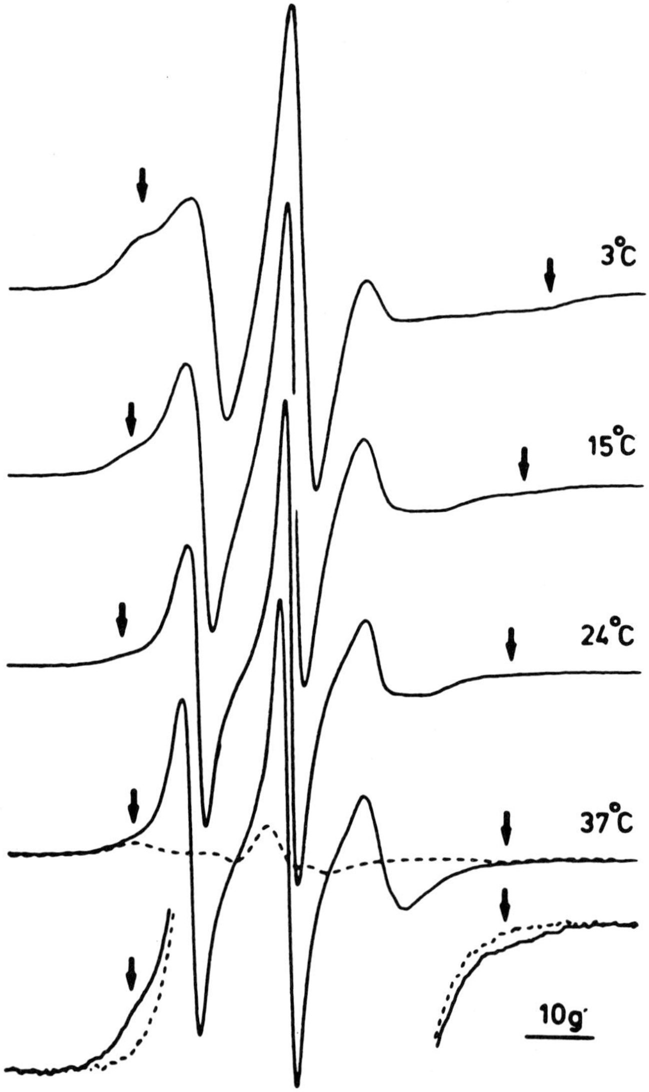

Figure 8. ESR spectra of 14NS in bovine rod outer segment membranes at 3, 15, 24, and 37°C. The outer wings of the 37°C spectrum are expanded 5 times to display the immobilized component, and the dotted lines in the expanded spectrum are from an extracted lipid spectrum which has the same spectral parameters and shape as the fluid membrane spectrum. The immobilized spectrum (dotted line) superimposed upon the 37°C membrane spectrum has an integrated intensity of 37% of the membrane spectrum. Arrows indicate the outermost wings of the immobilized spectra, which are absent in the extracted lipid spectra. (Reprinted with permission from Watts, A. et al., 1979).

mixture of monomer and oligomer units which were separated by ultracentrifugation. The spin labeled oligomers showed a two component spectrum composed of a strongly immobilized and a mobile spectrum but the monomers showed only mobile spectra. The strongly immbolized component is attributed to lipids trapped between the protein oligomers.

In membranes with non-aggregated proteins, the motionally restricted component of the spectrum is associated with the lipid layer in direct contact with the hydrophobic surface of the integral membrane proteins (boundary layer). If the hydrophobic interaction between the boundary layer lipids and the protein surface results in a reduced mobility for these lipids compared to the bulk membrane lipids then a two component spectrum will be observed. The rotational correlation time for boundary lipids must be slow compared to the time scale of the spectroscopic method being used ($\tau_c < 10^{-7}$ s for conventional EPR). Unlike the strongly immobilized spectrum found for lipids trapped by proteins, the boundary layer lipids exhibit a temperature dependent lineshape. An example is the spectrum for 14-DOXYL stearic acid in rod outer segment membranes at different temperatures (Watts et al., 1979 and 1981) Fig. 8. At 0°C (not shown) the hyperfine splitting, $2A_{\|}$ is 59 G whereas at 37°C it has decreased to 54 G. At 37°C the effective correlation time is about 10^{-8} s. According to Davoust and Devaux (1982) the spectrum at any temperature can be accounted for by assuming an exchange between the boundary layer and bulk membrane lipids. If the exchange rate is slow on the EPR time scale ($\sim 10^7 s^{-1}$), a two component spectrum will be observed. The hyperfine splitting and lineshape of the motionally restricted component will change as the rate increases until only a fast exchange spectrum results. Conversely, for slow exchange, as the exchange rate decreases (for example, with decreasing temperature) a more motionally restricted spectrum will result until the rate is $\leq 10^7 s^{-1}$ below which no further change in the conventional EPR spectrum will be observed.*

The boundary layer has been shown to be composed of only the first shell of lipids in contact with the protein surface. The lipids maintain a fixed stoichiometry in proportion to the square root of the protein molecular weight and independent of the lipid to protein ratio (Marsh, 1985). Thus, although the motion of the boundary layer lipids is temperature dependent, their stoichiometry is not.

*This concept of lipid exchange has been invoked to explain the absence of any detectable boundary layer lipids in the NMR spectra of membranes. A rate of $10^7 s^{-1}$ is fast on the NMR time scale.

Plant membrane studies

Relatively few spin label studies of plant membranes have been published (Vigh et al., 1979), (Singh and Miller, 1980) (Singh, 1981), (Boss, 1983), (Caldwell and Haug, 1981), (Briggs, Haug and Scheffer, 1982) and (Wright et al., 1982) in comparison to the extensive spin label literature on non-plant membranes. In most plant studies, stearic acid spin labels have been used to monitor changes in membrane fluidity and to calculate relative correlation times. Arrhenius plots of τ_c or $2A_{\parallel}$ were constructed in search of lipid phase transitions.

EPR of rye plasma membranes

The winter rye studies are part of an investigation into the strategies plants employ in adapting to environmental stress. Much has been written on the response of plants to environmental stress but the mechanisms underlying acclimation are largely unknown (Levitt, 1980). The plasma membrane is believed to play a critical part in the adaptive process because of its role as the interface between the cell and the external environment (Steponkus, 1984). Winter rye (*Sceale cereal* cv. L. Puma) was selected because both tender (non-acclimated) and hardy (cold acclimated) specimens can be obtained from the same species by growing it at different temperatures. Cold-hardy rye cells can survive to −67°C whereas tender rye cannot survive below −7°C. Spin labeling studies using 16NS and PD^{15}N-17NS were undertaken to investigate the alterations in the plasma membrane during cold acclimation. Spectra were obtained at temperatures between −40°C and 40°C on living cells, protoplasts (cells without cell walls), plasma membrane fractions and plasma membrane lipids. It was found that within this temperature range the mobility of the membrane lipids ranged from completely fluid (40°C) to completely immobilized (−40°C).

The stearic acid spin labels orient themselves in the membrane parallel to the phospholipid acyl chains with the carboxyl group near the lipid head group region and the nitroxide reporter group near the bilayer center (Hubbell et al., 1969). An illustration of a membrane lipid bilayer indicating the orientation of 16NS and 5NS is shown in Fig. 9. This figure also illustrates the concepts of chain ordering, chain flexibility and of the membrane flexibility gradient. Near the head group region, band A, the lipid acyl chains are more ordered and less flexible than toward the center of the bilayer, band B where chain flexibility and disorder increase reaching a maximum at the center, band C. Spectral linewidths of reporter groups located near the center will be narrow and the most sensitive to changes in lipid order and mobility. The effect of probe

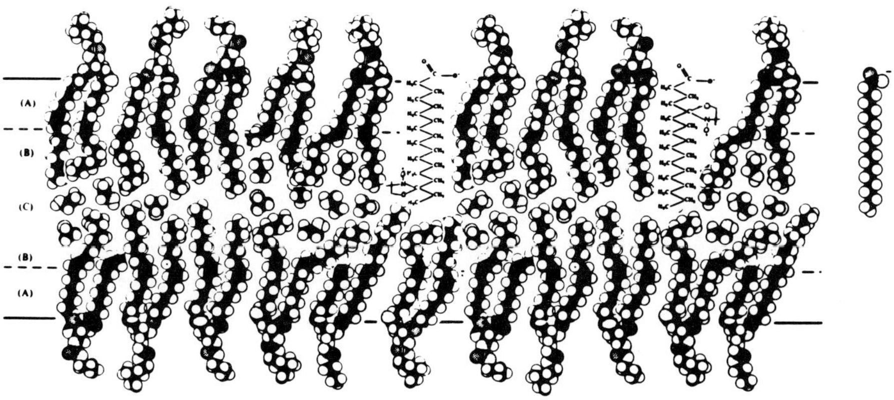

Figure 9. Diagram of phospholipid bilayer based on space-filling atomic models. Near the head-group region, band A, the acyl chains are more ordered and less flexible than toward the bilayer center, band B, where chain flexibility and disorder increase, reaching a maximum at the center, band C. Also shown to scale are the spin-labels, 16NS and 5NS, in their normal orientation in the bilayer. (Modified from Robertson, R.N., 1983).

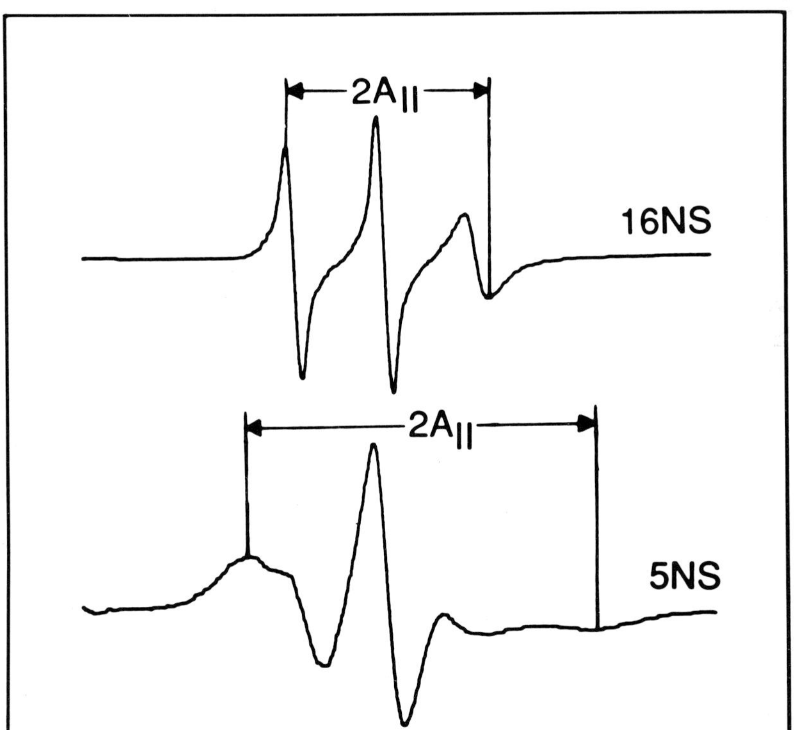

Figure 10. Conventional EPR spectra of the spin-labels, 16NS and 5NS in rye-protoplast plasma membranes. The greater fluidity of the environment around the 16NS-probe is evident from the sharper lines and smaller value of the motion parameter, $2A_{\parallel}$.

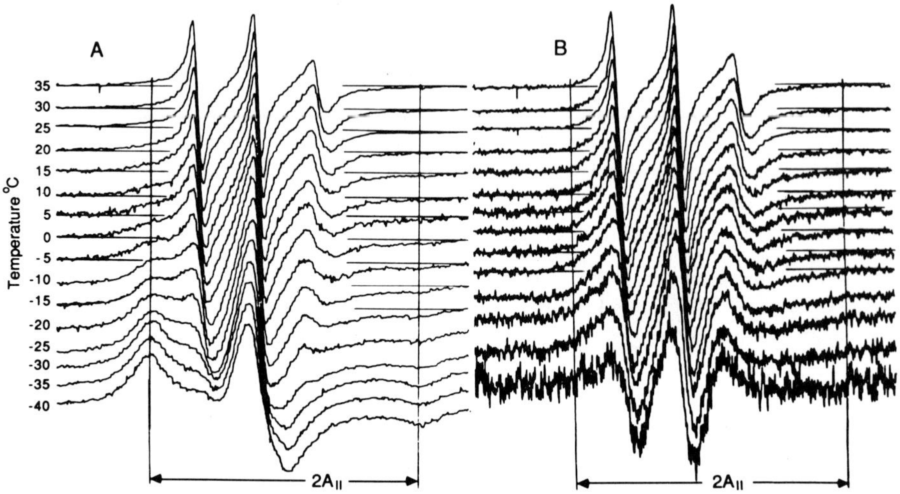

Figure 11. (a) Conventional EPR spectra for the spin-label, 16NS in the plasma membrane of cold-hardy rye whole cells for different temperatures. (b) The difference spectra (the fluid-lipid component) obtained by digital subtraction of the -40°C spectrum. The vertical lines indicating $2A_{\parallel}$ for the -40°C spectrum and the horizontal base lines have been included to help visualize the immobilized components.

position in the bilayer on the spectral lineshape is shown in Fig. 10, where the spectra of 16NS and 5NS in the plasma membrane of rye protoblasts are compared. The motion parameter, $2A_{\parallel}$, is inversely related to the mobility. The narrower lines and smaller value of $2A_{\parallel}$ for 16NS are indicative of a fluid-like environment. The results obtained with whole cells will be given as an example.

The effects of temperature on the specimen of 16NS in the plasma membrane of cold-hardy rye cells in shown in Fig. 11(a). At a temperature of 35°C a narrow, three-line spectrum is observed that is typical for this probe in a fluid-lipid environment. As the temperature is decreased, the fluidity of the lipids decreases and the mobility of the probe is reduced. For example, at 10°C there is a noticeable increase in the linewidths of the spectrum indicating a decrease in the probes' rotational diffusion rate. In addition, a broad shoulder is becoming evident on the low field line (left side of spectrum). As the temperature is lowered further, the broad shoulder grows in intensity and a negatively directed peak begins to appear on the high field side. By -5°C this peak is clearly evident, the broad, low-field peak and the negative, high-field peak are the extreme peaks of a motionally restricted spectrum. As the temperature continues to decrease, the linewidths of the fluid component continue to broaden and the amount of immobilized component increases until at -40°C only an immobilized component is present. A careful examination of the spectra shows that the immobilized component also is present at the higher temperatures. The vertical lines indicating the outer peak hyperfine splitting, $2A_{\parallel}$, and the horizontal base lines in Fig. 11(a) have been added to emphasize the presence of the immobilized component.* To a first approximation the $2A_{\parallel}$ splitting appears to be independent of temperature except perhaps for temperatures above about 25°C where $2A_{\parallel}$ may be decreasing slightly. (The high field immobilized peak is weak and difficult to detect at these temperatures). If it is assumed that the rye plasma membrane spectra are composed of only two components, and that the immobilized component can be represented by the −40°C spectrum, then by subtracting the immobilized component from the composite spectra, the spectra for the fluid component can be obtained. The subtraction is carried out digitally by adjusting the amplitude of the immobilized spectrum until the low and high field immobilized peaks disappear. If too much immobilized spectrum is subtracted, it will be evident by the appearance of a negative, low field and a positive, high field peak. This subtraction method is least accurate at the higher temperatures where the immobilized signal is weak. We have ascertained experimentally by subtracting an immobilized signal from a purely fluid spectrum not containing any immobilized component that an error of as much as 10% can be made in estimating the amount of immobilized component present. Alternative methods of subtracting a fluid signal cannot be applied because of the difficulty of accurately matching the fluid spectra over the temperature range studied. The percentages of immobilized component are obtained from the areas of the immobilized and composite spectra after double integration of each spectrum. The fluid spectra obtained by the subtraction are shown in Fig. 11(b). These spectra can be characterized by the hyperfine splitting parameter, $2A_{\parallel}$, which is inversely related to the mobility of the probe.

Membrane temperature profile

The above analysis provides two parameters with which to characterize the membrane: (1) the fraction of immobilized lipids present at any temperature and (2) $2A_{\parallel}$, a measure of the mobility of the fluid lipids as a function of temperature. Taken together these provide a *temperature profile* of the membrane. The temperature profile for the plasma membrane of cold-hardy, whole rye cells is shown in Fig. 12. The percent immobilized lipids is plotted on the right hand axis and the motion

*In previous plant membrane studies no reference has been made to the possibility of a complex spin-label spectrum. It is possible that the spectral analyses of the earlier studies may be subject to error by neglecting this possibility.

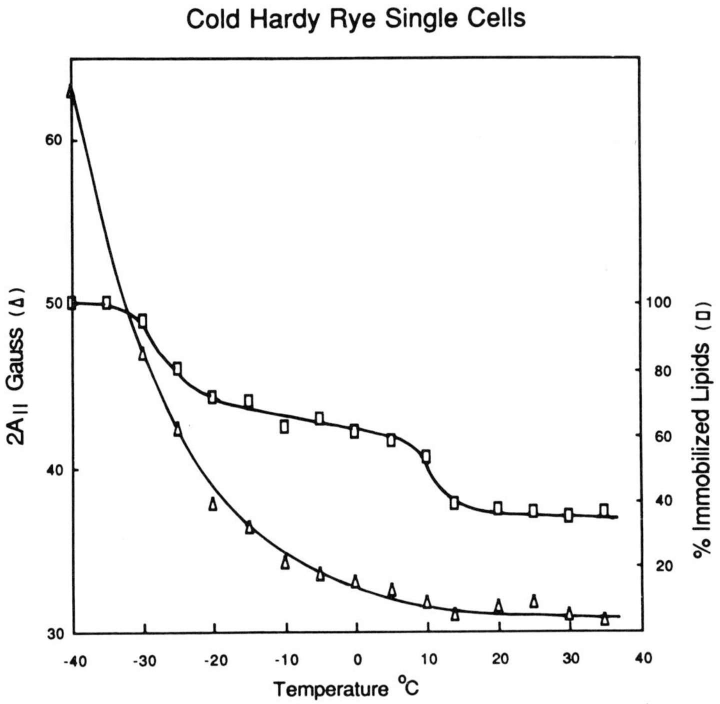

Figure 12. Temperature profile for cold-hardy rye whole cells labeled with 16NS. The percent immobilized component and the motion parameter for the fluid component, $2A_{\parallel}$, are plotted vs. temperature.

parameter for the fluid lipids, $2A_{\parallel}$, is plotted on the left hand axis. Consider first the percent immobilized lipids. At high temperatures the percentage remains essentially constant and equal to about 35% until about 15°C. These immobilized lipids are believed to be the boundary layer lipids around the hydrophobic surfaces of the integral membrane proteins. This result is similar to that obtained with frog rod outer segment disc membranes except that for the frog membranes the percentage of boundary layer lipids was greater ($\sim$ 37%) and remained unchanged over a wider temperature range (to 5°C) (Pates et al., 1985). As the temperature of the rye membrane decreased from 15° to 10°C an increase in the amount of immobilized lipids from 35% to 55% is observed. The reason for this sudden increase is not known, but one possiblity might be protein aggregation trapping some of the lipids. The hyperfine splitting of about 63 Gauss is in accord with this suggestion. As the temperature is decreased further, the percentage of immobilized lipids increases slowly. This broad, slow, rise is thought to be due to increasing fractions of bulk membrane lipids undergoing a liquid crystal to gel phase transition. No sharp phase transitions are to be expected in real membranes because of the diversity of sterols, lipids and proteins present. A series of lipid phase transitions broadened by sterol and protein interactions and merging together to give a continuously varying ratio of gel to fluid lipids is more likely to be the situation. Thus for rye plasma membranes we postulate two kinds of immobilized lipids: (1) lipids immobilized by interaction with membrane proteins and (2) bulk membrane lipids in the gel phase. The temperature profile for the fluid lipids is indicated by $2A_{\parallel}$. At high temperatures $2A_{\parallel}$ is small, about 30 Gauss, corresponding to a probe in a highly disordered fluid environment. Such an environment also is indicated by the narrow lines resulting from rapid rotational diffusion. Between 40° and 10°C the motion parameter does not change appreciably, but the line width is increasing. The probe is still in a mobile and disordered environment, but its rotational correlation time is increasing. Below 10°C, $2A_{\parallel}$ increases more rapidly indicating an increasing amount of chain ordering and loss of mobility. At 10°C only one half the lipids are in the fluid state and this fraction is decreasing as the immobilized fraction increases with decreasing temperature. At − 40°C, $2A_{\parallel} \cong 63$ Gauss, typical for an immobilized fatty acid spin label.

Plasma membrane lipids

A temperature profile for lipids isolated from cold-hardy rye plasma membranes is shown in Fig. 13. The temperature profile for the lipid sample is in accord with the above analysis for the whole cells. First, at high temperatures, the absence of any immobilized component in the pure lipid spectra supports the contention that the immobilized component in the whole cells was associated with membrane proteins, i.e., was a boundary layer. Second, from about 10°C and below, the lipid spectra became complex and were composed of an immobilized and a fluid component. The immobilized component increased as the fluid component decreased until at −40°C only the immobilized component was present. Since no proteins are present in the pure lipid sample, the immobilized component must have been due to gel phase lipids. The beginning of the gel phase transition at 10°C in the lipids coincided with the starting temperature of the proposed gel state in the whole cells. We have observed similar temperature profiles for mixed phospholipids from soybeans (unpublished results).

From an analysis of the temperature profiles we found a complex series of lipid-protein and lipid-lipid interactions occurring in rye plasma membranes within the physiological temperature range. By comparing such measurements from hardy and tender cells, we hope to learn more about the role of the plasma membrane in adaptation and acclimation.

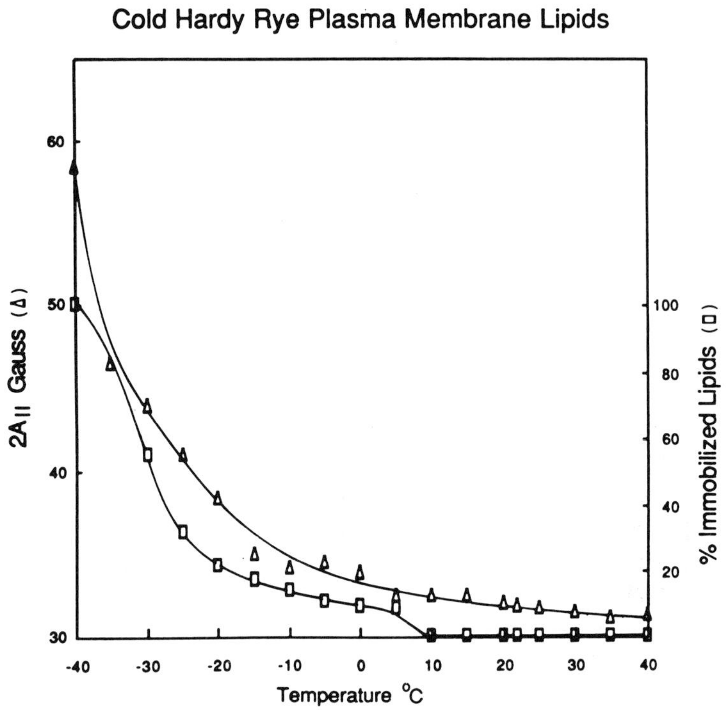

Figure 13. Temperature profile for cold-hardy rye plasma membrane lipids labeled with 16NS. The percent immobilized component and motion parameter for the fluid component, $2A_{\|}$ are plotted vs. temperature. Note the absence of any immobilized fraction above 10°C.

Spin labeling of living cells

In the application of the spin labeling method to living cells and protoplasts two problems are encountered. The first is to determine the distribution and location of the spin label in the specimen. The second is to cope with the biochemical reduction of the spin label and the concomitant loss of signal. When fatty acid and other lipid soluble spin labels are added to living cells or protoplasts, the labels are found to partition rapidly throughout the intracellular membranes (Briggs et al., 1982) and in many cases the labels become reduced (Kaplan et al., 1973). In such cases the spin label signal will be a mixture of all of the cellular membrane signals and may decay before the experiment is completed. Under such circumstances the results with cells and protoplasts will be of limited interest. We were confronted with both of these problems in the spin labeling of rye cells and protoplasts. To prevent reduction and decay of the signal we followed the common practice of adding an impermeant oxidizing agent such as ferricyanide to the system (Kaplan et al., 1973), (Singh et al., 1980) and (Boss, 1983). After decay of the signal, addition of ferricyanide regenerated 16% of the 16-DOXYL stearic acid signal and 88% of a sterol-like spin label, 3DA. Since ferricyanide is impermeant to the cells, the regenerated spectrum was presumed to be localized in the plasma membrane. Thus the reoxidized probe could be used as a plasma membrane-specific probe for living cells and protoplasts. This result raised the question of the mechanism by which ferricyanide can reoxidize a spin label inserted within a lipid bilayer. We investigated this problem and ruled out the possibility of direct interaction between ferricyanide and reduced spin label. We suspect that the probe is reoxidized *via* a transplasma membrane electron transport system (Windle and DuPont, manuscript in preparation).

Conclusions

Almost from its inception, EPR spectroscopy has played an important part in biological research. A significant advancement which greatly expanded the biological application of EPR was the development of spin labeling. The anisotropic characteristics of the nitroxide free radical made it an ideal reporter group for spin label studies of molecular motion in biological membranes. The development of saturation transfer spin labeling, ST-EPR extended the range of molecular correlation times to the slow motion region of 10^{-7} to 10^{-3} s, thus complementing conventional EPR which is sensitive to motion in the 10^{-11} to 10^{-7} s range. The ability to directly measure molecular motions over such a wide range of correlation times and in such complex structures as membranes in living cells makes EPR spectroscopy a powerful tool for biological research. EPR application to biology is expected to continue well into the 21st century as further advances in instrumentation and spin labeling techniques occur.

Acknowledgement

I thank Dr. Frances DuPont for many helpful comments and criticisms.

References

1. Alger, R.S. 1968. *Electron paramagnetic resonance. Techniques and Applications.* Interscience Publishers, New York. pp. 8–13.
2. Anderson, J.P., P. Fellmann, J.V. Moller and P.F. Devaux. 1981. *Immobilization of a spin label fatty acid chain covalently attached to a Ca^{2+} AtPase from sarcoplasmic reticulum suggests an oligomeric structure.* Biochemistry 18: 4928–4936.
3. Bales, B. 1980. *A simplified, accurate method of correcting for unresolved hyperfine broadening in the EPR of nitroxide spin probes to determine the intrinsic linewidth and Heisenberg spin exchange frequency.* J. Magn. Reson. 38: 193–205.

4. Beinert, H. 1985. *Electron-paramagnetic resonance spectroscopy in biochemistry: past, present and future*. Biochem. Soc. Trans. 13: 542–547.
5. Beringer, R. and J.G. Castle. 1950. *Magnetic resonance absorption in nitric oxide*. Phys. Rev. 78: 581–586.
6. Boss, W.F. 1983. *Poly (ethylene-glycol) induced fusion of plant protoplasts. A spin label study*. Biochim. Biophys. Acta 730: 111–118.
7. Briggs, S.P., A.R. Haug and R.P. Scheffer. 1982. *Localization of spin labels in oat leaf protoplasts*. Plant Physiol. 70: 662–667.
8. Caldwell, C.R. and A. Haug. 1981. *Temperature dependence of the barley plasma membrane-bound Ca^{2+}- and Mg^{2+} dependent ATPase*. Physiol. Plant. 53: 117–124.
9. Chignell, C.F. and D.A. Chignell. 1975. *A spin label study of purple membrane from Halobacterium halobium*. Biochem. Biophys. Res. Commun. 62: 136–143.
10. Commoner, B., J. Townsend and G. Pake. 1954. *Free radicals in biological materials*. Nature 174: 689–691.
11. Copeland, E.S. 1972. *Electron spin resonance studies in radiation biology*. In *Biological Applications of Electron Spin Resonance*. H.M. Swartz, J.R. Bolton and D.C. Borg eds. Wiley-Interscience, New York, p. 449–481.
12. Davoust, J. and P.F. Devaux. 1982. *Simulation of electron spin resonance spectra of spin labeled fatty acids covalently attached to the boundary of an intrinsic membrane protein. A chemical exchange model*. J. Magn. Reson. 48: 475–494.
13. Devaux, P.F. 1985. *Conventional ESR spectroscopy of membrane proteins. Recent applications*. In *The Enzymes of Biological membranes V. 1* A.N. Martonosi ed. Plenum Press, New York. pp. 259–285.
14. Devaux, P.F. and M. Seigneuret. 1985. *Specificity of lipid-protein interactions as determined by spectroscopic techniques*. Biochim. Biophys. Acta 827: 63–125.
15. Edmondson, D.E. 1985. *Electron spin resonance study of flavoenzymes*. Biochem. Soc. Trans. 13: 593–600.
16. Evans, M.C.N., R. Ford and Y.E. Atkinson. 1985 *Electron paramagnetic resonance studies of photosynthesis*. Biochem. Soc. Trans. 13: 585–588.
17. Froncisz, W. and J.S. Hyde. 1982 *The loop-gap resonator: A new microwave lumped circuit ESR sample structure*. J. Magn. Reson. 47: 515–521.
18. Griffith, O.H. and P.C. Jost. 1976. *Lipid spin labels in biological membranes*. In *Spin Labeling. Theory and Application*. L.J. Berliner ed. V. 1 Academic Press, New York, pp. 453–523.
19. Hidalgo, C. and D.D. Thomas, 1977. *Heterogeneity of SH groups in saroplasmic reticulum*. Biochem. Biophys. Res. Commun. 78: 1175–1182.
20. Hubell, W.L. and H.M. McConnell. 1969. *Orientation and motion of amphiphilic spin labels in membranes*. Proc. Natl. Acad. Sci. USA 64: 20–27.
21. Hyde, J.S. and L.R. Dalton. 1979. *Saturation transfer spectroscopy*. In *Spin Labeling. Theory and Applications*. L.J. Berliner ed. V. 2 Academic Press, New York pp. 1–70.
22. Hyde, J.S. and D.D. Thomas. 1973. *New EPR methods for the study of very slow motions. Application to spin labeled hemoglobin*. Ann. N.Y. Acad. Sci. 222: 680–692.
23. Johnson, M.E., L. Lee and LW-M. Fung. 1982. *Models for slow anisotropic rotational diffusion in saturation transfer electron paramagnetic resonance at 9 and 35 GHz*. Biochemistry 21: 4459–4467.
24. Jost, P.C. and O.H. Griffith. 1982. eds. *Lipid-Protein Interactions* V.2. John Wiley and Sons, New York. 307 pages.
25. Jost, P.C., O.H. Griffith, R.A. Capoldi, and G. Van der Kooi. 1973. *Evidence of boundary layer lipids in membranes*. Proc. Natl. Acad. Sci USA 70: 480–484.
26. Kaplan, J., P.G. Canonico and W.J. Caspary. 1973. *Electron spin resonance of spin labeled mammalian cells by detecting surface membrane signals*. Proc. Natl. Acad. Sci. USA 70: 66–70.
27. Levitt, J. 1980. *Response of Plants to Environmental Stresses*. 2d ed. V1 & V2. Academic Press, New York.
28. Marsh, D. 1980. *Molecular motion in phospholipid bilayers in the gel phase: long axis rotation*. Biochemistry 19: 1632–1637.

29. Marsh, D. 1981. *Electron spin resonance, spin labels.* in *Membrane Spectroscopy*, E. Grell, ed. Springer-Verlag, Berlin, pp 51–142.
30. Marsh, D. 1985a. *ESR probes for structure and dynamics of membranes.* in *Spectroscopy and Dynamics of molecular biological systems.* Academic Press Inc. London pp 209–238.
31. Marsh, D. 1985b. *ESR spin label studies of lipid-protein interactions.* In *Progress in Protein-lipid Interactions.* Watts/de Pont, eds. Elsevier Science Publisher B.V. pp. 143–172.
32. Ohnishi, S. and H.M. McConnell. 1965. *Interaction of the radical ion of chloropromazine with deoxyribonucleic acid.* J. Amer. Chem. Soc. 87: 2293.
33. Pates, P.D., A. Watts, R. Uhl and D. Marsh. 1985. *Lipid-protein interactions in frog rod outer segment disc membranes. Characterization by spin labels.* Biochim. Biophys. Acta 814: 389–397.
34. Robertson, R.N. 1983. *The lively membranes.* Cambridge University Press, Cambridge. 206 pp.
35. Singh, J. 1981. *Isolation and freezing tolerance of mesophyll cells from cold hardened and nonhardened winter rye.* Plant Physiol. 67: 906–909.
36. Singh, J. and R.W. Miller. 1980. *Spin-label studies of membranes in rye protoplasts during extra-cellular freezing.* Plant Physiol. 66: 349–352.
37. Steponkus, P.L. 1984. *Role of the plasma membrane in freezing injury and cold acclimation.* Ann. Rev. Plant Physiol. 35: 543–584.
38. Swartz, J.M., J.R. Bolton and D.C. Borg. 1972 eds. *Biological Applications of Electron Spin Resonance.* Wiley-Interscience, New York. 569 pp.
39. Thomas, D.D., L.R. Dalton and J.S. Hyde. 1976. *Rotational diffusion studied by passage saturation transfer electron paramagnetic resonance.* J. Chem. Phys. 65: 3006–3024.
40. Thomas, D.D. and C. Hidalgo. 1978. *Rotational motion of the sarcoplasmic reticulum* Ca^{2+}*-ATPase.* Proc. Natl. Acad. Sci. USA 75: 5488–5492.
41. Thomas, D.D., C.H. Wendt, W. Froncisz and J.S. Hyde. 1983. *Saturation transfer EPR spectroscopy on spin labeled muscle fiber using the loop-gap resonator.* Biophys. J. 43: 131–135.
42. Vigh, L., L.I. Horvath, D. Dudits and T. Farkas. 1979. *Protoplast plasmalemma fluidity of hardened wheats correlates with frost resistance.* FEBS Lett. 107: 291–294.
43. Vistnes, A.I., D.J. Wormald, S. Isachsen and D. Schmalbein. 1984. *An efficient digital phase sensitive detector for use in election spin resonance spectroscopy.* Rev. Sci. Instrum. 55: 527–532.
44. Watts, A., I.D. Volotovski and D. Marsh. 1979. *Rhodopsin lipid associations in bovine rod outer segment membranes. Identification of immobilized lipids by spin labels.* Biochemistry 18: 5006–5012.
45. Watts, A.J. Davoust, D. Marsh and P.F. Devaux. 1981. *Distinct states of lipid mobility in bovine rod outer segment membranes. Resolution of spin label results.* Biochim. Biophys. Acta 643: 673–676.
46. Windle, J.J. 1981. *Hyperfine coupling constants for nitroxide spin probes in water and carbon tetrachloride.* J. Magn. Reson. 45: 432–439.
47. Windle, J.J. and W.H. Thurston. 1957. *Electron spin resonance in a stable phenoxy radical.* J. Chem. Phys. 27: 1429–1430.
48. Wright, L.C., E.J. McMurchie, K.M. Pomeroy and J.K. Raison. 1982. *Thermal behavior and lipid composition of cauliflower plasma membranes in relation to ATPase activity and chilling sensitivity.* Plant Physiol 69: 1350–1360.
49. Zavoisky, E., 1945. *Paramagnetic relaxation of liquid solutions for perpendicular fields.* J. Phys. USSR 9: 211–216.

Recent developments in mass spectrometry

HENRY M. FALES
Laboratory of Chemistry, National Heart, Lung, and Blood Institute, National Institutes of Health, Bethesda, MD 20894, U.S.A.

Rather than the more classical attributes of resolution and sensitivity, recent developments in mass spectrometry appear to have addressed problems associated with sample involatility, complexity and high molecular weight. The first was solved originally by conversion to derivatives, first methyl esters for example, and later others, trimethylsilyl ethers in particular. In my opinion this method, in the hands of a chemist knowledgeable enough to be aware of its attendant complications, is still the most powerful of all. It allows use of labelled reagents and separation from byproducts and impurities by GC and LC. Quite often the elucidation of molecular weight and other structural features is then straightforward.

Particularly important where high molecular weight or complex polar structures are encountered is proper placement of the sample within the source. This was probably first emphasized by Reed during sterane studies as early as 1958, and then rediscovered in different contexts (direct C.I., in-beam EI., etc) later. On the other hand, when molecular complexity reaches a certain level, even this approach fails. The number of derivatizing groups required to reduce polarity, for example, may themselves begin to mask structural features and spectral interpretation is complicated.

Chemical ionization (C.I., Harrison 1982) was one early and very successful approach in such cases, especially in the form of "direct CI". This is nothing more than optimum placement of sample within the ion source and, importantly, its distribution on the outside of the probe as emphasized by Baldwin and McLafferty and later by Hunt et al. Our early prototype CI source built by SRI in Baltimore was, in fact constructed in this manner but we were not aware of the importance of placing the sample on the outside exposed to the reagent ion. Interestingly, many more recently designed CI sources such as the LKB 2791, lack this construction and can be profitably modified. Signal enhancements of several orders of magnitude are possible. Some spectra obtained in this mode even suggest that certain compounds (e.g. adenosine monophosphate, certain salts, etc.) may be ionized by processes other than simple volatilization. Of course, the real strength of CI is in enhancement of the molecular ion and the ability one has to select the energy transferred to the molecule via the choice of a specific reagent ion.

A related technique "in beam EI" discovered by Dell and then Ohashi seems to be, in reality, localized CI in an EI source, using the compound under study as its own reagent gas. This is a technique whose principle was well known to early mass spectrometrists who often increased sample "pressure" to elucidate the molecular weight by causing the appearance of an $M + H^+$ or (M + acyl$^+$) ion due to ion-molecule reactions.

Probably the first successful extension of mass spectrometry to truly involatile classes of compounds, such as arginine-containing peptides, involved the field desorption (FD) methods of Beckey and Schulten. Even though the far simpler FAB method has largely superceded FD systems today, I feel that the demonstration by these workers that certain classes of molecules were amenable to mass spectrometry broke an important psychological barrier.

Laser desorption mass spectrometry (Kistemaker 1980, Posthumus 1978, Cotter 1986) where the sample is volatilized by brief intense input of radiation has several distinct advantages. Today it is often combined with time-of-flight mass spectrometers, allowing a complete analysis of the results of each individual laser pulse. The ability to focus the laser spot on a specified area is important and the ability of being able to control duration of the pulse is useful too. We can hope for even more selective lasers in the future that will excite only certain vibrational frequencies, particularly those corresponding to bonds holding molecules to surfaces. Spectra obtained from the laser plume are most often the result of protonation or cationization by alkali metal, i.e., basically CI spectra, but their presence certainly does shed light on the processes obtained by other methods including Cf-252 plasma desorption.

The thermospray system of Blakely and Vestal et al., (1980) is remarkable in two regards. First, it is ideally suited to combination with a liquid chromatograph since it requires that the sample be dissolved in a liquid. Second, the work required to separate and ionize the sample molecules has already been done by the dissolution process. It is thus ideally suited to polar and ionic compounds. In this method the solution is injected from a hot needle into a vacuum where solvent is freed from the droplets whose multiple charge then renders them unstable. Breakup and more solvent is lost until finally only sample ions (and neutrals) are left. This is a very mild ionization technique indeed and again either CI type ions ($M + H^+$, $M + Na^+$) or even free quaternary ammonium ions are observed. The simplicity of the system, usually using a quadrupole spectrometer, further recommends it and I am sure it will continue to gain in popularity.

The secondary ion mass spectrometer, used originally for inorganic analysis is now applied to organic problems (Benninghoven 1978, Ens 1982) through the use of the so-called fast atom bombardment (FAB) technique. Barber, Bordoli and Sedgewick showed originally how use of a fast atom beam in place of the fast ion beam ordinarily used in SIMS could provide good spectra of very high molecular weight polar compounds. Then, by use of a liquid matrix (glycerol etc.) it was later seen that even a fast *ion* beam would provide good spectra. Today, nearly every mass spectrometer sold must lay claim to successful operation in this important "secondary ion" mode. Since the FAB adaption actually entails little modification of most spectrometers, and because it gives such elegant results, it is probably the most important of the new techniques I will discuss today.

Of course many of the new methods, while providing excellent ion intensity near the molecular weight, often give weak or poorly characterized fragment ions. One way around this problem is to make use of an ion cyclotron resonance spectrometer (Lehman and Bursey 1976) where individual ion types may be separated and studied either by their spontaneous decompositions or by activating them through collisions with target gases or by thermal excitation with RF energy. This is also the method employed in MS/MS analysis where individual ions are examined as they cascade through several fragmentations, (daughter, granddaughter ion analysis). These techniques are competing well with the older chemical methods involved in structure elucidation and this technique will be discussed at length in Professor Bieman's lecture. A price one pays appears to be instrument complexity, ranging from n-sector magnet systems down to relatively simple "ion traps". But the advantages are real and several recent papers have shown how parent ions may be selected from biological matrices using CI or FAB, then subsequently decomposed through collisions with inert gases to yield structural information or, at least, satisfactory "fingerprint" spectra.

Indeed, it seems that sometimes authors using these techniques have gone to extremes to avoid the simplest of extraction procedures, but perhaps this is just to emphasize the elegance of the method. Certainly, the less handling a biological sample undergoes the better, for obvious reasons. Still, in these days of budgetary constraints it is increasingly difficult to justify instrument time on a very expensive system merely to avoid a simple solvent extraction. The inherent power of these instruments certainly suggests that they will be acquired by more laboratories to solve important problems.

I would like to suggest further that artificial intelligence techniques may be applied to help interpret the wealth of ion decomposition pathways available from MS/MS instrumentation. Some of us had hoped earlier, perhaps naively, that AI would be useful in piecing together the structure of a molecule from the exact masses (formulae) of its fragments. But of course, in most ion decompositions, bonds are not only broken, but made, broken again, and remade in strange and wonderful ways, making interpretation by any method a formidable task.

In our laboratory, we are currently specializing in the technique known, somewhat obscurely, as plasma desorption mass spectrometry or Cf-252 plasma desorption mass spectrometry. In this method, the fission fragments from ^{252}Cf are used both to ionize and volatilize the sample and to trigger the detectors. This technique, discovered by Professor R. Macfarlane of Texas A & M. (see 1985 review) provides high abundance cationized molecule peaks that allow identification of very high molecular weight, complex substances. In a sense, it is like FAB, but especially because it uses a time-of-flight spectrometer, very little sample (10^{-6}) is actually used up in the process of taking a spectrum. Excellent results are obtained with a wide variety of inorganic salts, peptides, polysaccharides, dyes, glycosides, and other natural products. An especially important feature of the method, in my opinion, is the ease of running samples and the virtually maintenance-free operation of the spectrometer.

All of the above methods have certain advantages, and it must be admitted, certain disadvantages. Furthermore, a given problem may usually be solved by several different approaches, and it is often true that a direct solution involving a little chemical manipulation can cost less in time than that required for the maintenance and operation of these state-of-the-art techniques.

References

1. Baldwin and McLafferty. 1982. *Direct C.I.* Org. Mass Spec. 1353.
2. Barber, et al. 1982. *Fast Atom Bombardment.* Anal. Chem. 845A.
3. Beckey. 1970. *Field Desorption.* Field Ionization Mass Spectrometry, Pergamon.
4. Benninghoven, et al. 1978. *Secondary Ion Mass Spectrometry.* Anal. Chem. 1180.
5. Blakely, et al. 1980. *Thermospray.* Anal. Chem. 1636.
6. Chapman. 1985. *Soft Ionization Methods.* Practical Organic Mass Spectrometry. Wiley.
7. Cotter. 1980. *Laser Desorption.* Anal. Chem. 1767.
8. Dell, et al. 1958. *In-Beam EI.* J.A.C.S. 2497.
9. Ens, et al. 1982. *Secondary Ion Mass Spectrometry.* Anal. Chem. 960.
10. Harrison. 1982. *Chemical Ionization.* Mass Spectrometry, C.R.C. Press.
11. Hercules, et al. 1982. *Laser Desorption.* Anal. Chem. 280A.
12. Hunt, et al. 1977. *Direct C.I.* 1161.
13. Lehman and Bursey. 1976. *Ion Cyclotron Resonance.* Mass Spectrometry, Wiley.
14. Lyon. 1985. *Soft Ionization Methods.* Desorption Mass Spectrometry, ACS. 291.
15. Morris. 1981. *Soft Ionization Methods.* Mass Spectrometry, Heyden.
16. Ohashi, et al. 1976. *In-Beam EI.* Chem. Let. 439.

17. Posthumus, et al. 1978. *Laser Desorption.* Anal. Chem. 985.
18. Reed. 1958. *In-Beam EI.* J. Chem. Soc. 3432.
19. Schulten. 1979. *Field Desorption.* Int. J. Mass Spec. & Ion Physics. 97–283.
20. Sundqvist and Macfarlane. 1985. *Californium-252 Plasma Desorption (fission fragment mass spectrometry).* Mass Spectrometry Reviews. 421–460.

Recent developments in high mass and tandem mass spectrometry

KLAUS BIEMANN
Department of Chemistry, Massachusetts Institute of Technology, Cambridge, MA 02139, U.S.A.

Introduction

Over the past few years developments in the field of mass spectrometry occurred which opened it to areas of research previously completely inaccessible. First was the discovery by Barber et al. (1981) of a novel ionization technique, fast atom bombardment (FAB), which suddenly made it possible to ionize large, polar and thermally labile molecules. Prior to that time, such compounds had to be degraded into smaller molecules and/or chemically transformed to compounds that could be volatilized into the mass spectrometer. By using FAB it now became possible to ionize, generally in the form of protonated molecular ions, $(M + H)^+$, compounds of molecular weight of a few thousand daltons which, however, exceeded the mass range of the then commercially available mass spectrometers.

This situation spurred the second development, namely the design of mass spectrometers of expanded mass range, generally by modification of the ion optics, increasing the magnetic fields and radius of deflection. Instruments with a mass range of up to 14 500 dalton at full accelerating voltage (10 kV), and higher at proportionally lower acceleration, are now commercially available (Kammei et al., 1985).

The third development was prompted by the first. The spectra produced by FAB are characterized by an abundant $(M + H)^+$ ion, but there is little excess energy in this very stable ion. Therefore, relatively little fragmentation takes place, particularly if the molecule is large. Thus, while the molecular weight of a few thousand daltons can be easily determined to ± 1 or better, little structural information can be obtained on the compound because of the paucity (or absence) of structurally significant fragment ions.

It was well known that one can induce fragmentation by colliding a preselected ion with a neutral molecule and the resulting product (or daughter) ions can then be mass analyzed in a second mass spectrometer. Prior to the development of FAB, this tandem mass spectrometry (MS/MS) was chiefly used to study mass spectrometric fragmentation

Address correspondence to: Klaus Biemann, Dept. of Chemistry, Massachusetts Institute of Technology, Cambridge, MA 02139, U.S.A.

processes or analyze selected components of mixtures (for a review, see McLafferty, 1983). However, this approach has now become very useful to generate structurally specific fragment ions from the stable $(M + H)^+$ ions produced by FAB from large molecules. In order to obtain practically useful product ion spectra from compounds of a molecular weight range of a few thousand dalton, sufficient sensitivity and resolution is required so that one can preselect a single mass as the precursor ion in the first mass spectrometer (MS-1) and uniquely define (to better than one dalton accuracy) the mass of all the product ions in the second mass spectrometer (MS-2). This again required an improvement in instrumentation and led to the design of high performance tandem mass spectrometers which consist of two double focusing spectrometers in sequence.

The development of new strategies for the determination of the structure of peptides and proteins is important because of extreme structural variability and biological importance. These molecules are particularly suited for mass spectrometry because they represent linear arrangements of unique building blocks, the twenty common amino acids. The following discussion is therefore devoted to this area of research, but it should be kept in mind that similar approaches can be applied to other so-called biopolymers as well, such as carbohydrates, oligonucleotides, etc.

Mass spectra at high mass

There are many instances where reliable molecular weight information (accurate to better than one dalton) is sufficient to solve a certain problem, particularly since the mass spectrometer allows this to be done even for components present in a mixture. For example, the various insulins differ by the replacement of one or two amino acids which causes their molecular weights to differ accordingly. Figure 1 shows the FAB spectrum obtained from porcine and human insulin, respectively. They differ in the last amino acid of the β-chain, which is alanine in the former and threonine in the latter. Therefore, the two insulins differ by 30 daltons, which is very easily measured. The figure also shows another problem one encounters with organic compounds of high mass. The chances that a molecule containing more than 250 carbon atoms would consist entirely of ^{12}C-atoms is low, in fact the statistically most likely species is, in this case, that containing three ^{13}C-atoms. For compounds of this size, or even larger ones, it is thus not necessarily best to try to resolve these isotope clusters but determine the center of mass of the unresolved multiplet, i.e., the abundance-weighted average mass of all isotopic molecules. This average mass then equals the sum of the average

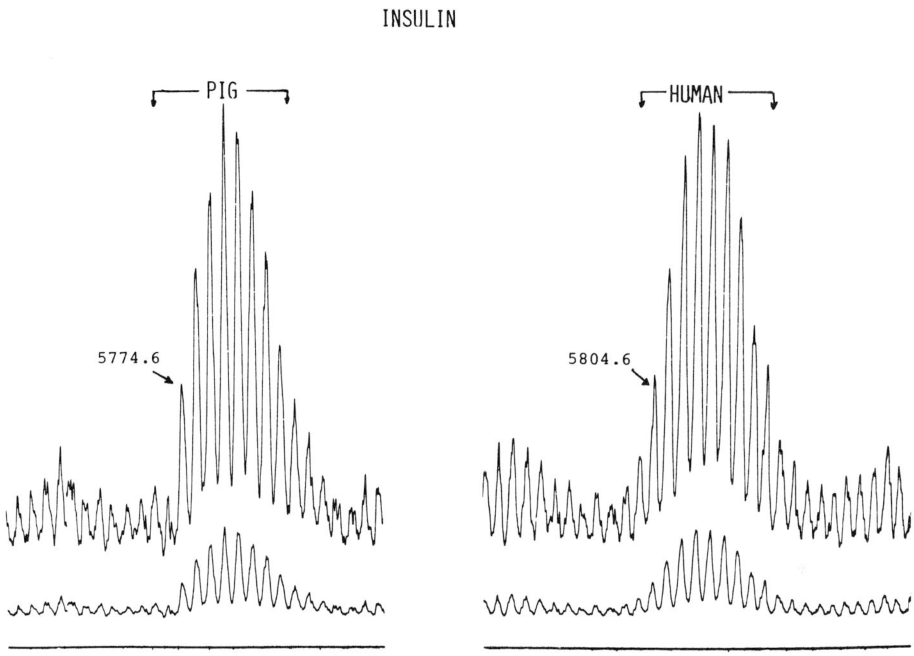

Figure 1. The molecular ion region for porcine insulin ($C_{256}H_{381}N_{65}O_{76}S_6$, monoisotopic $(M + H)^+$ = *m/z* 5774.6) and human insulin ($C_{257}H_{383}N_{65}O_{77}S_6$, monoisotopic $(M + H)^+$ = *m/z* 5804.6).

mass of all atoms, i.e., their "chemical" atomic weights which also represent the abundance weighted mass of all isotopes of the particular element (McNeal & Macfarlane, 1981; Yergey et al., 1983). Reducing resolution causes an increase in sensitivity and this allows one to obtain a measurable signal for the $(M + H)^+$ ion of compounds of even higher molecular weight. For instance, the average molecular weight of a small protein, thioredoxin isolated from *E. coli*, was found to be 11 673.4 based on the average mass of the $(M + H)^+$ ion (*m/z* = 11 674.4; calculated from known structure: 11 672.1) (Biemann et al., 1986). Such molecular weight determinations will be very useful as the final confirmation of the correctness and completeness of a protein structure.

There are many other uses for simple molecular weight determinations. As the example of the two insulins implies, amino acid substitutions in peptides or small proteins can be detected or verified. With the importance of biogenetically modified proteins – amino acids changed either by natural mutation, recombinant DNA techniques, or fully synthetic modification of genes – easy detection and verification is of great utility.

One such example is from the investigation of the biological properties

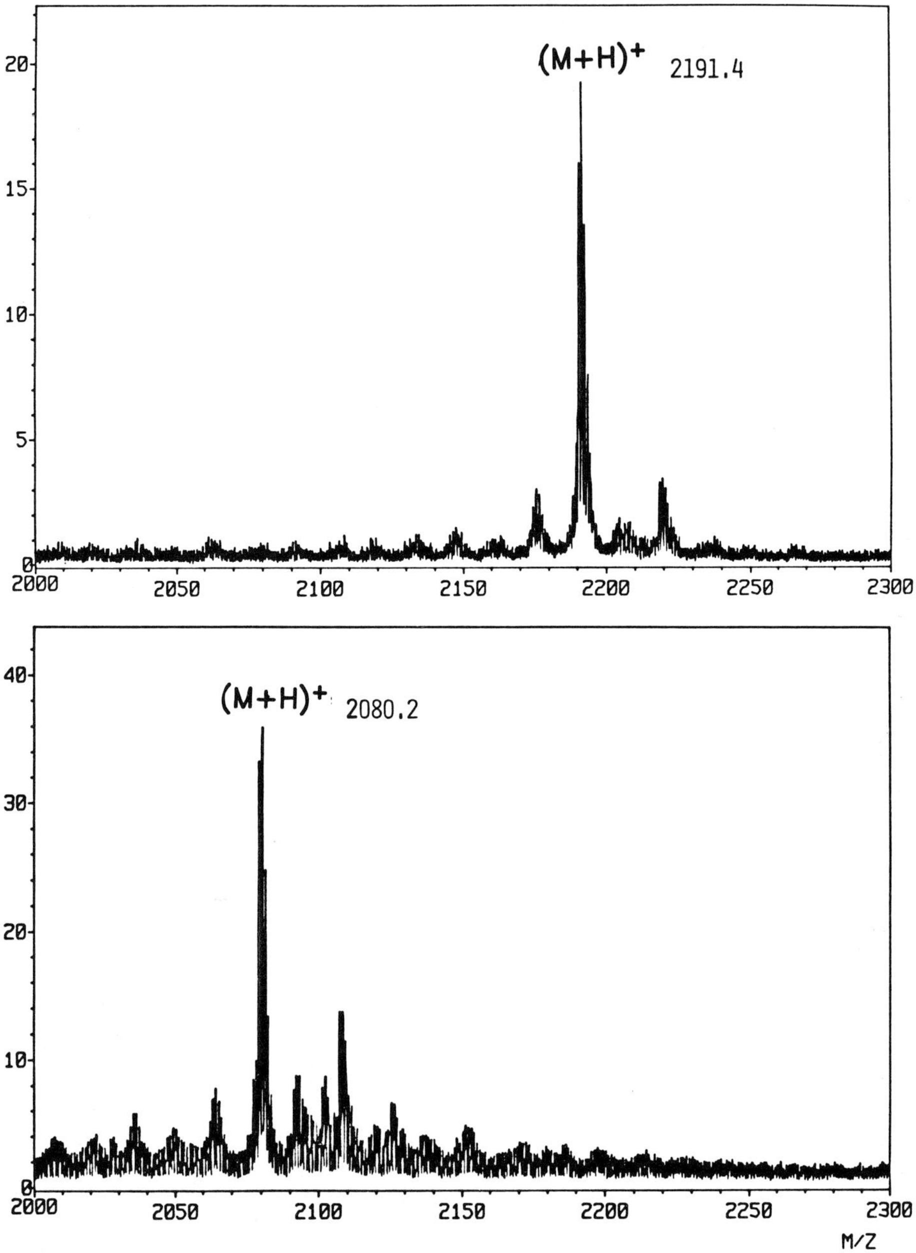

Figure 2. Molecular ion region of the *N*-terminal cyanogen bromide peptide of native bacteriorhodopsin (top) and of a modified analog (bottom).

of bacteriorhodopsin molecules which had undergone site-specific amino acid replacements by chemical modification of the base sequence of the gene coding for this protein, which plays an important role in the light-induced transfer of protons across cell membranes (Khorana, 1986). It contains 248 amino acids and is therefore too large for direct molecular weight measurements with sufficient accuracy. However, cleavage with cyanogen bromide at the ten methionines present along the amino acid chain generates eleven peptides of much more manageable size. Their molecular weights can be measured by mass spectrometry with good accuracy. Comparison with those obtained by cleavage of native bacteriorhodopsin reveals in which region of the molecule a change has taken place and the difference in mass is a clue to which amino acid has been modified. An example is presented in Fig. 2 which shows the molecular weight region of the FAB spectrum obtained from the *N*-terminal CNBr peptide of native (top) and of biosynthetically modified (bottom) bacteriorhodopsin. The difference of 111 daltons uniquely indicate that the *N*-terminal pyroglutamic acid of the native protein is absent in the modified one (Allmaier et al., 1986).

Considering the future commercial importance of proteins produced by recombinant techniques and the need for ascertaining their structure and integrity, such rapid and efficient analytical methodology should gain quick, general acceptance and may even become a regulatory requirement.

Peptide sequencing by tandem mass spectrometry

Up to this point, cases were discussed where molecular weight information was sufficient, because much was known about the structure of the peptide or protein under scrutiny and only a specific change was to be identified. For the determination of a previously unknown primary structure of a peptide or protein, fragmentation along the various peptide bonds must be achieved to generate ions the mass of which is indicative of the sum of the amino acids therein. Furthermore, the set of ions must be sufficient (i.e., cleavage must occur at each peptide bond) to allow the unambiguous interpretation of the resulting spectrum in terms of a unique amino acid sequence.

As alluded to in the Introduction, this fragmentation can be accomplished by collision of the protonated peptide ion with an inert gas. Collision induced dissociation (CID) leads to the desired fragment ions.

In the determination of the primary structure of a protein, the latter is usually cleaved into polypeptides by a very specific reaction (Allen, 1981). Most common is digestion with trypsin, a proteolytic enzyme that cleaves at the *C*-terminal side of lysine or arginine. Fortunately, these two amino acids are generally of medium abundance in proteins and trypsin produces, therefore, a mixture of peptides the vast majority of

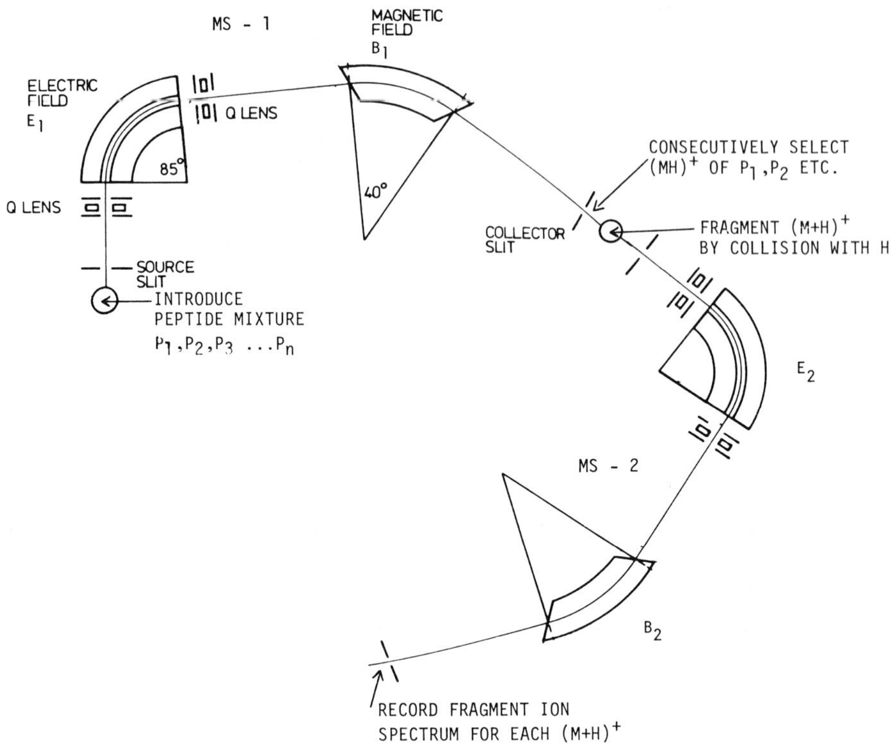

Figure 3. Schematic diagram of a magnetic deflection tandem mass spectrometer (JEOL, HX110/HX110) indicating the ionization of a mixture, consecutive selection of $(M + H)^+$ ion of components in MS-1 for fragmentation by collision with a neutral gas (He) and mass analysis of the product ion is MS-2.

which have molecular weights between 500 and 2500 daltons. This is just the right size for CID because the collision process – at least in the presently used form – does not cause complete fragmentation of larger peptides.

As alluded to earlier, in order to obtain clearly interpretable product ion spectra, one must use a tandem mass spectrometer that allows the selection of a single mass as the precursor ion and the product ions must be sufficiently well focused and resolved so that their mass can be unambiguously determined (which requires an accuracy of $< \pm 0.3$ dalton).

The required performance can be achieved by the combination of two double focusing mass spectrometers, separated by a field free region containing the collision cell filled with an inert gas, such as helium, at low pressure (10^{-3} to 10^{-4} torr). Such an arrangement is shown in Fig. 3. It is important to reiterate that it is not necessary to separate the tryptic

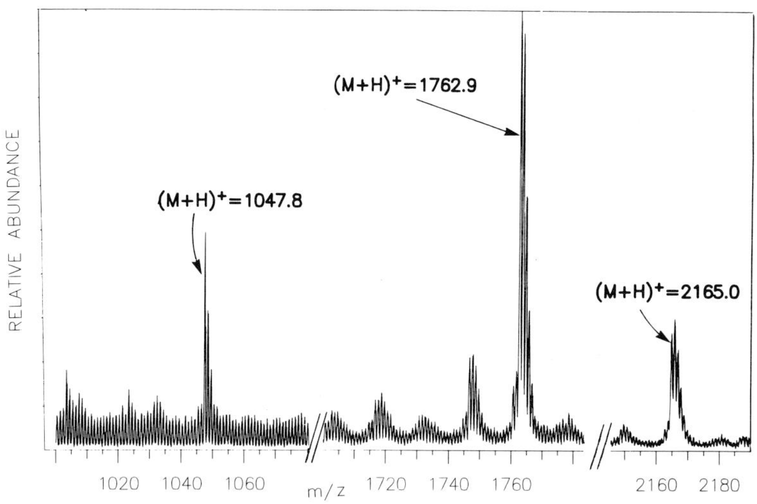

Figure 4. Segments of the FAB mass spectrum of HPLC Fraction # 10 of a tryptic digest from the thioredoxin isolated from *Chromatium vinosum*. It indicates the presence of three peptides in the mass range of m/z 1000 to m/z 2200.

digest into the individual peptides because the first mass spectrometer also separates any peptide from any other as long as their molecular weights differ by at least one dalton, which is practically always the case. For technical reasons, it is still advisable to fractionate the often quite complex mixture of tryptic peptides into less complex ones containing two to five peptides each, which can be easily accomplished by high performance liquid chromatography (HPLC).

We are presently using this approach for the determination of the primary structure of a number of small proteins, such as thioredoxins and serine protease inhibitors. The former have molecular weights in the range of 11 000–12 000 (Holmgren, 1985) and have a lysine-arginine content to result in about 10–15 peptides upon digestion with trypsin. An example of the high mass regions of the FAB spectrum obtained from an HPLC fraction of the tryptic digest of thioredoxin isolated from the anaerobic photosynthetic sulfur bacterium *Chromatium vinosum* is shown in Fig. 4. Each of the monoisotopic (^{12}C only) $(M + H)^+$ ions of the three peptides present in this fraction, m/z = 1047.8, 1762.9 and 2165.0 is consecutively focused on the collector slit of MS-1 and passed through into the collision cell. The emerging fragment ions enter MS-2 and their m/z values are recorded by scanning MS-2 in the so-called linked mode which involves simultaneous decrease of the electric field E_2 and magnetic field B_2 while keeping their ratio constant, rather than scanning only B while E remains constant, as is done when recording a

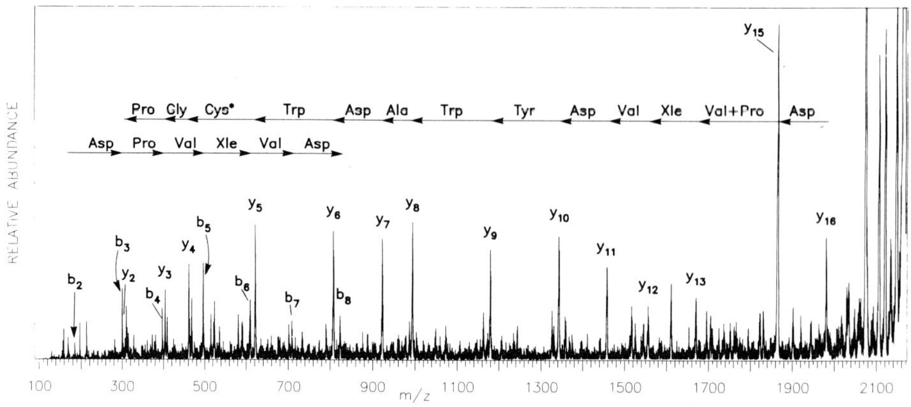

Figure 5. Tandem mass spectrum of the tryptic peptide of $(M + H)^+ = m/z$ 2165.0 (Fig. 4) after collision and mass analysis in MS-2. See Scheme I for an explanation of the notation identifying the fragment ions. Cys* refers to carbamidomethylated cysteine.

normal mass spectrum (such as is the case in MS-1). This is because the fragment ion produced upon collision carries only that fraction of the kinetic energy of the precursor ion which corresponds to the ratio of the mass of product ion to precursor ion.

The MS/MS spectrum obtained from $(M + H) = m/z$ 2165.0 (Fig. 4) is shown in Fig. 5. It clearly consists of two sets of peaks where the numbers of each set differ by the residue mass (NH–CHR–CO) of an amino acid. These values differ for all amino acids except for the isomers leucine and isoleucine (mass 113) and the isobars lysine and glutamine (mass 128). Their differentiation will be discussed later. It may not always be obvious from a casual inspection of Fig. 5 why these particular peaks were chosen over others of sometimes similar intensity. It is only these two sets which represent internally consistent mass differences. These differences, when expressed as the corresponding amino acids, reveal that they are due to sequences of opposite directions and this indicates that one set of fragments retained the *N*-terminus, the other the *C*-terminus of the peptide. The former are acyl ions and the latter ammonium ions, both formed by cleavage of a CO–NH bond retaining the charge on the –CO group and the NH group (as H_3N^+–), respectively, as shown in Scheme I (the nomenclature is a variant of that proposed by Roepstorff & Fohlman, 1984).

Close inspection of Fig. 5 reveals that there is no clear indication of the nature of the first two amino acids, except that the sum of their masses must be consistent with that of the b_2 ion (m/z 185). This could be either alanine + leucine, or serine + proline. Such ambiguities can be resolved by subjecting the peptide fraction to a one-step manual

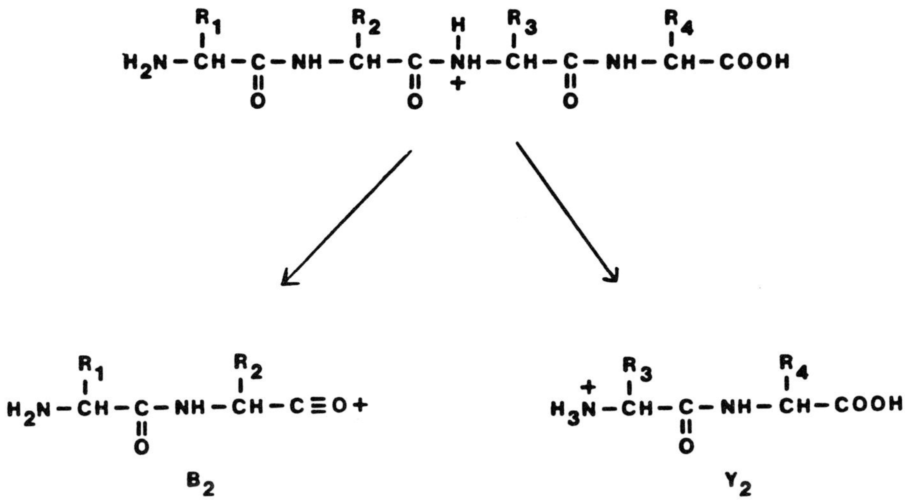

Scheme I. Two of the most common fragmentation processes along the peptide bond.

Edman degradation followed by redetermination of the FAB spectrum which then reveals the molecular weights of the truncated peptides, each of which having lost the original *N*-terminal amino acid. In this case, it was found to be serine and the sequence must therefore be Ser-Pro-. The same reaction can be used to distinguish lysine from glutamine, because only the former adds the Edman reagent (phenylisothiocyanate) which increases the molecular weight by 135 daltons.

The amino acid sequence of the other ten tryptic peptides (all were smaller than that discussed above) from this thioredoxin was determined in the same manner. In order to arrange them in the correct final sequence, one needs to generate a second set of peptides with an enzyme or chemical reaction which is specific for another amino acid. *S. aureus* protease cleaves at glutamic acid and the molecular weights of the resulting peptides must, therefore, match that particular arrangement of the tryptic peptides which results in the proper distribution of the glutamic acid residues. This experiment concluded the determination of the primary structure of this thioredoxin which turned out to 107 amino acids long (Johnson & Biemann, 1987). The only unresolved question was the differentiation of leucine from isoleucine, which can be done only if there are abundant ions of type a_n (which correspond to those of type b_n but have lost the CO group). Such ions eliminate C_2H_4 (28 daltons) if the *C*-terminal amino acid of the fragment is isoleucine, but if it is leucine, C_3H_6 (42 daltons) is lost (Martin & Biemann, 1986).

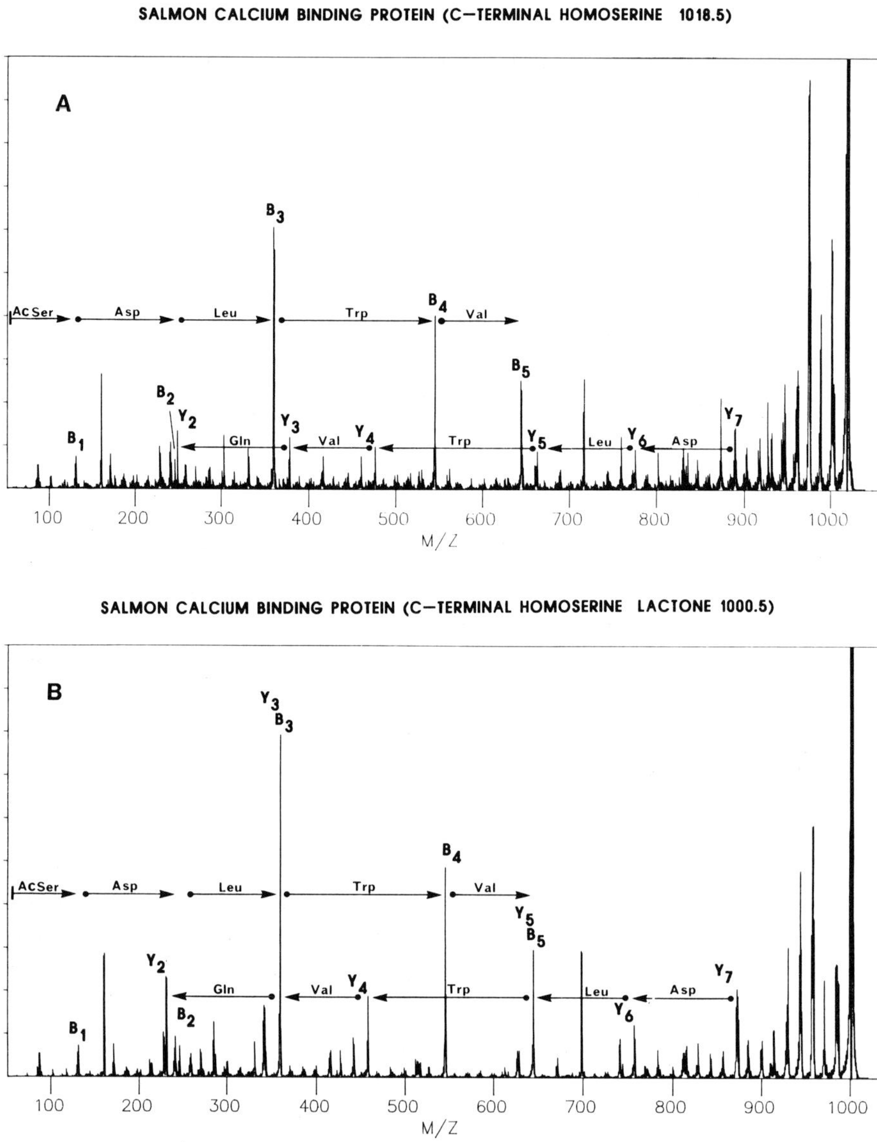

Figure 6. CID mass spectrum of an *N*-blocked peptide (generated by cyanogen bromide cleavage from a salmon calcium binding protein) ending in free homoserine (a) or homoserine lactone (b).

Post-translationally modified proteins

Frequently the biologically active gene product does not correspond to that synthesized among the mRNA chain but the primary protein is modified enzymatically thereafter. Most prevalent is removal of one or more amino acids from the *N*-terminus and this is often accompanied by acylation of the newly exposed amino group. This sequence of events results in a protein that is not amenable to the sequential Edman degradation, the conventional method for the sequencing of peptides or proteins (Hewick et al., 1981).

Such a modification presents no problem for the tandem mass spectrometer, which merely sees an *N*-terminal amino acid of different residual mass. The nature of the blocking group is deduced from the difference between the mass of the *N*-terminal fragment and the residual mass of any of the amino acids (which must also be the *N*-terminal one). As an example, the CID spectrum of a peptide produced by cyanogen bromide cleavage of a calcium binding protein is shown in Fig. 6(a). It clearly reveals the sequence

CH_3CO-Ser-Asp-Leu-Trp-Val-Gln-Lys-Hse

where Hse stands for homoserine, $H_2N–CH(CH_2CH_2OH)COOH$, the conversion product of methionine upon cleavage by CNBr. The lysine preceding Hse is deduced from the mass of the y_2 ion which must correspond to the sum of the two *C*-terminal amino acids. Figure 6(b) represents the CID spectrum of a by-product, the peptide in which the homoserine is present as its lactone. Therefore, the molecular weight is 18 daltons less and the same holds for all fragment ions that contain the *C*-terminus, a feature which facilitates the unambiguous assignment of all fragment ions in both spectra.

This example suffices to demonstrate that not only modifications at the *N*-terminal amino acid but anywhere along the protein chain can be identified from the change of the residue mass of the modified amino acid.

Conclusion

The structural information provided by tandem mass spectrometry is unique and nicely complements that obtainable by other techniques. It is particularly suitable for the detection and characterization of structural modifications. Furthermore, it opens a way for an entirely new strategy for the determination of the primary structure of proteins. Although it requires at present more material (10–25 nmol of total protein) than the gas-phase Edman method (Hewick et al., 1981), future improvement in the sensitivity may overcome this obstacle, which is only significant when the protein is in very short supply. While this discussion dealt exclusively with peptides and proteins, the same principle applies to other biopolymers and other complex, preferably linear, molecules.

It should also be noted that tandem mass spectrometry is not limited to magnetic deflection mass spectrometers. Multiple quadrupole mass spectrometers have been used for peptide sequencing (Hunt et al., 1980), although their resolution and mass range is lower than that of the magnetic instruments. Fourier-transform mass spectrometers are also, in principle, capable of high sensitivity and resolution, and are well suited for tandem operation (in this case, consecutive in time rather than space) (Cody et al., 1985). At present their practical performance and quality of spectra are apparently not good enough to reliably deduce the previously unknown sequence of a peptide or protein. However, FTMS is most likely to become an excellent method for this purpose in the not too distant future, when some of the practical and operational problems of present instruments have been overcome.

Acknowledgement

The author is indebted to his associates J.E. Biller, C.E. Costello, R.S. Johnson, S.A. Martin and H.A. Scoble for their contributions to the work discussed, which was supported by grants from the National Institutes of Health (GM05472 and RR00317).

References

1. Allen, G. (1981). *Sequencing of Proteins and Peptides.* Elsevier: North-Holland Biomedical.
2. Allmaier, G., B.H. Chao, H.G. Khorana and K. Biemann. (1986). *The Determination of the Location and Nature of Amino Acid Substitutions in Biosynthetic Bacterio-opsin*, presented at the 34th Annual Conference on Mass Spectrometry and Allied Topics, Cincinnati, OH, pp. 308–309.
3. Barber, M., R.S. Bordoli, R.D. Sedgwick and A.N. Tyler. (1981). *Fast Atom Bombardment of Solids (F.A.B.): A New Ion Source for Mass Spectrometry.* J. Chem. Soc. Chem. Comm. 7: 325–327.
4. Biemann, K., S.A. Martin, H.A. Scoble, R.S. Johnson, I.A. Papayannopoulos, J.E. Biller and C.E. Costello. (1986). *How to Obtain and How to Use Mass Spectral Data at High Mass.* In *Mass Spectrometry in the Analysis of Large Molecules*, C.J. McNeal, Ed., John Wiley & Sons, Sussex, England, pp. 131–149.
5. Cody, R.B., I.J. Amster and F.W. McLafferty. (1985). *Peptide Mixture Sequencing by Tandem Fourier-Transform Mass Spectrometry.* Proc. Natl. Acad. Sci. USA 82: 6367–6370.
6. Hewick, R.M., M.E. Hunkapillar, L.E. Hood and W.J. Dreyer. (1981). *A Gas-Liquid Solid Phase Peptide and Protein Sequenator.* J. Biol. Chem. 256: 7990–7997.
7. Holmgren, A. (1985). *Thioredoxin.* Ann. Rev. Biochem. 54: 237–271.
8. Hunt, D.F., J.R. Yates III, J. Shabanowitz, S. Winston and C.R. Hauer. (1986). *Protein Sequencing by Tandem Mass Spectrometry.* Proc. Natl. Acad. Sci. USA 83: 6233–6237.
9. Johnson, R.S. and K. Biemann. (1987). *The Primary Structure of Thioredoxin from* Chromatium vinosum *Determined by High Performance Tandem Mass Spectrometry.* Biochem., 26: 1209–1214.
10. Kammei, Y., Y. Itagaki, E. Kubota, H. Kunihiro and M. Ishihara. (1985). *The Design of a New Type Double Focusing Mass Spectrometer for High Mass Analysis*, presented at the 33rd Annual Conference on Mass Spectrometry and Allied Topics, San Diego, CA, pp. 855–856.

11. Khorana, H.G. (1986). *Rhodopsin, the Visual Pigment and Bacteriorhodospin.* Ann. New York Acad. Sci. 471: 272–288.
12. Martin, S.A. and K. Biemann. (1986). *Tandem Mass Spectrometry of Oligopeptides*, presented at the 34th Annual Conference on Mass Spectrometry and Allied Topics, Cincinnati, OH, pp. 854–855.
13. McLafferty, F.W., Ed. (1983). *Tandem Mass Spectrometry.* John Wiley & Sons, New York.
14. McNeal, C.J. and R.D. Macfarlane. (1981). *Observation of a Fully Protected Oligonucleotide Dimer at* m/z *12637 by* ^{252}*Cf-Plasma Desorption Mass Spectrometry.* J. Am. Chem. Soc. 103: 1609–1610.
15. Roepstorff, P. and J. Fohlman. (1984). *Proposal for a Common Nomenclature for Sequence Ions in Mass Spectra of Peptides.* Biomed. Mass Spectrom. 11: 601.
16. Yergey, J., D. Heller, G. Hansen, R.J. Cotter and C. Fenselau. (1983). *Isotopic Distributions in Mass Spectra of Large Molecules.* Anal. Chem. 55: 353–356.

Prospective trends in quantitative mass spectrometry

JAMES S. HOLLER
Division of Environmental Health Laboratory Sciences, Center for Environmental Health, Centers for Disease Control, Public Health Service, U.S. Department of Health and Human Services, Atlanta, GA 30333, U.S.A.

Abstract Solving most analytical problems requires information about the identity of compounds in a sample and some estimate of the amounts. Mass spectrometry, alone and in combination with chromatographic techniques, often provides the most satisfactory quantitative answer in terms of specificity and sensitivity. Monitoring human populations with exposure to elevated levels of toxic chemicals illustrates the type of analytical problem the chemist can solve by using existing technologies. Gas chromatography/mass spectrometry is the best method for measuring the extent of environmental contamination from toxic waste sites. Mass spectral methods are often used in regulatory situations to validate routine analytical methods for biomedical measurement. Measuring isotope ratios by mass spectrometry can provide information on the origins of certain compounds, as demonstrated in tracer studies for inorganic elements and for the determination of adulteration for several natural products. These analytical techniques will become more important in quantitative studies. New, low cost mass spectrometers and computerized data systems will be used in routine studies, and developments in chromatographic interface technology will facilitate specialized instrument systems designed for particular applications. A hope for future technological development is improved source design, with increased sensitivity and the ability to use samples more efficiently.

Introduction

To answer an analytical problem, estimating accurately the amount of compound present is often almost as important as identifying the chemical itself. The quantity of chemical present determines whether a toxic substance will result in: (1) acute toxicity, (2) a situation of long term chronic exposure with small biological effects, or (3) a level too low to induce any health effect to a particular organism. Mass spectrometry is one of the most useful techniques in organic analysis because of its ability to study large numbers of compounds whose mass spectrum may relate back to original chemical compound identity. A whole subdiscipline, that of quantitative mass spectrometry (Millard, 1978), has evolved in which the main emphasis is on answering the question, "How much of a compound is present?"

A brief review of mass spectroscopic instrumentation is in order. The mass spectrometer receives the sample in some form, generates an ion beam, and conducts a mass to change analysis. Sample introduction technqiues include gas or liquid chromatographic systems and a variety of probe devices. Ionization techniques include electron impact, chemical ionization, or, alternatively, a variety of modes dictated by special sample requirements, such as fast atom bombardment, field desorption, or chemical desorption. Types of analyzers include quadrupole systems, double-sector high resolution instruments, triple- and four-sector instruments. The spectra obtained may be that of the electron impact fragmentation; the $[M + H]^+$ ion with a few daughter ions in chemical ionization; or such newer approaches as parent or precursor scans, daughter scans, and metastable scans.

The mass spectrum produced from such study is a series of ion intensities at various mass-to-charge values or m/z values. The spectrum obtained can be either the result of one observation or the summation of a repetitive series of scans, but the x axis is one of m/z. This spectral presentation is good for interpretation when qualitative information is desired. Several modifications are made, both in data acquisition and presentation, when quantitative information is desired. Although it is possible to quantify, based on a single primary ion in a single mass spectrum, measuring several ions over a time period can improve precision and accuracy of quantification. These ion intensities can be compared with similar measurements for corresponding analytical standards. Because of the spread of sample concentration into the ion source, no matter which sample introduction is used, the presentation of this data mimics that of a gas chromatographic peak in appearance and in plotting observed ion intensity versus time. Using this approach, both qualitative and quantitative data are obtained in a single analysis. Precision of 5% to 10% and detection limits of 1–10 ng are possible using this approach.

Unfortunately, these detection limits are not sufficient to meet the needs for ultra-trace analysis. Several modifications in operational approach allow mass spectroscopic detection at 1 pg or below. First, the analytical question is changed from "What compound is present and how much?" to "Is compound x present above a concentration, and, if so, how much is present?" Under these conditions, the mass spectrometer is operated to monitor only the most prominent or important ions in the mass spectrum of x. Again, the data mimics a gas chromatographic peak in its appearance. This technique, denoted selected ion recording (SIR) or multiple ion detection (MID), result in an analytical method with precision of 5–10% and detection limits of 1–10 pg.

Unlike other types of spectroscopy, mass spectrometry is not precise in its measurement of absolute signals. The reproducibility of sample introduction into a high vacuum system, ionization, and mass analysis is not usually good enough to quantify compounds based upon signal intensity alone. The internal standard method alleviates the need for precise reproducibility in introduction or ionization because measurement of the ratio of two ions, (one due to the target compound and one for the internal standard) determines the quantity of target compound present. To conduct the internal standard method, a known amount of internal standard is added to the sample first before any preparation. Chromatography or other pretreatment is then conducted, and mass spectroscopic analysis is performed. Quantification is performed, using a calibration curve of the ratio of ion intensities of target compound to internal standard versus moles of target compound to moles of internal standard. A second approach is the calculation of a response factor:

$$RF_x = \frac{A_x}{A_{is}} \times \frac{Q_{is}}{Q_x}$$

RF_x = response factor for x; A_x = ion intensity corresponding to x; A_{is} = ion intensity for internal standard, Q_x = quantity of x, Q_{is} = quantity of internal standard initially added.

The concentration may then be determined by using the response factor:

$$C_x = \frac{A_x}{A_{is}} \times \frac{Q_{is}}{RF_x} \times \frac{1}{W}$$

C_x = concentration of x, W = sample weight.

This quantitative technique also relies on the fact that the internal standard will mimic the target compound in sample pretreatment and in analysis, thus compensating for less than 100% recovery. The ideal internal standard is one that is labeled with stable isotopes and hence is not differentiated in the method until it is detected by the mass spectrometer. Such an approach is described as isotope dilution-mass spectrometry.

Dioxin in people. An excellent example of the current capabilities of quantitative mass spectrometry is the measurement of dioxin in people. The manufacturing impurity, 2,3,7,8-tetrachlorodibenzodioxin (TCDD), which occurs in 2,4,5-trichlorophenol, is carried over to the final herbicide products, notably 2,4,5-T. Vietnam veterans and a number of groups of Missouri residents, who have possibly been exposed to the 2,3,7,8-TCDD, expressed apprehension upon learning of the toxicity observed in one animal species. As with many questions related to toxic chemicals, public policy is based in part on the perception of risk rather than on demonstrated hazards. The concern over the dioxin issue has supported considerable research. A significant portion of this research is devoted to the analytical determination of this and related compounds in various matrices.

The analytical method for measuring TCDD (Patterson et al., 1986) relies on a mass spectrometer operated in the selected ion recording mode at high resolution conditions, i.e., resolving power of 10 000. The high resolution approach increases selectivity by reducing the number of potential interfering compounds. This approach also appears to increase sensitivity by reducing background chemical noise. Using a suitable capillary column in the gas chromatographic introduction, which permits an isomer-specific analysis, provides additional specificity. The $^{13}C_{12}$ labeled TCDD functions as the internal standard for quantification. An extensive sample pretreatment includes isolation of the target compound through five different column systems. This analytical method is able to measure down to 1 part per trillion (ppt, i.e., 10 pg TCDD in a 10 g tissue sample). Based upon the analysis of quality control materials, a coefficient of variation (CV) of 19% is observed at the 1.6 ppt level and a CV of 11% at the 8.4 ppt level. The analysis of spiked samples is in good agreement with that target value – usually within 5% with a 5 ppt spike and within 15% with a 1 ppt spike. Tissue samples analyzed in duplicate generally agree within 10%. This reproducibility is remarkable because it includes biological variability in the samples.

Successfully conducting large studies in body-burden determinations requires that a number of specialists must be joined as a team approach in problem solving. Trained epidemiologists establish valid field studies and enroll volunteers in the study. This measurement currently requires 10 g of omental tissue, which can be obtained through surgical procedures that require a local anesthetic. Obtaining volunteers for such procedures is difficult. Careful statistical analysis of the data is critical in determining trends and differences between exposed and control groups. Chemists who are specially trained in sample preparation and others who are trained in instrument operation actually conduct the analyses. Public health officials must review the observed trends and estimate the risks associated with environmental or other sources.

The conclusion to date from human studies is that the presence of 2,3,7,8-TCDD as well as some of the other dioxin isomers is ubiquitous. Like PCBs and DDT metabolites, these compounds are found in all humans today. A main source of these compounds is municipal incineration. Human exposure to this and other sources has resulted in a mean tissue value approaching 7 ppt. Recent studies, soon to be published, demonstrate that exposure to high concentrations of dioxin does result in elevated body burden. Additionally, abnormal health effects have been clearly observed in this exposed population. More research is required, unfortunately, to understand the public health significance of these findings.

As a note on the technology supporting these studies, such methods would have been quite difficult to conduct only ten years ago. Certainly the magnitude of effort needed in monitoring hundreds of human samples would not have been possible. One of the technological advances is the commercially available fused silica capillary column. This is essential for achieving isomer specific separation. Electronic systems have improved to the degree of instrument stability necessary for this measurement process. The evolution of computerized data systems permits continuous interaction with the instrument system in adjusting operating parameters, and a variety of options in processing the data after acquisition (Chapman, 1978).

Several trends in instrumentation are apparent from these studies. There is increasing specialization in the treatment of the sample and the types of instruments used for the measurement process. Analytical difficulty in conducting these studies requires a highly specialized approach. Concurrent with this focused approach, each project team must include specialists in chromatography, in instrumentation and in data refinement.

This work also readily points to future needs for instrumentation. The continued search for zero will require increased instrument sensitivity while retaining the current selectivity (Crummett, 1979). Incremental improvements in instrument systems will retain the present selectivity, but additional sensitivity will be required to address the dioxin and other chemical issues. One of the problems in assessing the dioxin problem for humans has been the sample requirement of 10 g tissue obtained from volunteers by surgery. From the volunteers' point of view a better sample is a 10 ml serum sample. However, because of the unfavorable distribution of dioxin between fat tissue and serum, there is an increased sensitivity requirement of $\times$ 500 for an analytical method to measure "normal" value in serum. Such analytical characteristics are likely to be obtained only through a quantum leap in instrument source improvement. The changes in mass spectrometers in the last two to three decades have improved instrument transmission of ions to within a factor of 10 to the theoretical limit. Interestingly enough, the key to sensitivity is sample introduction and ionization, which occurs in the instrument source. However, source design has not changed significantly over the last 30 years and remains basically the same from manufacturer to manufacturer. A radical approach in sample handling and ionization must occur to meet needs in the next decades.

Monitoring hazardous substances in the environment. The extent of man-made chemical pollution of the environment is another area of significant public interest. Toxic chemicals throughout our environment can lead to extensive human exposure. The growth of activities by the U.S. Environmental Protection Agency and the passage of Superfund legislation are evidence of the increased interests of both the public and government in programs that will monitor and control such chemical pollution. Support for such programs continues to be strong, and thus these environmental programs are likely to continue for some time.

Quantitative mass spectrometry has played a predominant role in providing data necessary to evaluate the extent of pollution in many landfill sites and natural water systems (Rushneck, 1986). In the late 1970s, a group of methods, the so-called priority pollutant analyses, were developed and established as the acceptable methods for measuring a number of important chemicals in water. This procedure uses standard extraction techniques and, in many cases, detection by gas chromatography/mass spectrometry (GC/MS). Samples are run in the "full scan" mode for qualitative identification, and intensities of selected ions are used for quantification. These procedures have been modified to use new capillary columns in the analytical scheme and have been the main basis for the Superfund Contract Laboratory. After five years of operation under Superfund, this laboratory program now includes over 50 laboratories. It is estimated that over 60 000 mass spectroscopic analyses were conducted during FY '85.

Recent studies suggest that the Superfund Contract Laboratory program will significantly influence production of analytical mass spectrometer systems (Stafford, Kelley and Bradford, 1983). The advent of a second generation of low cost mass spectrometer systems could revise analytical approaches to regulatory environmental matters. In the past, official methods often relied on two gas chromatographic procedures, with alternate types of detectors and columns. With low cost mass spectrometer systems, the analyses can still supply the specificity needed while saving on capital and labor costs. Thus, we can anticipate the widespread availability of instruments capable of conducting these analyses, analytical standards for the target compounds, and software facilitating automated processing and research supporting this type of work. Because of the general approach taken by the method design, these procedures and instruments may be readily applied to analytical problems of a similar nature in biological monitoring, clinical analyses, and other disciplines.

A major shortcoming in environmental monitoring is translating the large volume of data produced by the laboratory into meaningful information. The day has passed when a computer printout of numbers is sufficient to meet the needs of a monitoring program. Protocols are being established today for routine electronic transmission of analytical data (Almich, Budde and Shobe, 1986). We also need new approaches that will insure valid data. These include new approaches to examine quality control trends, detect and eliminate artifacts, and present the final data in a format where trends are readily discerned. Certainly the evolution of statistical programs such as SIMCA and the advances in artificial intelligence will play major rôles in translating data into useful information.

One interesing trend evolving in mass spectrometry is the use of personal computers (PCs) as an accessory to instrument data systems. Now data generated and calculated by the data system is downloaded to the PC. Once there, it can be manipulated using any spread sheets or other general software programs. Because of the ease of networking PCs, this route may be the easiest route for electronic transmissions of analytical results, particularly in network systems.

Reference and definitive analytical methods. Quantitative mass spectrometry also forms the basis for definitive or reference methods in areas of clinical research and regulatory action. A hierarchy of analytical methods has been proposed to insure accuracy in a measurement system (Tietz, 1979). The routine method is one based upon low cost, high sample throughput, and the degree of accuracy required for the study being conducted. Unfortunately such methods often exhibit significant bias, (up to $\pm 25\%$), and this is cause for concern, particularly with clinical analyses. Reference and definitive methods were developed to evaluate routine clinical methods, monitor obvious problem areas, and encourage manufacturers of diagonstic kits used to improve any products used in routine analysis. Mass spectrometry is generally accepted as bias free when used as an isotope-dilution technique, and it is applicable to a number of organic compounds.

Our development of a candidate definitive method for serum cortisol (Patterson et al., 1984) is an example of the type of effort required to demonstrate a bias-free measurement for an analyte in biological matrix. Serum cortisol is an indicator of adrenocortical function, and it has been included for several years in laboratory proficiency testing programs. Our main analytical approach was to (1) initially spike the serum sample with the commercially available cortisol-d_3 compound, (2) extract with methylene chloride, (3) form the trimethyl silyl derivative, and (4) analyse by gas chromatography/mass spectrometry in the multiple ion monitoring approach. Analytical standards were made to bracket the target value and analyze the sample under study. The overall coefficient of variation (CV) for this method was 0.69% based upon 32 measurements.

We used a number of alternate approaches in sample preparation and mass spectroscopic analysis to demonstrate validity in this determination. One group of samples used high performance liquid chromatography (HPLC) as the initial separation technique, followed by derivatization and mass spectroscopic analysis. A second set of samples used solid phase (C_8) extraction, followed by derivatization and instrumental analysis. We also pursued alternate instrumental approaches. A group of samples were quantified using the molecular ions for the cortisol and the internal standard, $m/z = 636$ and $m/z = 639$. Another set of samples were quantified in the negative chemical ionization mode using the primary ions at $m/z = 459$ and 462. All the studies described support the original quantitative determination.

A suitable reference material is needed to transfer this accuracy base to reference and routine methods. We prepared a serum pool by ultra filtration, dispensing into five-milliliter aliquots and then lyophilizating. Such materials have been shown to be stable for years when maintained at $-40°C$.

Reference methods are applied in a similar manner to verify measurements obtained using routine methods. Reference methods do not require the intensive verification that definitive methods demand. Reference methods have often been required in studies of residue analysis, new drug applications, and other areas that require official approval or sanction. Confirmation of routine analytical results is particularly important where the sample matrix is complex, as is found in environmental and biological samples. A significant problem may exist with false positive determinations in such studies (Holler, Alley and Yert, 1983). For the foreseeable future, quantitative mass spectrometry is the method of choice for such confirmation because of its general applicability to many compounds and its inherent high selectivity.

Applications of naturally occurring $^{13}C/^{12}C$ isotope ratios. Carbon occurs in nature primarily as two stable isotopes, with 1.104% as ^{13}C and with 98.892% (existing) as ^{12}C. The formation of biochemical materials occurs through photosynthesis in plants, and, depending on the species involved, a specific fraction of these isotopes occurs. The two routes involve a three-carbon intermediate phosphoglycorate or a four-carbon dicarboxylic acid intermediate. The incorporation of ^{12}C is greater in the

C_3 route than in the C_4 route. With δ^{13} defined as the ratio of ^{13}C to ^{12}C times 1000, atmospheric carbon dioxide is observed as -7, average C_3 plants as -26.5, and C_4 plants as -12.5 (Fernelius, 1984). The mass spectroscopic measurement of this isotopic ratio has been recently proposed as an analytical method for determining the basis for adulteration of honey (Lucchesi, 1979). The method is based upon the fact the two different sources of sugar are formed in different biochemical pathways. Bees gather honey from C_3 plants, while corn syrup comes from a C_4 plant. Measurement of the $^{13}C/^{12}C$ ratio, therefore, will indicate whether adulteration has taken place. A similar analysis can indicate adulteration of orange juice (Doner and Bills, 1981) or apple juice (Doner and Phillips, 1981). This analytical approach has been used to investigate the type of diets involving in archaeological studies of early human civilization based upon the observed isotope ratios in bone collagen (Lewin, 1983). A number of other disciplines, using carbon, hydrogen, or nitrogen ratios (Bricout, 1982) have successfully applied this technique in environmental studies.

A very specialized type of mass spectrometer is required for these quantitative determinations. For carbon isotope ratio measurement, the sample is first combusted in a special inlet system to produce carbon dioxide. This gas is then introduced into a mass spectrometer system, and the isotope ratios are measured simultaneously with multiple collectors as detection devices. Calibration of the instrument system is based upon the analysis of a sample of carbon dioxide obtained from the analytical standard used for this determination.

This analytical approach to study natural processes is likely to increase in the future. This technique is applicable because of the occurrance of a number of different isotopes for individual elements in nature and because of the discrimination between isotope exhibited by many biochemical systems. As with other analytical techniques, the capacity to analyze smaller samples is needed to study very rare samples or to examine for in homogenity in larger samples. A number of needs have been described for instrumentation for this work (Wright, 1984). These include an update of instrument design and a novel means of generating and handling the sample gas from the sample. Chemical contamination of the sample becomes a significant problem as smaller samples are used. One possible approach is the use of ion probe techniques combined with the appropriate mass spectroscopic detector. The use of tandem mass spectrometry or mass spectrometry/mass spectrometry is an analytical approach which is just coming of age in analytical problem-solving. This technique uses two separate mass analyzers to conduct a determination. After introduction into the instrument and appropriate ionization, the resultant ion beam is passed through the first mass analyzer and a key ion selected. This ion may be the molecular ion, the $[M + H]^+$ ion, or a key fragment ion. A collision chamber is then used to impart extra energy to the chosen ion. This added energy causes the selected ion to fragment in a fashion characteristic of the chemistry of this ion. The fragment ions, called daughter ions, are then observed by the second mass analyzer, and hence the term tandem mass spectrometry. This approach can be much more selective than conventional mass spectroscopic analysis. The increased selectivity permits a significant reduction in chromatographic pretreatment of the sample, and reduces analysis time. The systems have yet to reach the estimate limit of 1000 quantitative determinations a day (McLafferty, 1981a). In a study quantifying hexachlorobenzene and 2,3,4-trichlorophenol in serum and urine (Johnson and Yost, 1985), better detection limits were observed and sample throughput increased by a factor of 30 obtained over the conventional capillary gas chromatography/mass spectrometry approach. Such speed of analysis, coupled with the ability to tolerate reduced sample cleanup, is very valuable when accurate, repetitive analyses are needed for a particular study. The appearance of lower-cost, second-generation commercial instruments will further facilitate the widespread applications of tandem mass spectrometry. This success of instrumentation in trace analytical applications has been related to the "4 Ss of trace analysis": sensitivity, selectivity, speed and $ (McLafferty, 1981b). Many developing trends in mass spectrometric instrument design address one or more of these characteristics. A number of new techniques in sample handling and introduction should increase selectivity and reduce sample pretreatment. The development of workable liquid chromatograph/mass spectrometer (LC/MS) interface (Garteiz and Vestal, 1985) promises a number of new methods for polar nonvolative compounds. Supercritical fluid chromatography (SFC), when combined with mass spectrometry, has a position intermediate between that of gas chromatography and liquid chromatography. Problems still remain for obtaining a workable interface to the high vacuum system and a wider choice of solvent systems. Lasers are finding increased use as a desorption and ionization mechanism. Recently, an instrument system uses a laser microprobe to examine the surface components of samples (McMahon, 1985). Unfortunately such a system is limited to qualitative determination because none of the traditional quantitative approaches apply.

A number of specialized mass spectrometer systems have a promising future. Fourier transform/mass spectrometry (FT/MS) (Johlman, White and Williams, 1983) has excellent selectivity potential because of its high resolution ability. This instrument also operates under complete computer control during the course of an experiment. A key requirement for high resolution operation is high vacuum at the source, unfortunately a situation somewhat incompatable with sample introduction. A successful sample-handling system has yet to evolve that is widely applicable to quantitative problems. A new generation of mobile mass spectrometers has been introduced commercially. Such a system could revolutionize the analytical approach to environmental monitoring. The reduction in sample handling and the facility for long-term monitoring studies could increase our knowledge of toxic chemicals in the environment. The new field of inductively coupled plasma/mass spectrometry (ICP/MS) could bring a new dimension in accuracy to inorganic quantitative determination by increased selectivity.

Acknowledgements

Donald Patterson was a key researcher in the development of quantitative steroids analysis, and he has played a critical role in the evolution of dioxin analysis at CDC. Other significant contributors to the dioxin quantitative program were Chester Lapeza, Lewis Alexander, and S. Jay Smith. I appreciate the patience of Quin Long in the preparation of the manuscript.

References

1. Almich, B.P., W.L. Budde and W.R. Shobe. 1986. *Waste monitoring*. Environ. Sci. Technol. 20: 16–21.
2. Bricout, J. 1982. *Isotope ratios in environmental studies*. Int. J. Mass Spectrom. Ion Phys. 45: 195–205.
3. Chapman, J.R. 1978. *Computers in mass spectrometry*. Academic Press, New York, N.Y.
4. Crummett, W.B. 1979. *Fundamental problems related to validation of analytical data elaborated on the example of TCDD*. Toxicol. Environ. Chem. Rev. 3: 61–71.
5. Doner, L.W. and D.D. Bills. 1981. *Stable carbon isotope ratios in orange juice*. J. Agric. Food Chem. 29: 803–804.
6. Doner, L.W. and J.G. Phillips. 1981. *Detection of high fructose corn syrup in apple juice by mass spectrometric* $^{13}C/^{12}C$ *analysis: collaborative study*. J. Assoc. Off. Anal. Chem. 64: 85–90.
7. Fernelius, W.C. 1984. *Is sugar from sugar beets the same as sugar from sugar cane?* J. Chem. Ed. 61: 249.
8. Fjeldsted, J.C. and M.L. Lee. 1984. *Capillary supercritical fluid chromatography*. Anal. Chem. 56: 619A–628A.
9. Garteiz, D.A. and M.L. Vestal. *Thermospray LC/MS interface: principles and applications*. L.C. 3: 334–346.
10. Holler, J.S., C.C. Alley and L.W. Yert. *Role of mass spectroscopic confirmation for routine methods in environmental analysis*. Presented at the 31st Annual Conference on Mass Spectrometry and Allied Topics; Boston, Massachusetts; May 8–13, 1983.
11. Houk, R.S. 1986. *Mass spectrometry of inductively coupled plasmas*. Anal. Chem. 58: 97A–105A.
12. Johlman, C.L., R.L. White and C.L. Williams. 1983. *Applications of Fourier transform mass spectrometry*. Mass Spectrometry Reviews 2: 389–415.

13. Johnson, J.V. and R.A. Yost. 1985. *Tandem mass spectrometry for trace analysis*. Anal. Chem. 57: 758A–768A.
14. Lewin, R. 1983. *Isotopes give clues to past diets*. Science 220: 1369.
15. Lucchesi, C.A. 1979. *Assuring the quality of honey. Is it honey or syrup?* Anal. Chem. 51: 224A–232A.
16. McLafferty, F.W. 1981. (a). *Tandem mass spectrometry*. Science 214: 280–287.
17. McLafferty, F.W., 1981. (b). *Tandem mass spectrometry in trace toxicant analysis*. Biomed. Mass Spectro. 8: 446–448.
18. McMahon, J.M. 1985. *Laser microprobe MS of inorganic and organic species*. Amer. Lab 17(6): 50–57.
19. Millard, B.J. 1978. *Quantitative mass spectrometry*. Heyden & Sons, Ltd., Philadelphia, PA.
20. Patterson, D.G., J.S. Holler, L.R. Lapeza, Jr., L.R. Alexander, D.F. Groce, R.C. O'Connor, S. Jay Smith, J.A. Liddle and L.L. Needham. 1986. *High-resolution gas chromatographic/high-resolution mass spectrometric analysis of human adipose tissue for 2,3,7,8-tetrachlorodibenzo-p-dioxin*. Anal. Chem. 58: 705–713.
21. Patterson, D.G., M.B. Patterson, P.H. Culbreth, D.M. Fast, J.S. Holler, E.J. Sampson and D.D. Bayes. 1984. *Determination of steroid hormones in a human-serum reference material by isotope dilution-mass spectrometry: a candidate definitive method for cortisol*, Clin. Chem. 30: 619–626.
22. Rushneck, D.R. 1986. *Environmental analysis with mass spectrometry*. Spectra, Vol. 10, No. 4.
23. Stafford, G.C. Jr., P.E. Kelley and D.C. Bradford. 1983. *Advanced ion trap technology in an economical detector for GC*. Amer. Lab. 15(6): 51–57.
24. Tietz, N.W. 1979. *A Model for a Comprehensive Measurement System in Clinical Chemistry*. Clin. Chem. 25: 833–839.
25. Wright, I.P. 1984. *$\delta^{13}C$ measurements of smaller samples*. Trends Anal. Chem. 3: 210–215.

Microcomputer interfacing in the analytical laboratory

T.C. O'HAVER
Department of Chemistry and Biochemistry, University of Maryland, College Park, MD 20742, U.S.A.

High-level interfacing

It is convenient in dealing with interfacing to distinguish between *high-level* and *low-level* interfacing, much as we distinguish between the high-level and low-level programming languages. High-level interfacing is employed on "computer-ready" instruments which already have a standard computer interface built in, most commonly an RS-232 or an IEEE-488 interface. In such cases the actual interfacing often involves nothing more that obtaining the proper cable with suitable connectors and an interface card for the computer. Some computers have a built-in RS-232 or IEEE-488 port, but in most cases it will be necessary to buy an interface card which plugs into the computer's bus and connects by cable to the instrument. RS-232 and IEEE-488 interface cards are available for most popular brands of microcomputers. Of the two, IEEE-488 is more standardized and likely to require little or no "fine tuning" to get it to work. On the other hand, there are quite a number of ways that RS-232 can be set up, both with respect to the wiring of the cable and connectors (especially the "handshaking lines") and with respect to the "communication parameters" such as baud rate, parity, number of data and stop bits. A book on data communications (1) will be essential for understanding and working out these details.

The RS-232 and IEEE-488 interfaces are *character-oriented*, meaning that programming them is simply a matter of sending and receiving alphanumeric characters. Thus, sending a command to the instrument through the interface is like sending output to a printer, which is also a character-oriented device. For example, you would simply execute a **PRINT** or **OUTPUT** in BASIC or a **Writeln** in Pascal. Similarly, getting input from the instrument through the interface is like getting input from the keyboard; you would execute and **INPUT** or **ENTER** command in BASIC or a **Readln** in Pascal. All high-level languages support character-oriented input and output, although the details naturally depend on the system; in particular there has to be some way to distinguish which interface card you are communicating with. In general, the interface card will come with documentation which will include program examples in popular languages such as BASIC and Pascal, as well as assembly language, for that model of computer.

How do you know what commands to send to the instrument? This depends only on the instrument itself and must be found in the manufacturer's documentation. For example, the Shimadzu model 250 u.v.-vis spectrophotometer can be obtained with an IEEE-488 interface which recognizes a large set of simple alphanumeric compounds, a sub-set of which is listed in Table 1.

For example, to set the spectrophotometer to a wavelength of 500 nm, you send the characters "W500". To read the absorbance, you send the letter "D" and then read the input as a string. For example, here is a simple routine, written in HP BASIC, to set wavelength to 500 nm, then read the absorbance and store it in the variable "A".

```
OUTPUT 709; "W500"
OUTPUT 709; "D"
ENTER 709; B$
LET A = VAL (B$, 2)
```

The 709 comes about as follows: 7 is the interface number (the HP 85 has a built-in IEEE port) and the 09 is the device address of the spectrophotometer (given in the manufacturer's documentation). The VAL (B$, 2) function strips off two extra leading characters which are sent by the spectrophotometer along with the absorbance.

Low-level interfacing

The RS-232 and IEEE-488 interfaces are convenient, but may not be flexible enough for some purposes and are not available on all instruments, particularly older models. In many cases only a analog voltage output signal may be available, such as might be displayed on a strip-chart or X–Y recorder. In such cases you will have to use a analog-to-digital converter (ADC). ADC's are commercially available for most popular brands of microcomputers either as plug-in boards or as part of a complete data acquisition hardware system.

Table 1. Shimadzu IEEE-488 commands

W	Set wavelength
S	Set slit width
D	Read absorbance
V	Set mode
J	Set scan speed
A	Begin scan
H	Set start wavelength
I	Set end wavelength

A block diagram of a typical commercial converter product, shown in Fig. 1, includes a multiplexer to provide multiple input channels and a variable gain amplifier. The important characteristics include:

1. *Resolution,* expressed in terms of the number of bits. The most commonly encountered resolutions and the approximate corresponding reading precisions are:

8 bits: $\approx$ 1/250
12 bits: $\approx$ 1/4000
14 bits: $\approx$ 1/16000
16 bits: $\approx$ 1/65000

Higher resolution gives better precision and greater dynamic range but are more costly. Generally, 12 bits is considered adequate for most laboratory work.

2. *Speed,* expressed as the conversion time in μsec or msec, determines the maximum number or readings that may be taken per second, although in most cases the speed of execution of the software is the limiting factor. Typical conversion times range from a few μsec to several msec.

3. *Number of input channels.* Most commercial ADC's have 8 or 16 input channels which can be connected to different signal sources. They can

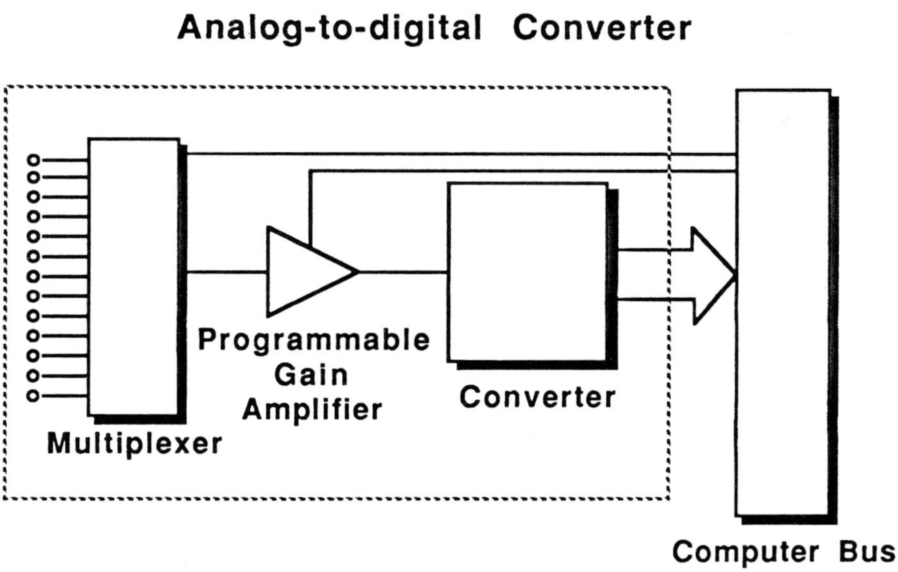

Figure 1. Block diagram of typical analog-tp-digital converter board for microcomputer data aquisition.

actually read only one at a time, but switching from one input to another can be done very quickly so as to give the effect of monitoring several signals simultaneously.

4. *Input sensitivity* refers to the input signal voltage which gives a full-scale response on the ADC. Ideally this should be able to be changed by the user in order to accomodate different instrument output voltages. Some ADC's set the input sensitivity by means of hardware jumpers or switches, but it is more convenient if the input sensitivity is software programmable. That way, the software can be written to switch ranges automatically as needed to handle large changes in input voltage level.

5. *Type of interface.* Commercial ADC's are either circuit cards which plug into the computer's bus or they are separate boxes which connect usually by means of a parallel or a high-speed serial connection such as RS-232 or RS-422. The circuit cards are less expensive but are specific for one brand of computer. External boxes which use a standard interface (RS-232, RS-422 or IEEE-488) may be more easily adapted to different computer models.

In some cases it is desired to have the computer generate a variable output voltage which can be used to control some part of the experiment. This can be done with a digital-to-analog converter (DAC). These are basically the reverse of ADC's and, like ADC's, come in various resolutions. They are usually faster and cheaper than ADC's of the same resolution, and are available from the same companies that make ADC's.

Digital interfacing

Parallel digital inputs and outputs are commonly available on computers, either built-in or as an optional card. Each wire of a parallel interface carries a binary (off/on) signal which can be used individually to control or sense an event which is naturally off/on in nature. Most commonly these utilize so-called "TTL" levels, meaning that they take *off* to be a low voltage (typically between 0 and 0.2 volts) and *on* to be a voltage between 3 and 5 volts. If required, the off/on sense can be reversed with a very simple logic circuit element called an *inverter.*

A common requirement in automating equipment is to have the computer activate a switch or pushbutton on an instrument. Figure 2 shows a few approaches which can be used. First you must use a voltmeter or oscilloscope to determine the voltage between each terminal of the switch and ground. If one end of the switch is grounded and the

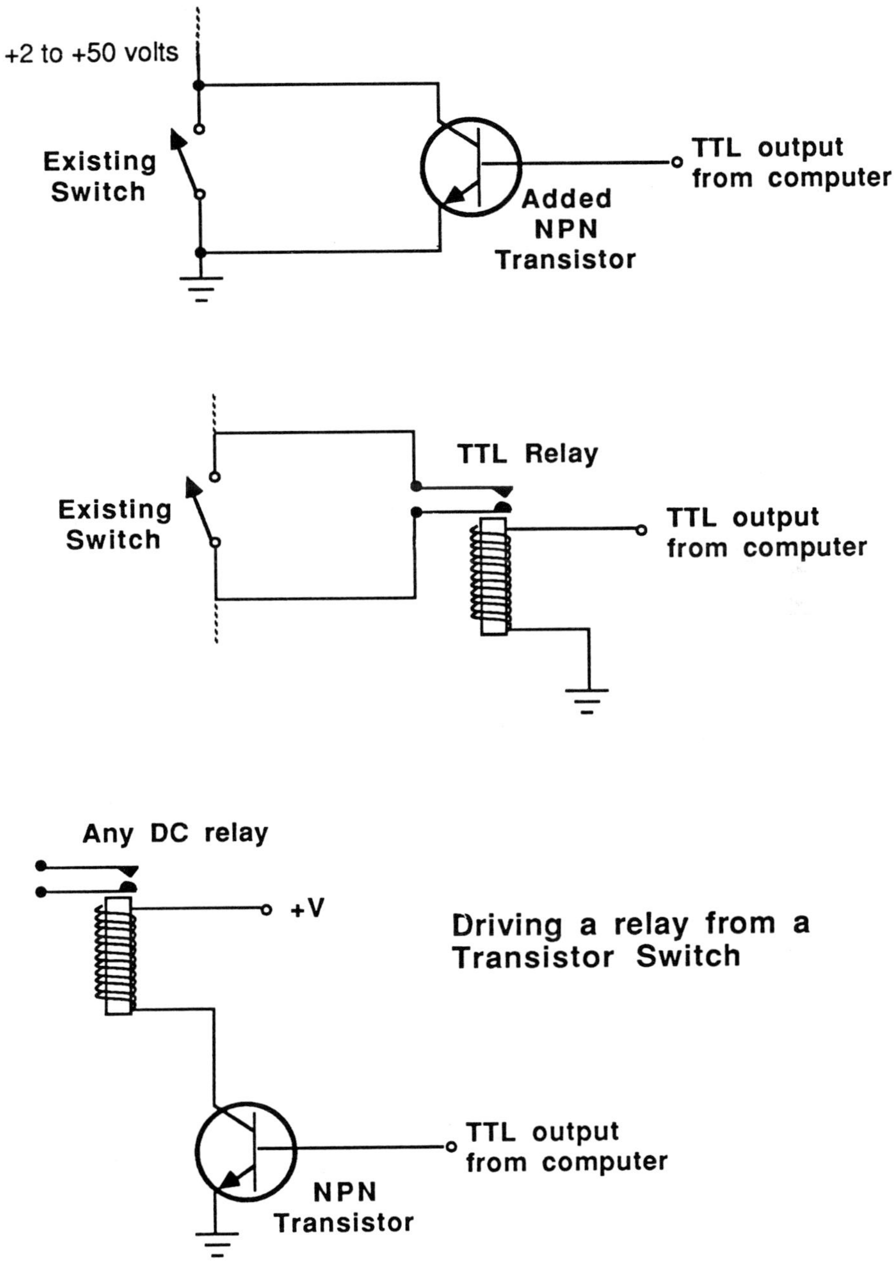

Figure 2. Adding computer control to an existing instrument.

other end is positive with respect to ground by no more than 50 volts or so when the switch is open, then you can use a simple NPN transistor as shown. Just about any NPN silicon transistor will work. Be sure to connect the grounds of the instrument and the computer together. A more general solution, which will work under more circumstances, involves the use of a relay. Relays which operate directly from TTL voltage levels are available, or a transistor can be used to drive the relay, as shown at the bottom of Fig. 2.

Commercial data acquisition systems

A number of data acquisition systems, which incorporate digital-to-analog conversion, signal conditioning, digital-to-analog conversion and digital input/output, are commercially available for popular microcomputers. A few of these are listed in Table 2.

Most of these systems are rather general purpose devices and can be used for a wide range of laboratory interfacing projects in spectroscopy, chromatography, electrochemistry, etc. The Oriel system has a number of additions designed specifically for spectroscopy, including a 6-decade, computer controlled nanoammeter with dark current subtraction, a high-voltage supply for photomultiplier tubes, and drivers for stepper-motor controlled monochromators. Most vendors supply software for their systems, which varies from rather general-purpose data logging programs, to program language enhancements which allow customers to design their own application software, to stand-alone application programs which support various spectroscopy and chromatography experiments. It is useful to determine whether the source code for the software is available and whether it is possible for the user to modify the programs if necessary.

Data file interfacing: transferring data between computers

Most laboratories have more than one computer, and sooner or later the need arises to transfer some data acquired on one computer to another computer or to the same computer running a different software package. Thus data file compatability becomes an issue. In a few cases one particular program has become so popular that its data file structure has become almost a standard which can be read by other programs; the spreadsheet program Lotus 1-2-3 is an example. But in the majority of cases, the most universal file type is the plain text file, that is, a file

Table 2. Some commercial data acquisition systems for microcomputers

Interactive Mircoware Inc., Box 139, State College, PA 16804 (Apple, IBM-PC)
Keithley-DAS, 349 Congress St., Boston, MA 02210 (Apple, IBM-PC)
Dynamic Solutions Corp., 61 Lake Ave., Suite 305, Pasadena, CA 91101 (Apple, IBM)
Oriel Corp., 250 Long Beach Blvd, Box 872, Stratford, CT 06497 (Apple)
GW Instruments, 3 Ames St., Cambridge, MA 02139 (Macintosh)

containing only ASCII alphanumeric characters. Most programs can generate and read text files. A text file may not contain all the formatting and other special information contained in the program-specific file types, but it will contain most of the essential data. Data acquisition programs measure and store numerical data, such as spectra or chromatograms, which can be represented as a table of numbers. If these programs can generate text files containing the numeric data in a simple format, then it may be a simple matter to transfer the data to other programs which can read text files.

A typical example of this situation occurs when it is desired to transfer data from a data acquisition program to an independent word processing program to include the data in a printed report, to a spreadsheet or statistics program to perform further data reduction and analysis which is beyond the capabilities of the data acquisition program, to a data plotting program to prepare presentation-quality graphics, or to a data base program to store the data in a structured database along with data from other sources. The problem is eased when the other software runs on the same type of computer and under the same operating system; then it may be possible for the receiving program to read a text file generated by the data acquisition program. However, here may be some file format incompatibilities, such as the presence (or absence) of column headers, carriage returns, line feeds, or tab characters which are used as delimiters to separate the different elements in the data file. In that case some editing may be necessary, but that still may be easier than typing in the entire data set into the receiving program. Editing can be done by reading in the data file into your word processor (any decent word processor can read plain text files), or the process can be automated by writing a simple program in any convenient language (e.g. BASIC) which acts as a "filter" by reading the text file, reformatting it, and writing a text file in a different format.

A less convenient situation occurs when you wish to send data to a different computer or operating system; in that case the direct reading of files off the disk is probably out of the question. However, it is still possible to transfer text data between different computers by using telecommunication software (e.g. terminal programs) on both computers. Telecommunication programs are available for all computers and can invariably read and write plain text files. The two computers can be connected in one of two ways, either directly by means of an RS-232 connection or indirectly over the telephone lines by means of a modem. A direct connection is feasible if the two computers are not too far apart and if they both have, or can be provided with, RS-232 ports. The communications programs will have to be configured to recognize the

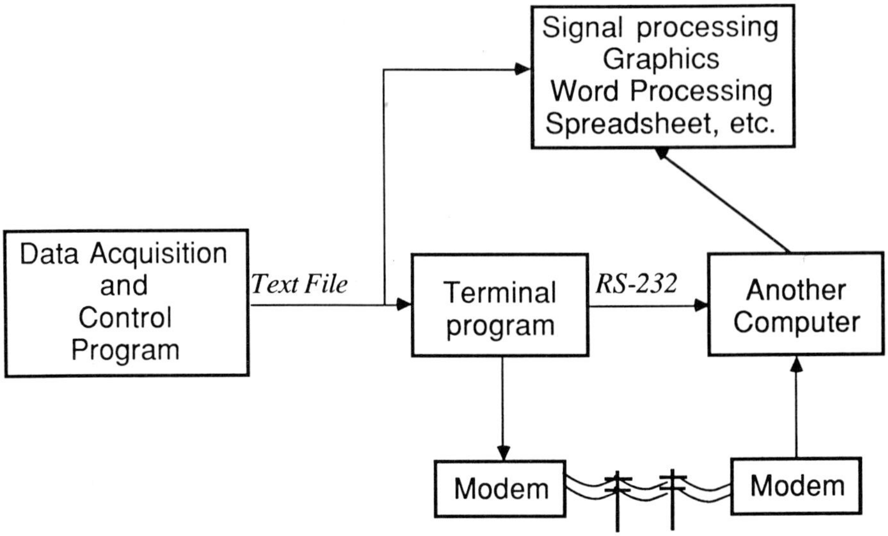

Figure 3. Data file interfacing.

RS-232 ports and to use the same communication parameters (baud rate, parity, number of data and stop bits). Also, a suitable interconnecting cable will have to be obtained, probably (but not necessarily) a "null modem" cable, which is wired specifically for a direct connection between two computers. If a direct connection is not practical, then a modem transfer over the phone lines may be possible. The disadvantages of a modem transfer are that it is much slower than a direct connection (typically 30 to 120 characters per second versus 480–1900 characters per second by direct connection) and that it may be less reliable, due to noise and distortion on the phone lines. Most telecommunicaiton software supports some kind of error-detection and correction protocol which will automatically correct communication errors (popular ones are XMODEM and KERMIT), but for this to work the same protocol must be used at both ends. Telecommunications is a bit technical and often hard to get working right the first time. Still, the modem and phone lines are the "great equalizer" of the computer world, allowing data transfer between even the most mismatched computer systems.

The user interface

A different aspect of interfacing which has been receiving much attention lately is the *user interface*, that is, the environment within which the user interacts with the computer system. New operating systems, and extensions of older operating systems, have been developed based on the idea that the operating system should encourage a standardized user interface, rather than requiring each independently developed program to develop its own from scratch (2, 3). In this way, standard unit operations such as opening and saving a file, transferring blocks of data, inserting and deleting, scrolling, etc., would be uniform across all available software for that system. These ideas were developed at the Xerox Palo Alto Research Center and popularized in the microcomputer world by the Apple Macintosh, Microsoft's Windows, and Digital Research's GEM. A characteristic of these systems is that they operate in a totally graphics-oriented environment (2). This offers greater flexibility than the traditional text-oriented systems and yields important benefits for the scientist and technical user. For example, special text characters and symbols are readily produced, such as boldface, italics, underlining, subscripts, superscripts, Greek and other non-English-language characters, and math symbols needed for equations. Graphics, such as chemical structures, are easily imbedded in text. A graphical user interface has proven to greatly increase the ease of learning for the new or casual user, provide a more consistent operation between independent programs, and allow easy transfer of text, numeric, and graphic data between different application programs. In many applications, actions are indicated and registered by the direct manipulation of on-screen graphic elements (buttons, icons, scroll bars, click boxes, etc) which have strong suggestive power and allow the new user to work intuitively with much less study of user manuals than is ordinarily required.

These developments have already had a major impact in office productivity software and are just beginning to be felt in the area of laboratory applications. Already there are examples in the area of chromatography data acquisition and gradient control software, and it seems likely that future laboratory systems will be increasingly concerned with user interface, as well as with the instrument interface. The new graphic-based operating environments are very demanding of computer power, memory, and screen resolution, but these needs are being met by improvements in technology. The newest personal computer models, which feature 32-bit processors, high-resolution bit-mapped screens, standard 1 megabyte memory, and floppy disk capacities of at least 800 K bytes, are rapidly approaching the level of professional graphic workstations (4, 5).

References

1. E.A. Nichols, J.C. Nichols and K.R. Musson. (1982). *Data Communications for Microcomputers*. McGraw-Hill, New York.

2. T.C. O'Haver. (1985). *Trends in User Interface Technology*. Trends in Anal. Chem. 4: 191.
3. T.C. O'Haver. (1986). *User Environments for PC Workstations*. American Lab. 18: 22.
4. Larry Press. (1985). Tomorrow's Personal Computer is Here Today, Abacus 2: 42.
5. J.P. Crecine. (1986). *The Next Generation of Personal Computers*. Science 231: 935.

Computer aided interpretation of spectra

J.T. CLERC
University of Berne, Switzerland

Introduction

Identification and structure elucidation of organic compounds is today mostly done with spectroscopic methods, the most often used methods include infrared spectroscopy (IR), mass spectrometry (MS), nuclear magnetic resonance spectroscopy (NMR), and spectroscopy in the ultraviolet and visible region of the electromagnetic spectrum (UV).

Even though the fundamental theory of the techniques mentioned are well known, calculation of spectra from first principles is generally not feasible. Reliable values for the various parameters required for such calculations are in most cases not available, the calculations use excessive amounts of cpu-time, and the precision of the results is not adequate for practical applications in analytical chemistry. Thus, the analyst still has to rely heavily on the use of (at best semi-) empirical methods, based on previously accumulated reference data from compounds of known structure.

The high throughput of modern instruments, the impositions of legislation and environmental protection, and the costs of qualified manpower make sophisticated procedures for automated evaluation of spectroscopic data increasingly important. Thus, considerable effort is currently invested into the development of such systems. All systems hitherto described in the scientific literature base directly or indirectly on complications of reference data. This paper will discuss some of the more successful approaches to computer aided spectra interpretation and, in particular, the important role played by the reference data base.

The mathematical model

There is obviously a functional relationship between the chemical structure of a compound and its spectrum. This may abstractly be formulated as [1]:

$$\text{Spectrum} = F\,(\text{structure}) \qquad (1)$$

This deceptively simple formulation immediately suggests how the structure of an unknown compound might be elucidated: Invert the

Send proofs to: J.T. Clerc, Pharmazeutisches Institut, Baltzerstr. 5, CH-3012 Berne, Switzerland function F to F^{-1} and apply it to the spectrum of the unknown compound to get

$$\text{Structure} = F^{-1}\,(\text{spectrum}) \tag{2}$$

However, this is in general not possible due to several reasons. The two most important limitations are the following. The function F is generally not known, and neither is the function F^{-1}. Furthermore, the forward function F is a projection type function, its application to a chemical structure results in a loss of information. Thus, the function would not be invertible, even if the function F would be known and simple.

The human analyst thus uses an approach, where the extremely complicated function F, which relates the total structure to the total spectrum is partitioned into a set of more simple functions G_l, relating partial structures to partial spectra:

$$\text{Partial spectrum } i = G_\text{i}\,(\text{partial structure } i) \tag{3}$$

The functions G_i are selected to be invertible. Thus, operating with G_i^{-1} on a suitably selected partial spectrum can recover the respective partial structure.

$$\text{Partial structure } i = G_\text{i}^{-1}\,(\text{partial spectrum}) \tag{4}$$

The various partial structures identified by the application of (4) are then combined to give the total structure of the unknown compound at hand

$$\text{total structure} = \text{Sum}_\text{i}\,(\text{partial structure}_\text{l})$$

This formalism seems very abstract and mathematical, and far removed from the analyst's day-to-day work. However, all parts of it may easily be identified with tools familiar to the practical chemist. The spectrometer is an analog computer, that implements the function F of eqn. (1). The partial functions G_i and their inverses G_i^{-1} are represented in spectra correlation tables, correlation charts, additivity rules, and other rules of thumb.

The general algorithm

With the partition of the function F into simpler partial functions G_i the context information is lost. One no longer looks at the full structure, only isolated substructural elements are perceived. In a real compound, their spectral properties will depend on the neighbouring groups and may thus vary considerably. This results in the G_i functions being fuzzy. Spectra correlation tables do not give one single value for the spectral

Table 1. General algorithm for spectra interpretation

1. Partial structures. Apply the G_i^{-1} functions to compile (a) a list of substructures possibly present (good list) and (b) a list of substructures definitely not present (bad list) in the compound at hand.
2. Consistent sets. From the lists compiled in step 1 select all sets of structural elements possibly present simultaneously in the compound at hand which are self consistent and not at variance with non-spectroscopic evidence.
3. Structure assembly. From all sets generated in step 2 generate all possible full structures that are in accordance with the rules of chemistry.
4. Spectra prediction. Predict the spectral data for the next tentative structure from the list of step 3, using the approximate G-functions and taking into account the structural context.
5. Spectra comparison. Compare the predicted spectrum with the experimental spectrum. If the two spectra are similar, keep the tentative structure as a possible solution. Then repeat steps 4 and 5 until the list from step 3 is exhausted. Then stop. The list generated in step 5 now contains all possible solutions to the problem.

parameters associated with a given partial structure, they rather give a range. Consequently, application of the inverted functions G_i^{-1} results not in a unique solution, one rather finds, that a given subspectrum, i.e. a given subset of spectral parameters may correspond to several different partial structures. On the other hand, if one uses G_i functions to predict the spectral properties expected for a partial structure embedded in a total structure, the context information is available and may be used to narrow down the range. Thus, prediction of spectral data from structures has an inherently higher precision than the prediction of the structure from spectral data. Therefore, to derive the structure of an unknown compound from its spectral data, one uses the stepwise procedure depicted in Table 1.

The key elements for the five steps in the general algorithm and the type of problem to be solved are now easily identified. In step 1, the selection of appropriate set of G_i^{-1} functions is critical. It has to be complete. A substructural element not represented among the G_i^{-1} functions will never be taken into consideration and thus will never be identified. The functions have to have an appropriate degree of fuzziness. If the range is too narrow, a substructure element in an exotic environment may be missed. If, on the other hand, the range is made too wide, too many solutions will be found and step 3 and/or step 2 will combinatorically explode.

Step 2 calls for the solution of a conceptually simple set of logical equations. However, due to the size of the problem and due to the wide variety of non-spectroscopic constraints, the procedure generally becomes quite involved.

Step 3 is a combinatorical problem. The methods for handling chemical structures with the computer are not too well developed. Also, to translate the rules of structural chemistry into precise, unambiguous and universally applicable computer programs is by no means easy.

In step 4, the quality of the model used for spectra prediction is critical. The prediction has to be precise. If the precision of the G_i-functions used in this step is not significantly superior to the precision of their inverses G_i^{-1} used in step 1, no reduction of the list from step 3 will result. Furthermore, the prediction has to be accurate. If there are errors in the predicted values, acceptable solutions (and, according to Murphy's law, the correct solution) will be missed in step 5.

The key word in step 5 is "similar". The relation "similar" has to be defined so as to adequately take into account the instrumental and technical artifacts (i.e. the overall reproducibility of spectra registration), as well as the precision and accuracy of the G-functions used in step 4.

All known computerized spectra interpretation systems (e.g. [2–15], for a review see [16]) work according to this general algorithm. They differ mainly by the way the five steps are implemented and by the degree of sophistication used for their computer implementation.

Structure inference

There are two pure strategies one may follow in step 1. On one hand one may define the spectral properties of the selected structural features so narrow, that no overlap occurs. In this case the list of substructures identified will contain only elements really present in the target compound. Consequently, consistency check, spectra prediction, and spectra comparison are hardly required. However, in most cases some partial structures present in the compound at hand will be absent from the list, and thus the result will not be correct. On the other hand one may use windows wide enough to identify all structural features of the target molecule. In this case, however, there will be many overlaps and the list will contain many structure elements not present in the target molecule. The structure inference step then becomes a noisy operation. The steps consistency check, spectra prediction and comparison are then needed to remove incorrectly identified partial structures.

In most of today's systems the designers have followed a strategy close to the first alternative, carefully minimising overlaps in the definitions. Thus, consistency checks are in most cases not too involved. Furthermore, no attempt is generally made to use the context information recovered in the structure assembly step for spectra prediction with improved precision. The two final steps, namely spectra prediction and spectra comparison are thus often not realized.

In the most straightforward systems the functions G_i^{-1} used to predict the spectral features of the target compound are part of the program. This approach is used for example in McLafferty's STIRS system for the interpretation of mass spectra [7] or in Sasaki's CHEMICS system [12], which uses data from various spectral techniques. Also the interpretation system described by Gribov [4] (using predominantly IR data) uses this approach. One problem here is that improving and extending the spectra-structure correlations is tantamount to making changes in the program. This problem may be solved by using a table of spectral characteristics for the set of substructures considered in the program (e.g. the IR system designed by Munk [8], or by using a rule generator program that accepts statements in (almost) natural language. This solution [17, 18] has been realized in the PAIRS system [15] described by Woodruff.

Still another solution is used in Bremser's ^{13}C-NMR system [2]. Here the inference of partial structures is based on a large reference data collection. For each line in the spectrum of the unknown the system retrieves from the reference library all substructures (atom centred fragments) which give raise to a line close to the respective line of the unknown. Of course, this approach is possible only where a strict one-to-one relation between lines and atoms exists. Furthermore, the reference spectra have to be interpreted, i.e. all reference lines have to be assigned to a well defined substructure. In ^{13}C-NMR these requirements can be met. It should also be possible in ^{1}H-NMR, but it has, to our best knowledge, never been realized.

Another problem in step 1 is the fact, that with a simple approach a given substructure element is considered either as possibly present in the target structure or as definitely absent, i.e. the inference rules are rigidly defined. Their results are either true or false with no intermediates. If a compound does not behave in accordance with rules, the system will break down. Some systems use a more relaxed approach by assigning a probability to the occurrence of a given substructure rather than a strict yes/no answer. For example systems described by Hippe [5] and by Woodruff [15] do this explicitly, where the STIRS-system by McLafferty [7] uses this strategy implicitly.

Consistency check

This step eliminates false hits generated in the structure inference step. The degree of sophistication required varies with the precision realized in step 1. If the precision is perfect, this step is not needed at all. However, extreme precision goes parallel with low recall, and low recall in the inference step cannot be corrected in later steps. Thus, some consistency checking will always be required. Nonspectroscopic information, in most cases available from the source and history of the sample is introduced into the system at this point.

In many cases step 2 is integrated into step 1 and performed concurrently. A trivially simple example from IR spectroscopy, originally given by Gribov shall illustrate the general procedure.

We consider the functional groups CH_3, CH_2, and C=C only. From experience the following 3 statements may be formulated:

CH_2 always gives a band at $1460\,cm^{-1}$

CH_3 always gives bands at $1460\,cm^{-1}$ and $1380\,cm^{-1}$

C=C sometimes gives a band at $1650\,cm^{-1}$

These facts may now be cast into G functions, relating structural data to spectral data:

G_1: CH_3 implies 1380 and 1460

G_2: CH_2 implies 1460

G_3: C=C implies 1650 or $\overline{1650}$

where the bar is used to indicate negation.

Inversion of the G_1 functions leads to

G_1^{-1}: 1380 implies CH_3

G_2^{-1}: 1460 implies CH_2 or CH_3

G_3^{-1}: 1650 implies C=C

Both sets of functions may now be combined in a set of logical equations, which all have to have the value true for a consistent set:

$T_1 = \overline{1380}$ or CH_3

$T_2 = \overline{1460}$ or CH_3 or CH_2

$T_3 = \overline{1650}$ or C=C

$T_4 = \overline{CH_3}$ or (1380 and 1460)

$T_5 = \overline{CH_2}$ or 1460

$T_6 = \overline{C{=}C}$ or 1650 or $\overline{1650}$

Obviously, T_6 is always true (it represents a tautology) and may be dropped. Then, for a consistent set we have the condition

$T = T_1$ and T_2 and T_3 and T_4 and T_5 = true

has to be met.

Structure assembly

In this step the structure fragments resulting from the previous step are combined into meaningful chemical structures. If the aim of a system is to identify functional groups only (e.g. [13]), this step is not required.

In all other cases it serves to eliminate sets of structural fragments which cannot be combined into a meaningfull chemical structure. It thus complements step 2 by providing a filter using non-spectroscopic constraints. This important function is well represented in the program CASE [9, 10] developed in Munk's group and by Sasaki's program [19]. The CASE program is designed to generate from a set of partial structures a complete and non-redundant list of total structures, taking into account various structure constraints specified by the user.

If combined with a suitable module to convert the connectivity information into reasonably nice looking structural diagrams, it allows for a presentation of the results easily perceived by the chemist. Some systems use highly sophisticated algorithms to generate good looking structure diagrams for display of the results. Particularly nice drawings are produced by the systems of Shelly [20], Bremser [2] and Opferkuch [27].

A second important function of step 3 is to recover the context information lost in step 1. After step 3 is performed, the full structure of the candidate structure is available. Thus, the functional groups and partial structures inferred in step 1 are no longer seen in isolation. They are rather put into structural context. Therefore, it is now possible to predict their characteristic spectral data with greatly enhanced precision. Furthermore, ambiguity and overlap are eliminated, as the forward functions G_i may now be employed rather than the inverted G_i^{-1} functions.

Spectra prediction and comparison

In this step the spectral data for the candidate structures generated in step 3 are predicted, using the forward functions G_i. These functions, relating partial structures to spectral features need not to be the same as those used in inverted form in step 1. They may make full use of the context information (i.e. the surrounding structure) and need not to be invertible. Thus greatly improved accuracy and precision may be realized by using a second set of G_i-functions, more sophisticated than those used in the structure inference step. Most systems, however, do not make use of this possibility. Exceptions are the Bremser ^{13}C-NMR system [2] and the CASE program [9, 10].

The Bremser system uses a library of reference spectra as a pointwise representation of the G_i-functions. For every carbon atom in the candidate structure a search is made in the library for carbon atoms having the same set of first neighbours. This is the exact inverse of the procedure used in step 1 to infer possible substructure elements. In the spectra prediction step, however, further searches are performed taking into account more distant neighbours. Finally, the reference substructure most similar to the respective substructure in the candidate structure is used to predict the expected chemical shift value. This value is then compared to the actually measured value in the spectrum of the unknown. If there is a large deviation, the respective candidate structure is eliminated.

A conceptually simple but quite efficient procedure is used in the CASE system. The candidate structure is analysed for symmetry relations between carbon atoms. This information is then used to set an upper limit to the number of distinct lines expected in the (noise decoupled) ^{13}C-NMR spectrum. Again candidate structures exhibiting too many lines are dropped.

Discussion

Present systems using an algorithmic approach to the computer aided interpretation of spectra for the identification and structure elucidation of organic compounds may be compared within the framework of the general procedure given in Table 1.

In the first step a list of substructures possibly present in the target compound is assembled. The system has to look at isolated parts of the spectra, it cannot yet take into account the surroundings of the functional group under consideration. Thus, wide windows for the spectral data have to be used in order to be able to correctly identify substructures even if they are in an exotic environment. A functional group missed in this step leads always to an unrecoverable error. Thus, step 1 should emphasize recall rather than precision, even at the expense of an extensively long list of possible substructures. Furthermore, the list of substructures tested for in this step has to be essentially complete. This can be achieved either by using a short list of small and simple fragments or by using an extensive list of larger and more complex fragments. Practical realizations will fall somewhere between this two extreme cases. The fragment selection is guided by the type of spectroscopy used. Best performance is achieved with fragments for which presence as well as absence from the target structure can be unambigously inferred from the spectroscopic data. If this is not possible, it seems to be advantageous to give priority to fragments for which absence from the target structure can be proved. If the respective spectral features are missing in the spectrum of the unknown, the fragment can be omitted from the list. If, however, only presence can be proved, the fragment has to be put on the list even though the respective spectral features are not present in the spectrum.

The next two steps, consistency check and structure assembly work entirely in the structure domain, i.e. they are fully independent from the spectroscopic technique used. One should therefore keep them strictly isolated from the other parts of the program in order to retain their universal applicability.

The last two steps, spectra prediction and spectra comparison have received the least amount of attention. Here significant improvements seem possible. An information gain is realized only if the prediction rules are significantly more precise as the rules used in step 1. This is not too difficult to achieve.

What one needs here is a set of rules that approximate the G_i-functions, relating one selected spectral datum to a set of structure parameters, as e.g. the well known additivity rules routinely used in NMR spectroscopy do [21]. In step 1, however, one needs a rule relating one selected structure parameter to a set of spectral features. This is much harder to realize.

When inferring partial structures from spectral data, the values measured for the unknown at hand are compared to the ranges expected for a given substructure. If the data from the unknown fall within these ranges, the respective substructure may be present in the target structure. The ranges used have to be wide enough to cover the variability of the spectral parameters due to the unknown structural context. In the spectra comparison context, however, the range needs only to cover the uncertainities of the prediction rule and the experimental errors. Thus, checking a candidate structure against the experimental spectrum is a far more efficient filter for the elimination of noise structures then checking the experimental spectrum against a list of partial structures.

Library search systems

Overview

From a computer science point of view algorithmic spectra interpretation systems, often classified under the heading "artificial intelligence", are undoubtedly very interesting. Research work in this field has lead to a better understanding of the problems involved and many new and useful concepts for handling structural and spectral data with the computer have been developed. Up to now, however, algorithmic systems have not yet found widespread application in the chemical laboratory. Library search systems (e.g. [7, 22–32], on the other hand, are routinely used in many laboratories for the identification and structure elucidation of organic compounds.

Library search systems do not use explicitly formulated interpretation rules, they rather rely on a large library of reference spectra from compounds of known structure. The spectrum of the unknown at hand is compared to all library spectra. The structures of those reference compounds exhibiting spectra closely similar to the spectrum of the unknown are selected and put out to the user. It is then up to the user to decide whether one of the top hits is identical to the unknown or if they just have some structural features in common. Library search systems also follow the pattern given in Table 1, however, they make some clever shortcuts, thereby avoiding most of the problems associated with algorithmic systems. Step 1, partial structures, is skipped in library search systems, the spectrum of the unknown at hand is in no way interpreted. Thus, step 2, consistent sets, in not applicable. In step 3, structure assembly, all structures represented in the reference library are selected as candidate structures. Spectra prediction, as called for in step 4 then becomes trivial. The reference library includes the real spectra of all tentative structures selected in step 3.

This amounts to using the exact version of the *F*-function for spectra prediction. Reliability, accuracy and precision are thus determined by the respective parameters of the reference library and do not depend on the quality of a simplified model. Step 5, spectra comparison, is the only major processing step left. The quality of the similarity measure used in step 5 is of utmost importance for the performance of library search systems. For algorithmic systems this is generally much less critical. Provided the structure generator steps are not completely irrational, the number of tentative structures generated is limited, and the structures have a reasonable relation to the spectral data of the unknown. If some of these structure proposals manage to pass the similarity filter, not too much harm is done to the result. In library search systems, however, all entries in the library are indiscriminately taken into consideration. If the similarity filter does a bad job, the hit list will be dominated by noise structures, bearing no relation at all to the unknown at hand.

The model

The most important thing to realize when dealing with library search systems is that when we compare spectra, we really mean structures. We may look at the system in the following way. Each chemical compound is characterized by a set of structural descriptors and by a set of spectral descriptors. We may now set up two independent multidimensional coordinate systems, where to each coordinate axis is assigned one descriptor from the respective set. Every conceivable chemical compound can be mapped into a point in each of these multidimensional spaces, using the values of the respective descriptors as coordinate values. We refer to these two multidimensional spaces as structure space and spectra space respectively.

Comparing the spectra of two compounds amounts to measuring the distance between their respective points in spectra space. The degree of similarity between the two spectra is inversely proportional to this distance. Library systems now base on the assumption, that two points close together in spectra space will also be next to each other in structure space. Consequently, for a library system to be useful, the two spaces have to be similarily structured.

The descriptors for the structure space and the distance measure to be used are not defined in a library search system. They are implied by the user of the system and are determined by tradition and by the type of problem to be solved. If e.g. the user is predominantly concerned about chemical reactions, he will most probably consider compounds exhibiting the same functional groups as similar and not pay much attention to the remainder of the molecule. In another case he may be more interested in the skeleton of a compound. Then he will base his similarity measure for structures mainly on the skeleton and give low weight to the number and type of functionalities present.

The spectral features defining the spectra space and the distance measure to be used in this space is, however, an integral part of the library search system. The goal in defining these descriptors is to model in the spectra space the structure of the spectra space as closely as possible. As the structure space is not part of the system and exists only in the views of the user, it is impossible to do this perfectly.

Spectral feature selection

The spectral features used to span the spectra space determine how well the spectra space will conform to the structure space. We want the description of the compounds to be sensitive to differences in the chemical structure, but at the same time insensitive to instrumental and technical artifacts. The type of spectral features to be selected also depends on the search strategy used.

One may assume that the unknown at hand is documented in the reference library. The goal is then to retrieve from the library the reference compound identical to the unknown, despite the fact that the sample may be impure and that the spectra recording conditions may be different. We refer to this strategy as identity search.

On the other hand one may assume that the unknown is a new

compound which is not documented in the library. The system is then expected to put out to the user a list of reference compounds structurally similar to the unknown. Furthermore, one tacitly assumes that the sample is reasonably pure. This strategy is referred to as similarity search.

Sensitivity to sample preparation techniques and to recording techniques can often be minimized by using suitably defined relative values for the spectral parameters. In mass spectrometry for instance, signal intensities may be related to the local total ion current rather than to the strongest signal. This effectively compensates for mass discrimination. In infrared spectroscopy it has sometimes proved adantageous to use integrated absorption intensities rather than maximum absorbance values and differences in absorption band positions rather than absolute values. This often helps to compensate for matrix effects. The most simple and straight forward way is, however, to use a large enough window.

To select for appropriate sensitivity to structure requires insight into the theory of, and much practical experience with the spectroscopic technique involved. It is here that spectroscopic knowledge may enter into the system. For an identity search system, any type of structure sensitive features may be used, but features changing their values significantly even with minor structure changes are preferred. Spectroscopic knowledge is not necessarily required. For example an infrared library search system which compares two spectra by summing the squares of the difference between the two curves at every wave number will make a fairly good identity search system. If instead of the curves their first derivative is used, baseline shifts will be eliminated, and the system becomes a good identity search system. If, however, no compound identical to the unknown is in the library, the results tend not to be too useful.

Similarity systems require a more sophisticated approach to feature selection. Any arbitrary complex comparison can be split up into a set of binary comparisons. This also implies that similarity in fact means identity on a majority of the (binary) features considered. A highly structure sensitive feature will change its state when the structure is but minimally changed. A feature of low structure sensitivity changes its state only upon a large change in structure. An identity search system works well with a collection of highly structure sensitive features. A powerful similarity search system, however, requires a carefull mix between features of high and low spectral sensitivity. It is via this selection that spectroscopic knowledge and experience enter into the system.

Implementation

When comparing the spectrum of an unknown compound to the spectra stored in a reference library the following operations are performed.

The spectrum of the unknown is converted to a series of binary values, each corresponding to the state of a predetermined spectral feature. The same conversion is done with every reference spectrum from the library. The two binary strings are then combined with the exclusive or function and the number of bits set is counted. The result is the number of (binary) spectral features the two compounds have in common. If the spectral features have been well selected, this number will be roughly proportional to the number of structural features common to both compounds and thus is a measure of structural similarity. This comparison is repeated for all compounds in the reference file, and the structures of the reference compounds getting the highest scores are put out to the user.

There are two important facts to note. First, the type of spectroscopy used enters the system only via the definitions of the spectral features to be used in the comparison. The result of the encoding process is a bit string, in which the spectroscopic information is no longer evident. Bit strings derived from different types of spectroscopy may thus easily be combined by concatenation of the respective bit strings and processed together. Thus integrated library search systems using spectral data from different spectroscopic techniques can be realized without any problems.

Secondly, the reference library needs to be encoded only once at the time the system is initialized. The bit strings generated are collected into a library of binary coded features which is then used in the comparison process. Thus, even if the definitions of the spectral features are very complex and the evaluation of their states requires significant amounts of computer time, the actual library search operation will still be fast. Only the spectrum of the unknown has to be encoded. The time required for this operation is insignificant compared to the time required to access and process the large number of bit strings from the reference library.

Composition of the reference data base

How many compounds are needed to make a library search system useful in practical applications? What compounds should be selected? For a system strictly limited to identity search the answer to these questions is trivial: We need all compounds we want to be able to identify. The larger the library, the more compounds the system can successfully deal with. For a similarity search system, however, these two questions touch on a fundamental problem. One thing is obvious: if, for a given unknown, there is no suitable reference compound in the library, no search system, however sophisticated, can ever produce a useful response. Thus, the contents of the spectral library set an upper limit to the performance of a search system. This seems to call for a comprehensive library containing the spectra of all known chemical compounds, or at least as many as one can get hold of. However, a moment's reflection shows that this is not a feasible goal. In the real world the number of compounds represented in a spectral library will always be a small percentage of nearly 10^7 chemical compounds described and characterized in the scientific literature. Thus, the probability of having a randomly selected compound in the library approaches zero. Moreover, in the most challenging case of a new compound the probability is exactly zero. Thus, every search system explicitly or implicitly aims at retrieving reference compounds similar (as opposed to identical) to the unknown at hand. One therefore expects as an answer, reference compounds belonging to the same compound class as the unknown (i.e. exhibiting common spectral or structural features). As compound classes overlap, there will be many different reference compounds similar to the unknown in question, when looked at from different points of view. If the library contains too many closely similar reference compounds from the same compound class a large number of them will be selected for retrieval by virtue of an unknown compound falling into this category. They will occupy the top places in the hit list, displacing reference compounds from other categories. For practical reasons the list of retrieved compounds has to be limited in length, therefore these second choice references will not even be presented to the user. This amounts to a serious information loss. Only the first entry in the hit list tells the user something new. All the following lines just repeat the same message. Thus, too large and unbalanced a library will lead to a deterioration in performance. One should therefore aim at an unbiased library of limited size, rather than spend time, work and money on futile attempts at universality.

Ideally, every compound class is represented by just a few carefully selected prototypes. Lengthy series of homologous compounds should be avoided. Except in special circumstances there is no need for storing several hundred spectra of, for example, long chain fatty acids. One or two model compounds will suffice to convey the structural concept "long-chain fatty acid" to the user. If, however, a given user's field of interest happens to be limited to long-chain fatty acids, he may well benefit from having many slightly different specimens available in his library. In general, the structure space should be sparsely and evenly populated with a few prototype compounds from a wide variety of compound classes. The user's region of interest, however, should be more densely populated. The general part of the library may well be acquired from a professional data compilation. For the special part, mapping the user's field of interest, only the user himself can provide an optimal selection of reference spectra. The compounds coming into his laboratory to have their spectra recorded are exactly those which are relevant to his work. Thus, it will not generally be possible to buy a ready-made spectral library optimal for one's own use.

The general part of a spectra library may well be available from outside sources, but the special part is best provided by the user himself. Thus, the software and hardware facilities to enrich a data compilation with one's own spectra are essential prerequisites. A new data set under consideration for insertion is first run as an unknown against the library. The compound should only be added to the library if the hit list does not give any reasonably similar reference compound. It may, however, replace a similar reference compound if the quality of the spectra is definitely superior.

It is difficult to give an estimate for the optimal size of a spectral library to be used in structure elucidation work. Our own experience would indicate the ideal number to be a few thousand rather than tens or hundreds of thousands. Most commercially available spectral collections are much too large and heavily biased due to the fact that everything available is indiscriminately added to the compilation. However, as long as the prospective customer's first question is "how many spectra do you have?" the professional data compilers are not solely to blame.

A third important aspect is the reliability, or more precisely the credibility, of the data. The analyst will consult the data bank if he is in doubt about the meaning of some piece of structural or spectral data. This is precisely the situation in which he cannot check the significance of the anwer he gets from the system. He has no option but to believe it. It is thus of utmost importance that the data are correct. It is virtually impossible to assemble an error-free spectra compilation. However, endeavours to keep the error rate at the lowest level economically attainable would do more to improve performance than hunting for the spectra of some exotic compounds.

References

1. Clerc, J.T. and Koenitzer, H. 1981. In *Data Processing in Chemistry*, Z. Hippe Ed., Elsevier Sci. Publ. Comp., Amsterdam, p. 151.
2. Bremser, W. and Fachinger, W. 1985. Magn. Res. Chem. 23: 1056.
3. Dubois, J.E. and Sobel, Y. 1985. J. Chem. Inf. Comput. Sci. 25: 326.
4. Serov, V.V., Gribov, L.A. and Elyashberg, M.E. 1985. J. Mol. Struct. 129: 183.

5. Jamroz, M., Latek, Z. and Hippe, Z. 1986. Anal. Chim. Acta, in press.
6. Hippe, Z. 1985. J. Chem. Inf. Comput. Sci. 25: 344.
7. McLafferty, F.W. and Stauffer, D.B. 1985. J. Chem. Inf. Comput. Sci. 25: 245.
8. Trulson, M.O. and Munk, M.E. 1983. Anal. Chem. 55: 2137.
9. Munk, M.E., Shelly, C.A., Woodruff, H.B. and Trulson, M.O. 1983. Fresenius Z. Anal. Chem. 313: 473.
10. Lipkus, A.H. and Munk, M.E. 1985. J. Chem. Inf. Comput. Sci. 25: 38.
11. Von der Lieth, C.W., Seil, J., Koehler, I. and Opferkuch, H.J. 1950. Magn. Res. Chem. 23: 1048.
12. Sasaki, S. and Kudo, Y. 1985. J. Chem. Inf. Comput. Sci. 25: 252.
13. Farkas, M., Markos, J., Szepesvary, G., Bartha, G., Szalontay G. and Simon, Z. 1981. Anal. Chim. Acta 133: 19.
14. Sazlontay, G., Csapo, Z. and Recsey Zs. 1982. Acta Chim. Acad. Sci. Hung. 111: 249.
15. Woodruff, H.B. and Smith, G.M. 1980. Anal. Chem. 52: 2321.
16. Gribov, L.A. and Elyashberg, M.E. 1979. CRC Crit. Reviews Anal. Chem. 8: 111.
17. Tomellini, S.A., Hartwick, R.A., Stevenson, J.M. and Woodruff, H.B. 1984. Anal. Chim. Acta 162: 227.
18. Tomellini, S.A., Hartwick, R.A. and Woodruff, H.B. 1985. Appl. Spectr. 39: 331.
19. Abe, H., Okuyama, T., Fujiwara, I. and Sasaki, S. 1984. J. Chem. Inf. Comput. Sci. 24: 220.
20. Shelly, C.A. 1983. J. Chem. Inf. Comput. Sci. 23: 61.
21. Pretsch, E., Clerc, J.T., Seibl, J. and Simon, W. 1983. Tables of Spectral Data for the Structure Determination of Organic Compounds. Springer-Verlag, New York.
22. Ruprecht, M. and Clerc, J.T. 1985. J. Chem. Inf. Comput. Sci. 25: 241.
23. Heller, S.R. 1985. J. Chem. Inf. Comput. Sci. 25: 224.
24. Henneberg, D. 1980. In *Advances in Mass Spectrometry*. A. Quale, Ed., Heyden and Son, London, p. 1511.
25. Lowry, S.R. and Huppler, D.A. 1983. Anal. Chem. 55: 1288.
26. Lowry, S.R., Huppler, D.A. and Anderson, D.A. 1985. J. Chem. Inf. Comput. Sci. 25: 235.
27. Zippel, M., Mowitz, J., Koehler, I. and Opferkuch, H.J. 1982. Anal. Chim. Acta 140: 123.
28. Coates, J.P. and Hannah, R.W. 1984. In *Fourier Transform Infrared Spectroscopy*. T. Theophanides Ed., D. Reidel Publ. Comp. p. 167.
29. Schaarschmidt, K. 1982. Z. Chem. 22: 438.
30. Shelly, C.A. and Munk, M.E. 1982. Anal. Chem. 54: 516.
31. Zupan, J. and Munk, M.E. 1985. Anal. Chem. 57: 1609.
32. Zupan, J. 1986. In *Computer Supported Data Bases*. J. Zupan, Ed., Ellis Horwood, Chichester, in press.

Computed tomography: the use and interpretation of imaged data in animal research

KREG A. LEYMASTER
U.S. Department of Agriculture, Roman L. Hruska U.S. Meat Animal Research Center, Clay Center, NE 68933, U.S.A.*

Abstract Animal scientists often conduct research that requires knowledge of body constituents, e.g., chemical composition or growth rates of tissues. X-ray computed tomography is a nondestructive procedure that presents imaged anatomical information. The potential contributions of computed tomography to animal science research and to animal production are currently being evaluated at several agricultural research institutions. Preliminary results reported in this paper are from an experiment conducted at the Agricultural University of Norway. Pigs weighing 30.6 ($\pm$ 5.2) kg live weight were scanned and the chemical composition of the carcass soft-tissue was determined. Chemical composition was also estimated using the tomographical data. Different statistical methods to summarize the imaged data were presented and used to estimate chemical composition. The methods varied with regard to precision of estimation and also biological interpretability.

Introduction

The primary objectives of animal science research are to reduce the costs of animal production and to improve product quality. These objectives are compatible with each other and relevant to problems facing the livestock industry today. Livestock producers face long-term industry threats from real or perceived concerns by consumers of the dietary-health questions presently surrounding red meats. Consumers seek nutritious, wholesome meat that is low in saturated fat and cholesterol content. Excessive fat is economically inefficient to produce and detrimental to product acceptability. Consequently, the reduction of fat is paramount to the livestock industries.

Therefore, it is appropriate that considerable efforts have been devoted to evaluation of numerous methods to estimate composition of living animals. The search for simple nondestructive methods to estimate composition has been intense. Levels of accuracy and precision that can be achieved vary considerably among methods. Experimental inferences often are limited to differences among treatments rather than individual animals due to the magnitude of imprecision. This limitation severely

*Mention of a trade name, proprietary product or specific equipment does not constitute a guarantee or warranty of the product by the USDA and does not imply its approval to the exclusion of other products that may also be suitable.

restricts the nature of hypotheses that can be addressed in all disciplines of animal science.

The technique of X-ray computed tomography, commonly referred to as CAT scanning, is universally accepted as a diagnostic tool in human medicine. Under the leadership of Professor Harald Skjervold, the Animal Breeding and Genetics Department of the Agricultural University of Norway recognized the potential opportunities of computed tomography to advance animal science research and production. This was the first agricultural institution to use computed tomography. Based on the encouraging results of a pilot trial (Skjervold et al., 1981), the Agricultural University of Norway acquired an X-ray computer tomograph in May of 1982. The purpose of this paper is to provide information concerning general principles of computed tomography, experimental procedures, biological interpretation of data and analytical methods. Preliminary results from an experiment conducted by the author while in Norway will be presented.

Principles of computed tomography

The following discussion of principles involved in X-ray computed tomography is based on specifications of the Somatom 2 computer tomograph manufactured by Siemens.

Computed tomography has been described as "a presentation of anatomical information by computed synthesis of an image from X-ray transmission data obtained in many different directions through the plane under consideration" (Hounsfield, 1979). The critical parameter defining the attenuation property of tissue is called the linear attenuation coefficient which is a function of numerous interactive processes between the energy of radiation and matter. As an initial level of radioactive energy (I_0) is transmitted through tissue of a specific thickness (d), the attenuation property of the tissue (μ) determines the intensity (I) of radioactive energy that passes through the tissue. This concept is illustrated in Fig. 1.

The X-ray tube of the Somatom 2 is positioned opposite of 512 detectors that monitor received intensities of X-rays. As this assembly rotates around the subject, pulses of radiation are generated at 12 msec intervals. During a single scan, 240, 360, or 720 pulses of radiation are generated, corresponding to scanning times of approximately 3, 5 or 10 s. With each pulse, 512 measurements of radioactive energy that have been transmitted through the subject are collected by the X-ray detectors. Each detector generates a current that is proportional to the received intensity of radioactive energy during each pulse. An analog to digital converter translates this information into a binary number that is stored

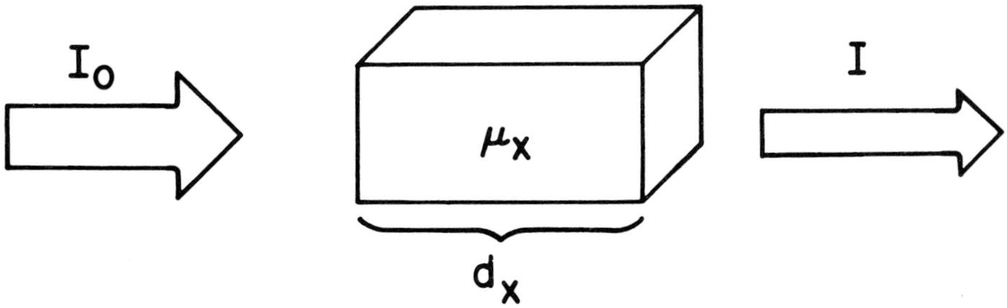

$$\text{where } I = I_0\, e^{-\int_0^D \mu_x\, d_x}$$

Figure 1. Function describing relationships among initial (I_0) and final (I) levels of radioactive energy, the linear attenuation coefficient (μ_x) and the thickness of tissue (d_x).

in the memory of the computer for later processing. This procedure generates 122880, 184320, or 368640 data from 3, 5 or 10 s scanning intervals, respectively. Each datum contribute to a set of simultaneous equations that upon solution yield the reconstructed image of the cross-sectional slice through the subject.

Data on radiation transmitted in many directions through a specific volume of tissue permit the linear attenuation coefficient to be estimated. An iterative approach involving a back-projection method is used to solve the simultaneous equations. Marginal totals of approximated values are compared with the measured marginal totals and discrepancies are then back-projected to produce a modified approximation. The set of simultaneous equations is solved in approximately 3 s by a computer dedicated to the system.

The linear attenuation coefficients are transformed into Hounsfield units (CT values) such that air, water and solid bone have values of – 1000, 0 and 1000, respectively. Muscle tissues have positive CT values, whereas adipose tissues have negative CT values. To the extent that CT values usefully measure differences in densities of tissue, computed tomography has potential to estimate compositional differences among individual animals.

A reconstructed image therefore can be regarded as a matrix of CT values, each value representing the density of a volume of tissue potentially as small as 1 mm^3. The image is transferred to magnetic media for storage and also displayed on the monitor at the operator's console. The reconstructed image represents a matrix of dimension 256 × 256 or 65536 CT values. An example of a reconstructed image taken at the 13th thoracic vertebra of a market weight pig is shown in Fig. 2.

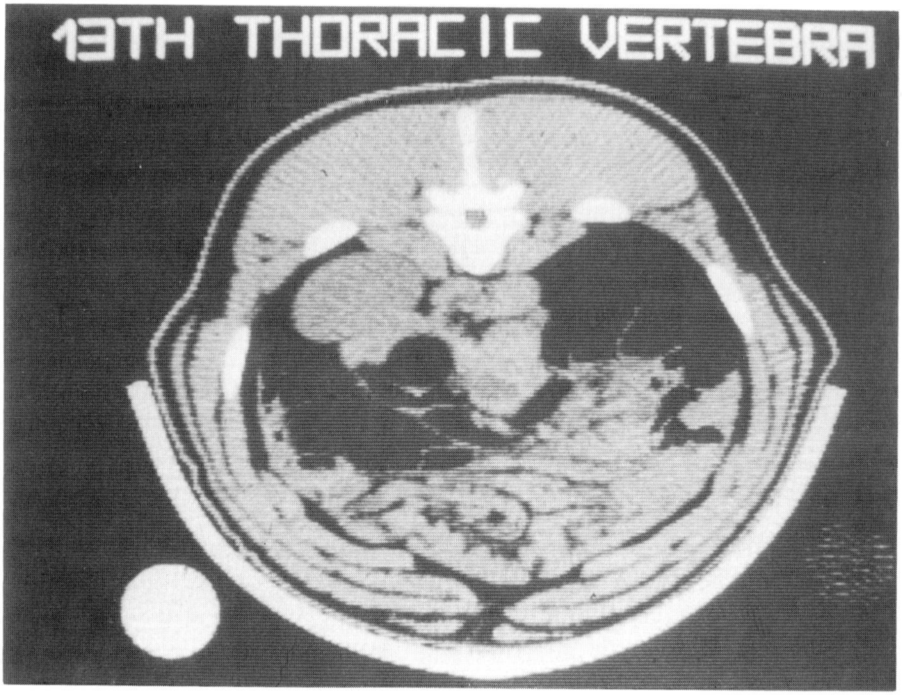

Figure 2. Example of a reconstructed image taken at the 13th thoracic vertebra of a 114 kilogram pig.

Experimental procedures

A general description of experimental procedures developed at the Agricultural University of Norway is presented. Requirements of specific experiments may deviate from those indicated below.

Animals were fasted at least 16 h prior to scanning. Pigs were given a tranquilizer followed by an anesthetic in order to reduce movement during scanning. It was not necessary to anesthetize sheep. Animals were placed on a plexiglass cradle, weighed and restrained prior to positioning the cradle on the patient support table. As X-rays were emitted from the temporarily nonrotating X-ray tube, the animal passed through the scanning gantry. This produced a horizontal profile of the animal so that standardized scanning locations could be identified by anatomical markers, e.g., the 13th thoracic vertebra. Animals were repositioned and scans recorded at the desired locations. Low quality scans, exhibiting artifacts that originated from animal movement, were repeated immediately after appearance of the reconstructured image. After scanning was completed, the reconstructed images were stored as data on permanent files for later analyses.

Following scanning, animals were slaughtered and carcasses chilled overnight. Carcasses were split and one side of each carcass was separated into bone, soft tissue and remainder (feet, head, kidney, skin, and kidney fat). The soft tissue was ground and a sample was chemically analyzed to determine proportions of water, fat, and protein.

Biological interpretation

The calibration of equipment for estimation of composition is a costly task, in terms of human labor and loss of carcass value. Yet, the scientific literature contains literally hundreds of regression equations describing numerous methods to estimate composition. The search for statistically robust equations has often been guided by numerical adventurism rather than by sound scientific methods. It is suggested that an understanding of biological principles underlying the growth and development of animals would facilitate more efficient use of the limited resources available for research. Furthermore, a consideration of biological factors might permit inferences to be extended effectively to specific experimental situations not previously tested.

As stated above, each reconstructed image (Fig. 2) illustrates 65536 CT values. In the experiment reported below, scans were taken at 11 different locations, yielding/720896 CT values per animal. This volume of data presents a statistical challenge. Software provided by the manufacturer was not adequate for quantifying compositional differences among animals. Programs were developed so that CT values representing the carcass could be separated from CT values representing the non-carcass components (air, organs, plexiglass cradle, and bodily gases). This was accomplished with the use of an electronic pen and manually drawing boundary lines on the reconstructed images. The desired CT values were then summarized as a frequency histogram and subsequently used as the analytical unit (Fig. 3). The histogram presented in Fig. 3 has been truncated to delete CT values corresponding to bone and will be referred to hereafter as the composite distribution.

The bimodal, composite distribution shown in Fig. 3 suggests the presence of two underlying component distributions corresponding to muscle and fat tissues. To examine this hypothesis, the 1448 CT values representing a longissimus muscle were defined on a reconstructed image and the resulting histogram was observed (Fig. 4). The histogram has distributional properties similar to normally distributed data, with apparently some degree of kurtosis. It has a mean CT value of about 58 Hounsfield units and a standard deviation of 15 Hounsfield units. If one assumes that the muscle distribution is symmetrical, then the right half

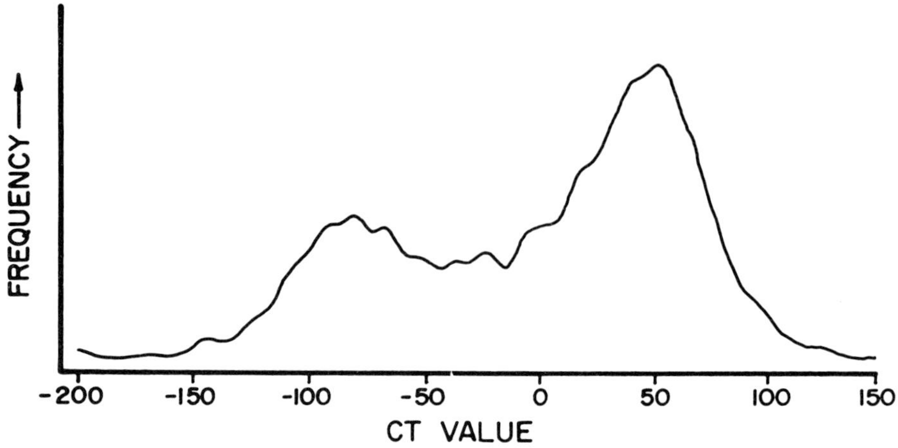

Figure 3. Representative distribution of carcass soft-tissue at the 13th thoracic vertebra of a pig.

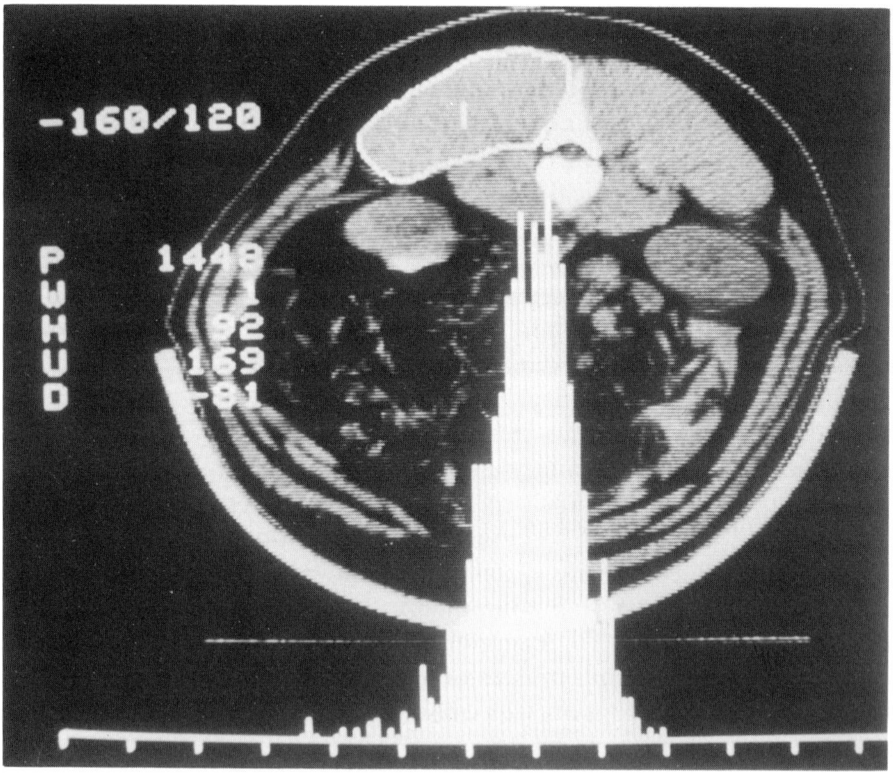

Figure 4. Distribution of CT values representing the left longissimus muscle.

of a muscle distribution can be mirror-imaged about the mean value to estimate the left half of the muscle distribution. This approach seems logical as numerous CT values less than the mean of muscle tissue represent volumes of tissue containing varying proportions of fat in addition to muscle (Fig. 3). The resulting hypothetical muscle distribution is superimposed on a composite distribution as shown in Fig. 5 and accounts for about 55% of the observed CT values for this particular composite distribution.

In an analogous manner, the component distribution of fat tissue was also estimated and superimposed on the same composite distribution (Fig. 6). About 30% of the total CT-values in this composite distribution were attributable to the fat distribution. If one considers the muscle and fat distributions jointly, approximately 15% of the composite distribution is unexplained. This residual portion is illustrated in Fig. 7 and apparently corresponds to CT values representing a mixture of muscle and fat tissues as described above.

If this biological interpretation of composite distributions is reasonably appropriate, then the implied concepts can be used to partly explain physiological differences among animals for developmental growth. As an example, composite distributions are illustrated in Fig. 8 for obese (solid line) and lean (dashed line) pigs of equal live weight. The obese pig had greater frequencies of CT values in the range corresponding to fat tissue and reduced frequencies of CT values in the range corresponding to muscle tissue than did the lean pig. These differences arose from

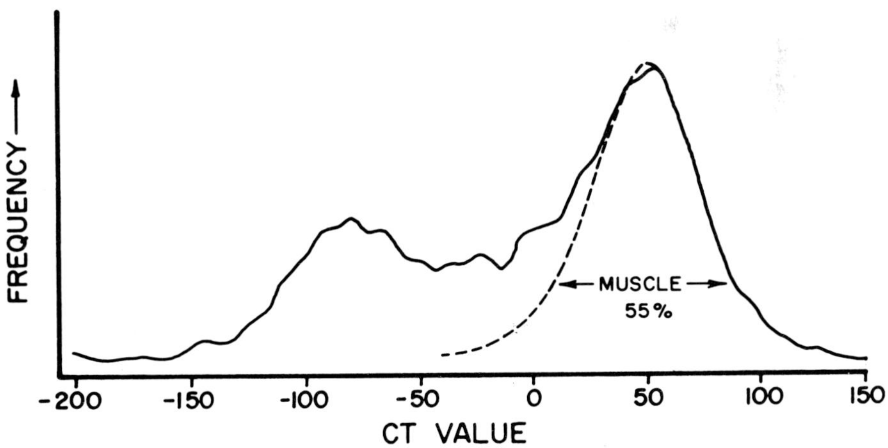

Figure 5. Underlying component distribution of muscle tissue representing 55% of the composite distribution.

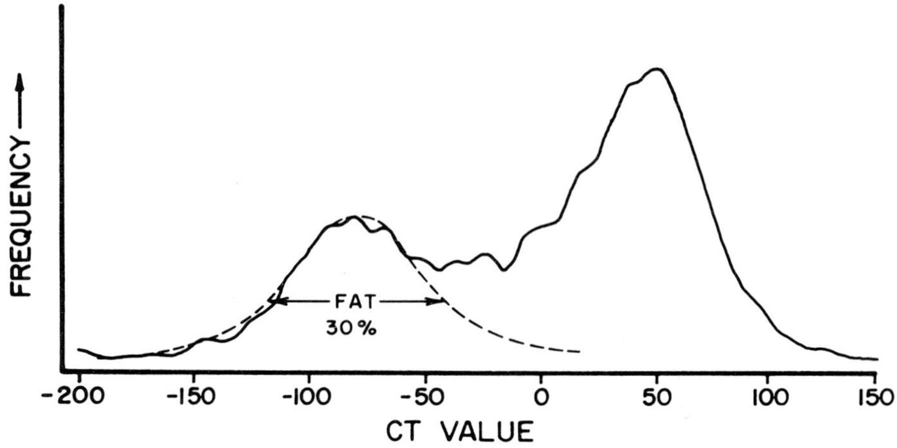

Figure 6. Underlying component distribution of fat tissue representing 30% of the composite distribution.

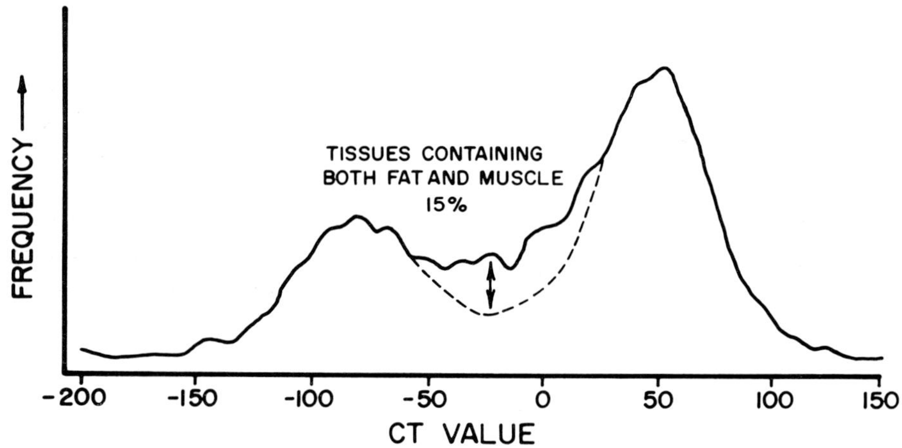

Figure 7. Residual distribution, after fitting component distributions of muscle and fat tissues, representing 15% of the composite distribution.

intentional selection for slow growth and increased backfat depth in the obese line of pigs. The consequences of selection, while obviously changing the relative amounts of fat and muscle tissue, affected the mean of the fat distribution but not the mean of the muscle distribution. That is, the obese and lean pigs exhibited similar means for their muscle distributions. However, the mean of the fat distribution for the obese pig was less than the mean of the fat distribution for the lean pig. As adipocytes develop, the content of water decreases suggesting that the fat tissue of the obese pig was more mature – a greater negative CT value – than the fat tissue of the lean pig.

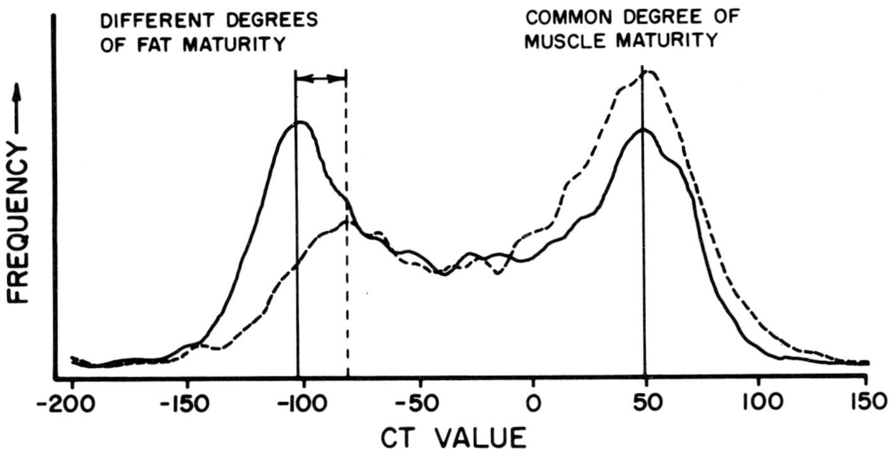

Figure 8. Composite distributions of obese (solid line) and lean (dashed line) pigs illustrating the similar means of muscle distributions, but different means of fat distributions.

Analytical methods

The tomography data can be used as independent variables in standard multiple regression equations to estimate composition. A simple approach is to calculate the mean CT value for the histogram determined at each scanning location. Alternatively, one can partition the histogram into intervals (e.g. 10 CT values) and use the corresponding frequencies as independent variables. Using Fig. 3 as an example, one could form 35 independent variables; the frequencies from -200 to -191, from -190 to -181, and so on through the last interval, 140 to 149. Stepwise regression procedures could be applied to identify the specific intervals jointly providing the most "useful" predictive value.

The data from different scanning locations may be analyzed separately to determine the most informative locations. This question is of practical merit because one desires to limit the number of scans necessary to predict composition with an acceptable level of precision. Another approach is to pool data from several locations together into a single histogram and use either the overall mean or the frequencies of CT intervals. Some of these approaches have been applied previously by Sehested (1983), Allen and Vangen (1983), Vangen, Standal and Walach-Janiak (1984), Sehested (1984), Vangen (1984), Allen (1985), Allen and Leymaster (1985), and Vangen, Sehested and Standal (1985).

As indicated by Sehested (1983, 1984) the mean of CT values is not as informative as using frequencies of CT intervals. This seems logical in light of the discussion presented above. The mean CT value would be relatively more useful if composition were determined by only two constituents rather than three (water, protein, and fat). The use of frequencies of CT intervals in stepwise regression analyses allows data from specific regions of the histogram to be exploited. For example, in estimating kilograms of carcass fat, CT intervals corresponding to the component distribution of fat (discussed above) are predominant relative to intervals in the muscle distribution (Sehested, personal communication). Another analytical method, not yet tested, would be to estimate the parameters (mean, standard deviation, and kurtosis) of the fat and muscle component distributions and use these estimates as independent variables. This method is appealing because such estimates provide greater biological interpretability as compared to previously published approaches.

Preliminary experimental results

Preliminary results will be presented from an experiment designed to develop estimation equations based on data collected on 50 Norwegian Landrace boars weighing about 30 kilograms live weight. This weight corresponds to the weights of boars when performance testing is initiated in the Norwegian national swine improvement program. Experimental procedures were similar to those described above. Pigs were scanned in batches of ten on five different dates. Scans were recorded at 11 locations on each pig: the midpoints of the 9th, 10th, 11th, 12th and 13th thoracic vertebrae and the 1st, 2nd, 3rd and 4th lumbar vertebrae as well as the narrowest point of the pelvic opening and the anterior end of the femur. Data collected on two pigs were deleted because complete sets of scans were not available.

The means, standard deviations, and coefficients of variation for growth rate, scanning age and weight, soft-tissue weight, and chemical composition are presented in Table 1. Note the relatively low variation for percent water and protein in this sample of young pigs. Correspond-

Table 1. Means, standard deviations and coefficients of variation of growth rate, scanning age and weight, soft-tissue weight and chemical composition.

Trait	Mean	Standard deviation	Coefficient of variation (× 100)
Growth rate[a], g/d	423	66	15.7
Scanning age, d	88.3	10.3	11.7
Scanning weight, kg	30.6	5.2	17.0
Soft-tissue weight, kg	14.2	3.1	22.2
Wet-tissue composition, g/100 g			
Percent water	71.90	1.43	2.0
Percent protein	17.49	0.73	4.2
Percent fat	9.96	1.67	16.8

[a]Postweaning rate of growth from 35 days of age to scanning at about 88 days of age.

ing statistics for mean CT values at each scanning location are given in Table 2. The histograms were truncated to include frequencies of CT values ranging from -90 to 89 Hounsfield units. As scanning locations range from the anterior to the posterior of the pig, the CT means increase, indicating a gradual shift from fat to muscle. The standard deviations tend to decrease from anterior to posterior locations, resulting in decreasing coefficients of variation also.

Compositional traits – percent protein, fat, and water – were estimated using three statistical models. The first (base) model fit the fixed effects of group (slaughter date) as a discrete variable and of scanning weight as a covariate. Subsequent models added effects of CT means or intervals to the base model so that inferences were to pigs of equal live weight scanned on the same day. The second model also fit the effect of CT mean at each location. Relevant statistics are reported in Tables 3, 4 and 5.

The decomposition of sums of squares for percent protein indicated that the base model accounted for 22.2% of the variation (Table 3). The addition of CT means, considered singly, only explained 11% or less of the base model residual variation. Consequently, the residual standard deviation was reduced significantly ($P < 0.05$) only by the effect of the CT mean at the third lumbar vertebra. The partial regression coefficients averaged about 0.03% per Hounsfield unit. These results suggest little, if any, relationship between CT means and percent protein when measured on young pigs of equal live weight.

The analyses of percent fat and water yielded results that were more encouraging. The base model explained 38.7% of the variation in percent fat (Table 4). The addition of CT means accounted for significant base-model residual variation at each scanning location. The partial

Table 2. Means, standard deviations and coefficients of variation of mean CT-values at each scanning location[a]

Location	Mean	Standard deviation	Coefficient of variation (× 100)
Thoracic vertebrae			
9th	25.6	5.0	19.5
10th	27.3	5.4	19.7
11th	28.8	5.2	18.2
12th	29.7	5.7	19.0
13th	30.9	5.0	16.3
Lumbar vertebrae			
1st	32.3	4.3	13.3
2nd	33.7	5.0	14.8
3rd	34.5	4.7	13.5
4th	34.4	4.8	13.9
Mid-pelvis	36.9	3.5	9.5
Femur	39.0	3.3	8.6

[a]Data are measured in Hounsfield units.

Table 3. Coefficients of determination relative to observed (R_0^2) and base model (R_b^2) variation, residual standard deviations (σ_e) and regression coefficients (b) for predicting percent protein on a wet-tissue basis

Model	R_0^2 (× 100)	R_b^2 (× 100)	σ_e	b
Base[a]	22.2		0.686	
Base plus CT data[b]				
Thoracic vertebrae				
9th	22.4	0.3	0.693	0.009
10th	22.2	0.1	0.694	−0.003
11th	23.9	2.3	0.686	0.022
12th	26.7	5.9	0.673	0.032
13th	23.7	1.9	0.687	0.021
Lumbar vertebrae				
1st	24.3	2.7	0.684	0.029
2nd	28.3	7.9	0.666	0.042†
3rd	30.5	10.7	0.656	0.053*
4th	28.3	7.9	0.666	0.047†
Mid-pelvis	24.7	3.3	0.682	0.036
Femur	28.1	7.6	0.667	0.058†

[a]The base model includes the fixed effects of group (discrete) and scanning weight (continuous).
[b]CT data are measured in Hounsfield units.
$*P < 0.05$, $^\dagger P < 0.10$.

Table 4. Coefficients of determination relative to observed (R_0^2) and base model (R_b^2) variation, residual standard deviations (σ_e) and regression coefficients (b) for predicting percent fat on a wet-tissue basis

Model	R_0^2 (× 100)	R_b^2 (× 100)	σ_e	b
Base[a]	38.7		1.382	
Base plus CT data[b]				
Thoracic vertebrae				
9th	59.3	33.6	1.140	−0.184***
10th	58.7	32.6	1.149	−0.163***
11th	59.0	33.0	1.145	−0.167***
12th	62.4	38.7	1.096	−0.165***
13th	61.8	37.7	1.105	−0.191***
Lumbar vertebrae				
1st	60.0	34.8	1.130	−0.207***
2nd	64.9	42.7	1.059	−0.196***
3rd	61.4	36.9	1.111	−0.198***
4th	54.6	25.8	1.205	−0.172***
Mid-pelvis	49.9	18.2	1.265	−0.172**
Femur	45.2	10.6	1.323	−0.137*

[a]The base model includes the fixed effects of group (discrete) and scanning weight (continuous).
[b]CT data are measured in Hounsfield units.
$***P < 0.001$, $**P < 0.01$, $*P < 0.05$.

regression coefficients averaged −0.18% per Hounsfield unit indicating that percent fat decreased as the CT mean increased. Similar results were obtained for percent water (Table 5) with the partial regression coef-

Table 5. Coefficients of determination relative to observed (R_0^2) and base model (R_b^2) variation, residual standard deviations (σ_e) and regression coefficients (b) for predicting percent water

Model	R_0^2 (× 100)	R_b^2 (× 100)	σ_e	b
Base[a]	47.2		1.099	
Base plus CT data[b]				
Thoracic vertebrae				
9th	64.1	32.1	0.917	0.143***
10th	66.9	37.4	0.880	0.139***
11th	62.0	28.0	0.944	0.122***
12th	64.2	32.9	0.915	0.120***
13th	64.6	33.1	0.910	0.142***
Lumbar vertebrae				
1st	64.8	33.4	0.907	0.161***
2nd	63.7	31.2	0.926	0.133***
3rd	58.8	21.9	0.983	0.121**
4th	58.6	21.6	0.985	0.125**
Mid-pelvis	53.8	12.4	1.041	0.113*
Femur	49.9	5.1	1.083	0.076

[a]The base model includes the fixed effects of group (discrete) and scanning weight (continuous).
[b]CT data are measured in Hounsfield units.
***$P < 0.001$, **$P < 0.01$, *$P < 0.05$.

ficients averaging about 0.12% per Hounsfield unit. Taken altogether, the results suggest that as the CT mean increased, there was a decrease in percent fat ($P < 0.001$), an increase in percent water ($P < 0.001$), and there tended to be a slight nonsignificant increase in percent protein. Apparently, the proportion of protein is relatively stable with fat and water substituted for one another.

The final statistical model used frequencies of 18 intervals as independent variables, where the frequencies were summed over histograms recorded at all 11 scanning locations. Stepwise procedures were not applied in these preliminary analyses. Due to the high degree of colinearity between interval traits, the partial regression coefficients are of questionable interpretative value. Also, if stepwise procedures were applied, it seems likely that the chosen intervals would be rather arbitrary and perhaps lowly repeatable. Nonetheless, use of this procedure accounted for 65.3, 55.1 and 71.6% of the variation in percent water, protein, and fat previously unexplained by the base model.

The superiority of the interval approach relative to using CT means confirms the findings of Sehested (1984) and supports the logic argued in the section on biological interpretation. It is recognized that the interval approach is an approximation to using estimates of parameters from the muscle and fat component distributions discussed above. Unfortunately, the latter method has not yet been tested, but will be pending the development of appropriate software.

Conclusion

The utility of tomography data to estimate composition has been rigorously tested in this experiment for the following reasons: (1) the use of immature animals which exhibited limited compositional differences, (2) compositional traits were defined as proportions rather than masses, and (3) evaluation of tomography data followed removal of variation due to base-model effects. In spite of this, it has been shown that computed tomography data can be used to predict composition with a relatively high degree of accuracy. As researchers gain experience with this technique and develop more useful software, it is anticipated that precision of estimation can be improved further.

Acknowledgements

I thank Professor Harald Skjervold and the staff of the Department of Animal Genetics and Breeding of the Agricultural University of Norway for their support, assistance, and inspiration. In particular, I thank Nils Standal, Petter Heyerdahl, Erling Sehested, Kjell Petterson, Knut Dalen and Odd Vangen. The many excellent contributions of Paul Allen, Dunsinea Research Center, Dublin, Ireland, are gratefully acknowledged.

References

1. Allen, P. 1985. *Developments in the use of computed tomography for live body evaluation.* Proc. Eur. Assoc. Anim. Prod., Khailkidhiki, Greece.
2. Allen, P. and K. Leymaster, 1985. *Machine error in X-ray computer tomography and its relevance to prediction of in vivo body composition.* Livestock Production Sci. 13: 383–398.
3. Allen, P. and O. Vangen. 1983. *X-ray tomography of pigs – some preliminary results.* In: *In vivo measurement of body composition in meat animals*, Ed. D. Lister, Elsevier, pp. 52–66.
4. Hounsfield, G.N. 1979. *Computed medical imaging.* J. Computer Assisted Tomography 4: 665–674.
5. Sehested, E. 1983. *Computerized tomography of sheep.* In: *In vivo measurement of body composition in meat animals*, Ed. D. Lister, Elsevier, pp. 67–74.
6. Sehested, E. 1984. *Evaluation of carcass composition of live lambs based on computed tomography.* Proc. Eur. Assoc. Anim. Prod., The Hague, Netherlands.
7. Skjervold, H., K. Gronseth, O. Vangen and A. Evensen. 1981. *In vivo estimation of body composition by computerized tomography.* Zeitschrift für Tierzüchtung und Züchtungsbiologie 98: 77–79.
8. Vangen, O. 1984. *Evaluation of carcass composition of live pigs based on computed tomography.* Proc. Eur. Assoc. Anim. Prod., The Hague, Netherlands.
9. Vangen, O., E. Sehested and N. Standal. 1985. *Protein and fat tissue changes in boars during fasting and performance testing, estimated by computerized tomography.* Proc. Eur. Assoc. Anim. Prod. Khailkidhiki, Greece.
10. Vangen, O., N. Standal and M. Walach-Janiak. 1984. *Tissue deposition rate in genetically lean and fat pigs estimated by computed tomography.* Proc. Eur. Assoc. Anim. Prod., The Hague, Netherlands.

Interactive microprocessor graphics as aid in physiological modeling: two examples

F. EUGENE YATES and TIMOTHY POSTON
Crump Institute for Medical Engineering, University of California, Los Angeles, CA 90024-1654, U.S.A.

Abstract To dramatize the often counterintuitive dynamic effects of clamping (fixing, holding constant or tightly bounding or constraining) an endogenous physiological variable such as a hormonal secretion rate or level, we have used a microcomputer to display vector field portraits of system trajectories under a wide variety of initial conditions or perturbations, and under interactive (moving-cursor) control with a mouse. These graphic views of behavior illustrate the occasionally profound consequences of freezing out a degree of freedom in a dynamical system. Stability can change in surprising ways. Although the stability analysis can be done without graphical aids, nevertheless the computed displays seem unmistakably to increase comprehension and provide a feeling for the mathematized organism (model), particularly for nonmathematicians. As further demonstration of the potential usefulness of a graphical expression of nonlinear dynamics, we have explored the behavior of a minimal model of the glucose/insulin system. Because this model has been used in clinical diagnosis, it is important to explore its behaviors.

Neither physiological models nor microprocessors are new – yet the combination of the microprocessor technology with the needs of modeling physiological problems, though much discussed, still remains exploratory in most biological and clinical fields. This paper emphasizes applications in which the interactive feature of the graphics provides the attractiveness of the approach and its tutorial effectiveness.

Simulations: two models

The place of modeling in physiological studies was properly fixed by Waggoner [1] who remarked that to embrace simulators is to return to the Dark Ages of endless debate, but to ignore simulators is to reject a new mode of ordering experiments, and so the wise biologist should take the middle course. These comments were meant to be a sermon on a very old subject – after all, in science the rules for good modeling are just the rules for good thinking. Perhaps the first sermon on the subject, from which all others derive was published by Francis Bacon in 1620 in his Novum Organum, as aphorism 95:

> The men of experiment are like the ant; they collect and use; the reasoners resemble spiders, who make cobwebs out of their own substance. But the bee takes a middle course, it gathers its material from the flowers of the garden and of the field, but transforms and digests it by a power of its own.

Invited paper, to appear in: Proceedings, XI Beltsville Symposium in Agricultural Research, U.S. Dept. of Agriculture. "Research Instrumentation for the 21st Century" (May 4–8, 1986).

A simulator is an embodiment of a mathematical model, whether as a physical analogue, or as a program being run on a computer. The computer has brought theories, models and simulators very close together. A theory expressed in a computer language as a program becomes simultaneously a model and simulator when it is put on a computer and run. Of course, biology is weak on theory and in practice models of biological systems are judged primarily on their ability to fit data according to some criterion. They are models of data rather than models on theory. The requirements for a good simulator are usually that it fit data, make predictions, use plausible structures and parameters, and be itself describable and communicable to other biologists [2]. Guidelines for such simulating have been suggested [3].

Much of the emphasis in earlier discussions of the use of simulators in physiological modeling has addressed quantitative models in which system and parameter identification (assignment of structure and numerical values) were of paramount importance. In this paper we call attention to the value of studying the *qualitative* aspects of models and system dynamics. To do that we shall introduce two mathematical models, one describing a continuous, stirred-tank reactor [4] as is common in chemical engineering and representative of many physiological transport, distribution and reaction processes in biology. This primitive model is given by the following equations:

$$\begin{aligned} \dot{x} &= \begin{bmatrix} \frac{3}{2} \\ \frac{1}{6} \end{bmatrix} - \begin{bmatrix} x \\ y \end{bmatrix} + \begin{bmatrix} \frac{18}{65} y^3 - \frac{9}{5} xy^2 \\ \frac{9}{5} xy^2 - \frac{18}{65} y^3 \end{bmatrix} \quad (1) \\ \dot{y} &= \quad (2) \end{aligned}$$

inflow efflux reaction

We shall refer to this as Model 1.

The second model is one proposed by Bergman and Cobelli, and described in four publications [5–8]. Specifically, it is the nonlinear model VI in reference [5]:

$$\frac{\mathrm{d}G}{\mathrm{d}t} = (P_1 - X)\,G + P_4 \tag{3}$$

$$\frac{\mathrm{d}X}{\mathrm{d}t} = P_2X + P_3I \tag{4}$$

These equations constitute Model 2.

Each of the models will have its variables and parameters defined, and its structure discussed, below. Our purpose, however, is not to explain these established models in detail, but to comment on the exploration of their dynamics by means of interactive microprocessor computer graphics. One way or the other we must "model or muddle" [9] and our aim is to describe one of the newer tools to aid modeling.

Of course, microprocessors themselves are hardly new today. The urge to model also is not new and the use of graphical displays [10] to dramatize features of data is an ancient impulse. The desire of modelers to have on-line, interactive access while their models are running on computers is at least as old as the general availability of computers in this country, starting in the 1960s. The novelty is inexpensive digital machinery that runs at a comfortable speed suitable for human interaction – a convenience of speed not seen since analog computer days. The state of technology in the shops has wider impact on modeling than does the state of the art. This paper must be taken only as a modest expression of some of the strong trends of shop activities in our times. We are riding the wave, not making it.

The hardware and software

The simulations of the above Models 1 and 2, that constitute the basis of our discussion, were carried out on an IBM PC/AT running under DOS, 3.1, equipped with an EGA graphics board. The models and ancillary aspects of the simulations were programmed in lattice C, using the CGI generalized interface for easy transfer between and among displays, plotters, printers, mice, etc. The displays were presented on IBM Enhanced Color Display.

The enclosed figures showing results were drawn with a Hewlett-Packard plotter (Model 7550), by minor redirection of the CGI commands used for screen output, and still with the interactive convenience of mouse control in selecting curves. They, too, illustrate the "state of the shops."

Exploration of Model 1

The details of Model 1 have been published [4]. In this trimolecular, stirred-tank reactor we imagine that two chemicals, X and Y, undergo a reaction of the form

$$X + 2Y \underset{18/65}{\overset{9/5}{\rightleftharpoons}} 3Y \tag{5}$$

with the rate constants shown. An inflow of concentration $3/2x$ and $1/6y$ occurs at the rate 1, and fluid with the instantaneous, stirred concentrations x and y is withdrawn at the same rate. The model shown in Eqs. (1) and (2), above is arrived at by ordinary mass action arguments. Figure 1 shows a phase protrait for Model 1. All trajectories relax to (x_0, y_0) where these are the equilibrium values of the concentrations. Any

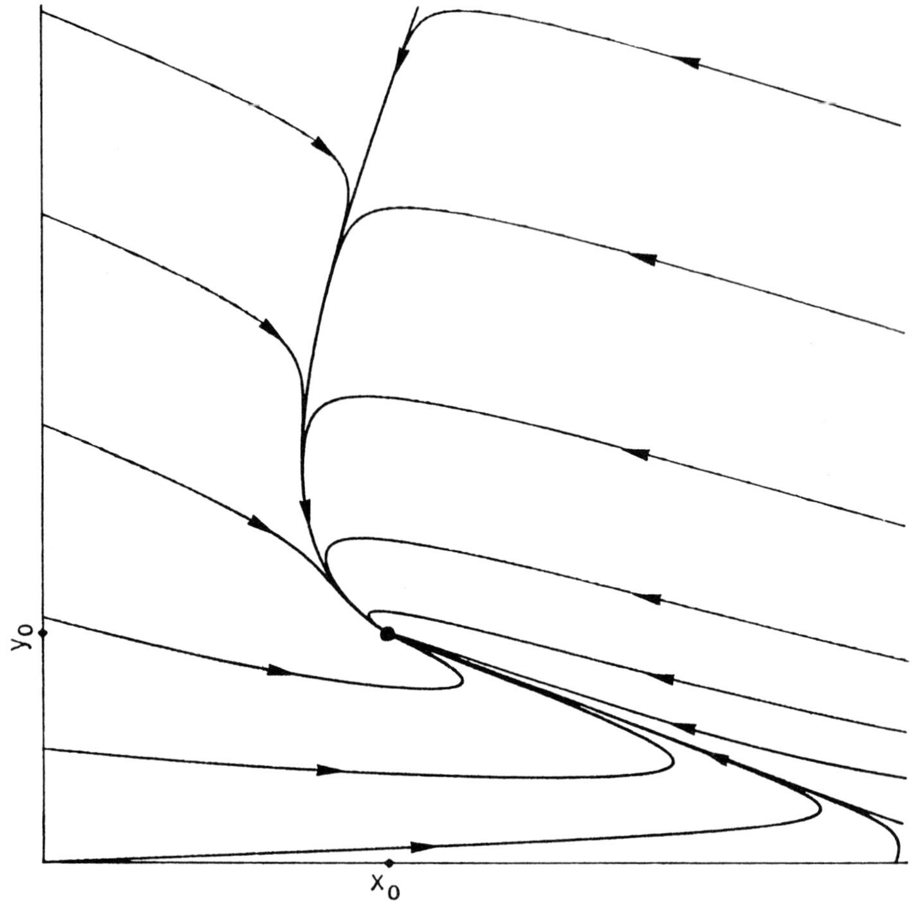

Figure 1. Phase portrait of model 1 (Eqns. (1) and (2)). In plots of this kind all possible coordinate pairs (x_i, y_j) are represented. The curving lines indicate some of the system trajectories from various initial conditions (x_n, y_m). All trajectories relax to an equilibrium point (x_0, y_0).

simple perturbation of the system, such as an increase in the concentration y, is followed by a smooth return to the value (x_0, y_0). Notice, however, that the return to equilibrium is not monotonic. Sometimes the immediate response to an increase in y is that y increases further, while x decreases.

Now we wish to explore the consequences of *constraining* x at its normal equilibrium value, x_0. When x is so constained and then there is a perturbation increasing y in the system, the preliminary decrease in x – essential to a decrease in y – no longer occurs, and the initial increase in y is not turned off. It continues as long as y is below a newly created equilibrium value $y_2 \approx 2.65$ (Fig. 2). Similarly, a small decrease from y_0 with fixed $x = x_0$ continues as long as $y > y_1 \approx 0.214$.

By freezing out a degree of freedom in this simple dynamic system we

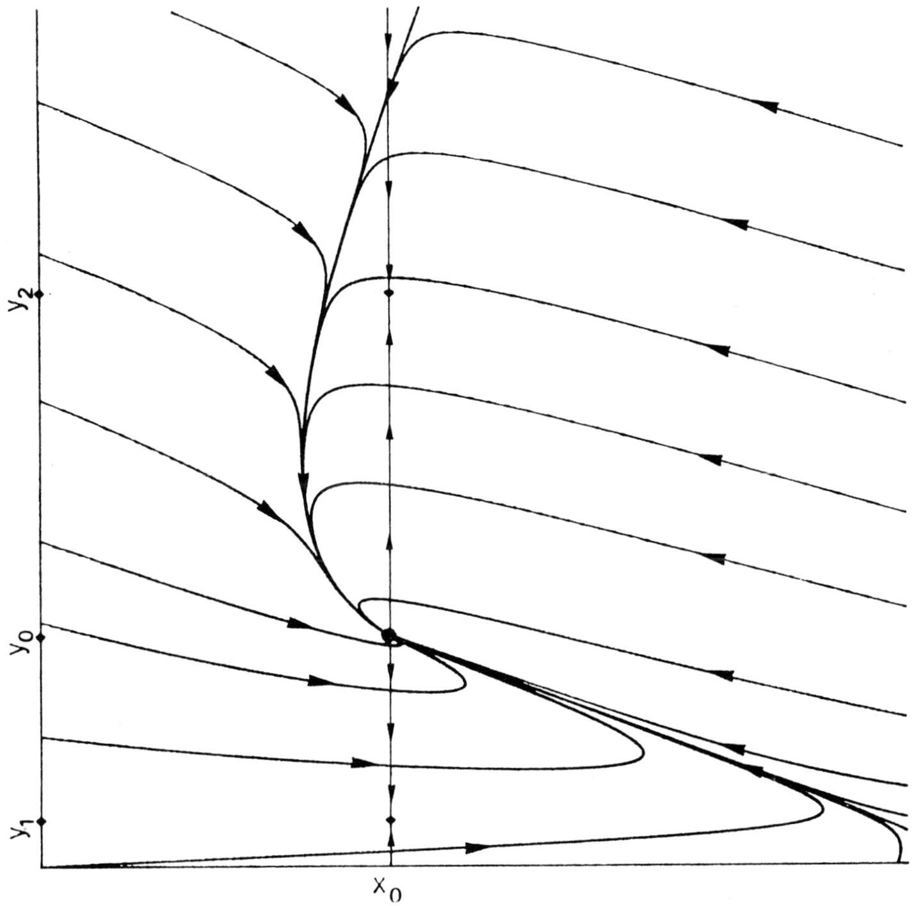

Figure 2. Behavior of variable y when variable x has been constrained to its equilibrium value $x = x_0$. Small arrows on the vertical line through x_0 show the up or down change of y at points (x_0, y), which under clamped conditions is the *only* change. Two new dynamical equilibria (x_0, y_1), (x_0, y_2) emerge from clamping x at its previous equilibrium value. The trajectories of the unclamped system (Fig. 1) are repeated for comparison.

have created new equilibria, and the system has become destabilized around the previously stable point (x_0, y_0). This destabilization by clamping x is not a result of the nonlinear character of this system: linearization of this system around (x_0, y_0) destabilizes in the y-direction in just this manner. The nonlinear terms merely allow other equilibria for y to exist (a linear or affine equation having at most one isolated zero), so that in the clamped model, y does not depart exponentially for $\pm\infty$.

Figure 3 shows a graph of $\dot{y}$ versus y in Model 1, where x is held fixed at the equilibrium value x_0: y increases where $\dot{y}$ is positive, decreases where it is negative, thus moving according to the arrows.

The importance of studying the dynamic consequences of clamping a

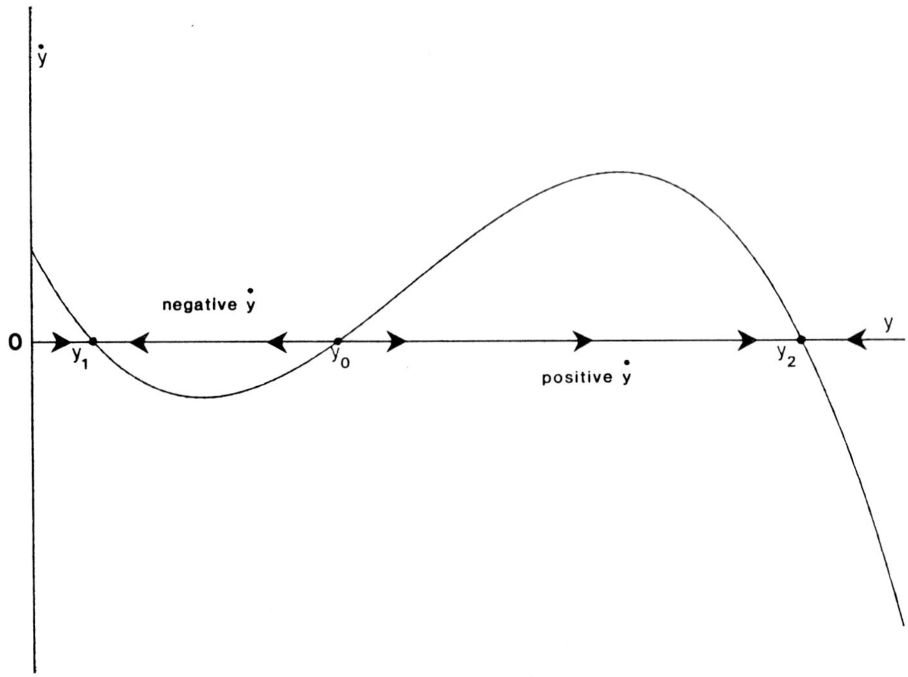

Figure 3. Graph of $\dot{y}$ versus y in Model 1, where x is held fixed at its equilibrium value x_0. The original equilibrium value for y (y_0 in Fig. 1) and the two new possible values after clamping $x = x_0$ (that is y_1 and y_2, Fig. 2) appear as zeroes for $\dot{y}$.

variable has relevance to medicine because administration of hormonal or other biological agents by rate-controlling delivery systems operating under open loop control with constant fluxes will lead to such "clamping" of the input rates if either the endogenous source is missing or it is shut off by closed loop internal controls. A consequence of clamping the input is near-constancy of the plasma concentration of the hormone – and that is usually the intended result. In many cases that intention is proper, justified, and effective. However, as the exploration of Model 1 shows, there can be surprising hazards and unexpected destabilizations. It is by no means intuitively obvious, when one clamps an endogenous variable that most of the time appears to be spontaneously constant (so as seemingly to replace constancy with constancy), that dramatic changes in system dynamics may occur. The use of vector field phase portraits such as shown in Figs. 1 and 2, especially when these are easily explored by use of a mouse-driven cursor, vividly illuminates the dynamical consequences of clamping. Unfortunately, in this paper we cannot demonstrate the "feeling" of exploring the model in this fashion. We can here show results only as static figures. In the actual simulations, the movements of y when x is free and when x is clamped, from a variety of initial conditions, can be easily followed and, we would say, "felt", with a deeper sense of reality than arises from mathematicians' marks on paper.

Exploration of Model 2

This model attempts to show certain aspects of the relations between the hormone insulin and the level of blood sugar. It was intended by its authors Bergman and Cobelli to represent a "minimal model" in the sense that it has very few parameters and all these are formally identifiable from a simple clinical, transient-response test – a modified, intravenous, "glucose tolerance" test. The relevant part of the simple model, for this discussion, is shown in Fig. 4 (from references [5, 6]), and in Eqns. (3) and (4).

The central part of the diagram is a "glucose space" which represents sugar in blood plasma and extracelllar fluids, in amount (or concentration) G. The glucose can enter this space only from the liver, where it can be synthesized; there is no allowance for input of sugar by the consumption of food in this case. (Intravenous pulses of glucose, not shown, are used to start the transient responses.) The sugar can leave the

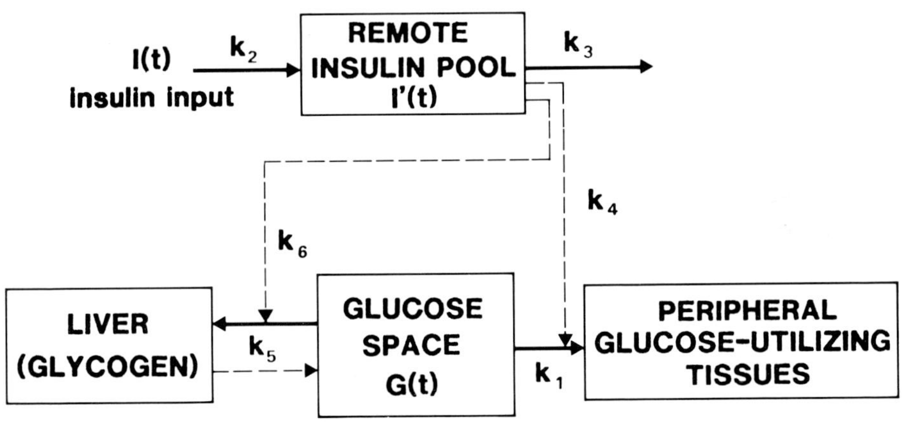

$$X(t) = (k_4 + k_6) I'(t) \quad \text{by definition}$$

Figure 4. Block diagram of glucose model. The diagram is modified from references [5, 6]. Solid arrows represent mass fluxes; dashed arrrows k_4 and k_6 are signal paths. $I(t)$ represents new insulin, above an initial condition established by a basal insulin secretion, not shown. This basal effect is buried in the values for P_4 and k_1 and k_5. As simulated here, k_5 addresses liver uptake of glucose. Under other conditions the liver can produce glucose – dashed arrow under k_5.

space by being taken up either by the liver or by other tissues. Some time-varying input rate I of the hormone insulin (above basal levels that are not shown) causes variation in the amount (or concentration) I' in a remote insulin pool from which insulin can be removed by enzymatic reactions that inactivate it. The influence of insulin on the flows of glucose is given by the nonlinear term GX (see Eqn. 3), where X is an informational, not mass flux, term proportional to the insulin in the remote pool, weighted separately for its influence on transports of glucose to (or from) liver and to other peripheral, glucose-utilizing tissues. The six constants shown in Fig. 4 were combined into three parameters:

$$P_1 = -(k_1 + k_5)$$

$$P_2 = -k_3$$

$$P_3 = k_2(k_4 + k_6)$$

In addition, a fourth parameter, P_4, was introduced to take account of the fact that under conditions in which there is no flow of insulin above basal insulin levels, some background transports of glucose occur. Finally, by definition, $X = (k_4 + k_6)I'$ (rescaling units already arbitrary), where X and I' reflect new insulin influences, above basal levels.

Like other models, this one has limitations and restrictions on its applicability. Our aim here is not to attack or defend it, but to explore it by means of interactive graphics. For the "normal" case, we set the parameters as follows:

$$P_1 = -2.5 \times 10^{-2}\ (\text{min}^{-1})$$

$$P_2 = -9.2 \times 10^{-2}\ (\text{min}^{-1})$$

$$P_3 = 5.7 \times 10^{-5}\ \text{min}^{-2}/(\mu\text{U/ml})$$

$$P_4 = 2.8\ (\text{mg}/100\,\text{ml})\ \text{min}^{-1}$$

For the "'obese" case we represented one type of diabetic pathology by decreasing P_3 to 10% of the above value, reflecting the hypothesis that some diabetic, overweight patients may have too few insulin receptors in the peripheral glucose-utilizing tissues. (These parameter values were chosen for illustrating the features of the model. They are not based on detailed clinical data.)

Insulin sensitivity – several definitions revealed by interactive graphics

Interactive graphics help disambiguate the common clinical term "insulin sensitivity". Regrettably, we are here limited to static figures to show the clarifications provided by our interactive explorations of the model. Figure 5 shows the phase portraits (analogous to Fig. 1) of the "normal" (a) and "obese" (b) dynamics (trajectories to basal-state equilibrium) of G and I from various nonbasal initial conditions, when $I(t) = 0$ (i.e., no new insulin is secreted or infused). The initial value of X, set by initial values of I', cannot be maintained while insulin is being destroyed by the k_3 process, and so the insulin-related signal X relaxes exponentially to a fixed, basal level X_b, because no new insulin comes in. The glucose level moves to its equilibrium value $-(P_4/P_1)$, which implicitly reflects the value of X_b.

Clamping a new input $I(t)$ to a non-zero, positive constant value I_c gives a new equilibrium value I'_c for I' and X_c for X, with a coresponding new equilibrium glucose level G_c. The collection of such combined *equilibrium points* (X_c, G_c) is marked in each figure as a dotted curve.

Because X_c is proportional to I'_c, which is in turn proportional to I_c, and may conveniently be again rescaled, to numerical equality, these dotted curves also serve as graphs of equilibrium glucose G_c against steady-insulin level I_c. The slope of the dotted curve is one possible measure of insulin sensitivity – one with strong appeal because of its intuitive clarity and obviousness. (However, it isn't easy to measure). We see that in the (modelled) normal case (Fig. 5(a)) a small increase of I_c from zero greatly decreases G_c, whereas in the "obese" case (Fig. 5(b)) the effect is small. The obese person is less insulin sensitive by this measure.

Less clear in a static plot, but vivid dynamically (with both initial (G, X) and time-dependent $I(t)$ under mouse control) is the transient response when I changes from one steady level, I_c, to another. The glucose level relaxes to the corresponding new equilibrium, with a time constant that might be taken as a second index of insulin sensitivity – in a "shift of insulin clamp level" experiment. But different parameters dominate this relaxation time in different regions of the figure. For low X, the relaxation constant $(P_1 - X)$ of G (recall P_1 is negative) is small, so that X re-equilibrates while G scarcely moves. Only if X is near the new X_c do we see the $(P_1 - X)$-controlled relaxation of G to its new equilibrium G_c. For high X, $(P_1 - X)$ is large enough to keep G always close to the equilibrium glucose for the current X; (G, X) then tracks the curve of equilibria, and the controlling time constant is P_2, governing the relaxation of X to equilibrium with the new level of I. The faster equilibration of G at high levels of X is thus invisible to this experiment, once it is two or three times faster than that of X. (Further increases are inevitably lost in measurement error, as (G, X) merely tracks the equilibrium curve even more closely.)

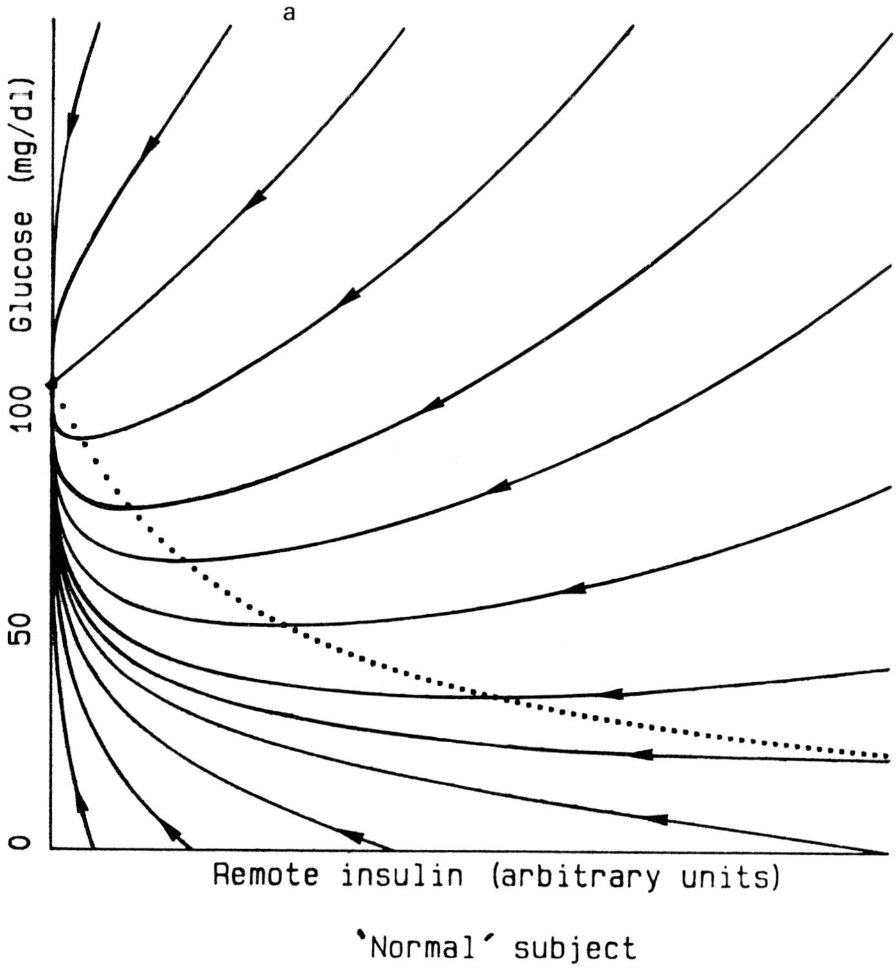

Figure 5. (a) Phase portraits with "normal" parameter settings for the glucose and remote insulin dynamics in Model 2 (Eqns. (3) and (4)), under conditions of zero new insulin input $I(t) = 0$. The trajectories start from any chosen initial condition (I', G) and relax to a non-zero, basal state equilibrium (I'_b, G_b). (A pre-existing basal insulin level is present, but not shown explicitly.) *Equilibrium values* I'_c, G_c at various other, non-zero levels of steady, incremental insulin input, are shown as a *dotted curve*. At equilibrium the "remote insulin" I'_c scales proportionally with I_c and X_c and so the abscissa for the dotted curves may be thought of as any of these three variables, with magnitudes increasing to the right, starting from 0 for I_c, and from non-zero, steady implicit basal levels I'_b, X_b for I'_c or X_c. (b) Model 2 modified to test an hypothesis about obesity-dependent diabetes. This phase portrait was generated by decreasing P_3 to 10 percent of its normal value, compared to the settings used to generate the data shown in Fig. 5(a). These parameter changes represent an hypothesis about the molecular defect in this type of diabetes. Again tne dotted line shows all the equilibrium values (I'_c, G_c), or, equivalently, (X_c, G_c) or (I_c, G_c).

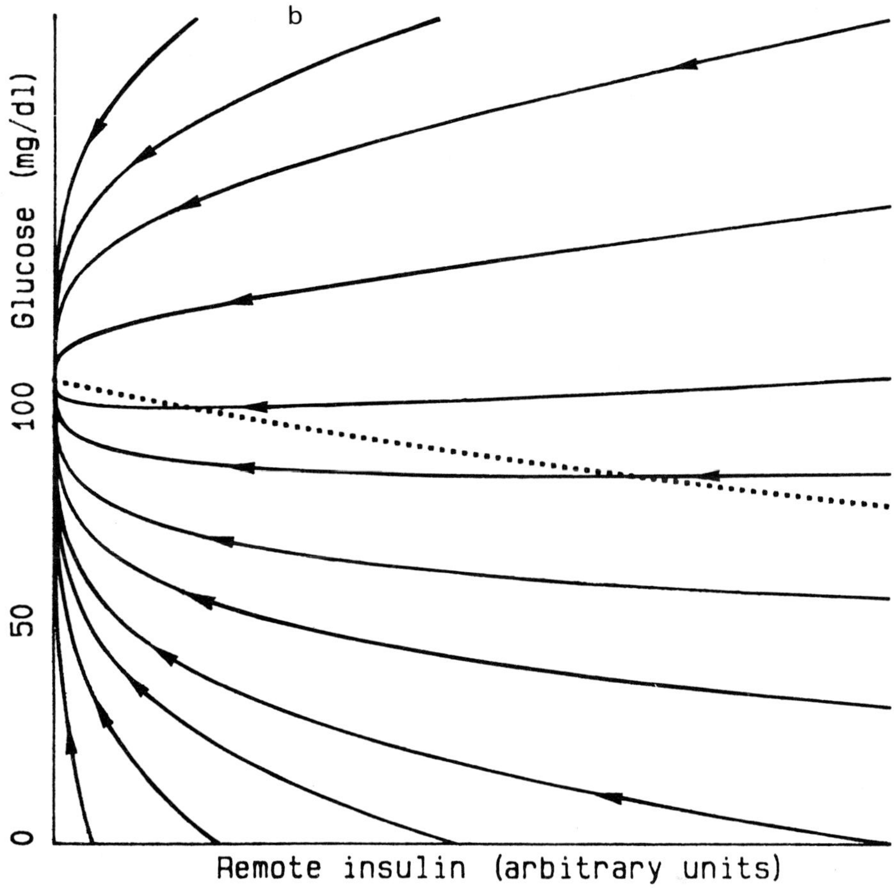

'Obese' subject

In the simulation, it is always possible to hold I constant during a transient response of G to an impulsive forcing by exogenous glucose. In real patients this clamp on insulin is possible only for Type I diabetics who make no insulin of their own, because G and I are normally coupled physiologically in a negative feedback loop. That loop is operating in the Type II (obesity) diabetic, but at lowered gain.

In the absence of an insulin clamp, in the regime of high $(P_1 - X)$ the relaxational time of a glucose tolerance test reflects only the responsiveness of I to G (not covered by this model) and of X to I, not of G to X. The clamped-I relaxational time of G can, however, be computed (within the model) using estimates of P_1, P_2, P_3 and P_4 made from the recorded time histories $I(t)$, $G(t)$ in such an unclamped test. (The possibility of such estimates was one of the criteria by which Bergman et al. selected this model.) The rate of change (not the magnitude) of this computed clamped-insulin glucose relaxation time, with insulin level, is the Bergman et al. definition of "insulin sensitivity". This rate of change of relaxation time with insulin level equals, analytically, $-(P_3/P_2)$. This is a third possibility for the meaning and one that can be calculated from transient response data from a single experiment. Thus, it is practical for clinical testing, but not intuitively obvious, nor close to the ordinary clinical impression about insulin sensitivity as this term appears informally in much of the medical literature we have examined.

These three different meanings of insulin sensitivity are easily conveyed with an interactive display, but detailed treatment without such methods requires more mathematical formalism.

Discussion

Model 1. A person sitting at a terminal exploring Model 1 through its graphical phase portrait, and using a mouse cursor to set initial conditions anywhere in the vector field, will see the trajectories of the motions of the point (x_i, y_j) at once on releasing the button on the mouse, and with it the dynamics. Under the conditions shown in Fig. 1, all motions are smooth, though not necessarily monotonic in x or in y. The motions stop at the point (x_0, y_0). If the cursor is on that point, but the model is constrained so that always $x = x_0$, the operator of the model feels how even very small errors in the placement of the cursor around the previous equilibrium point (x_0, y_0) lead to a motion of y either to the new equilibrium point y_1 or the new point y_2. The point (x_0, y_0) reveals itself as an equilibrium that is easy enough to control if the operator continues to make small adjustments to head off the effects of small fluctuations. But this equilibrium is unstable at any fixed level of the controller. The control operations needed to stabilize around (x_0, y_0) can literally be "felt" through a joystick while the operator visually closes a control loop on the dynamics of the system by trying to keep the cursor on the screen at that point.

Model 2. The operator can quickly assess the parameter-sensitivity of the model and test hypotheses about changes in insulin or glucose production rates, or uptake and removal rates in various disease states.

In the discussion above we have illustrated some of the insights to be gained and communicated this way, with simulations of glucose pulses and insulin production changes equally 'hands-on'. In fact, the operator can enter the loop as a 'pancreas' (insulin source), seeing the glucose level and responding through mouse or joystick with changes in insulin production. The loose or taut responsiveness of the system, in different test situations and parameter values, gives a feeling for its dynamics that is worth a thousand static pictures.

Envoi

Some features of modern personal computers that permit the style of modeling with interactive graphics that we are describing here, derive from the original work at the Xerox Corporation's Palo Alto Research Center (PARC). As decribed by Crecine [11] the PARC achievement "was the truly revolutionary operating system developed for the Alto, built around a graphically oriented, desk-top metaphor with icons representing files, documents, and programs, and a pointing device in the form of a "mouse" for selecting programs, documents, and complex commands from menus of choices. Further developments of high-resolution color graphics have brought us to the present stage." But no article such as that would be complete without speculations about the next generation of personal computers and the wonders they will bring. Crecine remarks: "Within two years the next generation of personal computer workstations will emerge. These computers will cost no more than fully-configured versions of current microcomputers, but they will be 5 to 10 times more powerful, with 10 to 20 times as much active memory and with the graphics capabilities previously available only on costly, specialized systems."

The potential of computer-aided, interactive, physiological modeling seems likely to be realized on a wide scale within the next five years. One of us (FEY) first encountered that nascent potential in the 1960s, working in collaboration with Robert Brennan on a dedicated IBM 360 system that we briefly had all to ourselves (in off hours). The power of interactive graphics as an aid to modeling, revealed under such extraordinary conditions, was never again within the reach of this ordinary professor of physiology, until now, 20 years later. The interactive feature now available with desk-top equipment such as that used in the simulations described here, is fundamental to the rhythm of the modeling process at its best.

Those of us who remember earlier, batch-process modeling, starting with a deck of punched cards and waiting through turnaround times of hours to days, recall also the difficulties of sustaining reflection, excitement and intellectual momentum when one's hands are not on the problem continuously. Model development and testing is best done in the interactive, graphic computer mode. That truism hardly needs emphasis today, yet we notice that exploration using dynamic phase portraits, such as we have shown here for two models, is not yet commonplace in physiological studies. In part, they are rare because biological phenomena are inherently difficult to mathematize, but no doubt there is also a sense of hardware and software limitations that most biologists have not yet overcome. As those limitations are lifted from us, physiological investigations and hypothesis testing will, in many cases, take a big step forward.

References

1. Waggoner, P.E. 1975. *Idols of the model or bringing home the Bacon.* In F.J. Vernberg, ed., *Physiological Adaptation to the Environment.* Intext Ed. Pub., New York, pp. 547–557.
2. Yates, F.E. 1979. *Physical biology: a basis for modeling living systems.* J. Cybernetics and Information Sci. 2: 57–70.
3. Yates, F.E. 1978. *Good manners in good modeling: mathematical models and computer simulations of physiological systems.* Am. J. Physiol.: Reg., Integrative and Comparative 3: R159–R160.
4. Yates, F.E. and T. Poston. 1986. *Rate-controlled delivery of endocrine agents: some paradoxical consequences of controlling the inputs.* In H.A.J. Struyker-Boudier, ed., *Principles and Applications of Rate-Controlled Drug Administration.* CRC Press, New York, pp. 247–272.
5. Bergman, R.N., Y.Z. Ider, C.R. Bowden and C. Cobelli. 1979. *Quantitative estimation of insulin sensitivity.* Am. J. Physiol.: Endocrine, Metabolism and Gastrointestinal Physiology 6: E667–E677.
6. Toffolo, G., R.N. Bergman, D.T. Finegood, C.R. Bowden and C. Cobelli. 1980. *Quantitative estimation of beta cell sensitivity to glucose in the intact organism, a minimal model of insulin kinetics in the dog.* Diabetes 29: 979–900.
7. Bergman, R.N., C.R. Bowden and C. Cobelli. 1981. *The minimal model approach to quantification of factors controlling glucose disposal in man.* In R.N. Bergman and C. Cobelli, eds., *Carbohydrate Metabolism – Quantitative Physiology and Mathematical Modelling.* John Wiley and Sons, New York, pp. 269–298.
8. Bergman, R.N., L.S. Phillips and C. Cobelli. 1981. *Physiologic evaluation of factors controlling glucose tolerance in man: measurement of insulin sensitivity and beta-cell glucose sensitivity from the response to intravenous glucose.* J. Clin. Invest. 68: 1456–1467.
9. Verveen, A.A. 1971. *In search of process: the early history of cybernetics.* Math. Biosci. 11: 5.
10. Tufte, E.R. 1983. *The Visual Display of Quantitative Information.* Graphics Press, Cheshire, Conn.
11. Crecine, J.P. 1986. *The next generation of personal computers.* Science 231: 935–943.

Index